T0395835

Second Edition

SYNCHRONOUS GENERATORS

Second Edition

SYNCHRONOUS GENERATORS

Ion Boldea

IEEE Life Fellow
University Politehnica Timisoara
Timisoara, Romania

CRC Press
Taylor & Francis Group
Boca Raton   London   New York

CRC Press is an imprint of the
Taylor & Francis Group, an **informa** business

CRC Press
Taylor & Francis Group
6000 Broken Sound Parkway NW, Suite 300
Boca Raton, FL 33487-2742

© 2016 by Taylor & Francis Group, LLC
CRC Press is an imprint of Taylor & Francis Group, an Informa business

No claim to original U.S. Government works

Printed on acid-free paper
Version Date: 20150817

International Standard Book Number-13: 978-1-4987-2356-5 (Hardback)

Visit the Taylor & Francis Web site at
http://www.taylorandfrancis.com

and the CRC Press Web site at
http://www.crcpress.com

Contents

8   Testing of Synchronous Generators

Preface to the Second Edition

The first edition of this single-author, two-book set was published in 2006. Since then, electric energy, "produced" mostly via electric generators, has become one of the foremost activities in our global economy world. The subject of Electric Generators (*Synchronous Generators* and *Variable Speed Generators* as two books) attracted special attention worldwide both from industry and academia in the last decade. Electric generators' design and control may constitute a new graduate course in universities with electric power programs.

Also, in the design and control of electric generators for applications ranging from energy conversion to electric vehicles (transportation) and auxiliary power sources, new knowledge and developments have been published in the last ten years. In the last ten years, in wind generators alone, the installed power has increased from some 40,000 MW to 300,000 MW (in 2014).

In view of these developments, we decided to come up with a new edition that

- Keeps the structure of the first edition to avoid confusion for users
- Keeps the style with many numerical worked-out examples of practical interest, together with more complete case studies
- Includes text and number corrections
- Adds quite a few new paragraphs in both books, totaling around 100 pages, to illustrate synthetically the progress in the field in the last decade

The new additions in the second edition are

Synchronous Generators

- Chapter 2 (Section 2.9): High Power Wind Generators, with less or no PM—an overview
- Chapter 4 (Section 4.15): PM-Assisted DC-Excited Salient Pole Synchronous Generators
        (Section 4.16): Multiphase Synchronous Machine Inductances via Winding Function Method
- Chapter 6 (Section 6.17): Note on Autonomous Synchronous Generators' Control
- Chapter 7 (Section 7.21): Optimization Design Issues
        (Section 7.21.1): Optimal Design of a Large Wind Generator by Hooke–Jeeves Method
        (Section 7.21.2): Magnetic Equivalent Circuit Population-Based Optimal Design of Synchronous Generators
- Chapter 8 (Section 8.10): Online Identification of SG Parameters
        (Section 8.10.1): Small-Signal Injection online Technique
        (Section 8.10.2): Line Switching (On or Off) Parameter Identification for Isolated Grids
        (Section 8.10.3): Synthetic Back-to-Back Load Testing with Inverter Supply

Variable Speed Generators

- Chapter 2 (Section 2.14): Ride-Through Control of DFIG under Unbalanced Voltage Sags
  (Section 2.15): Stand-Alone DFIG Control under Unbalanced Nonlinear Loads
- Chapter 5 (Section 5.8): Stand-Alone SCIG with AC Output and Low Rating PWM Converter
  (Section 5.10): Twin Stator Winding SCIG with 50% Rating Inverter and Diode Rectifier
  (Section 5.11): Dual Stator Winding IG with Nested Cage Rotor
- Chapter 6 (Section 6.8): IPM Claw-Pole Alternator System for More Vehicle Braking Energy Recuperation: A Case Study
- Chapter 8 (Section 8.12): 50/100 kW, 1350 –7000 rpm (600 N m Peak Torque, 40 kg) PM-Assisted Reluctance Synchronous Motor/Generator for HEV: A Case Study
- Chapter 9 (Section 9.11): Double Stator SRG with Segmented Rotor
- Chapter 10 (Section 10.16): Grid to Stand-Alone Transition Motion-Sensorless Dual-Inverter Control of PMSG with Asymmetrical Grid Voltage Sags and Harmonics Filtering: A Case Study
- Chapter 11 (Section 11.5): High Power Factor Vernier PM Generators

We hope that the second edition will be of good use to graduate students, to faculty, and, especially, to R&D engineers in industry that deal with electric generators, design control, fabrication, testing, commissioning, and maintenance. We look forward to the readers' comments for their confirmation and validation and for further improvement of the second edition of these two books: *Synchronous Generators* and *Variable Speed Generators.*

Professor Ion Boldea
IEEE Life Fellow
Romanian Academy
University Politehnica Timisoara
Timisoara, Romania

MATLAB® is a registered trademark of The MathWorks, Inc. For product information, please contact:

The MathWorks, Inc.
3 Apple Hill Drive
Natick, MA 01760-2098 USA
Tel: 508-647-7000
Fax: 508-647-7001
E-mail: info@mathworks.com
Web: www.mathworks.com

Preface to the First Edition

Electric energy is a key factor for civilization. Natural (fossil) fuels such as coal, natural gas, and nuclear fuel are fired to produce heat in a combustor and then the thermal energy is converted into mechanical energy in a turbine (prime mover). The turbine drives an electric generator to produce electric energy. Water potential and kinetic and wind energy are also converted to mechanical energy in prime movers (turbine) to drive an electric generator.

All primary energy resources are limited, and they have a thermal and chemical (pollutant) effect on the environment.

Currently, much of electric energy is produced in constant-speed-regulated synchronous generators that deliver electric energy with constant AC voltage and frequency into regional and national electric power systems, which further transport and distribute it to consumers.

In an effort to reduce environment effects, electric energy markets have been recently made more open, and more flexible distributed electric power systems have emerged. The introduction of distributed power systems is leading to an increased diversity and growth of a wider range power/unit electric energy suppliers. Stability, quick and efficient delivery, and control of electric power in such distributed systems require some degree of power electronics control to allow lower speed for lower power in electric generators to tap the primary fuel energy.

This is how *variable-speed* electric generators have come into play recently [up to 400 (300) MVA/unit], as for example, pump storage wound-rotor induction generators/motors have been in used since 1996 in Japan and since 2004 in Germany.

This book deals in depth with both constant- and variable-speed generator systems that operate in stand-alone and power grid modes.

Chapters have been devoted to topologies, steady-state modeling and performance characteristics, transients modeling, control, design, and testing, and the most representative and recently proposed standard electric generator systems.

The book contains most parameter expressions and models required for full modeling, design, and control, with numerous case studies and results from the literature to enforce the understanding of the art of electric generators by senior undergraduate and graduate students, faculty, and, especially, industrial engineers who investigate, design, control, test, and exploit electric generators for higher energy conversion ratios and better control. This 20-chapter book represents the author's unitary view of the multifacets of electric generators with recent developments included.

Chapter 1 introduces energy resources and fundamental solutions for electric energy conversion problems and their merits and demerits in terms of efficiency and environmental effects. In Chapter 2, a broad classification and principles of various electric generator topologies with their power ratings and main applications are presented. Constant-speed-synchronous generators (SGs) and variable-speed wound-rotor induction generators (WRIGs); cage rotor induction generators (CRIGs); claw pole rotors; induction; PM-assisted synchronous, switched reluctance generators (SRGs) for vehicular and other applications; PM synchronous generators (PMSGs); transverse flux (TF); and flux reversal (FR) PMSGs, and, finally, linear motion PM alternators are all included.

Chapter 3 treats the main prime movers for electric generators from topologies to basic performance equations and practical dynamic models and transfer functions.

Steam, gas, hydraulic, and wind turbines and internal combustion (standard, Stirling, and diesel) engines are dealt with. Their transfer functions are used in subsequent chapters for speed control in corroboration with electric generator power flow control.

Chapters 4 through 8 deal with SGs steady state, transients, control, design, and testing with plenty of numerical examples and sample results that cover the subject comprehensively.

This part of the book is dedicated to electric machines and power systems professionals and industries.

Chapters 9 through 11 deal with WRIGs that have a bidirectional rotor connected AC–AC partial rating PWM converter for variable-speed operation in stand-alone and power grid modes. Steady-state transients (Chapter 9), vector and direct power control (Chapter 10), and design and testing (Chapter 11) are treated in detail with plenty of applications and digital simulation and test results to facilitate in-depth assessment of WRIG systems currently built from 1 MVA to 400 MVA per unit.

Chapters 12 and 13 discuss cage rotor induction generators (CRIG) in self-excited modes used as power grid and stand-alone applications with small speed regulation by a prime mover (Chapter 12) or with full-rating PWM converters connected to a stator and wide-variable speed (Chapter 13) with ± 100% active and reactive power control and constant (or controlled) output frequency and voltage in both power grid and stand-alone operations.

Chapters 9 through 13 are targeted to wind, hydro, and, in general, to distributed renewable power system professionals and industries.

Chapters 14 through 17 deal with representative electric generator systems proposed recently for integrated starter alternators (ISAs) on automobiles and aircraft, all operating at variable speed with full power ratings electronics control. Standard (and recently improved) claw pole rotor alternators (Chapter 14), induction (Chapter 15), PM-assisted synchronous (Chapter 16), and switched reluctance (Chapter 17) ISAs are discussed thoroughly. Again, with numerous applications and results, from topologies, steady state, and transients performance, from modeling to control design and testing for the very challenging speed range constant power requirements (up to 12 to 1) typical to ISA. ISAs have reached the markets, used on a mass-produced (since 2004) hybrid electric vehicles (HEVs) for notably higher mileage and less pollution, especially for urban transport.

This part of the book (Chapters 14 through 17) is targeted at automotive and aircraft professionals and industries.

Chapter 18 deals extensively with radial and axial air gaps, surfaces, and interior PM rotor permanent magnet synchronous generators that work at variable speed and make use of full power rating electronics control. This chapter includes basic topologies, thorough field and circuit modeling, losses, performance characteristics, dynamic models, and bidirectional AC–AC PWM power electronics control in power grid and in stand-alone applications with constant DC output voltage at variable speed. Design and testing issues are included, and case studies are treated using numerical examples and transient performance illustrations.

This chapter is directed at professionals interested in wind and hydraulic energy conversion, generator set (stand-alone) with power/unit up to 3–5 MW (from 10 rpm to 15 krpm) and 150 kW at 80 krpm (or more).

Chapter 19 investigates with numerous case study designs two high-torque density PM synchronous generators (transverse flux [TFG] and flux reversal [FRG]), introduced in the last two decades that take advantage of non-overlapping multipole stator coils. They are characterized by lower copper losses/N m and kg/N m and find applications in very-low-speed (down to 10 rpm or so) wind and hydraulic turbine direct and transmission drives, and medium-speed automotive starter-alternators.

Chapter 20 investigates linear reciprocating and linear progressive motion alternators. Linear reciprocating PMSGs (driven by Stirling free-piston engines) have been introduced (up to 350 W) and are currently used for NASA's deep-mission generators that require fail-proof operation for 50,000 h. Linear reciprocating PMSGs are also pursued aggressively as electric generators for series (full electric propulsion) vehicles for power up to 50 kW or more; finally, they are being proposed for combined electric (1 kW or more) and thermal energy production in residences with gas as the only prime energy provider.

The author thanks the following:

- Illustrious people that have done research, wrote papers, books, patents, and built and tested electric generators and their control over the last decades for providing the author with "the air beneath his wings"
- The author's very able PhD students for electronic editing of the book
- The highly professional, friendly, and patient editors of CRC Press

Professor Ion Boldea
IEEE Life Fellow
University Politehnica Timisoara
Timisoara, Romania

Author

Ion Boldea:

- MS (1967), PhD (1973) in electrical engineering; IEEE member (1977), fellow (1996), and life fellow (2007)
- Visiting scholar in the United States (15 visits, 5 years in all, over 37 years), the United Kingdom, Denmark, South Korea
- Over 40 years of work and extensive publications (most in IEEE trans. and conferences and with IET (former IEE), London, in linear and rotary electric motor/generator modeling, design, their power electronics robust control, and MAGLEVs; 20 national and 5 international patents
- Eighteen books in the field, published in the United States and the United Kingdom
- Technical consultant for important companies in the United States, Europe, South Korea, and Brazil for 30 years
- Repeated intensive courses for graduate students and industries in the United States, Europe, South Korea, and Brazil
- Four IEEE paper prize awards
- Cofounding associate (now consulting) editor from 1977 for *EPCS Journal*
- Founding (2000) and current chief editor of the Internet-only international *Electronic Engineering Journal*: www.jee.ro
- General chair in 10 biannual consecutive events of the International Conference OPTIM (now IEEE-tech-sponsored, ISI and on IEEExplore)
- Member of the European Academy of Arts and Sciences located in Salzburg, Austria
- Member of the Romanian Academy of Technical Sciences (since 1997)
- Correspondent member of Romania Academy (2011)
- IEEE-IAS Distinguished Lecturer (2008–2009) with continued presence ever since
- IEEE "Nikola Tesla" Award 2015

1

Electric Energy and Electric Generators

1.1 Introduction

Energy is defined as the capacity of a body system to do mechanical work. Intelligent harnessing and control of energy determines essentially the productivity and, subsequently, the lifestyle advancement of society.

Energy is stored in nature in quite a few forms such as fossil fuels (coal, petroleum, and natural gas); solar radiation; and in tidal, hydro, wind, geothermal, and nuclear forms.

Energy is not stored in nature in electrical form. However, electric energy is easy to transmit over very long distances and complies with customer's needs through adequate control. More than 40% of energy is converted into electric energy before use, most of it through electric generators that convert mechanical energy into electric energy. Work and energy have identical units. The fundamental unit of energy is 1 J, which represents the work of a force of 1 N in moving a body through a distance of 1 m along the direction of force ($1\,J = 1\,N \times 1\,m$). Electric power is the electric energy rate, and its fundamental unit is $1\,W = 1\,J/s$. Electric energy is more commonly measured in kilowatt-hours (kWh):

$$1\,kWh = 3.6 \times 10^6\,J \tag{1.1}$$

Thermal energy is usually measured in calories (cal). By definition, 1 cal is the amount of heat required to raise the temperature of 1 g of water from 15°C to 16°C. The kilocalorie (kcal) is even more common ($1\,kcal = 10^3\,cal$).

As energy is a unified concept, as expected, the joule and calorie are directly proportional:

$$1\,cal = 4.186\,J \tag{1.2}$$

A larger unit for thermal energy is the British thermal unit (BTU):

$$1\,BTU = 1055\,J = 252\,cal \tag{1.3}$$

A still larger unit is the quad (quadrillion BTU):

$$1\,quad = 10^{15}\,BTU = 1.055 \times 10^{18}\,J \tag{1.4}$$

In the year 2000, the world used about 16×10^{12} kWh of energy, an amount above most projections (Figure 1.1).

An annual growth of 3.3%–4.3% was typical for world energy consumption in the period between 1990 and 2000. A slightly lower rate is forecasted for the next 30 years.

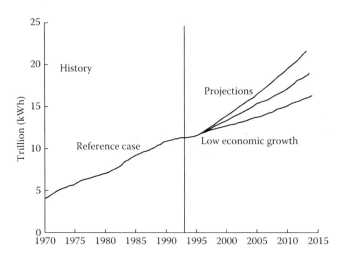

FIGURE 1.1 Typical world annual energy requirements.

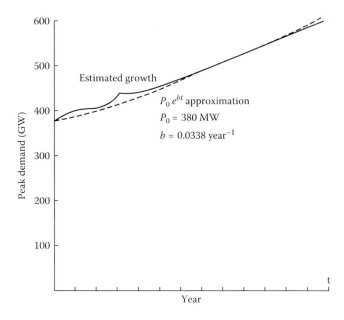

FIGURE 1.2 Peak electric power demand in the United States and its exponential predictions.

Besides annual energy usage (and growth rate), with over 30%–40% of total energy being converted into electric energy, it is equally important to evaluate and predict the electric power peaks for each country (region), as they determine the electric generation reserves. The peak electric power in the United States over several years is shown in Figure 1.2.

Peak power demands tend to be more dynamic than energy needs, and thus electric energy planning becomes an even more difficult task.

Implicitly, the transients and stability in electric energy (power) systems of the future tend to be more severe.

To meet these demands, we need to look at the main energy sources: their availability, energy density, the efficiency of the energy conversion to thermal to mechanical to electrical energy, and their secondary ecological effects (limitations).

1.2 Major Energy Sources

With the current annual growth in energy consumption, the fossil fuel supplies of the world will be depleted in, at best, a few hundred years, unless we do switch to other sources of energy or use energy conservation to tame energy consumption without compromising the quality of life.

The world estimated reserves of fossil fuel [1] and their energy density are shown in Table 1.1.

With a 14-year doubling time of energy consumption, if only coal would be used, the whole coal reserves will be depleted in about 125 years. Even if the reserves of fossil fuel were large, their predominant or exclusive usage is not feasible due to environmental, economical, or even political reasons.

Alternative energy sources are to be used more and more, with fossil fuels used slightly less, gradually, and more efficiently than today.

The relative cost of electric energy in 1991 from different sources is shown in Table 1.2.

Wind energy conversion is becoming cost-competitive while it is widespread, and it has limited environmental impact.

Unfortunately, its output is not steady, and thus very few energy consumers rely solely on wind power to meet their electric energy demands. As in general, the electric power plants are connected in local or regional power grids with regulated voltage and frequency, connecting of large wind generator parks to the power grids may produce severe transients that have to be taken care of using sophisticated control systems with energy storage elements in most cases.

By the year 2014, over 300,000 MW of wind power generators were in place, a good part in the United States.

The total wind power resources of the planet are estimated at 15,000 TWh, so much more work in this area is expected in the near future.

Another indirect means of using solar energy, besides wind energy, is the stream-flow part of the hydrological natural cycle.

The potential energy of water is transformed into kinetic energy using a hydraulic turbine, which then drives an electric generator. The total hydropower capacity of the world is about 3×10^{12} W. Only less than 9% of which is used today because many regions with greatest potential have economic problems.

However, as energy costs from water are low, resources are renewable and with limited ecological impact. Despite initial high costs, hydropower is up for a new surge.

Tidal energy is obtained by filling a bay, closed by a dam, during periods of high tides and emptying it during low-tide time intervals. The hydraulic turbine to be used in tidal power generation should be reversible so that tidal power is available twice during each tidal period of 12 h and 25 min.

Though the total tidal power is evaluated at 64×10^{12} W, its occurrence in short intervals only requires large rating turbine—generator systems that are still expensive. The energy burst cannot be easily matched with demand unless large storage systems are built.

TABLE 1.1   Estimated Reserves of Fossil Fuel

Fuel	Estimated Reserves	Energy Density (W-h)
Coal	7.6×10^{12} metric tons	937 per ton
Petroleum	2×10^{12} barrels	168 per barrel
Natural gas	10^{16} ft^3	0.036 per ft^3

TABLE 1.2   Cost of Electric Energy

Energy Source	Cents/kWh
Gas (in high efficiency combined cycle gas turbines)	3.4–4.2
Coal	5.2–6
Nuclear	7.4–6.7
Wind	4.3–7.7

These demerits make many of us still believe that the role of tidal energy to the world demand will be very limited, at least in the new future. However, exploiting submarine currents energy in wind-like low-speed turbines may be feasible.

Geothermal power is obtained by extracting the heat inside the earth. With a 25% conversion ratio, the useful geothermal electric power is estimated at 2.63×10^{10} MWh.

Fission and fusion are two forms of nuclear energy conversion to produce heat. Heat is converted to mechanical power in steam turbines that drive electric generators to produce electric energy.

Only fission-splitting nuclei of a heavy element such as uranium 235 are used commercially to produce a good percentage of electric power, particularly in developed countries.

As uranium 235 is in scarce supply, uranium 238 is converted into fissionable plutonium by absorbing neutrons. One gram of uranium 238 will produce about 8×10^{10} J of heat. The cost of nuclear energy is still slightly higher than that from coal or gas (Table 1.2). The environmental problems with used nuclear fuel conservation or with reactor potential explosions are more and more the problem of public acceptance.

Fusion power combination of light nuclei such as deuterium and tritium at high temperatures and pressures is scientifically feasible, but it is not yet technically proven for efficient energy conversion.

Solar radiation may be used either through heat solar collectors or through direct conversion to electricity in photovoltaic cells. From an average of 1 kW/m^2 of solar radiation, less than 180 W/m^2 could be converted to electricity with up-to-date solar cells. Small energy density and nonuniform availability (during mainly sunny days) lead to higher cents/kWh than from other sources.

1.3 Limitations of Electric Power Generation

Factors limiting electric energy conversion are related to the availability of various fuels, technical constraints, ecological, social, and economical issues.

Ecological limitations are due to low-temperature excess heat and carbon dioxide (solid particles) and of oxide of sulfur nitrogen emissions from fuel burning.

Low-temperature heat exhaust is typical in any thermal energy conversion. When too large, this heat increases the earth's surface temperature and, together with the emission of carbon dioxide and certain solid particles, have intricate effects on the climate. Global warming and climate changes appear to be caused by burning too much fossil fuel. Since the Three Mile Island, Chernobyl and Fukushima incidents, safe nuclear electric energy production has become not only a technical issue but also an ever-increasing social (public acceptance) problem.

Even hydro and wind energy conversion do pose some environmental problems, though much smaller than the fossil or nuclear fuel energy conversion does. We refer to changes in flora and fauna due to hydro-dams intrusion in the natural habitat. Big windmill farms tend to influence the fauna and are also sometimes considered "ugly" to the human eye.

Consequently, in forecasting the growth of electric energy consumption on earth, we must consider all these very complex limitation factors.

Shifting to more renewable energy sources (wind, hydro, tidal, solar, etc.) while using combined heat-electricity production from fossil fuels to increase the energy conversion factor, together with intelligent energy conservation, could be a complicated, but a potent, way of increasing material prosperity in more harmony with the environment.

1.4 Electric Power Generation

Electric energy (power) is produced by coupling a prime mover that converts the mechanical energy (called a turbine) to an electrical generator, which then converts the mechanical energy into electrical energy (Figure 1.3). An intermediate form of energy is used for storage in the electrical generator.

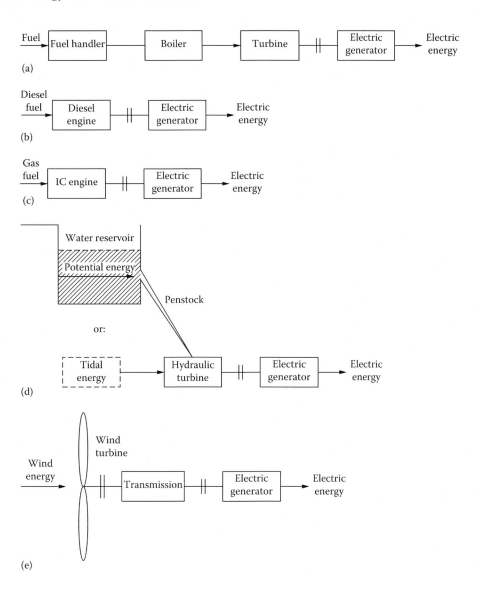

FIGURE 1.3 Most important ways to produce electric energy (a–e).

This is the so-called magnetic energy stored, mainly between the stator (primary) and the rotor (secondary). The main types of "turbines" or prime movers are as follows:

Steam turbines
Gas turbines
Hydraulic turbines
Wind turbines
Diesel engines
Internal combustion engines

Figure 1.3, which is self-explanatory, illustrates the most commonly used technologies to produce electric energy. They all use a prime mover that outputs mechanical energy. There are also direct ways of electric energy production that avoid the mechanical energy stage—such as photovoltaic, thermoelectric, and electrochemical (fuel cells) technologies. As they do not use electric generators, and still represent only a tiny part of all electric energy produced on earth, their treatment falls beyond the scope of this book.

The steam (or gas) turbines in various configurations practically make use of all fossil fuels from coal to natural gas and oil and nuclear fuel or geothermal energy inside earth.

Usually, their efficiency reaches 40%; but in combined cycle (producing heat and mechanical power), their efficiency has reached recently 55%–60%. Powers per unit go as high as 100 MW and more at 3000 (3600 rpm); but for lower powers, in the MW range, higher speeds are feasible to reduce weight (volume) per power.

Recently, low-power, high-speed gas turbines (with combined cycle) in the range of 100 kW at 70,000–80,000 rpm became available. Electric generators to match this variety of powers and speeds have also been produced. Such electric generators are also used as starting motors for jet plane engines on board of aircraft.

High speed, low volume and weight, and reliability are key factors for electric generators on board aircrafts. Power ranges are from hundreds of kWs to 1 MW in large aircraft. On ships or trains, electric generators are required either to power the electric propulsion motors or (and) for auxiliary multiple needs. Diesel engines (Figure 1.3b) drive the electric generators on board of ships or trains.

On vehicles, electric energy is used for various tasks for powers up to a few tens of klilowatts in general. The internal combustion (or diesel) engine drives an electric generator (alternator) directly or through a transmission (Figure 1.3c). The ever-increasing need for more electric power on vehicles to perform various tasks—from lighting to engine startup and from door openers to music devices and wind shield wipers and cooling blowers—imposes new challenges on electric generators of the future.

Hydraulic potential energy is converted to mechanical energy in hydraulic energy turbines. They in turn drive electric generators to produce electric energy. In general, the speed of hydraulic turbines is rather low—below 500 rpm—but in many cases below 100 rpm.

The speed depends on the water head and flow rate. High water head leads to higher speed, while high flow rate leads to lower speeds. Hydraulic turbines for low, medium, and high water heads have been perfected into a few most favored embodiments (Kaplan, Pelton, Francis, bulb type, Straflo, etc.)

With a few exceptions—in Africa, Asia, Russia, China, and South America—many large power/unit water energy reservoirs have been provided with hydroelectric power plants with large powers (in the hundreds and thousands of megawatt). Still by 1990 only 15% of the world's main 624,000 MW reserves have been put to work. However, very many smaller water energy reservoirs still remain untapped. They need small hydrogenerators with powers below 5 MW at speeds of a few hundred revolutions per minute. In many locations, tens of kilowatt micro hydrogenerators are more appropriate [2–5].

The time of small and micro hydroenergy has come finally, especially in Europe and North America where reserves left are less. Tables 1.3 and 1.4 show the world use of hydroenergy in terawatt in 1997 [6,7].

The World Energy Council estimated that by 1990 from total electric energy demand of 12,000 TWh, about 18.5% were contributed by hydroenergy. By 2020, the world electric energy demand is estimated to be 23,000 TWh. From this, if only 50% of all economically feasible hydroresources were put to work, the hydro would contribute 28% of the 2020s total electric energy demands.

These figures indicate that a new era of dynamic hydroelectric power development is to come soon, if the humanity desires more energy (prosperity for more people) but with a small impact on the environment (constant or less greenhouse emission effects).

TABLE 1.3   World Hydropotential by Region in Terawatt

	Gross	Economic	Feasible
Europe	5,584	2,070	1,655
Asia	13,399	3,830	3,065
Africa	3,634	2,500	2,000
America	11,022	4,500	3,600
Oceania	592	200	160
Total	34,231	13,100	10,480

TABLE 1.4   Proportion of Hydro Already Developed

Africa	6%
South and Central America	18%
Asia	18%
Oceania	22%
North America	55%
Europe	65%

Source: World Energy Council.

Wind energy reserves, though discontinuous and unevenly distributed, mostly around shores, are estimated at four times the electric energy needs of today.

To its uneven distribution, its discontinuity and some not insurmountable public concerns about fauna and human habitats, we have to add the technical sophistication and costs required to control, store, and distribute wind electric energy. These are the obstacles to the widespread use of wind energy. For comparison more than 100,000 MW of hydropower reserves are tapped today in the world.

But ambitious plans are ahead with European Union planning to install more than 10,000 MW from 2010 to 2020.

The power per unit for small hydropower increased to 4 MW and, for wind turbines, it increased up to 8 MW. More are being designed; but as the power per unit increases, the speed decreases to 10 rpm.

This poses an extraordinary problem: either use a special transmission and a high-speed generator or build a directly-driven low-speed generator. Both solutions have their own merits and demerits.

The lowest speeds in hydrogenerators are in general above 50 rpm but at much higher powers and thus much higher rotor diameters, which lead to still good performance.

Preserving high performance at 1–10 MW and less at speeds below 30 rpm in an electric generator poses serious challenges, but better materials, high energy permanent magnets, and ingenious designs are likely to facilitate in solving these problems.

It is estimated that wind energy will produce more than 10% of electric energy by 2020.

This means that wind energy technologies and business are apparently up to a new revival—this time with sophisticated control and flexibility provided by high-performance power electronics.

1.5 From Electric Generators to Electric Loads

Electric generators operate traditionally either in large power grids—with many of them in parallel to provide voltage and frequency stability to changing load demands—or they stand alone.

The conventional large power grid supplies most electric energy needs and consists of electric power plants, transmission lines, and distribution systems (Figure 1.4).

Multiple power plants, many transmission power lines and complicated distribution lines, and so forth constitute a real regional or national power grid.

Such large power grids with a pyramidal structure—generation—transmission—distribution and billing—are now in place and, to connect a generator to such a system, comply with strict rules.

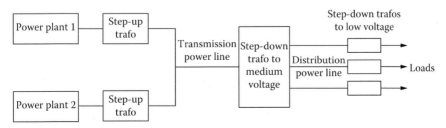

FIGURE 1.4   Single transmission in multiple power plant—standard power grid.

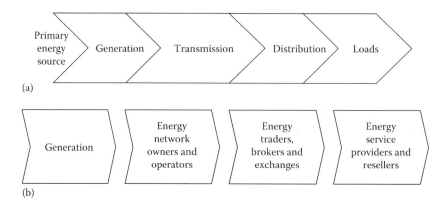

FIGURE 1.5 (a) Standard value chain power grid and (b) unbundled value chain.

The rules, standards, are necessary to provide quality power in terms of continuity, voltage, and frequency constancy, phase symmetry, faults treatment, and so forth.

The bigger, the more stable seems to be the driving force behind building such huge "machine systems."

The bigger the power/unit, the higher the energy efficiency was for decades, the rule that leads to steam generators of up to 1500 MW and hydrogenerators up to 760 MW.

However, investments in new power plants, redundant transmission power lines, and distribution systems did not always keep up with ever-increasing energy demands. This is how the blackouts have developed. Aside from extreme load demands or faults, the stability of power grids is limited mainly by the fact that existing synchronous electric generators work only at synchronism, that is, at a speed n_1 rigidly related to frequency f_1 of voltage $f_1 = n \times p$. Standard power grids are served exclusively by synchronous generators and have a pyramidal structure a called "utility" (see Figure 1.5). Utilities still run, in most places, the entire process from generation to retail settlement.

Today, the electricity market is deregulating at various pace in different parts of the world, though the process has been still considered in its infancy.

The new unbundled value chain (Figure 1.5b) breaks out the functions into the basic types: electric power plants, energy network owners, and operators; energy traders, brokers, and exchanges; and energy service providers and retailers [8,9]. This way, it is hoped to stimulate competition for energy costs reduction while also improving the power quality at the end users, by sustainable environment and more friendly technologies. Increasing the number of players needs clear rules of the game. In addition, the transients stresses on such power grid, with many energy suppliers entering, exiting, or varying their input, are likely to be more severe. To counteract such a difficulty, more flexible power transmission lines have been proposed and introduced in a few places (mostly in the United States) under the logo "FACTS": flexible AC transmission systems [10].

FACTS introduces controlled reactive power capacitors in the power transmission lines in parallel, for higher voltage stability (short-term voltage support), and (or) in series for larger flow management in the long term (Figure 1.6).

Power electronics at high power and voltage levels is the key technology to FACTS.

Additionally, FACTS includes the AC–DC–AC power transmission lines to foster stability and to reduce losses in energy transport over large distances (Figure 1.7).

The direct current (DC) high-voltage large power bus allows for parallel connection of energy providers with only voltage control, and thus the power grid becomes more flexible.

However, this flexibility occurs at the price of full-power high-voltage converters that take advantage of the power electronics technologies.

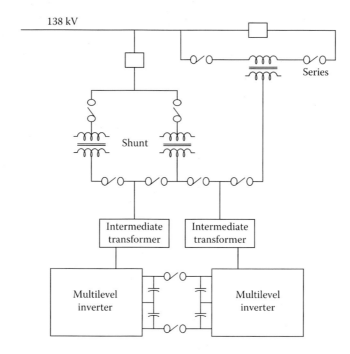

FIGURE 1.6 FACTS: series-parallel compensator.

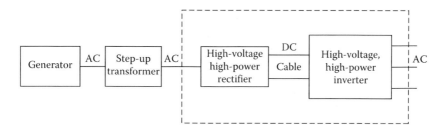

FIGURE 1.7 AC–DC–AC power cable transmission system.

Still, most electric generators are synchronous machines that need tight (rigid) speed control to provide constant frequency output voltage.

To connect such generators in parallel, the speed controllers (governors) have to allow for a speed droop in order to produce balanced output of all generators. Of course, frequency also varies with load, but this variation is limited to less than 0.5 Hz.

Variable speed constant voltage and frequency generators with decoupled active and reactive power control would make the power grids naturally more stable and more flexible.

The doubly fed induction generator (DFIG) with three-phase pulse width modulation (PWM) bidirectional converter in the three-phase rotor circuit supplied through brushes and slip rings does just that (Figure 1.8).

DFIG works as a synchronous machine and thus, fed in the rotor in AC at variable frequency f_2, and operating at speed n it delivers power at the stator frequency f_1:

$$f_1 = np + f_2 \tag{1.5}$$

where $2p$ is the number of poles of stator and rotor windings.

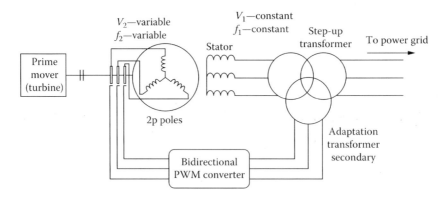

FIGURE 1.8 Variable speed constant voltage and frequency generator.

The frequency f_2 is considered positive when the phase sequence in the rotor is same as in the stator and negative otherwise. In the conventional synchronous generator $f_2 = 0$ (DC). DFIG is capable to work at $f_2 = 0$ but also for $f_2 <> 0$. With a bidirectional power converter, DFIG may work both as motor and generator with f_2 negative and positive, that is, at speeds lower and larger than that of the standard synchronous machine. Starting is initiated from the rotor, with stator temporarily short-circuited, then opened. Then the machine is synchronized and operated as motor or generator. The "synchronization" is feasible at all speeds within the design range (±20% in general). Therefore, not only generating but pumping mode is also available, besides flexibility in fast active and reactive power control.

Pump storage is used to store energy during peak-off hours and is then used for generation during peak hours at a total efficiency around 70% in large head hydropower plants.

DFIG units of up to 400 MW with about ±5% speed variation have been put to work in Japan, and also more recently (in 300 MW units) in Germany. The converter rating is about equal to the speed variation range, which limits the costs notably. Pump storage plants with conventional synchronous machines working as motors have been used in place for a few decades. DFIG/M, however, provides the optimum speed for pumping, which, for most hydroturbines, is different from that for generating.

While fossil-fuel DFIGs may be very good for power grids because of stability improvements, they are definitely the solution when pump storage is used and for wind generators above 1 MW per unit.

Will DFIG replace gradually the omnipresent synchronous generators in bulk electric energy conversion? Most likely, as the technology is here already up to 400 MW/unit. At the distribution stage (local) stage (Figure 1.5b), a new structure is getting ground: the distributed power system (DPS). It refers to low power energy providers that can meet or supplement some local power needs.

DPS is expected to either work alone or be connected at the distribution stage to existing systems.

DPS is to be based on renewable resources such as wind, hydro, biomass alone, or integrate also gas turbine generators or diesel engine generators, solar panels, and fuel cells.

Powers in the orders of 1–2 MW possibly up to 5 MW/unit of energy conversion are contemplated.

DPSs are to be provided with all means of control, stability, and power quality so typical to conventional power grids. With one big difference, they will make full use of power electronics to provide fast and robust active and reactive power control.

Here, besides synchronous generators with electromagnetic excitation, PM synchronous and cage-rotor induction generators and DFIGs, all with power electronics control for variable speed operation, are already in place in quite a few applications. But their wide spread usage is only about to take place.

Stand-alone electric power generation ties directly the electric energy generator to the load. Stand-alone systems may have one generator only (i.e., on board of trains and auto vehicles standby power groups) or 2–4 such generators such as on board of large aircraft or vessel. Stand-alone gas-turbine residential generators are also investigated for decentralized electricity production.

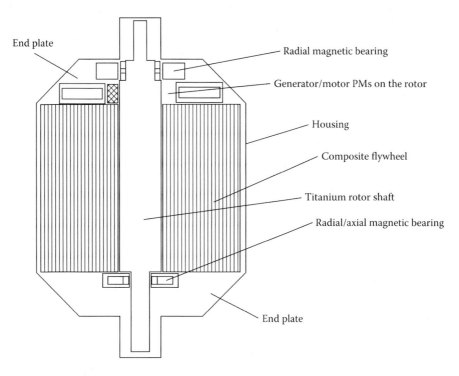

FIGURE 1.9   Typical flywheel battery.

Stand-alone generators and their control are tightly related to the application, from the design to the embodiment of control and protection.

Vehicular generators have to be lightweight and efficient, in this order.

Standby (backup) power groups for hospitals, banks, telecom, have to be quickly available, reliable, efficient, and environment friendly.

Backup power generators are becoming a must in public buildings as now all use a cluster of computers. Battery or fuel cell-based uninterruptible power supplies (UPS), all with power electronics control, are also used at lower powers especially. They do not include electric generators and fall beyond our scope here.

Electric generators or motors are also used for mechanical energy storage "inertial batteries" (Figure 1.9) in vacuum, with magnetic suspension to store energy for minutes to hours. Speeds up to 1 km/s (peripheral maximum speed with composite material flywheels) at the cost of 400–800 $/kW (50–100 $/kW for lead acid batteries) for an operation life of over 20 years (3–5 years for lead acid batteries) [11] are feasible today.

PM synchronous generator/motors are ideal for the scope at rotational speeds, preferably around 40 krpm, for the 3–300 kW range and less for the MW range.

Satellites, power quality systems (for active power control through energy storage), hybrid buses, trains (to store energy during braking), and electromagnetic launchers are typical applications for storage generator–motor systems. The motoring mode is used to reaccelerate the flywheel (or charge the inertia battery) via power electronics.

Energy storage up to 500 MJ (per unit) is considered practical for applications that (at 50 Wh/kg density or more) need energy delivered in seconds or minutes at a time, for the duration of power outage. As most (80%) of power line disturbances last for less than 5 s, flywheel batteries can fill up this time with energy as standby power source. Yet very promising electrochemical and superconducting coil energy storage fall beyond our scope here.

1.6 Summary

This introductory study leads to the following conclusions:

Electric energy demand is on the rise (at a rate of 2%–3% per annum), but so are the environmental and social constraints on the electric energy technologies.

Renewable resources input is on the rise—especially wind and hydro, at powers of up to a few MW per unit.

Single-value power grids will change to bundled valued chains with electric energy opening to markets.

Electric generators should work at variable speed but provide constant voltage and frequency output via power electronics with full or partial power ratings, to tap more energy from renewable and provide faster and safer reactive power control.

The standard synchronous generator, working at constant speed for constant frequency output, is challenged by the DFIG at high-to-medium power (from hundreds of megawatts up to 10 MW) and by the PM synchronous generator and the induction generator with full-power bidirectional power electronics in the stator up to 10 MW range.

Most variable speed generators, with bidirectional power electronics control, will also allow motoring (or starting) operation both in conventional or distributed power grids and in stand-alone (or vehicular) applications.

Home and industrial combined heat and electricity generation by burning gas in high-speed gas-turbines imposes special electric generators with adequate power electronics digital control.

In view of such a wide power/unit and applications range, a classification of electric generators seems in order and this is the subject of Chapter 2.

References

1. B. Sadden, Hydropower development in southern and southeastern Asia, *IEEE Power Engineering Review*, 22(3), 5–11, March 2002.
2. J.A. Veltrop, Future of dams, *IEEE Power Engineering Review* 22(3), 12–18, March 2002.
3. H.M. Turanli, Preparation for the next generation at Manitoba Hydro, *IEEE Power Engineering Review*, 22(3), 19–23, March 2002.
4. O. Unver, Southeastern Anatolia development project, *IEEE Power Engineering Review*, 22(3), 10–11, 23–24, March 2002.
5. H. Yang, G. Yao, Hydropower development in Southern China, *IEEE Power Engineering Review*, 22(3), 16–18, March 2002.
6. T.J. Hammons, J.C. Boyer, S.R. Conners, M. Davies, M. Ellis, M. Fraser, E.A. Nolt, J. Markard, Renewable energy alternatives for developed countries, *IEEE Transactions on Energy Conversion*, EC-15(4), 481–493, 2000.
7. T.J. Hammons, B.K. Blyden, A.C. Calitz, A.G. Gulstone, E.I. Isekemanga, R. Johnstone, K. Paleku, N.N. Simang, F. Taher, African electricity infrastructure interconnection and electricity exchanges, *IEEE Transactions on Energy Conversion*, EC-15(4), 470–480, 2000.
8. C. Lewiner, Business and technology trends in the global utility industries, *IEEE Power Engineering Review*, 21(12), 7–9, 2001.
9. M. Baygen, A vision of the future grid, *IEEE Power Engineering Review*, 21(12), 10–12, 2001.
10. A. Edris, FACTS technology development: An update, *IEEE Power Engineering Review*, 20(3), 4–9, 2000.
11. R. Hebner, J. Beno, A. Walls, Flywheel batteries come around again, *IEEE-Spectrum*, 39(4), 46–51, 2002.

2

Principles of Electric Generators

2.1 Three Types of Electric Generators

Electric generators may be classified in many ways, but the following are deemed fully representative:

- By principle
- By applications domain

The applications domain implies the power level. The classifications by principle unfolded here include commercial (widely used) types together with new configurations, which are still in the laboratory (although advanced) stages.

By principle, there are three main types of electric generators:

- Synchronous (Figure 2.1)
- Induction (Figure 2.2)
- Parametric (with magnetic anisotropy and permanent magnets)—Figure 2.3

Parametric generators have in most configurations doubly salient magnetic circuit structures; therefore, they may also be called "doubly salient electric generators."

Synchronous generators (SGs) [1–4] have, in general, a stator magnetic circuit made of laminations provided with uniform slots that house a three-phase (sometimes single- or two-phase) winding and a rotor. It is the rotor design that leads to a cluster of SG configurations as can be seen in Figure 2.1.

They are all characterized by the rigid relationship among speed n, frequency f_1, and the number of poles $2p_1$:

$$n = \frac{f}{p_1} \tag{2.1}$$

Those that are DC excited require power electronics excitation control, while those with permanent magnets (PMs) or (and) variable reluctance rotors have to use full power electronics in the stator to operate at adjustable speed. Finally, even electrically excited, SGs may be provided with full power electronics in the stator when they work alone or in power grids with DC high-voltage cable transmission lines.

Based on its principles, each of these configurations will be presented later on in this chapter.

For powers in the MW/unit range and less, induction generators have also been introduced. They are (Figure 2.2) as follows:

- With cage rotor and single stator winding
- With cage rotor and dual (main and additional) stator winding with different number of poles, in general
- With wound rotor

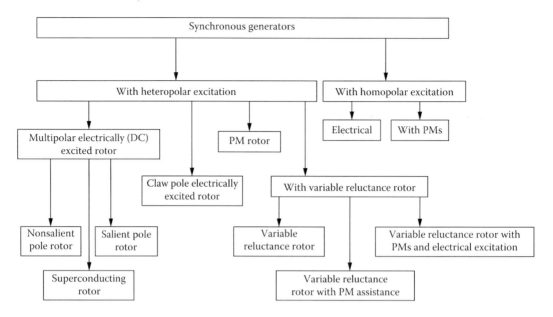

FIGURE 2.1 Synchronous generators.

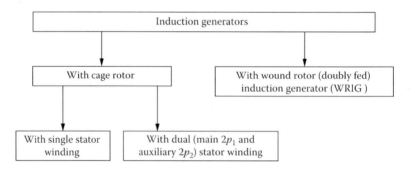

FIGURE 2.2 Induction generators.

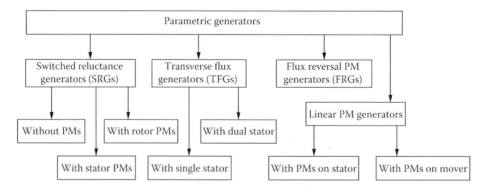

FIGURE 2.3 Parametric generators.

PWM converters are connected to the stator (for the single stator winding and, respectively, to the auxiliary stator winding for the case of dual stator winding)

The principle of the induction generator with single stator winding relies on the equation:

$$f_1 = p_1 n + f_2 \tag{2.2}$$

where

$f_1 > 0$ is the stator frequency
$f_2 <> 0$ is the slip (rotor) frequency
n is the rotor speed (rps)

f_2 may be either positive or negative in Equation 2.2, even zero, provided the PWM converter in the wound rotor is capable of supporting bidirectional power flow for speeds n above f_1/p_1 and below f_1/p_1.

Note that for $f_2 = 0$ (DC rotor excitation), the SG operation mode is reobtained with the DFIG.

The slip S definition is as follows:

$$S = \frac{f_2}{f_1} <> 0 \tag{2.3}$$

The slip is zero as $f_2 = 0$ (DC) for the SG mode.

For the dual stator winding, the frequency–speed relationship is applied twice:

$$\begin{aligned} f_1 &= p_1 n + f_2; \quad p_2 > p_1 \\ f_1' &= p_2 n + f_2' \end{aligned} \tag{2.4}$$

Therefore, the rotor bars experience, in principle, currents of two distinct (rather low) frequencies f_2 and f_2'. In general, $p_2 > p_1$ to cover lower speeds.

The PWM converter feeds the auxiliary winding. Consequently, its rating is notably lower than that of the full power of the main winding and is proportional to speed variation range.

As it may work in the pure synchronous mode too, the DFIG may be used up to highest levels of powers for SGs (400 MW units have been already in use for some years in Japan) and 2× 300 MW pump storage plant is being now commissioned in Germany.

On the contrary, the cage rotor induction generator (CRIG) is more suitable for powers in the megawatt and lower power range.

Parametric generators rely on the variable reluctance principle but may also use permanent magnets to enhance the power per volume and reduce generator losses.

There are quite a few configurations that suit this category such as the switched reluctance generator (SRG), the transverse flux PM generator (TFG), the flux reversal generator (FRG). In general, their principle relies on co-energy variation due to magnetic anisotropy (without or with PMs on rotor or on stator), in the absence of a pure traveling field with constant speed (f_1/p), so characteristic for synchronous and induction generators (machines).

2.2 Synchronous Generators

Synchronous generators (see Figure 2.1 for classifications) are characterized by an uniformly slotted stator laminated core, which hosts a three-, or two-, or one-phase AC winding and a DC current excited, or PM-excited or variable saliency, rotor [1–5].

As only two travelling fields—of the stator and rotor—at relative standstill interact to produce a ripple-less torque, the speed n is rigidly tied to stator frequency f_1, because the rotor-produced magnetic field is DC, typically heteropolar in SGs.

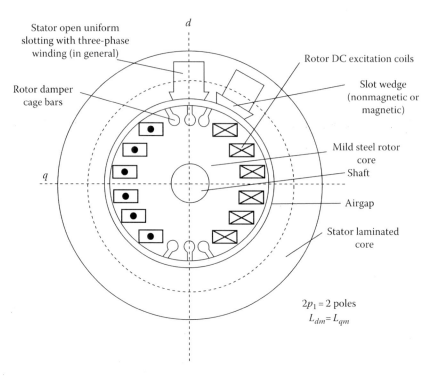

FIGURE 2.4 Synchronous generator with nonsalient pole heteropolar, DC distributed excitation.

They are built with nonsalient pole, distributed excitation rotors (Figure 2.4) for $2p_1 = 2$, 4 (i.e., high speed or turbogenerators) or with salient-pole concentrated excitation rotor (Figure 2.5) for $2p_1 > 4$ (in general, for low-speed or hydrogenerators).

As power increases, the rotor peripheral speed also increases. In large turbogenerators, it may reach more than 150 m/s (in a 200 MVA machine $D_r = 1.2$ m diameter rotor at $n = 3600$ rpm, $2p_1 = 2$, $U = \pi D_r$ $n = \pi \times 1.2 \times 3600/60 > 216$ m/s). The DC excitation placement in slots, with DC coil end connections protected against centrifugal forces by rings of highly resilient resin materials, thus becomes necessary. In addition, the DC rotor current air-gap field distribution is closer to a sinusoid.

Consequently, the harmonics content of the stator-motion-induced voltage (emf or no-load voltage) is smaller, thus complying with the strict rules (standards) of commercial, large power grids.

The rotor body is made of solid iron for better mechanical rigidity and heat transmission.

The stator slots in large SGs are open (Figures 2.4 and 2.5), and they are sometimes provided with magnetic wedges in order to further reduce the field space harmonics and thus reduce the emf harmonics content and rotor additional losses in the rotor damper cage. When $n = f_1/p_1$ and for steady-state (sinusoidal symmetric stator currents of constant amplitude), the rotor damper cage currents are zero. However, should any load or mechanical transients occur, eddy currents show up in the damper cage to attenuate the rotor oscillations when the stator is connected to a constant frequency and voltage (high power) grid.

The rationale neglects the stator magnetomotive force space harmonics due to the placement of windings in slots and also due to slot openings. These space harmonics induce voltages and thus produce eddy currents in the rotor damper cage even during steady state.

In addition, even during steady state, if the stator phase currents are not symmetric, their inverse components produce currents of $2f_1$ frequency in the damper cage.

Consequently, to limit the rotor temperature, the degree of current (load) unbalance permitted is limited by standards.

Nonsalient pole DC-excited rotor SGs have been manufactured for $2p_1 = 2$, 4 poles high speed turbogenerators that are driven by gas or steam turbines.

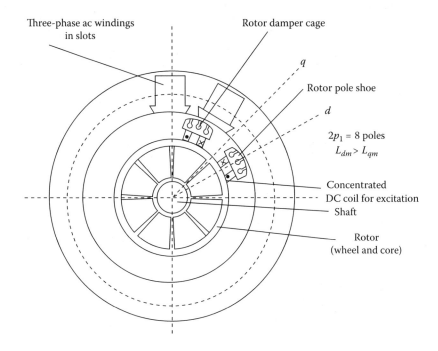

FIGURE 2.5 Synchronous generator with salient pole heteropolar DC-concentrated excitation.

For lower speed SGs with a large number of poles ($2p_1 > 4$), the rotors are made of salient rotor poles provided with concentrated DC excitation coils. The peripheral speeds are lower than for turbogenerators; even for high-power hydrogenerators (for 200 MW 14 m rotor diameter at 75 rpm, and $2p_1 = 80, f_1 = 50$ Hz, the peripheral speed $U = \pi \times D_r \times n = \pi \times 14 \times 75/60 > 50$ m/s). About 80 m/s is the upper limit, in general, for salient pole rotors. Still, the excitation coils have to be protected against centrifugal forces.

The rotor pole shoes may be made of laminations—to reduce additional rotor losses—but the rotor pole bodies and core are made of mild magnetic solid steel.

With a large number of poles, the stator windings are built for a smaller number of slot/pole: between 6 and 12, in many cases. The number of slots per pole and phase, q, is thus between 2 and 4. The smaller the q, the larger the presence of space harmonics in the emf. A fractionary q might be preferred, say 2.5, which also avoids the subharmonics and leads to a cleaner (more sinusoidal) emf, to comply with the current standards.

The rotor pole shoes are provided with slots that house copper bars short-circuited by copper rings to form a rather complete squirrel cage. A stronger damper cage is thus obtained.

DC excitation power on the rotor is transmitted by one of the following:

- Copper sliprings and brushes (Figure 2.6)
- Brushless excitation systems (Figure 2.7)

The controlled rectifier, whose power is around 3% of generator rated power, and that has a sizeable voltage reserve to force the current in the rotor quickly, controls the DC excitation currents according to the needs of generator voltage and frequency stability.

Alternatively, an inverted SG (whose three-phase AC windings and diode rectifier are placed on the rotor and the DC excitation in the stator) may play the role of a brushless exciter (Figure 2.7). The field current of the exciter is controlled through a low-power half-controlled rectifier. Unfortunately, the electrical time constant of the exciter generator notably slows down the response in the main SG excitation current control.

Still another brushless exciter could be built around a single-phase (or three phase) rotating transformer working at a frequency above 300 Hz to cut down its volume considerably (Figure 2.8).

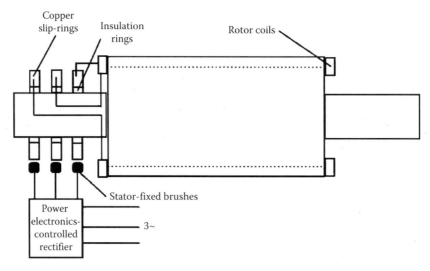

FIGURE 2.6 Slip-ring-brush power electronics rectifier DC excitation system.

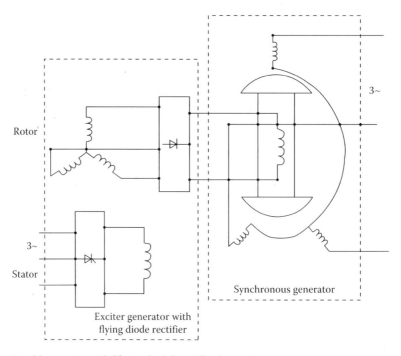

FIGURE 2.7 Brushless exciter with "flying diode" rectifier for synchronous generators.

An inverter is required to feed the transformer primary at variable voltage but at constant frequency.

The response time in the generator's excitation current control is short, and the size of the rotating transformer is rather small.

In addition, the response in the excitation control does not depend on speed and may be used from standstill.

Claw pole (Lundell) SGs are built now mostly as car alternators. The excitation winding power in p.u. is reduced considerably for the multiple rotor construction ($2p_1 = 10, 12, 14$) preferred to reduces external diameter and machine volume.

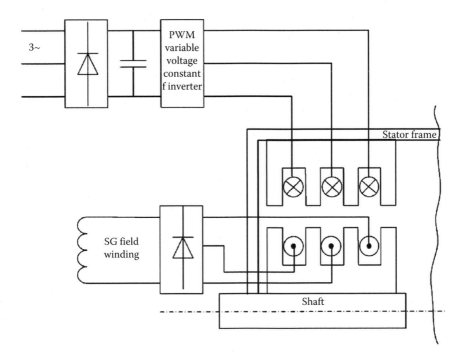

FIGURE 2.8   Rotating transformer with inverter in the rotor as brushless exciter.

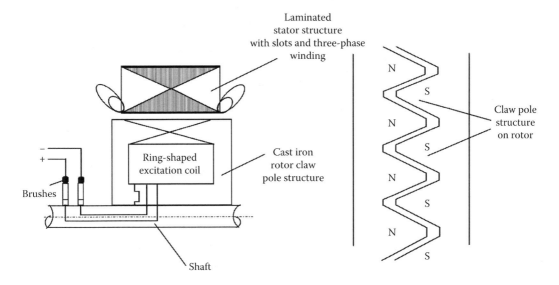

FIGURE 2.9   The claw pole synchronous generator.

The claw pole solid cast iron structure (Figure 2.9) is less costly to manufacture, while the single ring-shaped excitation coil produces a multipolar air-gap field (though with a three-dimensional field path) with reduced copper volume and DC power losses.

The stator holds a simplified three-phase single-layer winding with three slots per pole, in general. Though slip-rings and brushes are used, the power transmitted through them is small (in the order of 60–200 W for car and truck alternators), and thus small power electronics is used to control the output.

The total costs of the claw pole generator for automobiles including field current control and the diode full power rectifier is low and so is the specific volume.

However, the total efficiency, including the diode rectifier and excitation losses, is low at 14 V DC output: 70% at 2.5 kW and 6 krpm and below 55% at 12 krpm. To blame are the diode losses (at 14 V DC), the mechanical losses, and the eddy currents induced in the claw poles by the space and time harmonics of the stator currents mmf. Increasing the voltage to 42 V DC would reduce the diode losses in relative terms, while the building of the claw poles from composite magnetic materials would reduce notably the claw pole eddy current losses. A notably higher efficiency would result (80% at 8 kW), even if the excitation power might slightly increase, due to lower permability ($500\mu_0$) of the today's best composite magnetic materials. Also higher power levels might be obtained.

The concept of claw pole alternator may be extended to the megawatt range, provided the number of poles is increased (at 50/60 Hz or more) in variable speed wind and microhydrogenerators with DC-controlled output voltage of a local DC bus.

Though the claw pole SG could be built with the excitation on the stator too, to avoid brushes, the configuration is bulky and the arrival of high energy permanent magnets for rotor DC excitation has put it apparently in question.

2.3 Permanent Magnet Synchronous Generators

The rapid development of high energy PMs with a rather linear demagnetization curve has led to widespread use of PM synchronous motors for variable speed drives [6–10]. As electric machines are reversible by principle, the generator regime is also available and, for directly driven wind generators in the hundreds of kW or MW range (8 MW, 490 rpm), such solutions are being proposed. Super-high-speed gas turbine-driven PMSGs in the 100 kW range at 60–80 krpm are also introduced. Finally, PMSGs are being considered as starter-generators for the cars of the near future.

There are two main types of rotors for PMSGs:

- With rotor surface PMs (Figure 2.10)—nonsalient pole rotor (SPM)
- With interior PMs of Equation 2.11—salient pole rotor (IPM)

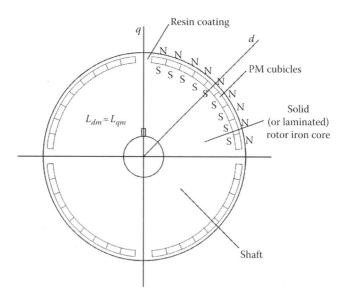

FIGURE 2.10 Surface PM rotor ($2p = 4$ poles).

The configuration in Figure 2.10 shows a PM rotor made with parallelepipedic PM pieces such that each pole is patched with quite a few of them, circumpherentially and axially.

The PMs are held tight to the solid (or laminated) rotor iron core by special adhesives, but also a highly resilient resin coating is added for mechanical rigidity.

The stator contains a laminated core with uniform slots (in general) that house a three-phase winding with distributed (standard) coils or with concentrated (fractionary) coils.

The rotor is practically isotropic from the magnetic viewpoint. There is some minor difference between d and q axis magnetic permeances because the PM recoil permeability ($\mu_{rec} \approx (1.04–1.07)\mu_0$ at 20°C) increases somewhat with temperature for NdFeB and SmCo high energy PMs.

Therefore, the rotor may be considered magnetically nonsalient (the magnetization inductances L_{dm} and L_{qm} are almost equal to each other).

To protect the PMs, mechanically, and produce reluctance torque, interior PM pole rotors have been introduced. Two typical configurations are shown in Figure 2.11a through c.

Figure 2.11a shows a practical solution for 2 pole interior PM (IPM) rotors. A practical $2p_1 = 4, 6,...$ IPM rotor as shown in Figure 2.11b has an inverse saliency: $L_{dm} < L_{qm}$ as usual with IPM machines. Finally, a high-saliency rotor ($L_{dm} < L_{qm}$), obtained with multiple flux barriers and PMs acting along axis q (rather than axis d), is presented in Figure 2.11c. It is a typical IPM machine but with large magnetic saliency. In such a machine, the reluctance torque is larger than the PM-interactive torque. The PMs field saturates the rotor flux bridges first and then overcompensates the stator-produced field in axis q. This way the stator flux along q axis decreases with current in axis q. For further flux weakening and torque control, the I_d current component is controlled. Wide constant power (flux weakening) speed range of more than 5:1 has been obtained this way. Starter/generators on cars are a typical application for this rotor.

As the PM role is limited, lower grade (lower B_r) PMs, at lower costs, may be used.

It is also possible to use the variable reluctance rotor with high magnetic saliency (Figure 2.11a)—without permanent magnets. With the reluctance generator, either power grid or stand-alone mode operation is feasible. For stand-alone operation capacitor self-excitation is needed. The performance is moderate, but the rotor cost is also moderate. Standby power sources would be a good application for reluctance SGs with high saliency $L_{dm}/L_{qm} > 4$.

PMSGs are characterized by high torque (power) density and high efficiency (excitation losses are zero). However, the costs of high energy PMs are still $100 and more per kilogram. In addition, to control the output, full power electronics is needed in the stator (Figure 2.12).

A bidirectional power flow PWM converter, with adequate filtering and control, may run the PM machine either as a motor (for starting the gas turbine) or as a generator, with controlled output at variable speed. The generator may work at the power grid mode or in stand-alone mode. These flexibility features, together with fast power-active and reactive-decoupled control at variable speed, may make such solutions a way of the future, at least in the tens and hundreds of kilowatt range, and more.

Many other PMSG configurations have been introduced, such as those with axial air gap. Among them, we will mention one that is typical in the sense that it uses the IPM reluctance rotor (Figure 2.11c), but it adds an electrical excitation. (Figure 2.13) [11].

Besides the reluctance and PM interaction torque, there will be an excitation interaction torque. The excitation current may be positive or negative to add or subtract from I_d current component in the stator.

This way at low speeds, the controlled positive field current will increase and control the output voltage; while at high speeds, a negative field current will suppress the emf, when needed to keep the voltage constant.

For DC-controlled generating only a diode rectifier is necessary, as the output voltage is regulated via DC current control in four quadrants. A low-power four-quadrant chopper is needed. For wide-speed-range applications, such hybrid excitation rotor may be a competitive solution. The rotor is not very rugged mechanically, but it can handle easily peripheral speeds of up to 50 m/s (10,000 rpm for 0.1 m diameter rotor).

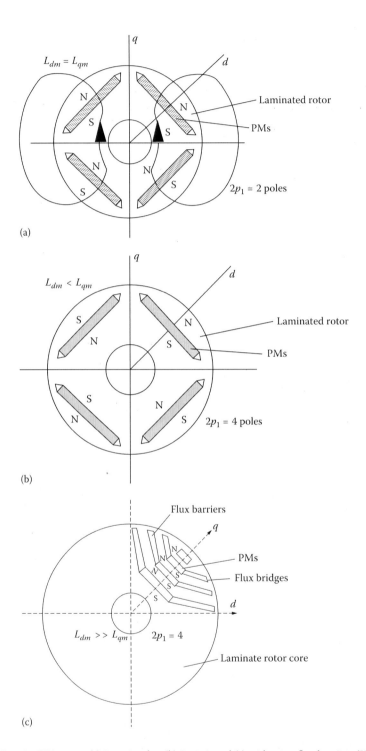

FIGURE 2.11 Interior PM rotors: (a) $2p_1 = 2$ poles, (b) $2p_1 = 4$, and (c) with rotor flux barriers (IPM reluctance).

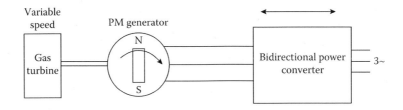

FIGURE 2.12 Bidirectional full-power electronics control.

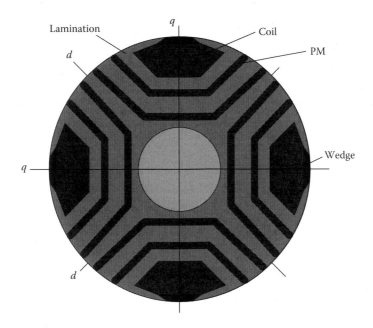

FIGURE 2.13 Biaxial excitation PM reluctance rotor generator (BEGA).

2.4 Homopolar Synchronous Generator

Placing both the DC excitation coils and the three-phase AC winding on the stator characterizes, the so-called homopolar (or inductor) synchronous machine (generator and motor)—Figure 2.14.

The rather rugged rotor with solid (even laminated) salient poles, and solid core is an added advantage. The salient rotor poles (segments) and interpoles produce a salient magnetic structure with a notable saliency ratio, especially if the air gap is rather small.

Consequently, the DC field current—produced magnetic field closes paths partially axially, partially circumpherentially, through stator and rotor, but it is tied (fixed) to the rotor pole axis.

It is always maximum in the axis of rotor poles and small, but of same polarity, in the axis of interpoles. An AC air-gap magnetic component is present in this homopolar distribution. Its peak value is ideally 50% of maximum air-gap field of the DC excitation current. Fringing reduces it to 35%–40% (Figure 2.14a through c), at best.

The machine is a salient pole machine with doubled air gap, but it behaves as a quasi-nonsalient pole rotor one and with rotor excitation.

Therefore, for same air-gap flux density fundamental B_{g1} (Figure 2.14), same mechanical air gap, the DC mmf of the field winding is doubled and the power loss quadruples. However, the ring-shaped coil

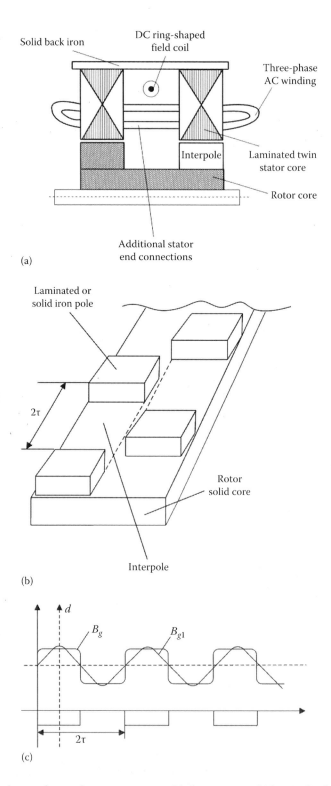

FIGURE 2.14 The homopolar synchronous generator: (a) the geometry, (b) longitudinal view, and (c) air-gap excitation field distribution.

reduces the copper weight and losses (especially when the number of poles increases) in comparison with a multipolar heteropolar DC rotor excitation system. The blessing of circularity comes into place here. We have to note also the additional end connection in the middle of the stator AC three-phase winding, between the two slotted laminated cores.

The rotor mechanical ruggedness is really superior only with solid iron poles when made in one piece. Unfortunately, stator mmf and slot space harmonics induce eddy currents in the rotor solid poles, reducing notably the efficiency, typically below 90% in a 15 kW, 15,000 rpm machine.

2.5 Induction Generator

The cage rotor induction machine is known to work as a generator, provided

- The frequency f_1 is smaller than $n \times p_1$ (speed $\times$ pole pairs): $S < 0$ (Figure 2.15a)
- There is a source to magnetize the machine

An induction machine working as a motor, supplied to a fixed frequency and voltage f_1, V_1 power grid, becomes a generator if it is driven by a prime mover above no-load ideal speed f_1/p_1:

$$n > \frac{f_1}{p_1} \tag{2.5}$$

Alternatively, the induction machine with cage rotor may self-excite on a capacitor at its terminals (Figure 2.15b).

For an induction generator connected to a strong (constant frequency and voltage) power grid, when the speed n increases (above f_1/p_1), the active power delivered to the power grid increases, but so does the reactive power drawn from power grid.

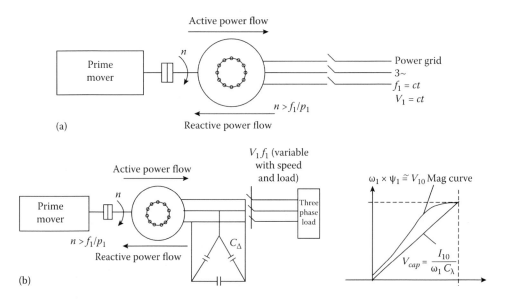

FIGURE 2.15 Cage rotor induction generator: (a) at power grid: $V_1 = ct$, $f_1 = ct$ and (b) standalone (capacitor excited): V_1, f_1, variable with speed and load.

Many existing wind generators use such CRIGs connected to the power grid. The control is only mechanical. The blade pitch angle is adjusted according to wind speed and power delivery requirements. However, such IGs tend to be rigid as they are stable only until n reaches the value:

$$n_{\max} = \frac{f_1}{p_1}\left(1 + |S_K|\right) \tag{2.6}$$

where S_K is the critical sleep, which decreases with rated power and is, anyway, below 0.08 for IGs in the hundreds of kilowatts. Additional parallel capacitors at power grid are required to compensate for the reactive power drained by the IG.

Alternatively, the reactive power may be provided by paralleled (plus series) capacitors (Figure 2.15b). In this case, we do have a self-excitation process that requires some remanent flux in the rotor (from previous operation) and the presence of magnetic saturation (Figure 2.15b). The frequency f_1 of self-excitation voltage (under no load) depends on the capacitor value and on the magnetization curve of the induction machine $\Psi_1(I_{10})$:

$$V_{10} \approx \psi_1(I_{10}) \cdot 2 \cdot \pi \cdot f_1 \approx V_{C_Y} = I_{10}\frac{3}{C_\Delta \cdot 2 \cdot \pi \cdot f_1} \tag{2.7}$$

The trouble is that on load, even if the speed is constant through prime mover speed control, the output voltage and frequency vary with load. For constant speed, if a frequency reduction under load of 1 Hz is acceptable, voltage control suffices. A three-phase AC chopper (variac) supplying the capacitors would do it, but the harmonics introduced by it have to be filtered out. In simple applications, a combination of parallel and series capacitors would provide rather constant (with 3%–5% regulation) voltage up to rated load.

Now, if variable speed is to be used, then, for *constant voltage and frequency*, PWM converters are needed. Such configurations are illustrated in Figure 2.16. A bidirectional power flow PWM converter (Figure 2.16a) provides both generating and motoring functions at variable speed. The capacitor in the DC line of the converter may lead not only to active but also reactive power delivery. Connection to the power grid without large transients is implicit, and therefore is fast, decoupled, active, and reactive power control.

The standalone configuration in Figure 2.16b is less expensive, but it provides only unidirectional power flow. A typical V_1/f_1 converter for drives is used. It is possible to inverse the connections, that is, to connect the diode rectifier and capacitors to the grid and the converter to the machine. This way, the system works as a variable speed drive for pumping, etc., if a local power grid is available. This commutation may be done automatically though it would take 1–2 min. For variable speed—in a limited range—an excitation capacitor in two stages would provide the diode rectifier with only slightly variable DC link voltage.

Provided the minimum and maximum converter voltage limits are met, the former would operate over the entire speed range. Now the converter is V_1/f_1 controlled for constant voltage and frequency.

A transformer (Y, Y_0) may be needed to accommodate unbalanced (or single-phase) loads.

The output voltage may be close-loop controlled through the PWM converter. On the other end, the bidirectional PWM converter configuration may be provided with a reconfigurable control system such that to work not only on the power grid but to separate itself smoothly to operate as standalone, or wait on standby and then be reconnected smoothly to the power grid. Thus, multifunctional power generation at variable speed is produced. As evident in Figure 2.16, full power electronics is required. For a limited speed range, say up to 25%, it is possible to use two IGs with cage rotor and different pole numbers ($2p_2/2p_1 = 8/6, 5/4, 4/3...$). The one with more poles, ($2p_2 > 2p_1$)

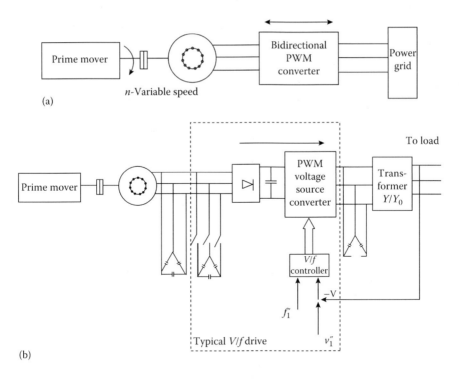

FIGURE 2.16 CRIGs for variable speed: (a) at power grid $V_1 = ct$, $f_1 = ct$ and (b) stand-alone $V_1 = ct$, $f_1 = ct$ (controlled).

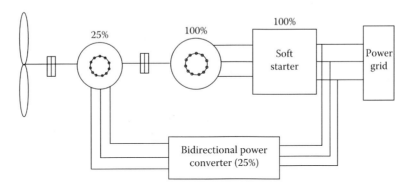

FIGURE 2.17 Dual IG system for limited speed variation range.

is rated at 25% of rated power and fed from a bidirectional power converter sized also at about 25%. The scheme works at the power grid (Figure 2.17).

The softstarter reduces the synchronization transients and disconnects the 100% IG when the power required is below 25%. Then the 25% IG remains alone at work, at variable speed ($n > f_1/p_1$), to tap the energy available from, for example, low-speed wind or from a low head microhydroturbine. In addition, above 25% load, when the main (100%) IG works, the 25% IGs may add power as generator or work in motoring for better dynamics and stability. Now we may imagine a single rotor-stator IG with two separate stator windings ($2p_2 > 2p_1$) to perform the same task.

The reduction in rating, from 100% to 25%, of the bidirectional PWM converter is noteworthy.

The main advantage of the dual IG or dual stator winding IG is lower costs, although for lower performance (low-speed range above f_1/p_1).

2.6 Wound-Rotor Doubly Fed Induction Generator

It all started between 1907 and 1913, with the Scherbius and Kraemer cascade configurations, which are both slip power recovery schemes of wound rotor induction machines. Leonhard analyzed it pertinently in 1928, but adequate power electronics for it was not available then. A slip recovery scheme with thyristor power electronics is shown in Figure 2.18a. Unidirectional power flow, from IG rotor to the converter, is only feasible because of the diode rectifier. A step-up transformer is necessary for voltage adaption, while the thyristor inverter produces constant voltage and frequency output. The principle of operation is based on the frequency theorem of traveling fields.

$$f_1 = np_1 + f_2; f_2 <> 0, \quad \text{and variable } f_1 = ct \tag{2.8}$$

Negative frequency means that the sequence of rotor phases is different from the sequence of stator phases. Now, if f_2 is variable, n may also be variable as long as Equation 2.7 is fulfilled.

That is, constant frequency f_1 is provided in the stator for adjustable speed. The system may work at the power grid or even as standalone, although with reconfigurable control. When $f_2 > 0$, $n < f_1/p_1$, we do have subsynchronous operation. The case for $f_2 < 0$, $n > f_1/p_1$ corresponds to hypersynchronous operation. Synchronous operation takes place at $f_2 = 0$ which is not feasible with the diode rectifier current source inverter, but it is with the bidirectional PWM converter.

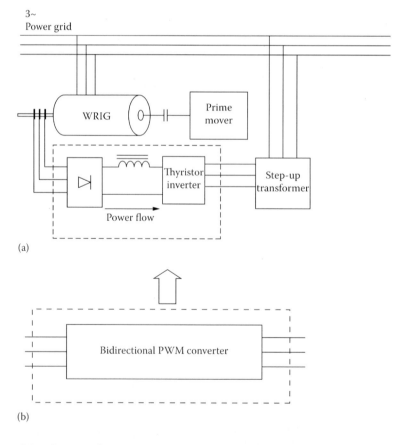

FIGURE 2.18 Wound rotor induction generator (WRIG): (a) with diode rectifier (slip recovery system) and (b) with bidirectional PWM converter.

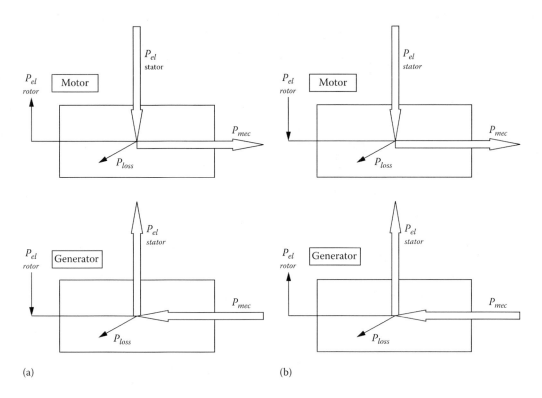

FIGURE 2.19 Operation modes of WRIG with bidirectional PWM converter (in the rotor): (a) $S < 0$ and (b) $S > 0$.

The slip recovery system can work as a subsynchronous ($n < f_1/p$) motor or as a supersynchronous ($n > f_1/p$) generator. The WRIG with bidirectional PWM converter may work as a motor and generator both for subsynchronous and supersynchronous speed.

The power flow directions for such a system are shown in Figure 2.19.

The converter rating is commensurable to speed range, that is, to maximum slip S_{max}:

$$\text{KVA}_{rating} = K \frac{f_{2max}}{f_1} \times 100[\%] \tag{2.9}$$

where $K = 1–1.4$ depending on the reactive power requirements from the converter.

Note that being placed in the rotor circuit, through slip rings and brushes, the converter rating is around $|S_{max}|$ in%. The larger the speed range, the larger the rating and the costs of the converter. Also the fully bidirectional PWM converter—as a back-to-back voltage source multilevel PWM converter system—may provide fast and continuous decoupled active and reactive power control operation even at synchronism ($f_2 = 0$, DC rotor excitation). And it may perform the self-starting as well. The self-starting is done by short-circuiting the stator first, previously disconnected from the power grid, and supplying the rotor through the PWM converter in the subsynchronous motoring mode. The rotor accelerates up to a prescribed speed corresponding to $f_2' > f_1(1 - S_{max})$. Then the stator winding is opened and, with the rotor freewheeling the stator no-load voltage, sequence and frequency are adjusted to coincide with that of the power grid, by adequate PWM converter control. Finally, the stator winding is connected to the power grid without notable transients.

This kind of rotor-starting requires $f_2' \approx (0.8 - 1)f_1$, which means that the standard cycloconverter is out of the question. Therefore, it is only the back-to-back voltage PWM multilevel converter or the matrix converter that are suitable for full exploitation of motoring/generating at sub- and supersynchronous speeds, so typical in pump storage applications.

2.7 Parametric Generators

Parametric generators exploit the magnetic anisotropy of both the stator and the rotor. PMs may be added on the stator or on the rotor. Single magnetic saliency with PMs on the rotor is also used in some configurations. The parametric generators use nonoverlapping (concentrated) windings to reduce end-connection copper losses on the stator.

As the stator mmf does not produce a pure traveling field, there are core losses both in the stator and in the rotor. The simplicity and (or) ruggedness of such generators made them adequate for some applications.

Among parametric generators, some of the most representative are detailed here:

- Switched reluctance generators (SRGs):
  - Without PMs
  - With PMs on stator or on rotor
- Transverse flux generators (TFGs):
  - With rotor PMs
  - With stator PMs
- Flux reversal generators (FRGs):
  - With PMs on the stator
  - With PMs on the rotor (and flux concentration)
- Linear motion alternators (LMAs):
  - With coil mover and PMs on the stator
  - With PM mover tubular or flat (with PM flux concentration)
  - With iron mover and PMs on the stator

The SRG [12] has a double saliency magnetic laminated structure—on the stator and the rotor—and concentrated coils on the stator (Figure 2.20). The stator phases are PWM voltage fed as long as the rotor poles are approaching them, one at a time, for the three-phase configuration. The phase inductances vary with rotor position (Figure 2.21) and, at least for the three-phase configuration, there is little magnetic coupling between phases.

Eventually, each phase is turned on around point A (in Figure 2.21); then it magnetizes the phase and the $dL/d\theta$ effect produces a motion-induced voltage (emf) which, in interaction with

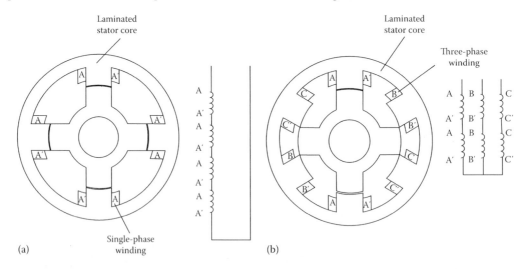

FIGURE 2.20 Switched reluctance generators (SRGs): (a) single-phase: 4/4 and (b) three-phase: 6/4.

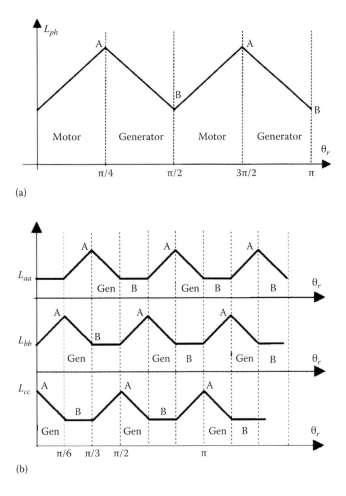

FIGURE 2.21   Phase inductance versus rotor position: (a) the single-phase 4/4 SRG and (b) the three-phase 6/4 SRG.

the phase current, produces torque. The phase is turned off around point B when the next phase is turned on. The current polarity is not relevant, and thus positive current flows through voltage PWM. The maximum voltage is applied until the phase current magnetizes to maximum admitted current.

As part of the AB interval available for generating is lost to the magnetization process, the latter takes up around 30% of energy available per cycle. For the single-phase machine, the torque, as expected, has notches as only the negative slopes of the inductance are adequate for generating.

It is the machine simplicity and ruggedness that characterize SRGs. High speed is feasible. Rotor higher temperature due to the local environment is also acceptable since these are no PMs or windings on the rotor. PMs may be added on the rotor (Figure 2.22a,b) [13,14]. In this situation, the current polarity has to change, and the torque production relies heavily on phase interaction through PMs. The reluctance torque is small.

Alternatively, PMs may be placed on the stator (Figure 2.22c) [15] with some PM flux concentration. Again the reluctance torque is reduced and PM torque prevails. The PM flux polarity in one phase does not change sign, therefore we may call it a homopolar PM excitation.

Other SRG configurations with homopolar excitation flux have been proposed but did not reach the markets thus far.

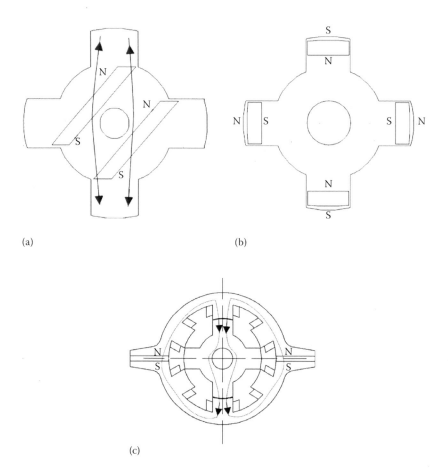

FIGURE 2.22 PM-assisted SRGs: (a) with long PMs on the rotor, (b) with short PMs on the rotor, and (c) with PMs on the stator.

2.7.1 Flux Reversal Generators

In these configurations, reliance is on PM flux switch (reversal) in the stator-concentrated coils (Figure 2.23a [16]).

The PM flux linkage in the stator coils of Figure 2.23a changes sign when the rotor moves 90° (mechanical) and does the same, for each phase in Figure 2.23b for the three-phase FRG, when rotor moves 22.5° mechanical degrees. We use "flux reversal" instead of "flux switch" in the hope for clarity.

In general, it is π/N_r (which corresponds to π electrical radius). The electrical and mechanical radians are related as follows:

$$\alpha_{el} = N_r\alpha_{mec}, \quad f_1 = N_r \cdot n \tag{2.10}$$

Therefore, the frequency of the electromagnetic force f_1 is as if the number of pole pairs on the rotor was N_r.

The three-phase FRG configuration [17] makes better use of stator and rotor core and the manufacturing process is easier than that for the single-phase, as the coils are inserted by the conventional technology. Premade stator poles, with coils on, may be mounted inside the stator back iron as done with rotor poles in salient poles in hydrogenerators. The main problem is the large flux fringing due to the juxtaposition of the N–S poles (there could be 2, 4, 6… of them on a stator pole). This reduces the useful flux to about 0.3–0.4 of its ideal value (in homopolar stator excitation this situation is quite normal).

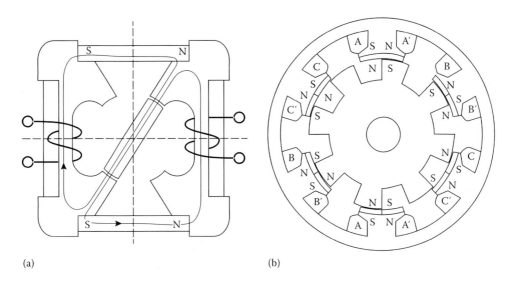

(a)   (b)

FIGURE 2.23 Flux reversal generators (FRGs) with stator PMs: (a) the single-phase 4/2 flux-switch alternator and (b) the three-phase 6/8 FRG.

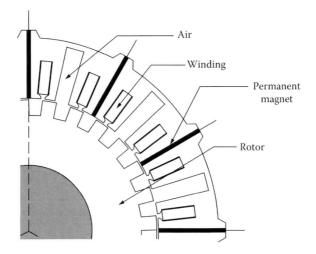

FIGURE 2.24 Three-phase reversal generator (FRG) with stator PM flux concentration.

Stator PM flux concentration should provide better torque density for same power factor. An example three-phase configuration is shown in Figure 2.24.

It is evident that the manufacturing of stator is a bit more complicated and the usage of stator core is partial, still PM flux concentration may increase the torque density without compromising too much the power factor. The phases are magnetically independent and thus high fault tolerance is expected.

For a better core utilization, the PMs with flux concentration may be placed on the rotor. An interior stator is added to complete the magnetic circuit (Figure 2.25).

The second (interior) windingless stator poses some manufacturing problems (the rotor also), but the higher torque/volume at acceptable power factor may justify it.

The power factor is mentioned here because it influences the converter KVAs through reactive power demands.

FRGs are, in general, meant mainly for low-speed applications such as directly driven wind generators, or vessel generator/motors, and so forth.

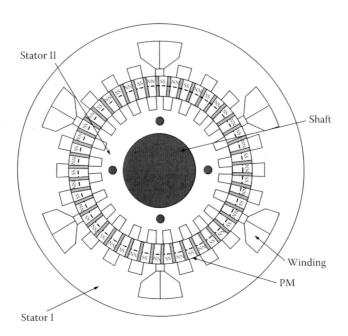

FIGURE 2.25 FRG with rotor PM flux concentration.

2.7.2 Transverse Flux Generators

The TFGs are built in single-phase configurations with ring-shaped stator coils and surface PMs on the rotor (Figure 2.26) or with PM rotor flux concentration (Figure 2.27) [17,18].

The double-sided (dual-stator) configuration in Figure 2.27 takes advantage of PM flux concentration on the rotor. In general, TFGs are characterized by low winding losses, due to the blessing of the ring-shaped coil.

A three-phase machine is built by adding axially three single-phase units, properly displaced tangentially by $2\pi/3$ with each other. It is evident that the stators could be best built from magnetic composite materials. However, this solution would reduce torque density since the permeability of such materials is below $500\mu_0$ under some magnetic saturation. The core losses would be reduced if frequency goes above 600 Hz with magnetic composite materials (magnetic powder).

The stator-PM TFG (originally called axial flux circumpherential current (AFCC) PM machine [19]) imposes the use of composite magnetic materials both on the stator and on the rotor due to its intricated geometry. Again, it is essentially a single-phase machine. PM flux concentration occurs along axial direction. Good use of PMs and cores is inherent in the axial–air gap stator PM FRG in Figure 2.28a.

FRG need more PMs than usual, but the torque density is rather good and compact geometries are very likely. The large number of poles on the rotor in most TFG leads to a good frequency, unless speed is not very low.

The rather high torque density (in Nm/m³, or (6–9) N/cm² of rotor shear stress) is inherent as the number of PM flux reversals (poles) in the stator ring-shaped coils per revolution is large. This effect may be called "torque magnification" [11].

2.7.3 Linear Motion Alternators

The microphone is the classical example of a linear motion alternator (LMA) with a moving coil. The loudspeaker illustrates its motoring operation mode.

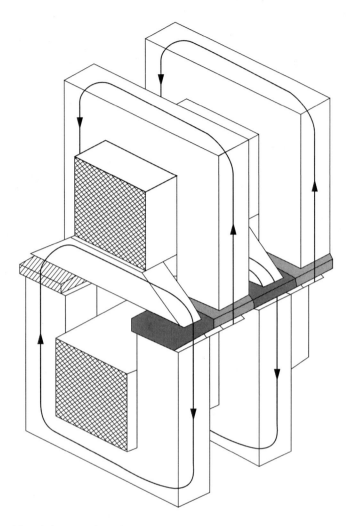

FIGURE 2.26 Double-sided TFG with surface PM rotor.

Though there are very many potential LMA configurations (or actuators), they all use PMs and fall into three main categories [20]:

- With moving coil (and stator PMs); Figure 2.29a
- With moving PMs (and stator coil); Figure 2.29b
- With moving iron (and stator PMs); Figure 2.29c

In essence, the PM flux linkage in the coil changes sign when the mover travels the excursion length l_{stroke}, which serves as a kind of pole pitch. Therefore they are, in a way, single-phase flux reversal machines. The average speed U_{av} is:

$$U_{av} = 2 \cdot l_{stroke} \cdot f_1 \qquad (2.11)$$

f_1 is the frequency of mechanical oscillations.

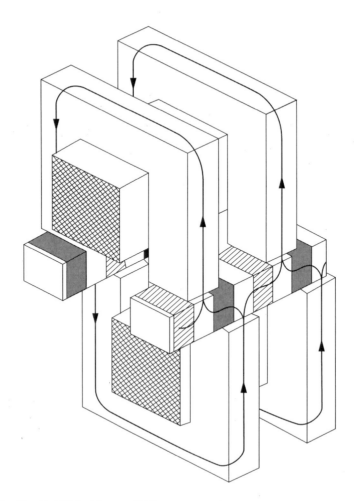

FIGURE 2.27   Double-sided TFG with rotor PM flux concentration.

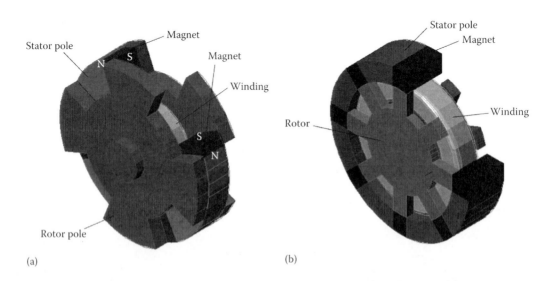

FIGURE 2.28   TFG with stator–PMs flux concentration: (a) with axial air gap and (b) with radial air gap.

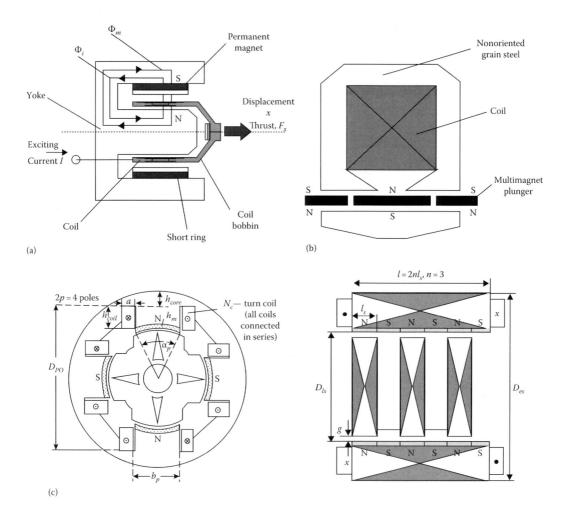

FIGURE 2.29 Commercial motion alternators with (a) moving coil, (b) moving PMs, and (c) moving iron.

To secure high-efficiency, beryllium–copper flexured springs (Figure 2.30) are used to store the kinetic energy of the mover at excursion ends. They also serve as linear bearings. The proper frequency of these mechanical springs f_m is good to be equal to electrical frequency:

$$f_e = f_m = \frac{1}{2\pi}\sqrt{\frac{K}{m}} \tag{2.12}$$

where
K is the spring rigidity coefficient
m is the moving mass

The current is in phase with speed for best operation.

The strokes involved in LMAs are in the order of 0.5–100 mm or so. Thus, their power, in general, is limited to 10–50 kW at 50 (60) Hz.

They are basically synchronous phase machines with harmonic motion and linear flux to position ideal variation.

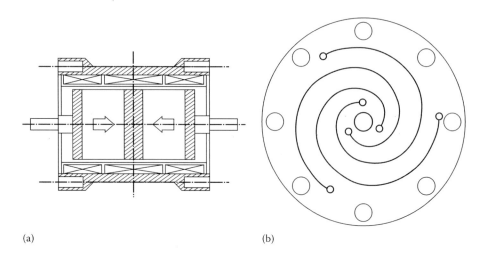

(a)   (b)

FIGURE 2.30   Tubular LMA with plunger supported on flexural springs: (a) cross-section and (b) mechanical flexure.

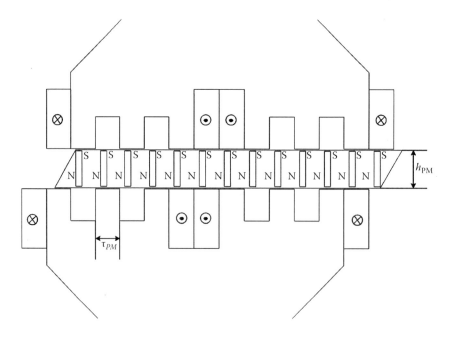

FIGURE 2.31   Flux reversal LMA with mover PM concentration.

Further increasing the power per volume requires—if average speed U_s is limited—configurations with PM flux concentration and three phases. Such a single-phase flat configuration—with moving PMs—is shown in Figure 2.31 [26].

Again it is a single-phase device and the PM flux reverses polarity when the mover advances one "small," stator, pole (tooth). The two twin stators are also displaced by one small stator tooth, to provide for optimal magnetic circuit completion. Large air gap PM flux densities of up to 1.25 T may be obtained under the stator teeth with 0.65 T left for armature reaction, to secure both high force (power) density and satisfactory power factor (or reasonable IX/E ≤ 0.5 ratio; X—machine reactance). The PM height h_{PM} per pole pitch τ_{PM} is $h_{PM}/\tau_{PM} \approx 1.5$–$2.5$ as all PMs are active all the time and full use of both stator and mover cores and copper is made.

Note: There are also LMAs that exploit progressive (rather than oscillatory) linear motion. Applications include auxiliary power LMAs on magnetically levitated vehicles (MAGLEVs) and plasma magnetohydrodynamic (MHD) linear motion DC generators with superconducting excitation (see Chapter 20).

2.8 Electric Generator Applications

The application domains for electric generators embrace almost all industries—traditional and new—with powers from mW to hundreds of MW/unit and more [20–25].

Table 2.1 summarizes our view of electric generator main applications and the competitive types that may suit each person and need. An overview on large power wind generators is offered as an example of the myriad of solutions for a single application field.

2.9 High-Power Wind Generators

2.9.1 Introduction

By high power, we mean here MW range (0.75–10) MW/unit wind generators, Figure 2.32a, and also by "less PM" the paper means less NdFeB kg/kW of wind generator or the use of ferrite PMs. "No PM" refers to wind generators without PMs, be them commercial (DC-excited SGs or CRIGs controlled by AC–AC PWM full power converters or wound rotor DFIG, controlled by partial (30%–40%) rating power AC–AC PWM converters), or solutions, under various development stages, which show good potential for the application. The present study was triggered by the steep rise in 2010–2012 (apparently here to stay) of NdFeB PM price (above 130 $/kg), as shown on Figure 2.32b, which may shift away the accent from the recently successful directly driven PMSGs with tone of NdFeB in the 3–10 MW range. Quite a few things have to be reconsidered on this occasion. Among them are brushless generator system configurations for lower maintenance effort, better handling of asymmetrical power grid voltage sags with PWM converters, practical protection schemes, and power/unit increase at higher efficiency

TABLE 2.1 Electric Generator Applications

Application	Large Power Systems (Gas, Coal, Nuclear Hydrogen)	Distributed Power Systems (Wind, Hydrogen)	Standby Diesel-driven EGs	Automotive Starter-Generators	Diesel Locomotives
Suitable generator	Excited rotor SGs, DFIGs (up to hundreds of MW/unit)	Excited rotor SGs, CRIGs, PMSGs. Parametric generators (up to 10 MW power/unit)	PMSGs CRIGs	IPM SGs, induction generators, TFGs	Excited-rotor SGs
Application	Home electricity production	Spacecraft applications	Aircraft applications	Ship applications	
Suitable generator	PMSGs and LMAs	LMAs	PM synchronous, CRIG or DFIGs (up to 500 kW/unit)	Excited SGs (power in the order of a few MWs)	
Application	Small-power telemetry-based vibration monitoring	Inertial batteries	Super high-speed gas-turbine generators		
Suitable generator	LMAs: 20–50 mW to 5 W	Axial–air gap PMSGs up to hundreds of MJ/unit	PMSGs up to 150 kW and 80,000 rpm (higher powers at lower speeds)		

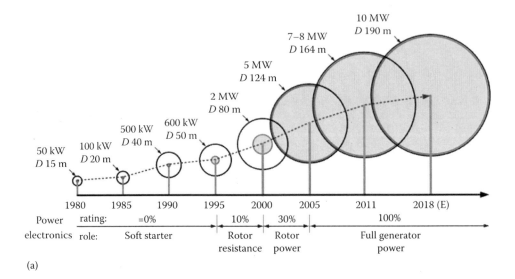

(a)

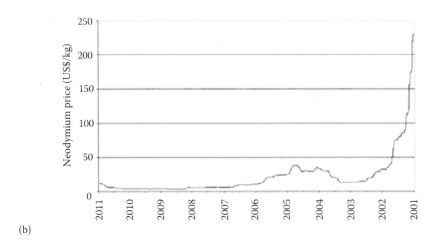

(b)

FIGURE 2.32 Evolution of power unit in wind generators: (a) Size development of Wind Turbine system power electronics (blue) and (b) Neodymium price versus time (© Kidd Capital Group).

multiplied by power factor in order to reduce the PWM converter kV A and costs. Even higher gear ratio for 3000–4000 rpm generators with modular (0.75 MW/unit) multiple units with smaller volume/kW and outer diameter or high voltage (>20 kV) generators will also be considered. All the above are to be considered first for large off-shore wind parks with on-ground DC–AC high voltage (50–200 MVA) inverter interfacing to the power grid.

Given the numerous large (MW range) wind generator/turbine powers/speeds available on the markets [28,29], up to 6 MW at around 10 rpm turbine speed with 7.5 MW (www.enercon.de) direct drive DC-excited synchronous and 8 MW, 9 rpm [1] with multibrid (500 rpm) PMSGs that are very close to the commercialization, a dual classification of existing solutions is done in Table 2.2.

Table 2.2 illustrates the four main commercial types of wind generators: CRIG, PMSG, dceSG, associated with full power AC–AC PWM converters and, by consequence, also with very good handling of voltage sags and high reactive power capability, and WRIG (DFIG) associated with partial rating

TABLE 2.2 Existing (Established) Wind Generators

	By Principle		
Cage Rotor Induction Generators (CRIGs)	PM Synchronous Generators (PMSGs)	Wound Rotor Induction Generators (WRIGs) (DFIGs)	DC-Excited Synchronous Generators (dce SGs)
Handling voltage sags			
Very good	Very good	Acceptable	Very good
Reactive power capability			
High	High	Medium-low	High
By gear rating			
250:1 (4G)	100:1 (3G)	(6–9):1 (1G)	Direct drive (DD)
None	CRIG DFIG	PMSG	PMSG (up to 5 MW), dce SG (up to 4.6 MW)

(20%–40%) AC–AC PWM converter connected to the wound rotor, through slip-rings and brushes. A generic wind generator system for DFIG, SG, and IG is shown in Figure 2.33. There are four speed ranges of wind generators in relation to the transmission ratio in Table 2.2, but, in reality, the high-speed (250:1) gear ratio class that leads to high frequency (max. 1–1.2 kHz) generators (PMSGs) (which require reasonably small PM quantities) is yet to be thoroughly evaluated; because it implies breakthroughs in the multi-stage mechanical transmission in terms of lower overall diameter, and volume to allow for a reasonable volume and weight of the nacelle, etc.

Detailed performance and active materials weight comparisons of various designs of the topologies in Table 2.2 for 3 MW, 15 rpm turbine generators are offered in Ref. [30]. In terms of total costs per kW or kW h, they are objectively in question, because the price of NdFeB magnets has increased sharply in the last 3 years from the 40$ considered in Reference [30]; in addition, the converter cost is given as 50 $/kW while it should be given, in $/kV A, because the machine power factor affects the peak kV A of the converter and thus its costs and losses.

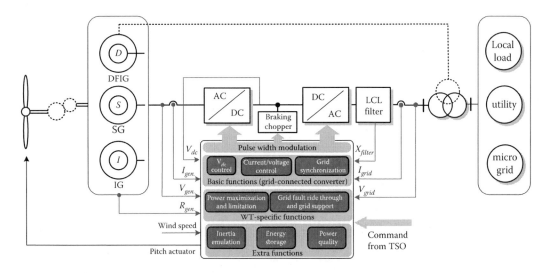

FIGURE 2.33 Generic wind generator system control.

TABLE 2.3   3 MW (15 rpm Turbine Speed) Wind Generator Design

	DD-dceSG	DD-PMSG	1G-PMSG	1G-DFIG	DFIG
Stator inner diameter (m)	5	5	1.8	1.8	0.42
Stack length (m)	1.2	1.2	0.4	0.6	0.75
Air gap (mm)	5	5	3.5	2	1
Pole pair p_1	40	80	56	40	3
Slot/pole/phase q_1	2	1	1	2	6
Generator active material					
Iron (*t*) (4 $/kg)	32.5	18.1	4.37	8.65	4.03
Copper (*t*) (20 $/kg)	12.6	4.3	1.33	2.72	1.21
PM (*t*)(33 $/kg)	—	1.7	0.41	—	—
Total (*t*)	45.1	24.1	6.11	11.37	5.25
Annual energy					
Total losses (MW h)	8.74	6.47	7.89	7.34	7.98
Energy yield (GW h)	7.74	7.89	7.70	7.76	7.69
Annual energy/total cost					
kW h/$ (1.33$ = 1 Euro)	2.76	3	3.075	3.17	3.09

Source: After from Richter, R., *Electrical Machines*, Vol. 2: *Synchronous Machines*, Verlag Birkhäuser, Basel, Switzerland, 1963 (in German).

Note: DD, direct drive; 1G, one stage transmission.

In addition, as the generator frame (construction) weight and cost, generator system cost and other wind turbine costs are much larger than the active materials costs in the generator, any error here (or any sharp progress in the near future: such as active magnetic suspension of the generators rotor) would render the conclusion of the study hardly practical in terms of kW h/$. Therefore, selecting from Reference [30] the results, in Table 2.3 have been obtained.

The differences in kW h/$ are up to 10%, but they may become more favorable to PM-less solutions (especially against direct-drive PMSG that needs 1.7 t NdFeB magnets). As the generator frame weight is rather proportional to the weight of its active materials, the reduction of the latter becomes even a more important challenge in the future; further gear weight reduction is also needed. As the total losses are minimum for DD-PMSG at reasonable weight (24.1 t), the following question arises: Can this solution be attempted with ferrite PMs (6 $/kg) for Br = 0.45 T by strong flux concentration? If so, at how much higher weight and at what power factor for similar efficiency with NdFeB machine; evidently, at lower active material costs. On the other hand, the even larger off-shore wind parks require turbine-generator units up to 10 MW. The results of such pitch-controlled turbine system designs with NdFeB PMSG [31] and WRIG (DFIG) direct drive [32] are summarized in Table 2.4, which suggests the following:

- The DD-DFIG seems capable of producing 10 MW at 10 rpm at about the same active material weight as the DD-PMSG, but at around (above) 10 m interior stator diameters, which is less practical.
- The overall cost of generator + frame, etc., is only 7% lower for the DD-DFIG.
- However, it should be noted that the air gap of DFIG design was considered to be 1 (2) mm as the centrifugal forces are small (peripheral speed is below 10 m/s), and a special flexible framing was FEM designed and tested for mechanical stress [32]. A 2 mm air gap would be more practical even in such a frame design, but that would also mean a lower efficiency.
- The DD-DFIG has been designed for unity power factor (however, at 50 Hz), while the DD-PMSG with full power AC–AC converter can deliver full rated reactive power (even when the turbine is idling, when a back-to-back capacitor DC link voltage source converter is used).
- The today's more than 100 $/kg price for NdFeB magnets considered in PMSG design, for 6 t of NdFeB magnets, means the DD-DFIG is even more attractive in terms of initial costs (Table 2.4).

TABLE 2.4 10 MW, 10 rpm, Direct-Drive Wind Generators

	DD-PMSG (After [4])	DFIG (After [5]) (50 Hz)
Stator inner diameter (m)	5	12
Stack length (m)	1.6	1.25 (1.3)
Air gap (mm)	10	1(2)!
Pole pair p_1/q_1	160/1	300/(1.8–2)
Generator material weight		
Iron (*t*) (4 \$/kg)	47	35
Copper (*t*) (20 \$/kg)	12	30
PM (*t*) (53 \$/kg)	6	—
Construction (*t*)	260	
Total weight (*t*)	325	
Total losses (GW h)	2.87 (4% loss)	n.a.
Energy yield (GW h)	48.4 (4% loss)	n.a.
Total generator and inverter cost		
	3.325*10⁶\$	3.192*10⁶\$

- The 25%–35% rated power transfer to the rotor through mechanical brushes for the DFIG also has to be considered together with DFIG inherent weakness in treating power grid deep voltage sags, and reactive power delivery limits.

In a further effort to sort out when, where, and which of the four standard wind generators are better, from 0.75 MW (38.6 rpm) to 10 MW (10 rpm). Reference 33 attempts a solid but analytical design for DD-PMSG, DD-dceSG, 1G-PMSG, 1G-DFIG and 3G-SCIG, 3G-DFIG, and 3G-PMSG again with a NdFeB PM price of 53.2 \$/kg. The results in Figure 2.34a and b [33] are safe-revealing, though again the cost of the converter is still considered in \$/kW.

Figure 2.34 suggests again that in terms of kW h/\$, the 3G-DFIG and the 1G-DFIG (multibrid) are better even at 10 MW (as it was for 3 MW), but the energy yield is smaller than at 3 MW. The DD-DFIG has not been considered in Reference [33]. In addition, it is notable that the 3G-CRIG is better than DD-PMSG or DD-dceSG in kW h/\$. The much higher than 53.2 \$/kg cost of NdFeB today makes the comparison even less favorable for the 1G and 3G PMSG.

As quite a few secondary effects such as magnetic saturation, losses, and torque pulsations have been indubitably coarsely treated in the even nonlinear analytical models in the optimal design [33], further 2(3)D FEM studies have to be run, even with a few ones included in the optimal design cycle, before industry-usable knowledge is obtained. Reaching this point with the overview, a few questions arise:

- If the DD-dceSG is capable of producing full active and reactive power control in a reliable configuration, which can stand severe voltage dips, why does it not become the norm in the MW range wind generators?: because it is too heavy, then the tower is heavier, etc., and the efficiency is moderately high probably 0.96 in the 7.5 MW dce DD-dceSG in the advanced development for a 220 t total generator weight (www.enercon.de); finally, the energy yield is lower at 10 MW, 10 rpm.
- If the 1G-DFIG (probably also DD-DFIG) produces the same energy yield for lower cost and weight than DD or 1G-PMSG (and is implicitly better than DD-dce SG), why does it not become the norm itself? The main reason is that the 30%–40% rated dual AC–AC converter connected to the DFIG rotor does not allow full (plus/minus) reactive power control and cannot handle easily deep voltage sags, as it controls only 20%–30% of the active power firmly. In addition, at 30%–40% power rating, 3 MW for 10 MW machines needs to be transferred through slip rings and brushes, which in off-shore sites may become a challenge.

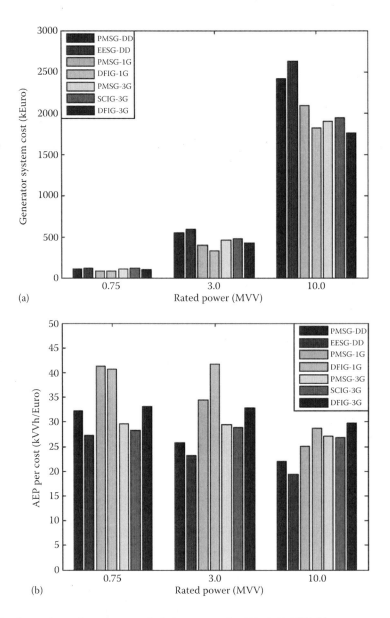

FIGURE 2.34 Comparison of seven types of wind generator for 0.75, 3, 10 MW: (a) generator system cost and (b) kWh/$. (From Hui, L.F. and Chen, Z., Design optimization and evaluation of different wind generator systems, Record of ICEMS, 2008, pp. 2396–2401.)

- If the 3G-SCIG, which is brushless and robust, operates at reasonable cost and energy yield (Figure 2.33), even at 10 MW, and by using a full power dual AC–AC converter can provide ±100% active and reactive power control, why it does not become the norm? Mainly because, it implies a 3G transmission that is bulky, requiring high diameter, heavy nacelle; it needs maintenance and, in case of oil spills, may produce sea water pollution in off-shore wind parks.

These questions led to the discussion in what follows, not only on the newly proposed solutions, but also to investigate variants and additions to DD-dceSG, 1G(DD) SCIG and brushless DFIG for off-shore high power units. In addition, it includes high speed (4G) on-shore solutions with dce SG, PMSG, for powers perhaps less than 3 MW/unit. The paper treats variants of DD-SG systems and 4G-SGs, less PM PMSGs,

BLDC—multiphase reluctance generators, brushless DFIGs, brushless doubly fed reluctance generators, SRGs, flux-switching ferrite PM stator generators. Finally, conclusions are drawn.

The choice of these innovative recent proposals is based on their assumed practicality, judged by the results obtained thus far. But also based on the fact that a high internal, synchronous reactance (above 1.00 pu) exerts a strong burden on the PWM converter connected to the generator, leading to high conductive losses in the converter full load and to high converter switching losses at low load, due the known "large voltage regulation syndrome" (this is why $q_s = 1$ windings may be preferred to $q_s < 0.5$ (fractionary) windings in PMSGs).

2.9.2 DC-Excited Synchronous Generator Systems

A typical DD-dce.SG with its generic power electronics grid interfacing is shown in Figure 2.35.

Besides, the bidirectional (in this embodiment) full-power PWM converter control through stator, an excitation converter of low power (less than 3% of the rating power) is provided to supply the excitation windings on the rotor through (in general) brushes and slip rings.

2.9.2.1 Brushless Excitation

To avoid brushes (especially for off-shore applications), a 500 Hz (or more) rotary transformer with a constant frequency and variable voltage inverter (3% power rating) in the primary side and a fast diode rectifier on SG rotor may replace the brush-based slip ring system within the same volume [34] and losses. Figure 2.36 shows the system.

The excitation current will be modified through the low power inverter voltage amplitude: the size of the rotary transformer is small at 500 Hz (or more) frequency and its small short-circuit reactance allows for fast field current control while its circular shape coils minimize the copper losses in it. The initial cost is still an issue.

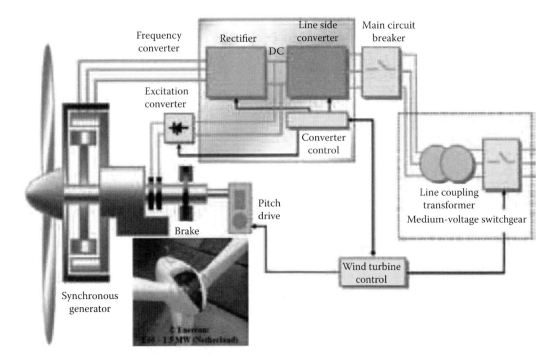

FIGURE 2.35 DD-dce.SG with full-power electronics grid interfacing.

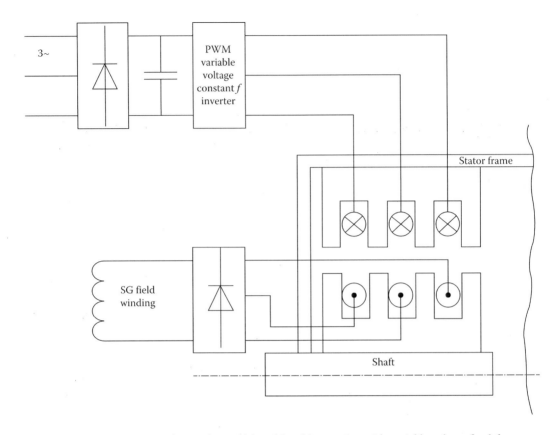

FIGURE 2.36 Rotating transformer (500 Hz)-based brushless exciter with variable voltage fixed frequency inverter (3% power rating).

2.9.2.2 Lower Size (Weight) by Optimal Design

With 41 t of active materials for DD at 3 MW and 15 rpm [30] and a total 220 t generator weight for 7.5 MW at 10 rpm (www.enercon.de), there is little available information about DD-dce.SGs on the market in terms of complete weight, cost and performance data. A dedicated optimal analytical design code with 2D FEM verifications was put in place and tried on a 5 MW, 12 rpm, 15 Hz, 6000 V, $q = 1$ slot/pole/phase DD-dce.SG. Table 2.5 summarizes key data on this endeavor.

The aggregated cost and loss objective function evolution during the optimization design is visible in Figure 2.37a with the losses seen in Figure 2.37b.

A few remarks are necessary:

- The active materials weight is reduced to 42.2 t at 5 MW, 12 rpm at the cost of a general rated efficiency (excitation losses included) of 93% ($P_{cun} = 114$ kW, $P_{iron} + P_{mec} = 115$ kW) by allowing a high rotor diameter $D_{ro} = 8.554$ m, and an air gap $g = 5$ mm.
- The machine synchronous (saturated) reactances are $x_d = 1.07$ pu and $x_q = 0.929$ pu. Consequently, the machine does not have the high voltage regulation problems, while the small saliency warrants rather pure i_q control ($i_d = 0$), which means a voltage boosting through the machine side converter as in PMSGs and CRIGs. If unity power factor control is considered, $i_d \neq 0$ under load and thus the copper losses will be larger than needed. In some simplified systems, a diode rectifier is used on the machine side with a DC link power controlled only by the field current. For small speed range cases (off-shore) such a compromise seems to be worth considering.

TABLE 2.5 Optimal Design of DD-dce.SG

Input data		
P_n	5000 (kW)	Rated power
f_n	15 (Hz)	Rated frequency
V_n	6000 (V)	Rated line voltage
q	1	Slots/pole/phase
p	150	Number of pole pairs
n_n	12 (rpm)	Rotor speed
Output data		
P_{cun}	114 (kW)	Stator copper rated losses
$P_{fe} + P_{mec}$	115 (kW)	Iron and mechanical total losses
$\eta/\cos\varphi$	0.93/1	Rated efficiency/power factor
D_{ro}	8554 (mm)	Rotor outer diameter
H_{ag}	5 (mm)	Length of the air gap
D_{so}	8885 (mm)	Stator outer diameter
R_w	15,346 (kg)	Rotor weight
S_w	26,900 (kg)	Stator weight
T_w	42,246 (kg)	Total weight

2.9.2.3 DD Superconducting Synchronous Generators

In efforts to reduce further the machine volume and weight, iron back core and air back core stator superconducting DC rotor excitation SGs have been vigorously investigated lately (Figure 2.38) [35,36].

About 20 t active materials designs for the air back core stator 10 MW, 0.833 Hz (10 poles), with low-temperature superconducting rotor SG inside a rotor diameter of 5 m and an efficiency of 96.4% have been proposed [37]. No data about the total cost of SC–SG systems are available thus far; but in terms of weight of active materials and efficiency, the solution is very interesting. However, it brings

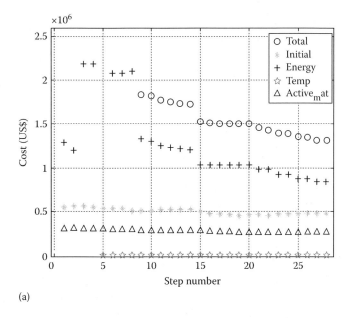

(a)

FIGURE 2.37 Optimization of a 5 MW, 12 rpm DD-dce.SG: (a) Objective function evolution.   (*Continued*)

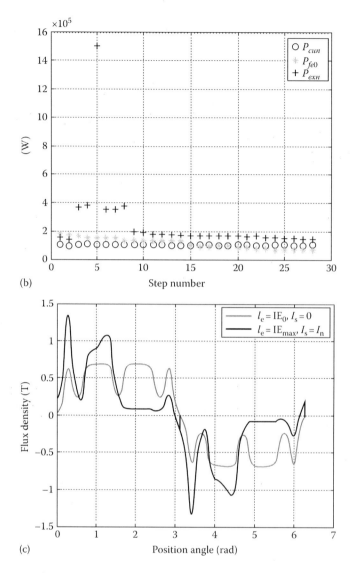

FIGURE 2.37 (*Continued*)   Optimization of a 5 MW, 12 rpm DD-dce.SG: (b) loss breakdown during optimal design and (c) 2D FEM extracted air-gap flux density distribution of radial component during optimal design.

along new challenges such as static power converters to operate properly at such a low fundamental frequency and at 10 MW of power. In addition, the lack of field current control for DC link power control, the rotor cooling and rotor SC replenishing with current periodically (daily) constitute additional problems to solve.

2.9.2.4 Claw Pole 1G-dce.SG (3 MW, 75 rpm)

The claw pole alternator makes most use of DC power excitation, which is, in general, three to four times smaller than in a machine with heteropolar DC excited, once the number of poles $2p_1 > 10$. It is investigated only for 1G–SG because the axial length is limited, and the configuration is most advantageous with a single rotor unit. A dedicated design methodology has been put in place for the claw pole SG and, for the 3 MW, 75 rpm, case study; the following results are shown in Table 2.6. Only a pre-optimization study was performed, but the results are worth considering.

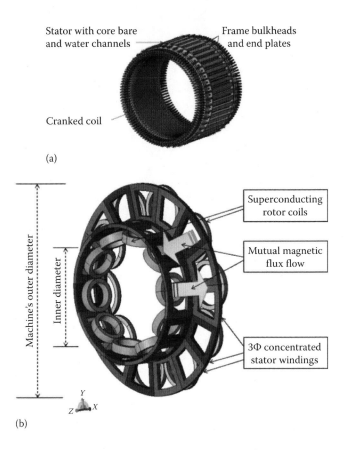

(a)

(b)

FIGURE 2.38 High-temperature superconducting direct drive synchronous generator: (a) with iron back core stator (8 MW, 11 rpm, 2.2 Hz!) and (b) with air core stator (10 MW, 10 rpm, 0.833 Hz). (From Karmaker, H. et al., Stator design concepts for an 8 MW direct drive superconducting wind generator, Record of ICEM, 2012, pp. 769–774; Quddes, M.R. et al., Electromagnetic design study of 10 MW-class wind turbine generators using circular superconducting field coils, Record of ICEMS, 2011, pp. 1–6.)

TABLE 2.6 3 MW, 75 rpm Claw Pole 1G-SG

Rated power (MW)	3
Rated speed (rpm)	75
Efficiency	0.9
Rated power factor	1
Number of pole pairs	40
Phase number	3
Number of current path	10
Air gap (mm)	10
Tangential force (N/cm²)	6
Inner stator diameter (m)	4.5
Fundamental flux density (T)	0.8
Phase voltage (V)	2900
DC voltage (V)	6436
Frequency at low speed (Hz)	33

Source: After from Boldea, I., Tutelea, L., and Blaabjerg, F., High power wind generator designs with less or no PMs: An overview, Keynote Address at *IEEE-ICEMS*, Hangazahou, China, 2014.

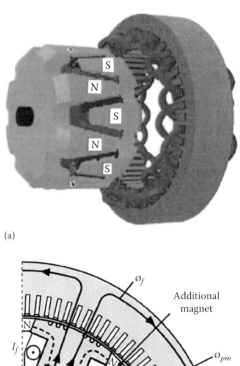

(a)

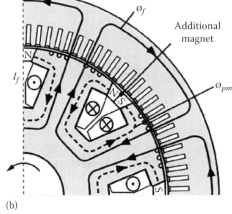

(b)

FIGURE 2.39   Ferrite PM-assisted SG (a) with claw pole rotor and (b) with salient rotor pole.

Ferrite PMs may be "planted" between the rotor poles of the SG or between the claw poles to focus better the excitation flux and de-saturate the rotor, thus bringing an additional 20%–25% output for same geometry and same losses [34,38,39] (see Figure 2.39).

2.9.2.5 Windformer

A high-voltage (40–60 kV) cable-stator-winding SG called "WINDFORMER" (*ABB Review*, vol. 3, 2000, pp. 31–37) has been proposed to eliminate the voltage matching transformer (and its full power switch) for the medium voltage required in power grids. These evident merits are partially compensated by the large insulation spaces inside the slots and to the frame, with a lower slot filling factor and thus larger copper losses, for a given stator geometry.

2.9.3 Less-PM PMSGs

As it has already been discussed thoroughly in the Introduction, the NdFeB PMSGs for both 1G and DD require large amounts of magnets (6 t for the DD-PMSG of 10 MW design at 10 rpm). An alternative will be investigated to reduce PM cost per kilowatt produced by PMSG. Among numerous proposals two are shown to have notable potential: the axial air-gap TF multiple rotor PMSG with circular stator coils and ferrite PM flux concentration rotor modules (Ferrite TF-PMSG) and the super-speed modular PMSG with NdFeB at rotational speeds of 4000 rpm (high frequency and 4G transmission [250:1 ratio]).

2.9.3.1 Ferrite TF-PMSG with Axial Air Gap

A 3D view of a section of one phase of an axial air-gap Ferrite TF-PMSG is shown in Figure 2.40 and was investigated by the authors, for the scope, through a dedicated optimal design software for 3 MW, 15 rpm and compared, again, with the same case study, but using NdFeB and, again, with flux concentration.

The main results obtained from the simulations are summarized in Table 2.7.

The results in Table 2.7 indicate that the NdFeB machine is slightly superior in performance but, in terms of active material weight, the multiple rotor system is much lighter (the stator is almost the same as that of the ferrite PM rotor). In terms of total active materials cost the two solutions are 34% apart, while a high 133 $/kg NdFeB price was already considered. Apparently, only the scarcity of NdFeB can rule here in favor of the ferrite-PM machine. On the other hand, the 0.7–0.75 power factor indicates for both machines a large synchronous reactance in p.u. machine, which leads to higher inverter losses and to an increased inverter kV A rating (and costs); a special, controlled, rectifier structure is claimed to yield higher efficiency than the standard PWM rectifier on the machine side [40].

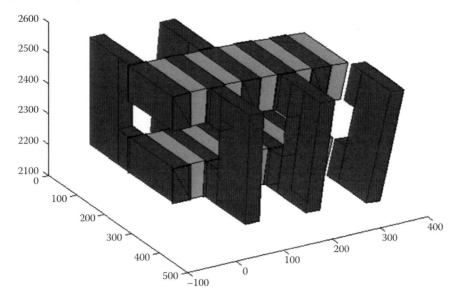

FIGURE 2.40 Axial air-gap ferrite TF-PMSG (3 MW, 15 rpm) design. (After from Boldea, I., Tutelea, L., and Blaabjerg, F., High power wind generator designs with less or no PMs: An overview, Keynote Address at *IEEE-ICEMS*, Hangazahou, China, 2014.)

TABLE 2.7 3 MW, 15 rpm TF-PMSG Performance

	Ferrite Rotor	NdFeB Rotor
External diameter Dext (m)	5.040	5.040
Number of poles, $2p$	216	216
Efficiency (%)	96.76	97
Power factor, $\cos(\varphi)$	0.7045	0.75
Active material weight (t)	36.961	16.597
PM weight (t)	1.983	1.764
Yield energy/year (MWh)	265.149	266.53
Generator active material ($)	380,014	543,905
Price: Ferrite 6 $/kg; NdFeB: 133.8 $/kg		

2.9.3.2 High-Speed Modular PMSG (4 × 0.75 MW, 4000 rpm)

For on-shore applications and probably up to 3 MW power, the generator may be split into a few (four) modules (340 kg each) that run at high speed (4000 rpm) [41]. But the required 260:1 ratio mechanical transmission has to be further improved by the reduction of volume and of outer diameter, to fit in to a reasonable diameter (weight) nacelle. The frequency of the generator may vary from 300 Hz to 1200 Hz (for 4 m/s to 16 m/s wind speed) [41] (see Figure 2.41).

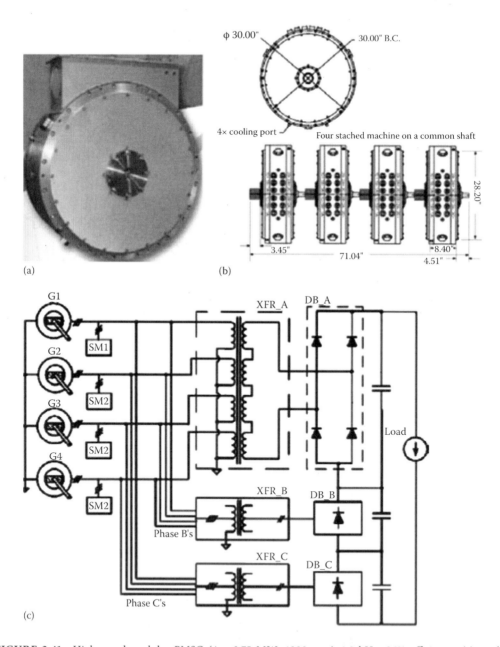

FIGURE 2.41 High-speed modular PMSG (4 × 0.75 MW, 4000 rpm), 1.2 kHz, 96% efficiency: (a) modular PMSG, (b) AC–DC conversion with switched embedded drive rectifiers (SMs), and (c) high-frequency transformer. *(Continued)*

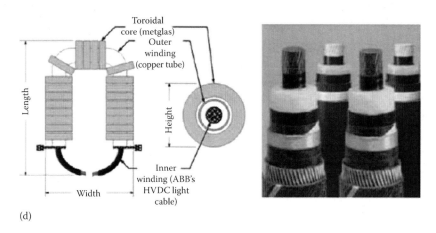

(d)

FIGURE 2.41 (*Continued*) High-speed modular PMSG (4 × 0.75 MW, 4000 rpm), 1.2 kHz, 96% efficiency: (d) small-size single-phase coaxial transformer (1.2 kHz). (From Prasai, A. et al., *IEEE Trans. Power Electron.*, PE-23(3), 1198, 2008.)

For such a high-speed solution, the PM weight per kilowatt is small and thus even NdFeB PMs could be cost-wise acceptable. The high frequency of PMSG voltage should provide low size in transformers while switched mode rectifiers (SMRs) perform the required control of the power delivered to the DC voltage link. A half bridge single-phase inverter is added to recover the capacitors' energy by connecting the generator modules and transformers neutral points. With a single diode output rectifier, the solution seems very tempting in performance/cost terms, provided the challenging 260/1 gear ratio transmission can be built.

2.9.3.3 Flux Reversal PMSGs

In yet another variant of ferrite PM rotor and non-overlapping coils stator, a flux reversal PMSG can be built either with axial or with radial air gap. Similar performance may be obtained (see Figure 2.42) [42].

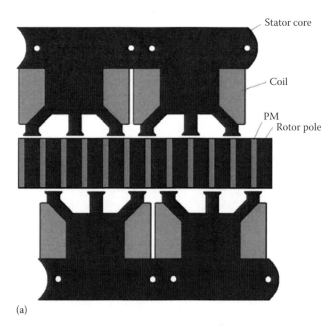

(a)

FIGURE 2.42 Flux reversal: (a) axial–air gap PMSG. (*Continued*)

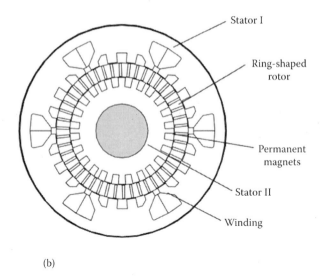

(b)

FIGURE 2.42 (*Continued*)   Flux reversal: (b) radial air gap.

It is easier to manufacture than the TF-PMSG, but it has a high reactance (in pu) and larger copper losses (TF-PMSG with circular shape coils is the best in terms of Joule of losses/Nm).

The radial air-gap configuration may allow for a two-stator single-rotor construction, which is easier to manufacture, while the copper losses per kilowatt are still low. As the ferrite PM rotor stands to be long axially (see Figure 2.42a), in order to limit the machine volume, the solution may be tried especially in 1G or even 3G PMSGs, where the torque per kilowatt is lower than in the direct drive wind generators.

2.9.3.4 The Vernier Machine

A radial air-gap, PM-flux concentration rotor dual stator, similar to the one in Figure 2.42, for the reversal flux PMSG, has been proposed in the Vernier machine topology [43] with claims of high power factor for distributed stator windings with $q = 2$ slots per pole per phase (Figure 2.43). It is yet to be tried for wind generator applications, but it is evident that the copper losses tend to be a bit smaller in the flux reversal PMSG (Figure 2.42), though the use of stator core is better in the Vernier machine (Figure 2.43).

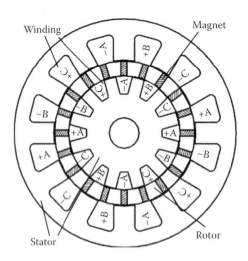

FIGURE 2.43   Vernier, dual-stator PM flux concentration rotor synchronous generator. (From Li, D. et al., High power factor vernier permanent magnet machines, Record of IEEE-ECCE, 2013, pp. 1534–1540.)

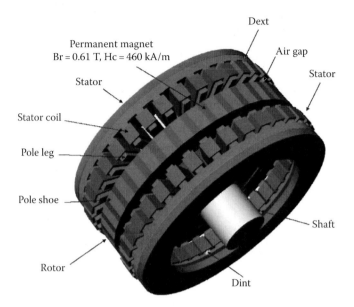

FIGURE 2.44 Tooth wound axial air-gap PMSG. (After from Boldea, I., Tutelea, L., and Blaabjerg, F., High power wind generator designs with less or no PMs: An overview, Keynote Address at *IEEE-ICEMS*, Hangazahou, China, 2014.)

This technology allows for the axial air-gap topologies when the two stators may be assembled easier, reducing the fabrication costs. Therefore, another variant is a dual stator axial–air gap, PMs flux concentration (spoke magnet) rotor PMSG [44,45], as shown in Figure 2.44. This configuration allows the stator poles insertion one by one, reducing the fabrication cost of the machine. With 0.6 T, 450 kA/m PMs, the in situ magnetization pole by pole may be attempted, to further reduce the manufacturing difficulties. But the copper losses are not small enough such that for a 3 MW, 15 rpm direct-driven PMSG the efficiency is moderate (0.93) for low power factor (0.54), which implies over-rated PWM converter on the machine side.

2.9.4 Multiphase Reluctance Generators (BLDC-MRG)

BLDC-MRGs (brushless multiphase reluctance generators) in two-pole configuration come from, say, a 12 slot DC brush machine single-layer winding rotor, where the excitation is stripped off and its brushes have been moved out from the neutral axis by a slot-pitch (to the stator pole corner [Figure 2.45a]). The equivalent BLDC-MRG with six phases is thus obtained if the mechanical commutation is replaced by electronic commutation (Figure 2.45b) and controlled as it is shown in Figure 2.45c,d.

Each current polarity (Figure 2.45c) has a field (i_c) coil role part (when the coils of that phase fall between the rotor poles) and a torque (i_T) coil role part, when the coils of that phase fall under rotor variable reluctance (salient) rotor [46–49]. The interaction of the excitation field (i_F) and torque current (by i_T) falls around 90 (close to ideal) to produce the torque T_e in the m phase machine (m_F—field phases, m_T—torque phase; $m = 6$, $m_F = 2$, $m_T = 4$). The average torque is basically

$$T_{eav} = k_\psi i_F i_T \left(m - m_F \right) \tag{2.13}$$

The flat-top currents i_F and i_T may be equal to each other on a rather optimal design for all operation regimes (or different from each other when the speed range has to be further extended and for generating mode). The high saliency ratio of rotors, which will be realized by a multiple flux barrier rotor or by an axially laminated anisotropic (ALA) rotor, is needed to secure a low armature reaction (higher torque

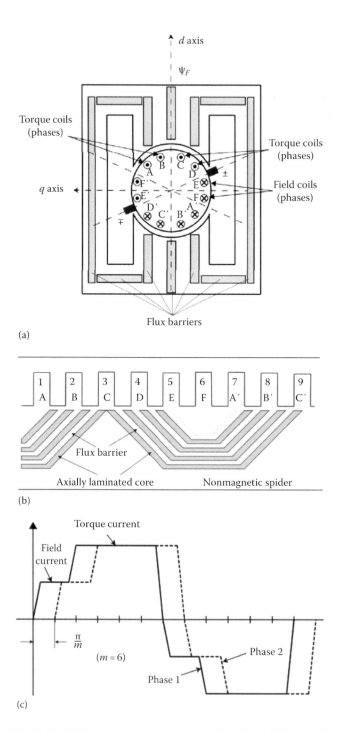

FIGURE 2.45 Brushless DC multiphase reluctance generators: (a) original DC brushed machine, (b) 6 phase BLDC-MRG, and (c) phase current waveform. (*Continued*)

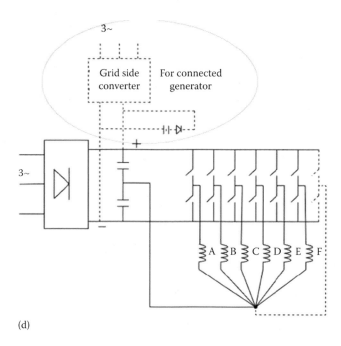

(d)

FIGURE 2.45 (*Continued*) Brushless DC multiphase reluctance generators: (d) inverter for control.

density for a given saturation level of flux density in the stator teeth facing rotor pole tip); a not-so-small air gap is good as a compromise between torque density, efficiency and current commutation quickness.

It is evident that the flat top (BLDC-type) current reveals a good use of the inverter voltage (lower "equivalent" reactive power) and, with a high rotor saliency, small voltage regulation. The number of phases may be 5, 6, 7, 9, 11, 15, and so forth, but the pole pitch increases with the number of phases. In addition, the multiphase inverter (Figure 2.45a) null current is smaller for an odd number of phases.

A proposal that divides the stator into four sectors, and uses 4 inverters that control the torque contribution, quicker, by the torque currents i_T and the air gap g_0, slower, by the field currents i_F provides for active magnetic levitation of the rotor, which should reduce drastically the generator structure (Figure 2.46 [42]).

A design case study for a 6 MW and 12 rpm direct-drive generator by using 2D FEM has led to FEM results as shown in Figure 2.47 with torque versus position as in Figure 2.46.

It is evident, as shown in Table 2.8, that the weight is acceptably low, while the copper losses make only 3% of output power: this means an efficiency by around 96% as the mechanical and iron losses ($f_{1N} = 14$ Hz) are considered to represent 1% of the output power.

Therefore, the performance looks good, except for the still relatively large torque pulsations (even after making the rotor of two pieces shifted by one slot pitch [skewing]). As the principle is sound, with all phases active and one phase-only commutating at any time, and the rotor has good manufacturability (the speed is below 10 rpm), the BLDC-MRG, taking advantage of multiphase inverters in recent developments, using its implicit notable fault tolerance capacity, seems to deserve attention both in direct drives, but especially, in 1G (single-stage gear) or 3(4)G solutions for wind generators.

2.9.5 DFIG: Brushless?

For a 10 MW DFIG about 2–3 MW of power has to be transferred to (from) the rotor through slip-rings and mechanical brushes. Besides the arch-occurrence-danger and maintenance problems that are far from trivial, this mechanical transfer of power implies also at least 0.5% of full power as losses. As a full power transformer would exhibit about the same losses (efficiency 99.5%) at grid frequency (50/60 Hz) it

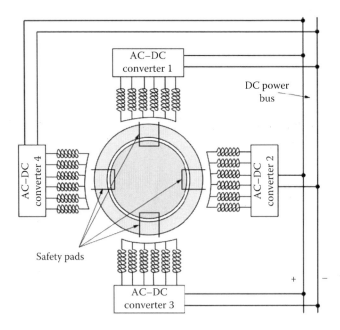

FIGURE 2.46 Brushless DC multiple phase reluctance generators with actively levitated rotor.

seems feasible to make the DFIG rotor as primary (full power) and then use the stator as secondary, fed through a dual PWM AC–AC converter system, designed at partial power rating (20%–30%). A rotary three-phase transformer with a low enough air gap (0.7–1 mm) designed for full power transfer to the rotor at grid frequency can retain an efficiency close to 99% for 1 kg/kW weight. The other transformer secondary placed on the stator (as the transformer primary) will feed the dual PWM converter that supplies the DFIG stator winding at "slip frequency" and power: $S_{max}S_n$; S_n-DFIG rating (Figure 2.48). At the same time, the rotary transformer primary—all transformer coils show minimum copper loss due the blessing of circularity—may be designed at higher voltage, above 36 kV, to eliminate the matching voltage existing transformer, needed for interfacing the typical 0.69–6 kV wind generators.

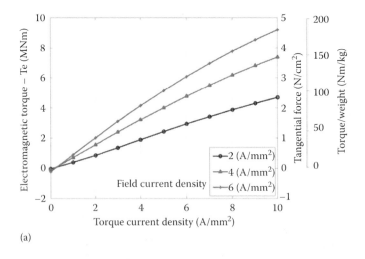

FIGURE 2.47 Brushless DC multiple phase reluctance generators: (a) torque production.   *(Continued)*

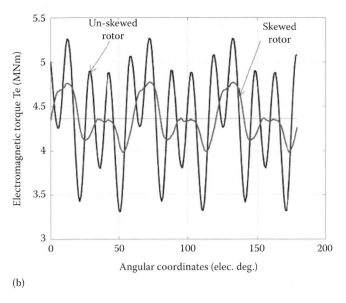

(b)

FIGURE 2.47 (*Continued*) Brushless DC multiple phase reluctance generators: (b) torque pulsations at nominal power. (After from Boldea, I., Tutelea, L., and Ursu, D., BLDC multiphase reluctance machines for wide range applications: A revival attempt, Record of EPE-PEMC-ECCE, 2012, Novisad, Serbia.)

TABLE 2.8 Brushless DC Multiple Phase Reluctance Generators (6 MW, 12 rpm)

Number of phases (m)	6
Pole pairs, p_1/f_1 (Hz)	70/14
Total active weight (t)	55
Air gap g (mm)	4
Inner stator diameter (m)	9.0
Stack length (m)	1.5
Copper losses (kW)	180
Predicted efficiency	96%

This brushless power transfer solution with full power transfer to the rotor at the grid frequency has the size of one transformer that would transfer the slip power ($|S_{max}S_n|$) at slip frequency $f_2 = S_{max}f_1$, but it may provide the DFIG transition through synchronous speed naturally (the rotary transformer has been initially proposed to transfer the slip power to the rotor without brushes [50], but power transfer at zero slip is not feasible).

Moreover, a recent study has proven that the wound rotor induction machine designed with the rotor as primary (at full power) and stator as secondary (at slip power in terms of slot area, etc.) may result in a 5%–6% size decrease and cost reduction [51]. The elimination of the brush system for full power transfer is not considered in [51]. For a 10 MW direct drive (DD) DFIG, where the active material weight is 65 t (by design), the addition of a 10 t (or less, with strong cooling) for the rotary transformer (for 10 MW at 50/60 Hz) may well be justified, especially if its primary is designed at above 30 kV, to provide direct interfacing with the middle-voltage local power grids.

2.9.6 Brushless Doubly Fed Reluctance (or Induction) Generators [52,53]

Another doubly fed machine, with dual stator winding (a power winding with $2p_1$ poles and a control winding with $2p_2$ poles) and a rotor, with $p_1 + p_2 = p_r$ poles, which has either a high reluctance

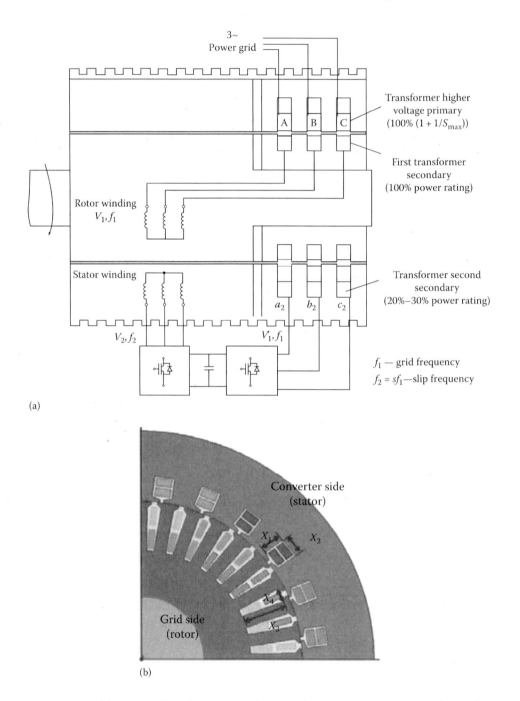

FIGURE 2.48 Brushless DFIG: (a) with rotary transformer and machine rotor as primary (designed at 100% power rating) and (b) cross-section. (From You, Y.M. et al., *IEEE Trans.*, MAG-48(11), 3124, 2012.)

ratio (multiple flux-barrier or ALA rotor [52,53]) or is a nested-cage rotor [54,55] has been proposed (Figure 2.49). It is essentially a brushless configuration, where the main winding is connected directly to the power grid, while the control winding (sized at 20%–30% power rating) is fed through an AC–AC PWM converter, much like in DFIG (Figure 2.49). As the torque density is paramount in wind generators (Nm/kg), it has to be mentioned that the magnetic coupling "between each of the

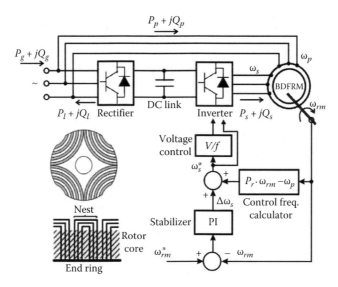

FIGURE 2.49 Brushless doubly fed reluctance generator with its control.

two windings and the rotor" takes place via an air-gap permeance space harmonic whose amplitude is less than 50% of its "DC" (main) level [53]. This means, inevitably, at best, moderate torque density. The magnetic coupling is thus weak and, consequently, each AC winding "reflects" an additional leakage-like reactance, which means a burden on the converter. A large air gap, inevitable in DD direct-drive wind generators, is not favorable for this machine. Therefore, it may be tried first for 3G low and medium-power wind generators, through design optimization studies in which the converter losses and cost have to be considered as a whole.

2.9.7 Switched Reluctance Generator Systems [37,56,57]

There are few switched reluctance generator (SRG) designs for wind energy, related to low-power case studies [28] and to direct-drive SRG [10] at 6 MW, 12 rpm. In order to secure a large efficiency and a large torque density, the SRG requires small air gap and large stator/rotor pole span. However, in DD-SRGs, a large number of rotor poles provide high torque copper losses, if the phase coils have a circular shape; and up to the point where the flux fringing increases ruins it, when the rotor pole span/air gap decreases too much.

2.9.7.1 DD-SRG

It has been calculated that in a DD-SRG design (using FEM) with 4 mm air gap, 9.0 m interior stator diameter, at a frequency of 175 Hz, the saliency ratio goes down to $L_{unaligned}/L_{aligned} = 0.6$ and thus the machine magnetization within 33% of the energy cycle time requires a 10/1 forcing voltage, which is unpractical [37]. This explains why for DD-SRG at 6 MW, 12 rpm power range DC excitation coils (which span three AC coils) on the stator are added, to produce a homopolar flux density that should vary from 1.5 T in aligned position and less than to 0.5 T for unaligned position. This is necessary for sufficient AC emf in the tooth-wound winding phase coils, which now are flowed by bipolar currents like in any stator excited SG. It has also been shown [37] that the SRG configuration (Figure 2.51a) is capable to produce 6 MW at 12 rpm, with an air gap of 4 mm at 9.0 m interior stator diameter and 1.5 m stack length and 80 t of active materials, but the total DC + AC losses are 1.5 MW, which means the solution is unpractical (80% efficiency). If a TF six-phase configuration is instead chosen (see Figure 2.50b,c) the design [37] will produce the required torque of 5.5 M N m (and more, see Figure 2.50c) for 240 kW of total (DC + AC coils losses).

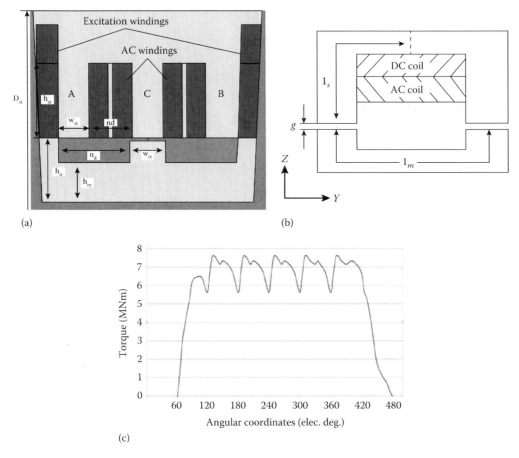

FIGURE 2.50 Direct-driven switched reluctance generator [10]: (a) standard DC + AC stator, (b) transverse flux circular coil DC + AC stator, and (c) FEM torque of the 6-phase circular coil DC + AC stator machine.

This leads to at least 94% of the total efficiency [37]. To avoid voltage induced by the AC coils in the DC coils [37] and thus reduce the voltage (and kV A) of the DC–DC converter that controls the DC (excitation) coils, the latter should first be connected in series. Unfortunately, the 4 mm air gap means a large p.u. reactance of the machine, which again means a burden of the converter kV A. In addition, the homo-polar character of the magnetic DC field in all stator poles implies higher than necessary normal (radial) forces. A lot is to be done in order to prove the practicality of the TF circular DC + AC coils stator for DD high-power wind generators (machine +converter +control) in terms of cost/performance, to prove their practicality.

2.9.7.2 High-Speed Wind SRG

On the other hand, for 1500–4000 rpm 3(4)G-SRG, the standard [56] or dual-stator segmented pole secondary [58] configurations (Figure 2.51) have already been proven to be capable of high torque density (45 Nm/l) at above 92% efficiency at 50 kW and more for HEV, and should be considered for high-speed wind generators for more reliable and lower cost and, at the same time, good torque density (reasonable weight). The 100(260)/1 ratio transmission has to be provided as usual for such systems, as in multi modular generator (4× 0.75 MW modules for 3 MW, as proposed with PMSGs [41]), which may be more feasible: "easy to say but difficult to do."

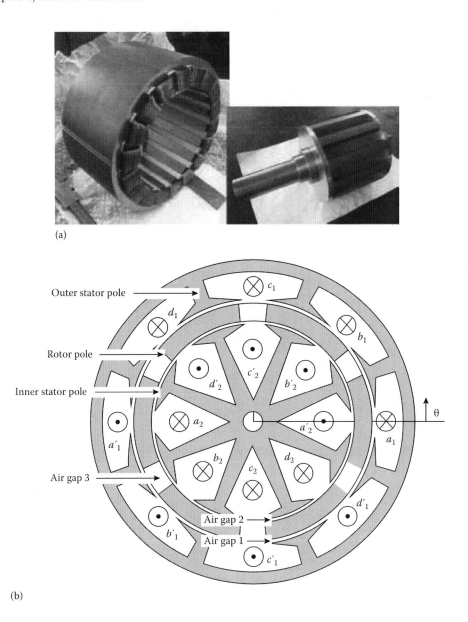

FIGURE 2.51 Standard switched reluctance generator [56]: (a) and double-stator segmented rotor and (b) high-speed SRGs for wind energy conversion. (After from Abbasian, M.A., Moallem, M., Fahimi, B., *IEEE Transactions on Energy Conversion*, EC-25(3), 589, 2010.)

2.9.8 Flux-Switch Ferrite PM Stator Generators

The stator PM flux switching double salient motors have been proven recently to produce a good torque density in industrial drives with stator NdFeB [59–61] or with hybrid stator excitation [62,63]. Already in Reference [63], it has been proposed to use the PM ferrite stator flux switch machine for wind generators (see Figure 2.52), even though in a low power design example.

Table 2.9 shows a comparison of the performance of a PMSG and a low-power PM ferrite flux switch generator [63] at 30 rpm.

Permanent magnet   Windings

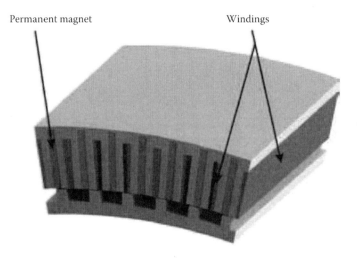

FIGURE 2.52   PM Ferrite stator flux switch generator: a view of a machine sector.

TABLE 2.9   Flux Switching Ferrite PM Stator Generator

	PM Machine "ALXION" 25	Ferrite
Turbine power (kW)	6.8	5
Maximum torque (Nm)	1049	1000
Efficiency (%)	79	77
Total mass (kg)	82	132
External radius (mm)	795	240
Air gap radius (mm)	689	208
Active length (mm)	55	476
Wind speed (m/s)	N/A	6
Rotor speed (rpm)	N/A	30
Turbine radius (m)	N/A	1

Note: The mass is given by manufacturer datasheet.

The efficiency is rather low (70%) at 30 rpm and 1000 Nm (1150 W losses for 5 kW of wind power). But extrapolating this design (using same torque) at 1500 rpm would mean, with doubled losses (due to higher core and some mechanical losses), an efficiency of about 96%. Also in this latter case, the weight of active materials corresponds to 1 kW/kg, which is a rather large value. Consequently, the solution deserves to be investigated further for 3(4)G-high speed modular wind generators. Not so, in our view, for high power DD wind generators, because the air gap is large and because the radial-long Ferrite PMs, needed in such a case for strong flux concentration, means too large an extra core. On the other hand, the ferrite-rotor double salient generator [64] may allow enough room for strong ferrite PM flux concentration with less extra weight in the rotor and thus yield a competitive direct drive wind generator along with this principle (Figure 2.53). It remains to be seen if the promises of a large torque density in Reference 64 may be duplicated in a large diameter (3 mm air gap) DD wind generators while the converter has to deal with bipolar currents whose summation is not zero.

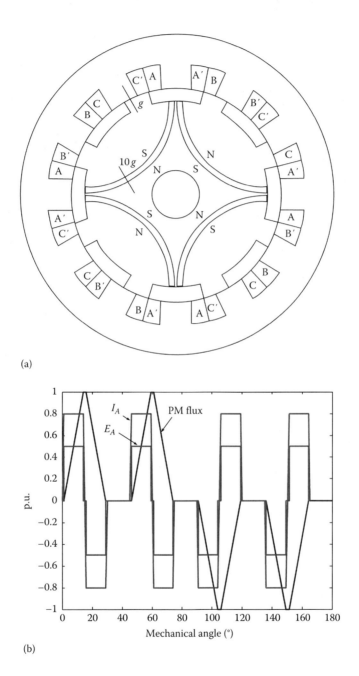

(a)

(b)

FIGURE 2.53 Ferrite PM double salient generator [64]: (a) construction and (b) PM flux current, emf.

2.10 Summary

In this chapter, we have presented some representative—in use and newly proposed—types of electric generators by principle, configuration, and application.

A few concluding remarks are in order:

- The power per unit range varies from a few mW to a few hundred megawatts (even 1500 MV A) per unit.
- Large power generators—above a few MW—are electrically excited on the rotor either DC, as in conventional synchronous generator (SG), or in three-phase AC, as in the wound rotor (doubly fed) induction generator (WRIG).
- While the conventional DC rotor-excited SG requires tightly controlled constant speed to produce constant frequency output, the WRIG may work with adjustable speed.
- The rating of the rotor-connected PWM converter in WRIG is about equal to the adjustable speed range (slip), in general around 20%. This implies reasonable costs for a more flexible generator with fast active and reactive power (or frequency and voltage) control.
- WRIG seems the way of the future in electric generation at adjustable speed for powers above a few megawatts, in general, per unit.
- PMSGs are emerging for kW, tenth of kW, even hundreds of kilowatts or 1–10 MW/unit in special applications such as automotive starter-alternators or super-high-speed gas turbine generators or DD wind generators, respectively.
- LMAs are emerging for power operation up to 15, even 50 kW for home or special series hybrid vehicles, with linear gas combustion engines and electric propulsion.
- Parametric generators are being investigated for special applications: SRGs for aircraft jet engine starter-alternators and TFGs/motors for hybrid or electrical bus propulsion or directly driven wind generators.
- Section 2.9 illustrates in detail the myriad of generators' topologies for wind (and small hydro) energy conversion; vehicular (stand-alone) generators on aircraft, ship, on-road vehicles, etc. are under very dynamic scrutiny, also.
- Electric generators are driven by different prime movers that have their own characteristics, performance, and mathematical models, which, in turn, influence the generator operation since at least speed control is enacted upon the prime mover. Chapter 3 dwells in some detail on most used prime movers with their characteristics, mathematical models, and speed control methods.

References

1. T. Bödefeld, H. Sequenz, *Elektrische Maschinen*, Springer, Vienna, Austria, 1938 (in German).
2. C. Concordia, Synchronous machines, in *Theory and Performance Book,* John Wiley & Sons, New York, 1951.
3. R. Richter, *Electrical Machines*, Vol. 2: *Synchronous Machines*, Verlag Birkhäuser, Basel, Switzerland, 1963 (in German).
4. M. Kostenko, L. Piotrovski, *Electrical Machines*, Vol. 2: *AC Machines*, Mir Publishers, Moscow, Russia, 1974.
5. J.H. Walker, *Large Synchronous Machines*, Clarendon Press, Oxford, U.K., 1981.
6. T.J.E. Miller, *Brushless PM and Reluctance Motor Drives*, Clarendon Press, Oxford, U.K., 1989.
7. S.A. Nasar, I. Boldea, L. Unnewher, *Permanent Magnet, Reluctance and Self-Synchronous Motors*, CRC Press, Boca Raton, FL, 1993.
8. D.C. Hanselman, *Brushless PM Motor Design*, McGraw Hill, New York, 1994.
9. D.R. Hendershot Jr., T.J.E. Miller, *Design of Brushless PM Motors*, Magna Physics Publishing/ Clarendon Press, Oxford, U.K., 1994.

10. J. Gieras, F. Gieras, M. Wing, *PM Motor Technologies*, 2nd edn., Marcel Dekker, New York, 2002.
11. I. Boldea, S. Scridon, L. Tutelea, BEGA: Biaxial excitation generator for automobiles, Record of OPTIM, Poiana Brasov, Romania, 2000, Vol. 2, pp. 345–352.
12. T. Miller, *Switched Reluctance Motors and Their Control*, OUP, Oxford, U.K., 1993.
13. Y. Liuo, T.A. Lipo, A new doubly salient PM motor for adjustable speed drives, *EMPS*, 22(3), 259–270, 1994.
14. M. Radulescu, C. Martis, I. Husain, Design and performance of small doubly salient rotor PM motor, *EPCS* (former EMPS), 30, 523–532, 2002.
15. F. Blaabjerg, I. Christensen, P.O. Rasmussen, L. Oestergaard, New advanced control methods for doubly salient PM motor, Record of IEEE-IAS, 1996, pp. 786–793.
16. S.E. Rauch, L.J. Johnson, Design principles of flux switch alternator, *AIEE Transactions*, 74(III), 1261–1268, 1955.
17. H. Weh, H. Hoffman, J. Landrath, New permanent excited synchronous machine with high efficiency at low speeds, *Proceedings of the ICEM*, Pisa, Italy, 1988, pp. 1107–1111.
18. G. Henneberger, I.A. Viorel, *Variable Reluctance Electric Machines*, Shaker Verlag, Aachen, Germany, 2001, Chapter 6.
19. L. Luo, S. Huang, S. Chen, T.A. Lipo, Design and experiments of novel axial flux circumpherentially current PM (AFCC) machine with radial airgap, Record of IEEE-IAS, 2001.
20. I. Boldea, S.A. Nasar, *Linear Electric Actuators and Generators*, Cambridge University Press, Cambridge, U.K., 1997.
21. J. Wang, W. Wang, G.W. Jewell, D. Howe, Design and experimental characterisation of a linear reciprocating generator, *Proceedings of the IEE*, 145-EPA(6), 509–518, 1998.
22. L.M. Hansen, P.H. Madsen, F. Blaabjerg, H.C. Christensen, U. Lindhard, K. Eskildsen, Generators and power electronics technology for wind turbines, Record of IEEE-IECON, 2001, pp. 2000–2005.
23. I. Boldea, I. Serban, L. Tutelea, Variable speed generators and their control, *Journal of Electrical Engineering*, 2(1), 2002 (www.jee.ro).
24. K. Kudo, Japanese experience with a converter fed variable speed pumped storage system, *Hydropower and Dams*, March 1994.
25. T. Kuwabata, A. Shibuya, M. Furuta, Design and dynamic response characteristics of 400 MW adjustable speed pump storage unit for Ohkawachi Power Station, *IEEE Transactions*, EC-11(2), 376–384, 1996.
26. T.-H. Kim, H.-W. Lee, Y.H. Kim, J. Lee, I. Boldea, Development of a flux concentration-type linear oscillatory actuator, *IEEE Transactions*, MAG-40(4), 2092–2094, 2004.
27. I. Boldea, L. Tutelea, F. Blaabjerg, High power wind generator designs with less or no PMs: An overview, Keynote Address at *IEEE-ICEMS*, Hangzhou, China, 2014.
28. Y. Amirat, M.E.H. Benbouzid, B. Bensaker, R. Wamkeue, H. Mangel, The state of the art of generators for wind energy conversion systems, *Proceedings of ICEM*, 2006, paper 2430.
29. M. Liserre, R. Cardenas, M. Molinas, J. Rodriguez, Overview of multi-MW wind turbines and wind parks, *IEEE Transactions on Industrial Electronics*, IE-58(4), 1081–1095, 2011.
30. H. Polinder, F.F.A. Van Der Pijl, G.-J. De Vilder, P.J. Tavner, Comparison of direct-drive and geared generator concepts for wind turbines, Record of IEEE-IEMDC, 2005, pp. 543–550.
31. H. Polinder, D.J. Bang, R.P.J.M. Van Rooij, A.S. McDonald, M.A. Mueller, 10 MW wind turbine direct-drive generator design with pitch or active speed stall control, Record of IEEE-IEMDC, 2007, pp. 1390–1395.
32. V. Delli Colli, F. Marignetti, C. Attaianese, Analytical and multiphysics approach to the optimal design of a 10-MW DFIG for direct-drive wind turbines, *IEEE Transactions on Industrial Electronics*, IE-59(7), 2791–2799, 2012.
33. L.F. Hui, Z. Chen, Design optimization and evaluation of different wind generator systems, Record of ICEMS, 2008, pp. 2396–2401.
34. I. Boldea, *Synchronous Generators*, CRC Press/Taylor & Francis, New York, 2006, pp. 2–7.

35. H. Karmaker, E. Chen, W. Chen, G. Gao, Stator design concepts for an 8 MW direct drive super-conducting wind generator, Record of ICEM, 2012, pp. 769–774.

36. M.R. Quddes, M. Sekino, H. Ohsaki, Electromagnetic design study of 10 MW-class wind turbine generators using circular superconducting field coils, Record of ICEMS, 2011, pp. 1–6.

37. C.B. Rauti, D.T.C. Anghelus, I. Boldea, A comparative investigation of three PM-less MW power range wind generator topologies, Record of OPTIM, 2012 (IEEE Xplore), Brasov, Romania.

38. L. Tutelea, D. Ursu, I. Boldea, IPM-claw pole alternator system for more vehicle breaking energy recuperation, 12(4), 2012. www.jee.ro.

39. K. Yamazaki, K. Nishioka, K. Shima, T. Fukami, K. Shirai, Estimation of assist effects by additional permanent magnets in salient-pole synchronous generators, *IEEE Transactions on Industrial Electronics*, IE-59(6), 2515–2523, 2012.

40. M.T. Kakhki, M.R. Dubois, High efficiency rectifier for a variable speed transverse flux permanent magnet generator for wind turbine applications, Record of EPE, 2011, Birmingham, U.K.

41. A. Prasai, J.-S. Yim, D. Divan, A. Bendre, S.K. Sul, A new architecture for offshore wind farms, *IEEE Transactions on Power Electronics*, PE-23(3), 1198–1204, 2008.

42. I. Boldea, L.N. Tutelea, M. Topor, Theoretical characterization of three phase flux reversal machine with rotor-PM flux concentration, Record of OPTIM, 2012 (IEEE Xplore).

43. D. Li, R. Qu, T. Lipo, High power factor vernier permanent magnet machines, Record of IEEE-ECCE, 2013, pp. 1534–1540.

44. W. Zhao, T.A. Lipo, B. Kwon, Design and analysis of a novel dual stator axial flux spoke-type ferrite permanent magnet machine, Record of IEEE-IECON, 2013, pp. 2714–2719.

45. O. Nymann, K.D. Larsen, U. Dam, Electrical motor/generator having a number of stator pole cores being larger than a number of rotor pole shoes, US Patent Us 7982352 B2, Published July 19, 2011, registered September 18, 2003.

46. R. Mayer, H. Mosebach, U. Schrodel, H. Weh, Inverter fed multiphase reluctance machine with reduced armature reaction and improved power density, Record of ICEM, 1986, Munchen, Part 9, pp. 1438–1441.

47. I. Boldea, Gh. Papusoiu, S.A. Nasar, A novel series connected SRM, Record of ICEM, 2000, MIT, Cambridge, MA, Part 3, pp. 1212–1217.

48. J.D. Law, A. Chertok, T.A. Lipo, Design and performance of the field regulated reluctance machine, Record of IEEE–IAS, 1992, Vol. 1, pp. 234–241.

49. I. Boldea, L. Tutelea, D. Ursu, BLDC multiphase reluctance machines for wide range applications: A revival attempt, Record of EPE-PEMC-ECCE, 2012, Novisad, Serbia.

50. M. Ruviaro, F. Runcos, N. Sadowski, I.M. Borges, Analysis and test results of a brushless doubly fed induction machine with rotary transformer, *IEEE Transactions on Industrial Electronics*, IE-59(6), 2670–2677, June 2012.

51. Y.M. You, T.A. Lipo, B. Kwon, Optimal design of a grid-connected-to-rotor type doubly fed induction generator for wind turbine systems, *IEEE Transactions*, MAG-48(11), 3124–3127, 2012.

52. M.G. Jovanovic, R.E. Betz, J. Yu, The use of doubly fed reluctance machines for large pumps and wind turbines, *IEEE Transactions on Industry Applications*, IA-38(6), 1508–1516, 2002.

53. A.M. Knight, R.E. Betz, D.G. Dorrell, Design and analysis of brushless doubly fed reluctance machines, *IEEE Transactions on Industry Applications*, IA-49(1), 50–58, 2013.

54. R.A. McMahon, P.C. Roberts, X. Wang, P.J. Tavner, Performance of BDFM as generator and motor, *IEE Proceedings on Electric Power Applications*, EPA-153(2), 289–299, 2006.

55. H. Gorginpour, H. Oraee, R.A. McMahon, A novel modeling approach for design studies of brushless doubly fed induction generator based on magnetic equivalent circuit, *IEEE Transactions on EC*, 28(4), 902–912, 2013.

56. M. Takeno, A. Chiba, N. Hoshi, S. Ogasawara, M. Takemoto, M.A. Rahman, Test results and torque improvement of the 50-kW switched reluctance motor designed for hybrid electric vehicles, *IEEE Transactions on Industry Applications*, IA-48(4), 1327–1334, 2012.

57. M.A. Mueller, Design and performance of a 20 kW, 100 rpm, switched reluctance generator for a direct drive wind energy converter, Record of IEEE-IEMDC-2005, 2005, pp. 56–63.

58. M.A. Abbasian, M. Moallem, B. Fahimi, Double-stator switched reluctance machines (DSSRM): Fundamentals and magnetic force analysis, *IEEE Transactions on Energy Conversion*, EC-25(3), 589–597, 2010.

59. J.T. Chen, Z.Q. Zhu, S. Iwasaki, R.P. Deodhar, A novel E-core switched-flux PM brushless AC machine, *IEEE Transactions on Industry Applications*, IA-47(3), 1273–1282, 2011.

60. W. Hua, Z.Q. Zhu, M. Cheng, Y. Pang, D.G. Howe, Comparison of flux-switching and doubly-salient permanent magnet brushless machines, Record of ICEMS, 2005, Vol. 1, pp. 165–170.

61. J. Ojeda, M.G. Simoes, G. Li, M. Gabsi, Design of a flux-switching electrical generator for wind turbine systems, *IEEE Transactions on Industry Applications*, IA-48(6), 1808–1816, 2012.

62. Y. Wang, Z. Deng, Hybrid excitation topologies and control strategies of stator permanent magnet machines for DC power system, *IEEE Transactions on Industrial Electronics*, IE-59(12), 4601–4616, 2012.

63. A. Zulu, B.C. Mecrow, M. Armstrong, A wound-field three-phase flux-switching synchronous motor with all excitation sources on the stator, *IEEE Transactions on Industry Applications*, IA-46(6), 2363–2371, 2010.

64. Y. Liao, T.A. Lipo, A new doubly salient permanent, magnet motor for adjustable speed drives, *EMPS Journal*, 22(2), 254–270, 1994.

3

Prime Movers

3.1 Introduction

Electric generators convert mechanical energy into electrical energy. Mechanical energy is produced by prime movers. Prime movers are mechanical machines that convert primary energy of a fuel or fluid into mechanical energy. They are also called "turbines or engines." The fossil fuels commonly used in prime movers are coal, gas, oil, and nuclear fuel.

Essentially, the fossil fuel is burned in a combustor and thus thermal energy is produced.

Thermal energy is then taken by a working fluid and converted into mechanical energy in the prime mover itself.

Steam is the working fluid for coal or nuclear fuel turbines, but it is the gas or oil in combination with air, in gas turbines, diesel, or internal combustion engines.

On the other hand, the potential energy of water from an upper-level reservoir may be converted into kinetic energy that hits the runner of a hydraulic turbine, changes momentum and direction, and produces mechanical work at the turbine shaft as it rotates against the "braking" torque of the electric generator under electric load.

Wave energy is similarly converted into mechanical work in special tidal hydraulic turbines. Wind kinetic energy is converted into mechanical energy by wind turbines.

A complete classification of prime movers is very difficult due to so many variations in construction, from topology to control. However, a simplified one is shown in Table 3.1.

In general, a prime mover or turbine drives an electric generator directly, or through a transmission (Figure 3.1) [1–3]. The prime mover is necessarily provided with a so-called speed governor (in fact, a speed control and protection system) that properly regulates the speed, according to electric generator frequency/power curves (Figure 3.2).

Note that the turbine is provided with a servomotor that activates one or a few control valves that regulate the fuel (or fluid) flow in the turbine, thus controlling the mechanical power at the turbine shaft.

The speed at turbine shaft is measured rather precisely and compared with the reference speed. The speed controller then acts on the servomotor to open or close control valves and control speed. The reference speed is not constant. In AC power systems, with generators in parallel, a speed droop of 2%–3% is allowed with power increased to rated value.

The speed droop is required for two reasons as follows:

1. With a few generators of different powers in parallel, fair (proportional) power load sharing is provided.
2. When power increases too much, the speed decreases accordingly signaling that the turbine has to be shut off.

The point A of intersection between generator power and turbine power in Figure 3.2 is statically stable as any departure from it would provide the conditions (through motion equation) to return to it.

TABLE 3.1   Turbines

	Fuel	Working Fluid	Power Range	Main Applications	Type	Observation
1.	Coal or nuclear fuel	Steam	Up to 1500 MW/unit	Electric power systems	Steam turbines	High speed
2.	Gas or oil	Gas (oil) + air	From watts to hundreds of MW/unit	Large and distributed power systems, automotive applications (vessels, trains and highway and off-highway vehicles), and autonomous power sources	• Gas turbines • Diesel engines • Internal combustion engines • Stirling engines	With rotary but also linear reciprocating motion
3.	Water energy	Water	Up to 1000 MW/unit	Large and distributed electric power systems, autonomous power sources	Hydraulic turbines	Medium- and low speed >75 rpm
4.	Wind energy	Air	Up to 10 MW/unit	Distributed power systems, autonomous power sources	Wind or wave turbines	Speed down to 10 rpm

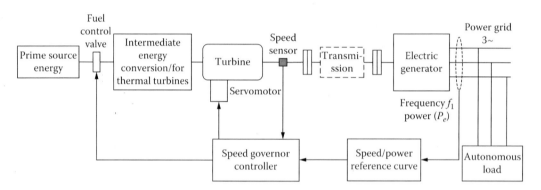

FIGURE 3.1   Basic prime mover generator system.

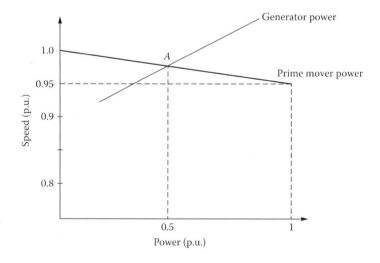

FIGURE 3.2   The reference speed (frequency)/power curve.

With synchronous generators operating in a rather constant voltage and frequency power system, the speed droop is very small, which implies strong strains on the speed governor—due to large inertia, etc. It also leads to not-so-fast power control. On the other hand, the use of doubly fed synchronous generators, or of AC generators with full power electronics between them and the power system, would allow for speed variation (and control) in larger ranges (±20% and more). That is, smaller speed reference for smaller power. Power sharing between electric generators would then be done through power electronics in a much faster and more controlled manner. Once these general aspects of prime mover requirements have been clarified, let us deal in some detail with prime movers in terms of principle, steady-state performance and models for transients. The main speed governors and their dynamic models are also included for each main type of prime movers investigated here.

3.2 Steam Turbines

Coal, oil, or nuclear fuels are burned to produce high pressure (HP), high temperature, steam in a boiler. The potential energy in the steam is then converted into mechanical energy in the so-called axial flow steam turbines.

The steam turbines contain stationary and rotating blades grouped into stages—HP, intermediate pressure (IP), low pressure (LP). The HP steam in the boiler is let to enter—through the main emergency stop valves (MSVs) and the governor valves (GVs)—the stationary blades where it is accelerated as it expands to a lower pressure (LP) (Figure 3.3). Then the fluid is guided into the rotating blades of the steam turbine where it changes momentum and direction thus exerting a tangential force on the turbine rotor blades. Torque on the shaft and, thus, mechanical power, are produced. The pressure along the turbine stages decreases and thus the volume increases. Consequently, the length of the blades is lower in the HP stages than in the lower power stages.

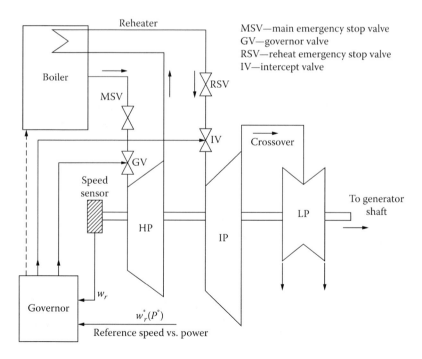

FIGURE 3.3 Single-reheat tandem-compound steam turbine.

The two, three, or more stages (HP, IP, and LP) are all, in general, on same shaft, working in tandem. Between stages the steam is reheated; its enthalpy is increased and thus the overall efficiency is improved, up to 45% for modern coal-burn steam turbines.

Nonreheat steam turbines are built below 100 MW, while single-reheat and double-reheat steam turbines are common above 100 MW, in general.

The single-reheat tandem (same shaft) steam turbine is shown in Figure 3.3.

There are three stages in Figure 3.3: HP, IP, and LP. After passing through MSV and GV, the HP steam flows through the HP stage where its experiences a partial expansion. Subsequently, the steam is guided back to the boiler and reheated in the heat exchanger to increase its enthalpy. From the reheater, the steam flows through the reheat emergency stop valve (RSV) and intercept valve (IV) to the IP stage of the turbine where again it expands to do mechanical work. For final expansion, the steam is headed to the crossover pipes and through the LP stage where more mechanical work is done.

Typically, the power of the turbine is divided as 30% in the HP, 40% in the IP, and 30% in the LP stages.

The governor controls both the GV in the HP stage, and the IV in the IP stage, to provide fast and safe control.

During steam turbine starting—toward synchronous generator synchronization—the MSV is fully open while the GV and IV are controlled by the governor system to regulate the speed and power. The governor system contains a hydraulic (oil) or an electrohydraulic servomotor to operate the GV and IV but also to control the fuel–air mix admission and its parameters in the boiler.

The MSV and RSV are used to stop quickly and safely the turbine under emergency conditions.

Turbines with one shaft are called tandem-compound while those with two shafts (eventually at different speeds) are called cross-compound. In essence, the LP stage of the turbine is attributed to a separate shaft (Figure 3.4).

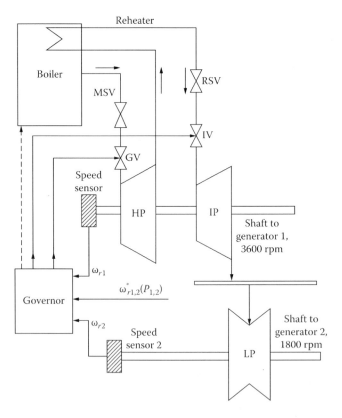

FIGURE 3.4 Single-reheat cross-compound (3600/1800 rpm) steam turbine.

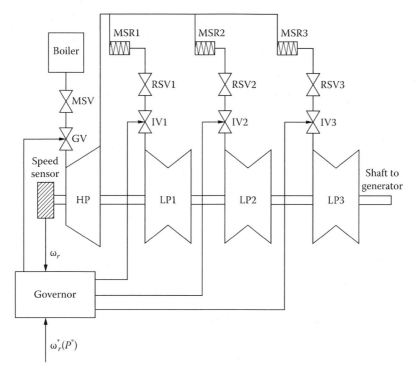

FIGURE 3.5 Typical nuclear steam turbine.

Controlling the speeds and powers of two shafts is rather difficult though it brings more flexibility. Also shafts are shorter. Tandem-compound (single shaft) configurations are more often used. Nuclear units have in general tandem-compound (single-shaft) configurations and run at 1800 (1500) rpm for 60 (50) Hz power systems.They contain one HP and three LP stages (Figure 3.5).

The HP exhaust passes through the moisture reheater (MSR) before entering the LP (LP 1,2,3) stages, to reduce steam moisture losses and erosion.The HP exhaust is also reheated by the HP steam flow.

The governor acts upon GV and IV 1,2,3 to control the steam admission in the HP and LP 1,2,3 stages while MSV, RSV 1,2,3 are used only for emergency tripping of the turbine.

In general, the governor (control) valves are of the plug-diffuser type, while the IV may be either plug or butterfly type (Figure 3.6). The valve characteristics are partly nonlinear and, for better control, are often "linearized" through the control system.

3.3 Steam Turbine Modeling

The complete model of a multiple-stage steam turbine is rather involved. This is why we present here first the simple steam vessel (boiler, reheater) model (Figure 3.7) [1–3], and derive the power expression for the single-stage steam turbine.

V is the volume (m³), Q is the steam mass flow rate (kg/s), ρ is the density of steam (kg/m³), and W is the weight of the steam in the vessel (kg).

The mass continuity equation in the vessel is written as follows:

$$\frac{dW}{dt} = V \frac{d\rho}{dt} = Q_{input} - Q_{output} \qquad (3.1)$$

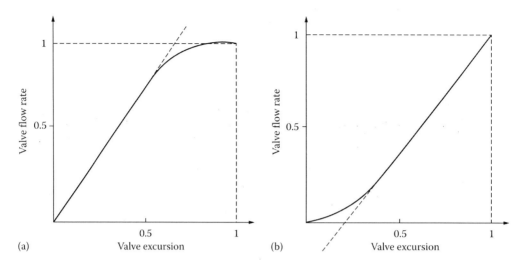

FIGURE 3.6   Steam valve characteristics: (a) plug-diffuser valve and (b) butterfly-type valve.

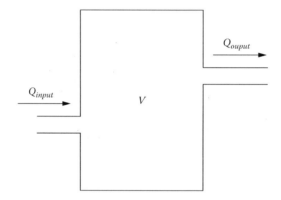

FIGURE 3.7   The steam vessel.

Let us assume that the flow rate out of the vessel Q_{output} is proportional to the internal pressure in the vessel:

$$Q_{output} = \frac{Q_0}{P_0} P \tag{3.2}$$

where
   P is the pressure (kPa)
   P_0 and Q_0 are the rated pressure and flow rate out of the vessel

As the temperature in the vessel may be considered constant:

$$\frac{d\rho}{dt} = \frac{\partial \rho}{\partial P} \cdot \frac{dP}{dt} \tag{3.3}$$

Steam tables provide $(\partial \rho / \partial P)$ functions.
   Finally, from Equations 3.1 through 3.3:

$$Q_{input} - Q_{output} = T_V \frac{dQ_{output}}{dt} \tag{3.4}$$

$$T_V = \frac{P_0}{Q_0} V \cdot \frac{\partial \rho}{\partial P} \qquad (3.5)$$

T_V is the time constant of the steam vessel. With $d/dt \rightarrow s$ the Laplace form of Equation 3.4 is written as follows:

$$\frac{Q_{output}}{Q_{input}} = \frac{1}{1 + T_V \cdot s} \qquad (3.6)$$

The first-order model of the steam vessel has been obtained. The shaft torque T_m in modern steam turbines is proportional to the flow rate:

$$T_m = K_m \cdot Q \qquad (3.7)$$

Therefore, the power P_m is

$$P_m = T_m \cdot \Omega_m = K_m Q \cdot 2\pi n_m \qquad (3.8)$$

Example 3.1

The reheater steam volume of a steam turbine is characterized by
$Q_0 = 200$ kg/s, $V = 100$ m³, $P_0 = 4000$ kPa, $\partial \rho / \partial P = 0.004$
Calculate the time constant T_R of the reheater and its transfer function.
We just use Equations 3.4 and 3.5 and, therefore Equation 3.6:

$$T_R = \frac{P_0}{Q_0} V \cdot \frac{\partial \rho}{\partial P} = \frac{4000}{200} \times 100 \times 0.004 = 8.0 \text{ s}$$

$$\frac{Q_{output}}{Q_{input}} = \frac{1}{1 + 8 \cdot s}$$

Now consider the rather complete model of a single-reheat, tandem-compound steam turbine (Figure 3.8). We will follow the steam journey through the turbine, identifying a succession of time delays/time constants.

The MVS and RSV (stop) valves are not shown in Figure 3.8 because they intervene only in emergency conditions.

The governor (control) valves modulate the steam flow through the turbine to provide for the required (reference) load (power)/frequency (speed) control.

The GV has a steam chest where substantial amounts of steam are stored, also in the inlet piping. Consequently, the response of steam flow to a change in GV opening exhibits a time delay due to the charging time of the inlet piping and steam chest. This time delay is characterized by a time constant T_{CH} in the order of 0.2–0.3 s.

The IV are used for rapid control of mechanical power (they handle 70% of power) during overspeed conditions, and thus their delay time may be neglected in a first approximation.

The steam flow in the IP and LP stages may be changed with the increase of pressure in the reheater. As the reheater holds a large amount of steam, its response time delay is larger. An equivalent larger time constant T_{RM} of 5–10 s is characteristic to this delay.

The crossover piping also introduces a delay that may be characterized by another time constant T_{CO}.

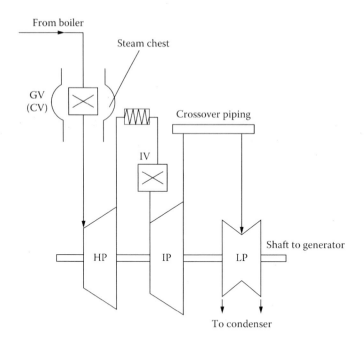

FIGURE 3.8   Single-reheat tandem-compound steam turbine.

We should also consider that the HP, IP, LP stages produce F_{HP}, F_{IP}, F_{LP} fractions of total turbine power such that:

$$F_{HP} + F_{IP} + F_{LP} = 1 \qquad (3.9)$$

We may integrate these aspects of steam turbine model into a structural diagram as in Figure 3.9.

Typically, as already stated: $F_{HP} = F_{IP} = 0.3$, $F_{LP} = 0.4$, $T_{CH} \approx 0.2\text{–}0.3$ s, $T_{RH} = 5\text{–}9$ s, $T_{CO} = 0.4\text{–}0.6$ s.

In a nuclear–fuel steam turbine, the IP stage is missing ($F_{IP} = 0$, $F_{LP} = 0.7$) and T_{RH} and T_{CH} are notably smaller.

As T_{CH} is largest, reheat turbines tend to be slower than nonreheat turbines.

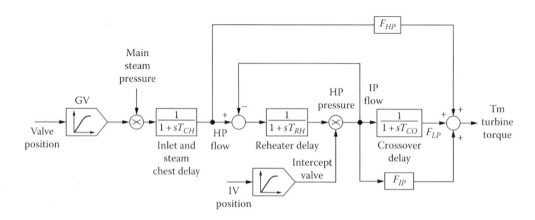

FIGURE 3.9   Structural diagram of single-reheat tandem-compound steam turbine.

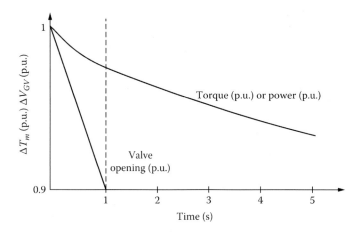

FIGURE 3.10 Steam turbine response to 0.1 (p.u.) 1 s ramp change of GV opening.

After neglecting T_{CO} and considering GV as linear, the simplified transfer function may be obtained:

$$\frac{\Delta T_m}{\Delta V_{GV}} \approx \frac{\left(1 + sF_{HP}T_{RH}\right)}{\left(1 + sT_{CH}\right)\left(1 + sT_{RH}\right)} \tag{3.10}$$

The transfer function in Equation 3.10 clearly shows that the steam turbine has a straightforward response to GV opening.

A typical response in torque (in PU)—or in power—to 1 s ramp of 0.1 (p.u.) change in GV opening is shown in Figure 3.10 for $T_{CH} = 8$ s, $F_{HP} = 0.3$, and $T_{CH} = T_{CO} = 0$.

Enhanced steam turbine models involving various details, such as IV more rigorous representation counting for the (fast) pressure difference across the valve, may be required to better model various intricate transient phenomena.

3.4 Speed Governors for Steam Turbines

The governor system of a turbine performs a multitude of functions such as the following:

- Speed (frequency)/load (power) control: mainly through GV
- Overspeed control: mainly through IV
- Overspeed trip: through MSV and RSV
- Start-up and shutdown

The speed/load (frequency/power) control (Figure 3.2) is achieved through the control of the GV to provide linearly decreasing speed with load, with a small speed droop of 3%–5%. This function allows for paralleling generators with adequate load sharing.

Following a reduction in electrical load, the governor system has to limit the overspeed to a maximum of 120%, in order to preserve the turbine integrity.

Reheat-type steam turbines have two separate valving groups (GV and IV) to rapidly control the steam flow to the turbine.

The objective of the overspeed control is set to about 110%–115% of rated speed to prevent overspeed tripping of the turbine in case a load rejection condition occurs.

The emergency tripping (through MSV and RSV; Figures 3.3 and 3.5) is a protection solution in case normal and overspeed controls fail to limit the speed below 120%.

A steam turbine is provided with four or more governor (control) valves that admit steam through nozzle sections distributed around the periphery of the HP stage.

In normal operation, the GVs are open sequentially to provide better efficiency at partial load. During the start-up, all the GVs are fully open and stop valves control the steam admission.

Governor systems for steam turbines evolved continuously, from mechanical–hydraulic to electro-hydraulic ones [4].

In some embodiments, the main governor systems activate and control the GV while an auxiliary governor system operates and controls the IV [4]. A mechanical–hydraulic governor contains, in general, a centrifugal speed governor (controller) whose effect is amplified through a speed relay to open the steam valves. The speed relay contains a pilot valve (activated by the speed governor) and a spring-loaded servomotor (Figure 3.11).

In electrohydraulic turbine governor systems, the speed governor and speed relay are replaced by electronic controls and an electric servomotor that finally activates the steam valve.

In large turbines, an additional level of energy amplification is needed. Hydraulic servomotors are used for the scope (Figure 3.12).

Combining the two stages—the speed relay and the hydraulic servomotor—the basic turbine governor is obtained (Figure 3.13).

For a speed droop of 4% at rated power, $K_{SR} = 25$ (Figure 3.13). A similar structure may be used to control the IV [2].

Electrohydraulic governor systems perform similar functions; but by using electronics control in the lower power stages, they bring more flexibility and faster and more robust response.

They are, in general, provided with acceleration detection and load power unbalance relay compensation.

The structure of a generic electrohyraulic governor system is shown in Figure 3.14.

We should notice the two stages in actuation: the electrohydraulic converter plus the servomotor and the electronic speed controller.

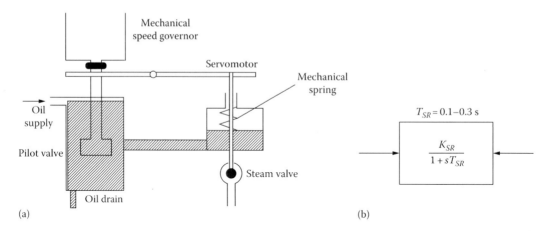

FIGURE 3.11   Speed relay: (a) configuration and (b) transfer function.

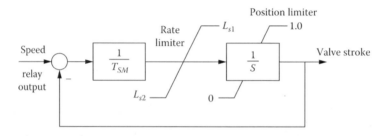

FIGURE 3.12   Hydraulic servomotor structural diagram.

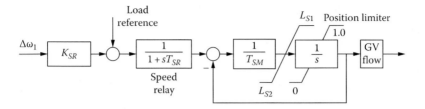

FIGURE 3.13 Basic turbine governor.

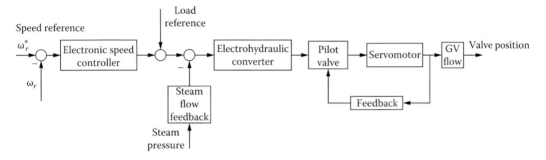

FIGURE 3.14 Generic electrohydraulic governing system.

The development of modern nonlinear control (adaptive, sliding mode, fuzzy, neural networks, H_∞, etc.) has led to a wide variety of recent electronic speed controllers or total steam turbine-generator controllers [5,6]. However, they fall beyond our scope here.

3.5 Gas Turbines

Gas turbines burn gas, whose thermal energy is converted into mechanical work. Air is used as the working fluid. There are many variations in gas turbine topology and operation [1], but the most commonly used seems to be the open regenerative cycle type (Figure 3.15).

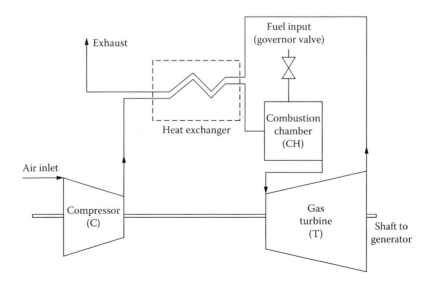

FIGURE 3.15 Open regenerative cycle gas turbine.

The gas turbine in Figure 3.14 consists of an air compressor (C) driven by the turbine itself (T) and the combustion chamber (CH).

The fuel enters the CH through the GV where it is mixed with the hot-compressed air from compressor. The combustion product is then directed into the turbine where it expands and transfers energy to the moving blades of the gas turbine. The exhaust gas heats the air from compressor in the heat exchanger. The typical efficiency of a gas turbine is 35%. More complicated cycles such as compressor intercooling and reheating or intercooling with regeneration and recooling are used for further (slight) improvements in performance [1].

The combined-gas and steam-cycle gas turbines have been proven recently to deliver an efficiency of 55% or even slightly more. The generic combined cycle gas turbine is shown in Figure 3.16.

The exhaust heat from the gas turbine is directed through the heat recovery boiler (HRB) to produce steam, which, in turn, is used to produce more mechanical power through a steam turbine section on same shaft.

With the gas exhaust exiting the gas turbine above 500°C and additional fuel burning, the HRB temperature may rise further the temperature of the HP steam and thus increase efficiency more.

Additionally, some steam for home (office) heating or process industries may be delivered.

Already in the tens of megawatts, combined cycle gas turbines are becoming popular for cogeneration and in distributed power systems in the megawatt or even tenth and hundreds of kilowatt per unit.

Besides efficiency, the short construction time, low capital cost, low SO_2 emission, little staffing, and easy fuel (gas) handling are all main merits of combined cycle gas turbines.

Their construction at very high speeds (tens of krpm) up to 10 MW range, with full power electronics between the generator and the distributed power grid, or in standalone operation mode at 50(60) Hz, make the gas turbines a way of the future in this power range.

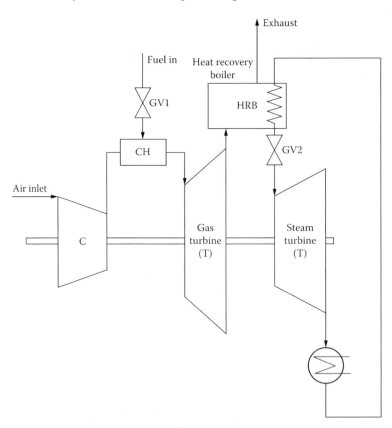

FIGURE 3.16 Combined cycle unishaft gas turbine.

3.6 Diesel Engines

Distributed electric power systems, with distribution feeders at 12 kV (or around it), standby power sets ready for quich intervention in case of emergency or on vessels, locomotives or series or parallel hybrid vehicles, or for power leveling systems in tandem with wind generators make use of diesel (or internal combustion) engines as prime movers for their electric generators. The power per unit varies from a few tenth of kilowatts to a few megawatts.

As for steam or gas turbines, the speed of diesel-engine generator set is controlled through a speed governor. The dynamics and control of fuel–air mix admission is very important to the quality of the electric power delivered to the local power grid or to the connected loads, in standalone applications.

3.6.1 Diesel Engine Operation

In four-cycle internal combustion engines [7] and diesel engine is one of them, with the period of one shaft revolution $T_{REV} = 1/n$ (n-shaft speed in rev/s), the period of one engine power stroke T_{PS} is

$$T_{PS} = 2T_{REV} \tag{3.11}$$

The frequency of power stroke f_{PS} is

$$f_{PS} = \frac{1}{T_{PS}} \tag{3.12}$$

For an engine with N_c cylinders, the number of cylinders that fire per each revolution, N_F, is

$$N_F = \frac{N_c}{2} \tag{3.13}$$

The cylinders are arranged symmetrically on the crankshaft such that the firing of the N_F cylinders is uniformly spaced in angle terms.

Consequently, the angular separation (θ_c) between successive firings in a four-cycle engine is

$$\theta_c = \frac{720°}{N_c} \tag{3.14}$$

The firing angles for a 12 cylinder diesel engine are illustrated in Figure 3.17a, while the two-revolution sequence is: intake (I), compression (C), power (P), and exhaust (E) (see Figure 3.17b).

The 12-cylinder timing is shown in Figure 3.18.

There are 3 cylinders out of 12 firing simultaneously at steady state.

The resultant shaft torque of one cylinder varies with shaft angle as in Figure 3.19.

The compression torque is negative while during power cycle it is positive.

With 12 cylinders, the torque will have much smaller pulsations, with 12 peaks over 720° (period of power engine stroke); Figure 3.20.

Any misfire in one or a few of the cylinders would produce severe pulsations in the torque that would reflect as flicker in the generator output voltage [8].

Large diesel engines are provided, in general, with a turbocharger (Figure 3.21) which influences notably the dynamic response to perturbations by its dynamics and inertia [9].

The turbocharger is essentially an air compressor that is driven by a turbine that runs on the engine exhaust gas.

The compressor provides compressed air to the engine cylinders. The turbocharger works as an energy recovery device with about 2% power recovery.

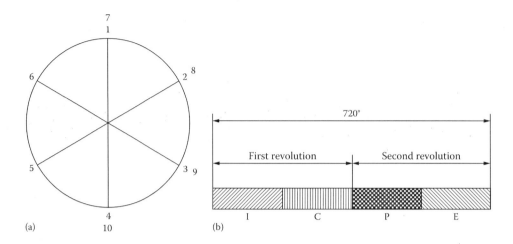

FIGURE 3.17 The 12-cylinder four-cycle diesel engine: (a) configuration and (b) sequence.

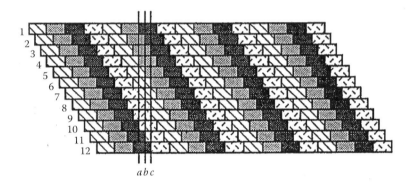

FIGURE 3.18 The 12-cylinder engine timing.

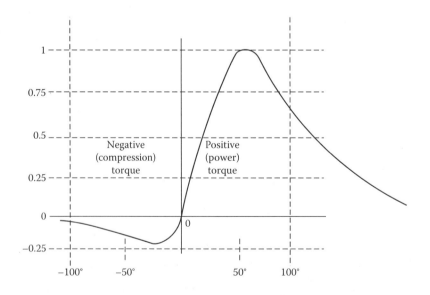

FIGURE 3.19 p.u. Torque/angle for one cylinder.

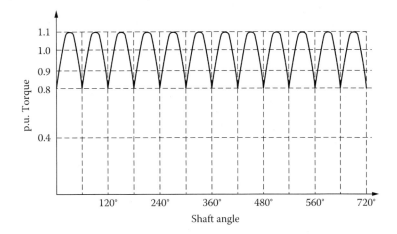

FIGURE 3.20 p.u. Torque versus shaft angle in a 12-cylinder ICE (internal combustion angle).

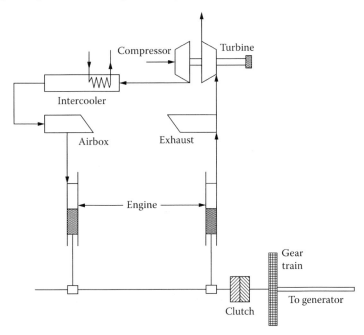

FIGURE 3.21 Diesel engine with turbocharger.

3.6.2 Diesel Engine Modeling

The general structure diagram of a diesel engine with turbocharger and control is shown in Figure 3.22. The most important components are as follows:

- The actuator (governor) driver that appears as a simple gain K_3.
- The actuator (governor) fuel controller that converts the actuator's driver into an equivalent fuel-flow Φ. This actuator is represented by a gain K_2 and a time constant (delay) τ_2 which is dependent on oil temperature, and an aging-produced back lash.
- The inertias of engine J_E, turbocharger J_T and electric generator (alternator) J_G.
- The flexible coupling that mechanically connects the diesel engine to the alternator (it might also contain a transmission).

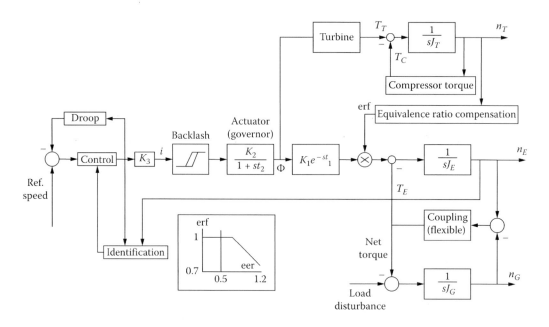

FIGURE 3.22 Diesel engine with turbocharger and controller.

- The diesel engine is represented by the steady-state gain K_1, constant for low fuel–flow Φ and saturated for large Φ, multiplied by the equivalence ratio factor (erf) and by a time constant τ_1.
- The erf depends on engine equivalence ratio (eer) which in turn is the ratio of fuel/air normalized by its stoichiometric value. A typical variation of erf with eer is also shown in Figure 3.22. In essence erf is reduced, because when the ratio fuel/air increases, incomplete combustion occurs, leading to low torque and smoky exhaust.
- The dead time of the diesel engine comprises three delays: the time elapsed until the actuator output actually injects fuel in the cylinder, fuel burning time to produce torque, and time until all cylinders produce torque at engine shaft:

$$\tau_1 \approx A + \frac{B}{n_E} + \frac{C}{n_E^2} \tag{3.15}$$

where n_E is the engine speed

The turbocharger acts upon the engine in the following ways:

- It draws energy from the exhaust to run its turbine; the more fuel in engine the more exhaust is available.
- It compresses air at a rate that is a nonlinear function of speed; the compressor is driven by the turbine, and thus the turbine speed and ultimate erf in the engine is influenced by the air flow rate.
- The turbocharger runs freely at high speed, but it is coupled through a clutch to the engine at low speeds, to be able to supply enough air at all speeds; the system inertia changes thus at low speeds, by including the turbocharger inertia.

Any load change leads to transients in the system pictured in Figure 3.22 that may lead to oscillations due to the nonlinear effects of fuel–air flow erf inertia. As a result, there will be either less or too much air in the fuel mix. In the first case, smoky exhaust will be apparent while in the second situation not enough torque will be available for the electric load, and the generator may pull out of synchronism.

This situation indicates that PI controllers of engine speed are not adequate and nonlinear controllers (adaptive, variable structure, etc.) are required.

A higher-order model may be adopted both for the actuator [11,12] and for the engine [13] to better simulate in detail the diesel engine performance for transients and control.

3.7 Stirling Engines

Stirling engines are part of the family of thermal engines: steam turbines, gas turbines, spark-ignited engines, and diesel engines. They all have already been described briefly in this chapter, but it is now time to dwell a little on the thermodynamic engines cycles to pave the way to Stirling engines.

3.7.1 Summary of Thermodynamic Basic Cycles

The steam engine, invented by James Watt, is a continuous combustion machine. Subsequently, the steam is directed from the boiler to the cylinders (Figure 3.23).

The typical four steps of the steam engine (Figure 3.23) are as follows:

1 → 2 The isochoric compression (1–1′) followed by isothermal expansion (1′–2). The hot steam enters the cylinder through the open valve at constant volume; then, it expands at constant temperature.

2 → 3 Isotropic expansion: once the valve is closed the expansion goes on till the maximum volume (3).

3 → 1 Isochoric heat regeneration (3–3′) and isothermal compression (3′–4)—the pressure drops at constant volume and then the steam is compressed at constant temperature.

4 → 1 Isentropic compression takes place after the valve is closed and the gas is mechanically compressed.

An approximate formula for thermal efficiency ηth is [13]:

$$\eta th = 1 - \frac{\rho^{K-1}(K-1)(1+\ln\rho)}{\varepsilon^{K-1}(x-1)+(K-1)\ln\rho} \tag{3.16}$$

where

$\varepsilon = V_3/V_1$ is the compression ratio
$\rho = V_2/V_1 = V_3/V_4$ is the partial compression ratio
$x = p_1'/p_1$ is the pressure ratio

For $\rho = 2$, $x = 10$, $K = 1.4$, $\varepsilon = 3$, and ηth = 31%.

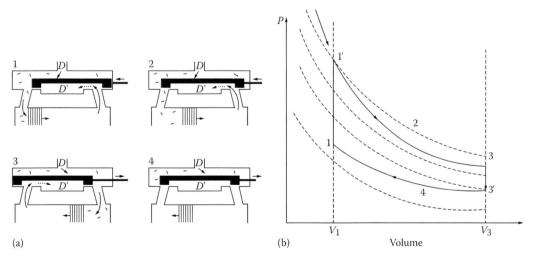

(a)   (b)

FIGURE 3.23 The steam engine "cycle": (a) the four steps and (b) PV diagram.

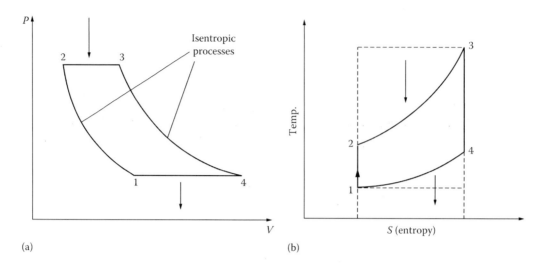

FIGURE 3.24 Brayton cycle for gas turbines: (a) PV diagram and (b) TS diagram.

The gas turbine engine fuel is also continuously combusted in combination with precompressed air. The gas expansion turns the turbine shaft to produce mechanical power.

The gas turbines work on a Brayton cycle (Figure 3.24).

The four steps include the following:

$1 \rightarrow 2$ Isentropic compression
$2 \rightarrow 3$ Isobaric input of thermal energy
$3 \rightarrow 4$ Isentropic expansion (work generation)
$4 \rightarrow 1$ Isobaric thermal energy loss

Similarly, with $T_1/T_4 = T_2/T_3$ for the isentropic steps and the injection ratio $\rho = T_3/T_2$, the thermal efficiency ηth is

$$\eta \text{th} \approx 1 - \frac{1}{\rho}\frac{T_4}{T_2} \tag{3.17}$$

With ideal, complete, heat recirculation,

$$\eta \text{th} \approx 1 - \frac{1}{\rho} \tag{3.18}$$

Gas turbines are more compact than other thermal machines; they are easy to start, have low vibrations, have low efficiency at low loads (ρ small), and tend to have poor behavior during transients.

The spark-ignited (Otto) engines work on the cycle in Figure 3.25.

The four steps are as follows:

$1 \rightarrow 2$—Isentropic compression
$2 \rightarrow 3$—Isochoric input of thermal energy
$3 \rightarrow 4$—Isentropic expansion (kinetic energy output)
$4 \rightarrow 1$—Isochoric heat loss

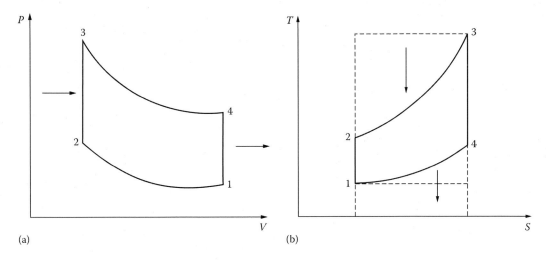

FIGURE 3.25 Spark ignition engines: (a) PV diagram and (b) TS diagram.

The thermal efficiency ηth is

$$\eta \mathrm{th} = 1 - \frac{1}{\varepsilon^{K-1}}; \quad \varepsilon = \frac{V_1}{V_2} \tag{3.19}$$

where

$$\frac{T_4}{T_3} = \frac{T_1}{T_2} = \left(\frac{V_3}{V_4}\right)^{K-1} = \frac{1}{\varepsilon^{K-1}} \tag{3.20}$$

for isentropic processes.

With a high compression ratio (say $\varepsilon = 9$) and the adiabatic coefficient $K = 1.5$ and ηth = 0.66.

The diesel engine cycle is shown in Figure 3.26.

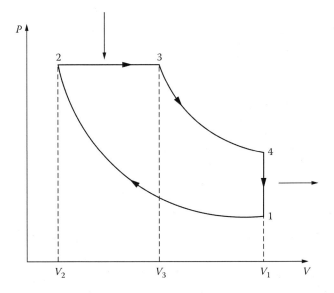

FIGURE 3.26 The diesel engine cycle.

During the downward movement of the piston, an isobaric state change takes place by controlled injection of fuel (2 → 3):

$$\rho = \frac{V_3}{V_2} = \frac{T_3}{T_2};$$

$$\eta_{\text{th}} = 1 - \frac{1}{\varepsilon^{K-1}} \cdot \frac{1}{K} \frac{\rho^K - 1}{\rho - 1}$$

(3.21)

Efficiency decreases when load ρ increases, in contrast to spark-ignited engines for same ε. Lower compression ratios (ε) than for spark-ignited engine are characteristic for diesel engines to obtain higher thermal efficiency.

3.7.2 Stirling Cycle Engine

The Stirling engine (born in 1816) is a piston engine with continuous heat supply (Figure 3.27).

In some respect, Stirling cycle is similar to Carnot cycle (with its two isothermal steps). It contains two opposed pistons and a regenerator in between.

The regenerator is made in the form of strips of metal. One of the two volumes is the expansion space kept at the high temperature $T_{\max}$, while the other volume is the compression space kept at low temperature $T_{\min}$. Thermal axial conduction is considered negligible. Let us suppose that the working fluid (all of it) is in the cold compression space.

During compression (1–2), the temperature is kept constant because heat is extracted from the compression space cylinder to the surroundings.

During the transfer step (2–3), both pistons move simultaneously; the compression piston moves toward the regenerator, while the expansion piston moves away from it. Thus, the volume stays constant. The working fluid is consequently transferred through the porous regenerator from compression to expansion space, and is heated from $T_{\min}$ to $T_{\max}$. An increase in pressure takes place also from 2 to 3. In the expansion step 3–4, the expansion piston still moves away from the regenerator, but the compression piston stays idle at inner dead point. The pressure decreases and the volume increases, but the temperature stays constant because heat is added from an external source. Then, again, a transfer step (4–1) occurs, with both pistons moving simultaneously to transfer the working fluid (at constant volume) through the regenerator from the expansion to the compression space. Heat is transferred from the working fluid to the regenerator, which cools at $T_{\min}$ in the compression space.

The ideal thermal efficiency η_{th} is

$$\eta_{\text{th}}^i = 1 - \frac{T_{\min}}{T_{\max}}$$

(3.22)

Therefore, it is heavily dependent on the maximum and minimum temperatures as the Carnot cycle is. Practical Stirling-type cycles depart from the ideal one. Practical efficiency of Stirling cycle engines is much lower: $\eta_{\text{th}} < \eta_{\text{th}}^i K_{\text{th}}$ ($K_{\text{th}} < 0.5$, in general).

Stirling engines may use any heat source and can use various working fuels such as air, hydrogen, or helium (with hydrogen the best and air the worst).

Typical total efficiencies versus HP/liter density are shown in Figure 3.28 [14] for three working fluids at various speeds.

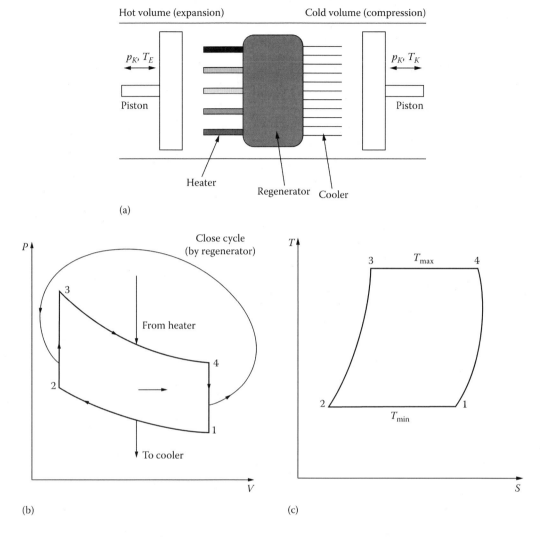

FIGURE 3.27 The Stirling engine: (a) mechanical representation and (b) and (c) the thermal cycle.

As the power and speed go up, the power density decreases. Methane may be a good replacement for air for better performance.

Typical power/speed curves of Stirling engines with pressure p are shown in Figure 3.29a. While the power of a potential electric generator, with speed, and voltage V as parameter, appear in Figure 3.29b.

The intersection at A of Stirling engine and electric generator power/speed curves looks clearly like a stable steady-state operation point. There are many variants for rotary-motion Stirling engines [14].

3.7.3 Free-Piston Linear-Motion Stirling Engine Modeling

Free-piston linear-motion Stirling engines were rather recently developed (by Sunpower and STC companies) for linear generators for spacecraft or for home electricity production (Figure 3.30) [15].

The dynamic equations of the Stirling engine (Figure 3.30) are

$$M_d \ddot{X}_d + D_d \dot{X}_d = A_d(P_p - P) \tag{3.23}$$

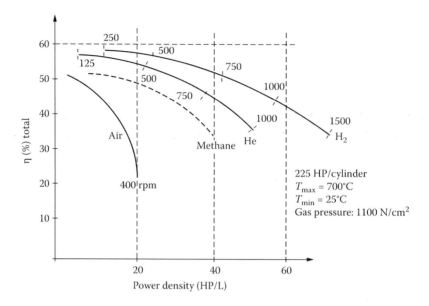

FIGURE 3.28 Efficiency/power density of Stirling engines.

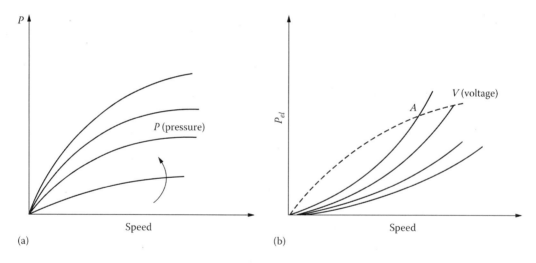

FIGURE 3.29 Power/speed curves: (a) the Stirling engine and (b) the electric generator.

For the normal displacer and:

$$M_p \ddot{X}_p + D_p \dot{X}_p + F_{elm} + K_p X_p + \left(A - A_d\right)\frac{\partial P}{\partial x_d} X_d = 0 \qquad (3.24)$$

for the piston, where

A_d is the displacer rod area in m²
D_d is the displacer damping constant in N/ms
P_d is the gas spring pressure in N/m²
P is the working gas pressure in N/m²
D_p is the piston damping constant (N/ms)
X_d is the displacer position (m)

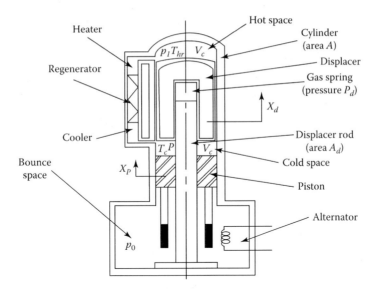

FIGURE 3.30 Linear Stirling engine with free-piston displacer mover.

X_p is the power piston position (m)
A is the cylinder area (m²)
M_d is the displacer mass (kg)
M_p is the power piston mass (kg)
F_{elm} is the electromagnetic force (of linear electric generator) (N)

Equations 3.23 and 3.24 may be linearized as follows:

$$M_d \ddot{X}_d + D_d \dot{X}_d = -K_d X_p - \alpha_p X_p \tag{3.25}$$

$$M_p \ddot{X}_p + D_p \dot{X}_p + F_{elm} = -K_p X_p - \alpha_T X_d$$

$$K_d = -A_d \frac{\partial P_d}{\partial X_d} - \frac{\partial P}{\partial X_d}; \quad \alpha_p = \frac{\partial P}{\partial X_d} A_d$$

$$K_p = \left(A - A_d\right) \frac{\partial P}{\partial X_p}; \quad \alpha_T = \left(A - A_d\right) \frac{\partial P}{\partial X_d}; \quad F_{elm} = K_e I \tag{3.26}$$

where I is the generator current

The electric circuit correspondent of Equation 3.25 is shown in Figure 3.31.

The free-piston Stirling-engine model in Equation 3.25 is a fourth-order system, with $X_d, \dot{X}_d, X_p, \dot{X}_p$ as variables. Its stability when driving a linear PM generator will be discussed in Chapter 20 dedicated to linear reciprocating electric generators. It suffices to say here that at least in the kilowatt range such a combination has been proven stable in stand-alone or power-grid-connected electric generator operation modes.

The merits and demerits of Stirling engines are as follows:

- Independence of heat source: fossil fuels, solar energy
- Very quiet
- High theoretical efficiency; not so large in practice yet, but still 35%–40% for T_{max} = 800°C and T_{min} = 40°C

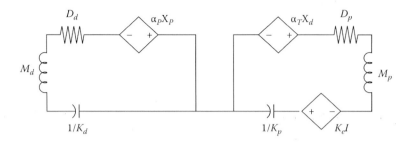

FIGURE 3.31   Free-piston Stirling engine dynamics model.

TABLE 3.2   Thermal Engines

Parameter Thermal Engine	Combustion Type	Efficiency	Quietness	Emissions	Fuel Type	Starting	Dynamic Response
Steam turbines	Continuous	Poor	Not so good	Low	Multifuel	Slow	Slow
Gas turbines	Continuous	Good at full loads, low at low load	Good	Reduced	Independent	Easy	Poor
Stirling engines	Continuous	High in theory, lower so far	Very good	Very low	Independent	NA	Good
Spark-ignited engines	Discontinuous	Moderate	Rather bad	Still large	One type	Fast	Very good
Diesel engines	Discontinuous	Good	Bad	Larger	One type	Rather fast	Good

- Reduced emissions of noxious gases
- High initial costs
- Conduction and storage of heat are difficult to combine in the regenerator
- Materials have to be heat resistant
- For high efficiency, a heat exchanger is needed for the cooler
- Not easy to stabilize

A general comparison of thermal engines is summarized in Table 3.2.

3.8 Hydraulic Turbines

Hydraulic turbines convert the water energy of rivers into mechanical work at the turbine shaft. River water energy or tidal (wave) sea energy are renewable. They are the results of water circuit in nature, and, respectively, are gravitational (tide energy).

Hydraulic turbines are one of the oldest prime movers that man has used.

The energy agent and working fluid is water: in general, the kinetic energy of water (Figure 3.32). Wind turbines are similar, but the wind air kinetic energy replaces the water kinetic energy. Wind turbines will be treated separately, however, due to their many particularities. Hydraulic turbines are, in general, only prime movers, that is motors. There are also reversible hydraulic machines that may operate either as a turbine or as pump. They are also called hydraulic turbine pumps. There are also hydrodynamic transmissions made of two or more conveniently mounted hydraulic machines in a single frame. They play the role of mechanical transmissions but have active control. Hydrodynamic transmissions fall beyond our scope here.

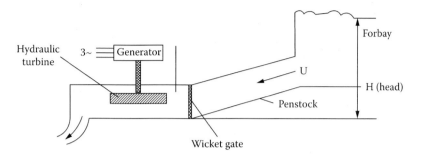

FIGURE 3.32 Hydropower plant schematics.

TABLE 3.3 Hydraulic Turbines

Turbine	Type	Head (m)	Inventor	Trajectory
Tangential	Impulse	>300	Pelton (P)	Designed in the transverse plane
Radial–axial	Reaction	<50	Francis (F)	Bented into the axial plane
Axial	Reaction (propeller)	<50	Kaplan (K), Strafflo (S), Bulb (B)	Bented into the axial plane

Hydraulic turbines are of two main types: impulse turbines for heads above 300–400 m and reaction turbines for heads below 300 m. A more detailed classification is related to main direction of the water particles in the rotor zone: bent axially or transverse to the rotor axis, or related to the inventor (Table 3.3). In impulse turbines, the run is at atmospheric pressure and all pressure drop occurs in the nozzles where potential energy is turned into kinetic energy of water that hits the runner.

In reaction turbines, the pressure in the turbine is above the atmospheric one; water supplies energy in both potential and kinetic form to the runner.

3.8.1 Basics of Hydraulic Turbines

The terminology in hydraulic turbines is related to variables and characteristics [16]. The main variables are of geometrical and functional type:

- Rotor diameter D_r (m)
- General sizes of the turbine
- Turbine gross head: H_T (m)
- Specific energy $Y_T = gH_T$ (J/kg)
- Turbine input flow rate: Q (m³/s)
- Turbine shaft torque: T_T (N m)
- Turbine shaft power: P_T, W (kW, MW)
- Rotor speed Ω_T (rad/s)
- Liquid (rotor properties): density ρ (kg/m³), cinematic viscosity ν (m²/s), temperature T (°C), and elasticity module E (N/m²)

The main characteristics of a hydraulic turbine are, in general, as follows

- Efficiency

$$\eta_T = \frac{P_T}{P_h} = \frac{T_T \cdot \Omega_r}{\rho g H_T Q} \qquad (3.27)$$

- Specific speed n_s

$$n_s = n \frac{\sqrt{P_T \cdot 0.736}}{H_T^{5/4}}, \text{rpm} \tag{3.28}$$

where
   n is the rotor speed in rpm
   P_T in kW
   H_T in m

The specific speed corresponds to a turbine that for a head of 1 m produces 1 HP (0.736 kW)
- the characteristic speed n_c:

$$n_c = \frac{n\sqrt{Q}}{H_T^{3/4}}, \text{rpm} \tag{3.29}$$

where
   n is the rotor speed in rpm
   Q is the flow rate in m³/s
   H_T in m

- Reaction rate γ

$$\gamma = \frac{p_1 - p_2}{\rho g H_T} \tag{3.30}$$

where
   p_1, p_2 are the water pressure right before and after turbine rotor
   $\gamma = 0$ for Pelton turbines $(p_1 = p_2)$—zero reaction (impulse) turbine and $0 < \gamma < 1$ for radial axial and axial turbines (Francis, Kaplan turbines)

- Cavitation coefficient σ_T:

$$\sigma_T = \frac{\Delta h_i}{H_T} \tag{3.31}$$

where Δh_i is the net positive suction head

   It is good for σ_t to be small, $\sigma_t = 0.01$–0.1. It increases with n_s and decreases with H_T
- Specific weight G_{sp}:

$$G_{sp} = \frac{G_T}{P_T}, \text{N/kW} \tag{3.32}$$

where G_T is the turbine mass $\times$ g, in N.

In general, $G_{sp} \approx 70$–150 N/kW.

In general the rotor diameter $D_r = 0.2$–12.0 m, the head $H_T = 2$–2000 m, the efficiency at full load is $\eta_T = 0.8$–0.96, the flow rate $Q = 10^{-3}$–10^3 m³/s, rotor speed $n \approx 50$–1000 rpm.

Typical variations of efficiency [16] with load are given in Figure 3.33.

The maximum efficiency [16] depends on the specific speed n_s and on the type of the turbine—Figure 3.34.

The specific speed is a good indicator for of the best type of turbine for a specific hydraulic site. In general, $n_{Sopt} = 2$–64 for Pelton turbines, $n_s = 50$–500 for Francis turbines and $n_s = 400$–1700 for

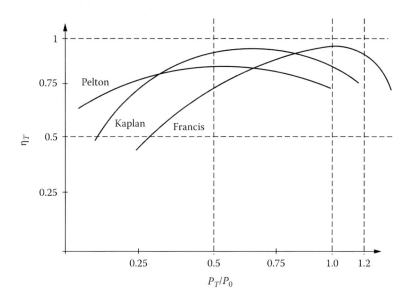

FIGURE 3.33 Typical efficiency/load for Pelton, Kaplan, and Francis turbines.

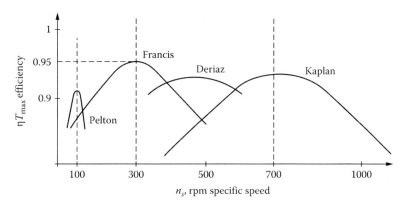

FIGURE 3.34 Maximum efficiency versus specific speed.

Kaplan turbines. The specific speed n_s could be changed by changing the rotor speed n, the total power division in multiple turbines rotors or injectors and the turbine head.

The tendency is to increase n_S in order to reduce turbine size, by increasing rotor speed, at the costs of higher cavitation risk.

As expected, the efficiency of all hydraulic turbines tends to be high at rated load. At part load Pelton turbines show better efficiency. The worst at part load is the Francis turbine. It is thus the one more suitable for variable speed operation. Basic topologies for Pelton, Francis, and Kaplan turbines are shown in Figure 3.35a–c.

In the high head, impulse (Pelton) turbine, the HP water is converted into high-velocity water jets by a set of fixed nozzles. The high-speed water jets hit the bowl-shaped buckets placed around the turbine runner and thus mechanical torque is produced at turbine shaft.

The area of the jet is controlled by a needle placed in the center of the nozzle. The needle is actuated by the turbine governor (servomotor).

In the event of sudden load reduction, the water jet is deflected from the buckets by a jet deflector (Figure 3.35a).

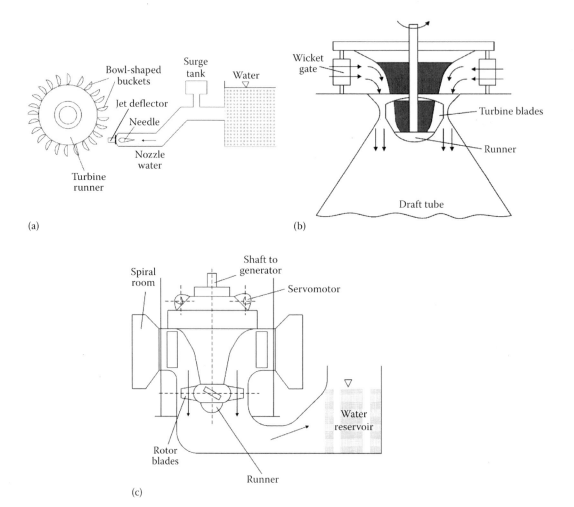

(a)

(b)

(c)

FIGURE 3.35 Hydraulic turbine topology: (a) Pelton type, (b) Francis type, and (c) Kaplan type.

In contrast, reaction (radial–axial) or Francis hydraulic turbines (Figure 3.35b) use lower head, high volumes of water and run at lower speeds.

The water enters the turbine from the intake passage or penstock, through a spiral chamber, passes then through the movable wicket gates onto the turbine runner and then, through the draft tube, to the tail water reservoir.

The wicket gates have their axis parallel to the turbine axis. In Francis turbines, the upper ends of the rotor blades are tightened to a crown and the lower ends to a band.

At even lower head, in Kaplan hydraulic turbines, the rotor blades are adjustable through an oil servomotor placed within the main turbine shaft.

3.8.2 First-Order Ideal Model for Hydraulic Turbines

Usually, in system stability studies [17], with the turbine coupled to an electrical generator connected to a power grid, a simplified, classical model of the hydraulic turbine is used. Such a lossless model assumes that water is incompressible, the penstock is inelastic, the turbine power is proportional to the product of head and volume flow (volume flow rate), while the velocity of water varies with the gate opening and with the square root of net head.

There are three fundamental equations to consider:

1. Water velocity U equation in the penstock
2. Turbine shaft (mechanical) power equation
3. Acceleration of water volume equation

According to the above assumptions the water velocity in the penstock U is

$$U = K_u G \sqrt{H} \tag{3.33}$$

where
  G is the gate opening
  H is the net head at the gate

Linearizing this equation and normalizing it to rated quantities ($U_0 = K_u G_0 \sqrt{H_0}$) yields

$$\frac{\Delta U}{U_0} = \frac{\Delta H}{2H_0} + \frac{\Delta G}{G_0} \tag{3.34}$$

The turbine mechanical power P_m writes

$$P_m = K_p HU \tag{3.35}$$

After normalization ($P_{m0} = K_p H_0 U_0$) and linearization, Equation 3.35 becomes

$$\frac{\Delta P_m}{P_0} = \frac{\Delta H}{H_0} + \frac{\Delta U}{U_0} \tag{3.36}$$

Substituting $\Delta H / H_0$ or $\Delta U / U_0$ from Equation 3.34 into Equation 3.36 yields

$$\frac{\Delta P_m}{P_0} = 1.5 \frac{\Delta H}{H_0} + \frac{\Delta G}{G_0} \tag{3.37}$$

and finally

$$\frac{\Delta P_m}{P_0} = 3 \frac{\Delta U}{U_0} - 2 \frac{\Delta G}{G_0} \tag{3.38}$$

The water column that accelerates due to change in head at the turbine is described by its motion equation:

$$\rho L A \frac{d\Delta U}{dt} = -A(\rho g)\Delta H \tag{3.39}$$

where
  ρ is the mass density
  L is the conduit length
  A is the pipe area, g is the acceleration of gravity

By normalization Equation 3.39 becomes:

$$T_W \frac{d}{dt} \frac{\Delta U}{U_0} = -\frac{\Delta H}{H_0} \tag{3.40}$$

where

$$T_W = \frac{LU_0}{gH_0} \tag{3.41}$$

is the water starting time. It depends on load, and it is in the order of 0.5–5 s for full load.

Replacing d/dt with the Laplace operator, from Equations 3.34 and 3.40 one obtains:

$$\frac{\Delta U/U_0}{\Delta G/G_0} = \frac{1}{1+\left(T_W/2\right)s} \tag{3.42}$$

$$\frac{\Delta P_m/P_0}{\Delta G/G_0} = \frac{1-T_W \cdot s}{1+\left(T_W/2\right)s} \tag{3.43}$$

The transfer functions in Equations 3.42 and 3.43 are shown in Figure 3.36.

The power/gate opening transfer function of Equation 3.43 has a zero in the right s plane. It is a non-minimum phase system whose identification may not be completed by investigating only its amplitude from its amplitude/frequency curve.

For a step change in gate opening, the initial and final value theorems yield

$$\frac{\Delta P_m}{P_0}(0) = \lim_{s \to \infty} s \frac{1}{s} \frac{1-T_w s}{1+\left(1/2\right)T_w s} = -2 \tag{3.44}$$

$$\frac{\Delta P_m}{P_0}(\infty) = \lim_{s \to \infty} s \frac{1}{s} \frac{1-T_w s}{1+\left(1/2\right)T_w s} = 1.0 \tag{3.45}$$

The time response to such a gate step opening is

$$\frac{\Delta P_m}{P_0}(t) = \left(1 - 3e^{-2t/T_W}\right)\frac{\Delta G}{G_0} \tag{3.46}$$

After a unit step increase in gate opening the mechanical power goes first to −2 p.u. value and only then increases exponentially to the expected steady state value of 1 p.u. This is due to water inertia.

Practice has shown that this first-order model hardly suffices when the perturbation frequency is higher than 0.5 rad/s. The answer is to investigate the case of the elastic conduit (penstock) and compressible water where the conduit of the wall stretches at the water wave front.

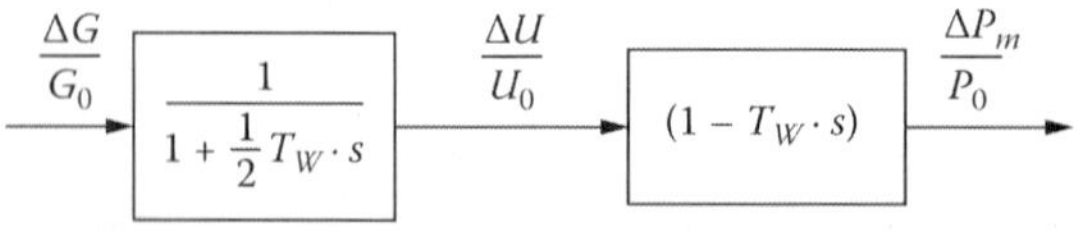

FIGURE 3.36   The linear ideal model of hydraulic turbines in p.u.

3.8.3 Second- and Higher-Order Models of Hydraulic Turbines

We start with a slightly more general small deviation linear model of the hydraulic turbine:

$$q = a_{11}h + a_{12}n + a_{13}z$$
$$m_t = a_{21}h + a_{22}n + a_{23}z \tag{3.47}$$

where
  q is the volume flow
  h is the net head
  n is the turbine speed
  z is the gate opening
  m_t is the shaft torque, all in p.u. values

As expected, the coefficients $a_{11}, a_{12}, a_{13}, a_{21}, a_{22}, a_{23}$ vary with load, etc. To a first approximation $a_{12} \approx a_{22} \approx 0$ and, with constant a_{ij} coefficients, the first-order model is reclaimed.

Now, if the conduit is considered elastic and water as compressible, the wave equation in the conduit may be modeled as an electric transmission line that is open circuited at the turbine end and short-circuited at forebay.

Finally, the incremental head and volume flow rate $h(s)/q(s)$ transfer function of the turbine is [2]:

$$\frac{h(s)}{q(s)} = -\frac{T_w}{T_e}\tan h(T_e \cdot s + F) \tag{3.48}$$

where
  F is the friction factor
  T_e is the elastic time constant of the conduit

$$T_e = \frac{\text{conduit\_length}:L}{\text{wave\_velocity}:a}; \quad a = \sqrt{\frac{g}{\alpha}} \tag{3.49}$$

$$\alpha = \rho g \left(\frac{1}{K} + \frac{D}{Ef}\right) \tag{3.50}$$

where
  ρ is the water density
  g is the acceleration of gravity
  f is the thickness of conduit wall
  D is the conduit diameter
  K is the bulk modulus of water compression
  E is the Young's modulus of elasticity for the pipe material

Typical values of α are around 1200 m/s for steel conduits and around 1400 m/s for rock tunnels. T_e is the order of fractions of a second, larger for larger penstocks (Pelton turbines).

If we now introduce Equations 3.42 and 3.43 into Equation 3.48, then the power $\Delta P_m(s)$ to gate opening $\Delta z(s)$ in p.u. transfer function can be obtained as follows:

$$G(s) = \frac{\Delta Pm}{\Delta z}(s) = \frac{1 - (T_W/T_e)\tan h(T_e \cdot s + F)}{1 + (T_W/2T_e)\tan h(T_e \cdot s + F)} \tag{3.51}$$

Alternatively, from Equation 3.47, we obtain the following:

$$G'(s) = \frac{\Delta Pm}{\Delta z}(s) = \frac{1 - q_p - (T_W/T_e)\tan h(T_e \cdot s + F)}{1 + 0.5q_p + (T_W/2T_e)\tan h(T_e \cdot s + F)} \tag{3.52}$$

where q_p is the friction

With $F = q_p = 0$, Equations 3.51 and 3.52 degenerate into the first order model provided $\tan h\, T_{es} \approx T_{es}$, that is for very low frequencies:

$$G_1(s) = \frac{1 - T_W \cdot s}{1 + (T_W/2) \cdot s} \tag{3.53}$$

The frequency response ($s = j\omega$) of Equation 3.51 with $F = 0$ is shown in Figure 3.36.

Now, we may approximate the hyperbolic function with truncated Taylor series [18]:

$$\tan h(T_e \cdot s) = \frac{T_e \cdot s}{1 + ((T_e \cdot s)/2)^2}$$

$$G_2(s) = \frac{(T_e \cdot s)^2 - 2T_W \cdot s + 2}{(T_e \cdot s)^2 + T_W \cdot s + 2} \tag{3.54}$$

Figure 3.37 shows comparative results for $G(s)$, $G_1(s)$ and $G_2(s)$ for $T_e = 0.25$ s and $T_w = 1$ s.

The second-order transfer function performs quite well to and slightly beyond the first maximum that occurs at $\omega = \pi/2T_e = 6.28$ rad/s in our case (Figures 3.37 and 3.38).

It is, however, clear that well beyond this frequency a higher-order approximation is required.

Such models can be obtained with advanced curve fitting methods applied to $G_2(s)$ for the frequency range of interest [19,20].

The presence of a surge tank (Figure 3.39) in some hydraulic plants calls for a higher-order model.

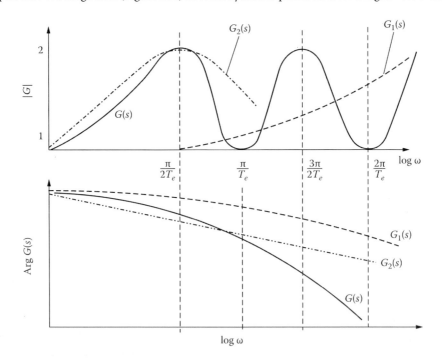

FIGURE 3.37   Higher-order hydraulic turbine frequency response.

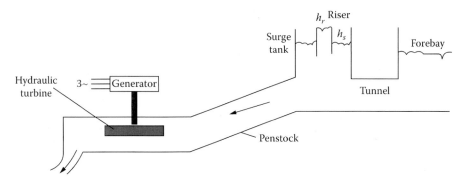

FIGURE 3.38 The second-order model of hydraulic turbines (with zero friction).

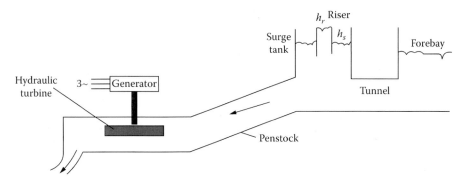

FIGURE 3.39 Hydraulic plant with surge tank.

The wave (transmission) line equations apply now both for tunnel and penstock. Finally, the tunnel and surge tank can be approximated to $F_1(s)$ [2]:

$$F_1(s) = \frac{q_c + s \cdot T_{WC}}{1 + s \cdot T_s \cdot q_c + s^2 T_{WC} T_S} = -\frac{h_S}{U_p}$$

$$\tan h(T_{ec} \cdot s) = T_{ec} \cdot s \tag{3.55}$$

$$Z_c = \frac{T_{WC}}{T_{ec}}$$

where

T_{ec} is the elastic time constant of the tunnel
T_{WC} is the water starting time in the tunnel
q_c is the the surge tank friction coefficient
h_S is the surge tank head
U_p is the upper penstock water speed
T_S is the surge tank riser time ($T_S \approx 600$–900 s)

Now for the penstock, the wave equation yields (in p.u.)

$$h_t = h_r \sec h(T_{ep} \cdot s) - Z_p U_t \tan h(T_{ep} \cdot s) - q_p U_t$$

$$U_p = U_t \cos h(T_{ep} \cdot s) + \frac{h_t}{Z_p} \sin h(T_{ep} \cdot s) \tag{3.56}$$

where

Z_p is the hydraulic impedance of the penstock ($Z_p = T_{Wp}/T_{ep}$)
q_p is thefriction coefficient in the penstock
T_{ep} is the penstock elastic time
T_{Wp} is the penstock water starting time
h_r is the riser head
h_t is the turbine head

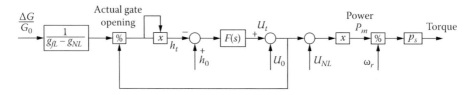

FIGURE 3.40   Nonlinear model of hydraulic turbine with hammer and surge tank effects.

The overall water velocity U_p to head at turbine h_t ratio is [2]

$$F(s) = \frac{U_t}{h_t} = -\frac{\left(1 + F_1(s) \times \tanh\left(T_{ep} \cdot s\right) / Z_p\right)}{q_p + F_1(s) + Z_p \tanh\left(T_{ep} \cdot s\right)} \tag{3.57}$$

The power differential is written as follows:

$$P_m = U_t h_t \text{ in p.u.} \tag{3.58}$$

$F(s)$ represents now the hydraulic turbine with wave (hammer) and surge tank effects considered.

If we now add Equation 3.33 that ties the speed at turbine head and gate opening to Equations 3.57 and 3.58, the complete nonlinear model of the hydraulic turbine with penstock and surge tank effects included (Figure 3.40) is obtained.

Notice that g_{fL} and g_{NL} are the full-load and no-load actual gate openings in p.u.

Also h_0 is the normalized turbine head, U_0 is the normalized water speed at turbine, U_{NL} is the no-load water speed at turbine, ω_r is the shaft speed, P_m is the shaft power, and m_t is the shaft torque differential in p.u. values.

The nonlinear model in Figure 3.40 may be reduced to a high-order (three or more) linear model through various curve fittings applied to the theoretical model with given parameters. Alternatively, test frequency response tests may be fitted to a third-, fourth-, and even higher-order linear system for preferred frequency bands [21].

As the nonlinear complete model is rather involved, the question arises as of when it is to be used. Fortunately, only in long-term dynamic studies it is mandatory.

For governor timing studies, as the surge tank natural period (T_s) is of the order of minutes, its consideration is not necessary. Further on, the hammer effect should be considered in general, but the second-order model suffices.

In transient stability studies, again, the hammer effect should be considered.

For small signal stability studies linearization of the turbine penstock model (second-order model) may be also adequate, especially in plants with long penstocks.

3.8.4 Hydraulic Turbine Governors

In principle, hydraulic turbine governors are similar to those used for steam and gas turbines. They are mechanohydraulic or electrohydraulic.

In general, for large power levels, they have two stages: a pilot valve servomotor and a larger power gate-servomotor.

An example of a rather classical good performance system with speed control is shown in Figure 3.41.

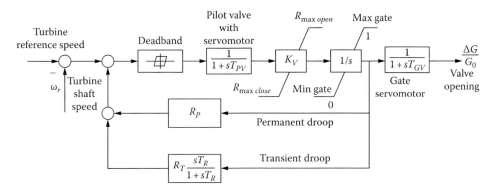

FIGURE 3.41 Typical (classical) governor for hydraulic turbines. T_{PV} is the pilot valve with servomotor time constant (0.05 s), T_{GV} is the main (gate) servomotor time constant (0.2 s), K_V is the servo (total) gain (5), $R_{max\ open}$ is the maximum gate opening rate $\approx$ 0.15 p.u./s, $R_{max\ close}$ is the maximum closing rate $\approx$ 0.15 p.u./s, T_R is the reset time (5.0 s), R_P is the permanent droop (0.04), and R_T is the transient droop (0.4).

Numbers in parenthesis are sample data [2] given only for getting a feeling of magnitudes. A few remarks on model in Figure 3.41 are in order:

- The pilot valve servomotor (lower power stage of governor) may be mechanical or electric; electric servomotors tend to provide faster and more controllable response.
- Water is not very compressible, and thus the gate motion has to be gradual; near the full closure even slower motion is required.
- Dead band effects are considered in Figure 3.41, but their identification is not an easy task.
- Stable operation during system islanding (stand-alone operation mode of the turbine—generator system) and acceptable response quickness and robustness under load variations are the main requirements that determine the governor settings.
- The presence of transient compensation droop is mandatory for stable operation.
- For islanding operation, the choice of temporary droop R_T and reset time T_R is essential; they are related to water starting time constant T_W and mechanical (inertia) time constant of the turbine/generator set T_M. Also the gain K_V should be high.
- According to Reference 2,

$$R_T = \left[2.3 - \left(T_W - 1.0\right)0.15\right]\frac{T_W}{T_M}$$

$$T_R = \left[5.0 - \left(T_W - 1.0\right)0.5\right]T_W \qquad (3.59)$$

$$T_M = 2H;$$

$$H = \frac{J\omega_0^2}{2S_0}(s)$$

where
   J (kg m²) is the turbine/generator inertia
   ω_0 is the rated angular speed (rad/s)
   S_0 is the rated apparent power (VA) of the electrical generator

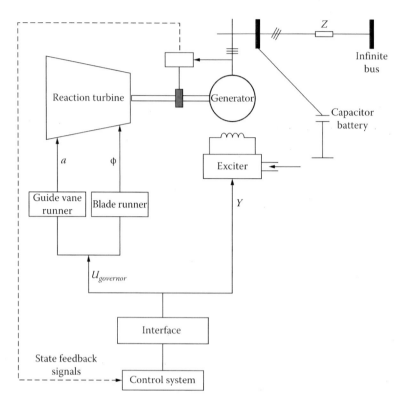

FIGURE 3.42   Coordinated turbine governor–generator–exciter control system.

- In hydraulic turbines where wickets gates (Figure 3.41) are also used, the governor system has to control their motion also, based on an optimization criterion.

The governing system thus becomes more involved. The availability of high-performance nonlinear motion controllers (adaptive, variable structure, fuzzy-logic, or artificial neural networks) and of various powerful optimization methods [22] puts the governor system control into a new perspective (Figure 3.42).

Though most such advanced controllers have been tried on thermal prime movers and especially on power system stabilizers that usually serve only the electric generator excitation, the time for comprehensive digital on line control of the whole turbine generator system seems ripe [23,24].

Still, problems with safety could delay their aggressive deployment; not for a long time, though, we think.

3.8.5 Reversible Hydraulic Machines

Reversible hydraulic machines are in fact turbines that work part time as pumps, especially in pump-storage hydropower plants.

Pumping may be required either for irrigation or for energy storage during off-peak electric energy consumption hours. It is also a safety and stability improvement vehicle in electric power systems in the presence of fast variations of loads over the hours of the day.

As up to 400 MW/unit pump storage hydraulic turbine pumps have been already in operation [25], their "industrial" deployment seems near. Pump-storage plants with synchronous (constant) speed generators (motors) are a well-established technology.

A classification of turbine-pumps is in order:

a. By topology
   - Radial–axial (Russel Dam) (Figure 3.43)
   - Axial (Annapolis) (Figure 3.44)
b. By direction of motion
   - With speed reversal for pumping (Figures 3.43 and 3.44)
   - Without speed reversal for pumping
c. By direction of fluid flow/operation mode
   - Unidirectional/operation mode (Figure 3.43)
   - Bidirectional/operation mode (Figure 3.44)

There are many topological variations in existing turbine pumps; it is also feasible to design the machine for pumping and then check the performance for turbining, when the direction of motion is reversible.

With pumping and turbining in both directions of fluid flow, the axial turbine pump in Figure 3.43 may be adequate for tidal-wave power plants.

The passing from turbine to pump mode implies the emptying of the turbine chamber before the machine is started by the electric machine as motor to prepare for pumping. This transition takes time.

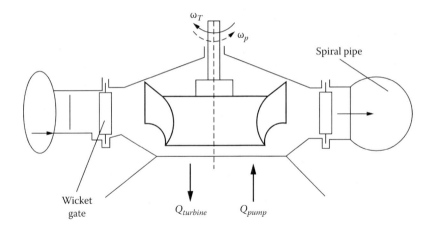

FIGURE 3.43 Radial–axial turbine/pump with reversible speed.

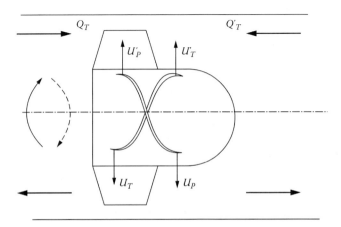

FIGURE 3.44 Axial turbine pump.

More complicated topologies are required to secure unidirectional rotation for both pumping and turbining, though the time to switch from turbining to pumping mode is much shorter.

In order to preserve high efficiency in pumping the speed in the pumping regime has to be larger than the one for turbining. In effect the head is larger and the volume flow lower in pumping. A typical ratio for speed would be $\omega_p \approx (1.12\text{--}1.18)\,\omega_T$. Evidently such a condition implies adjustable speed and thus power electronics control on the electric machine side. Typical head/volume flow characteristics [16] for a radial–axial turbine/pumps are shown in Figure 3.45.

They illustrate the fact that pumping is more efficient at higher speed than turbining and at higher heads, in general.

Similar characteristics portray the output power versus static head for various wicket gate openings [25] (Figure 3.46).

Power increases with speed, and higher speeds are typical for pumping. Only wicket gate control by governor system is used, as adjustable speed is practiced through instantaneous power control in the generator rotor windings, through power electronics.

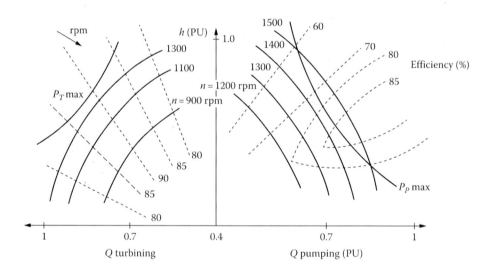

FIGURE 3.45 Typical characteristics pumping of radial–axial turbine + pumps.

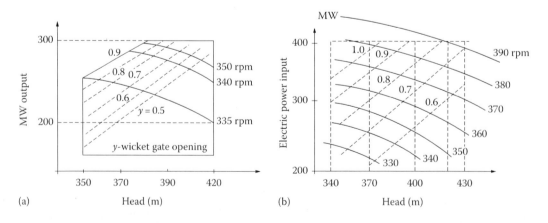

FIGURE 3.46 Turbine/pump system power/static head curves at various speeds: (a) turbining and (b) pumping.

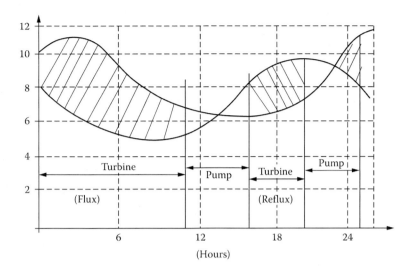

FIGURE 3.47 Head/time of the day in a tidal-wave turbine/pump.

The turbine governor and electric machine control schemes are rather specific for generating electric power (turbining) and for pumping [25].

In tidal-wave turbine/pumps, to produce electricity, a special kind of transit takes place from turbining to pumping in one direction of motion and in the other direction of motion in a single day.

The static head changes from 0% to 100% and reverses sign (Figure 3.47).

These large changes in head are expected to produce large electric power oscillations in the electric power delivered by the generator/motor driven by the turbine/pump. Discontinuing operation between pumping and turbining occurs and electric solutions based on energy storage are to be used to improve the quality of power delivered to the electrical power system.

3.9 Wind Turbines

Air pressure gradients along the surface of the earth produce wind whose direction and speed are highly variable.

Uniformity and strength of the wind are dependent on location, height above the ground, and size of local terrain irregularities. In general, wind air flows may be considered turbulent.

In a specific location, variation of wind speed along the cardinal directions may be shown as in Figure 3.48a.

This is an important information as it leads to the optimum directioning of the wind turbine, in the sense of extracting the largest energy from wind per year.

Wind speed increases with height and becomes more uniform. Designs with higher height/turbine diameter lead to more uniform flow and higher energy extraction. At the price of more expensive towers subjected to increased structural vibrations.

With constant energy conversion ratio, the turbine power increases approximately with cubic wind speed (u^3) up to a design limit, u_{max}. Above u_{max} (P_{rated}) the power of the turbine is kept constant by some turbine governor control to avoid structural or mechanical inadmissible overload (Figure 3.49).

For a given site, the wind is characterized by the so-called speed deviation (in p.u. per year). For example,

$$\frac{t}{t_{max}} = e^{-U^4} \tag{3.60}$$

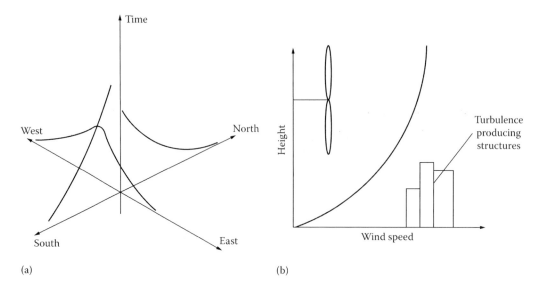

FIGURE 3.48 Wind speed vs. (a) location and (b) height.

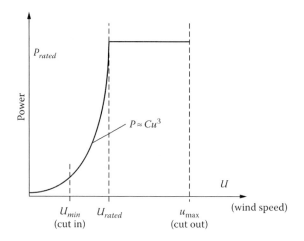

FIGURE 3.49 Wind turbine power vs. wind speed.

The slope of this curve is called speed/frequency curve f:

$$f(U) = -\frac{d\left(t/t_{\max}\right)}{dU}$$
(3.61)

The speed/duration is monotonous (as speed increases its time occurrence decreases), but the speed/frequency curve experiences a maximum, in general. The average speed U_{ave} is defined in general as follows:

$$U_{ave} = \int_{0}^{\infty} U f(U) dU$$
(3.62)

Other mean speed definitions are also used.

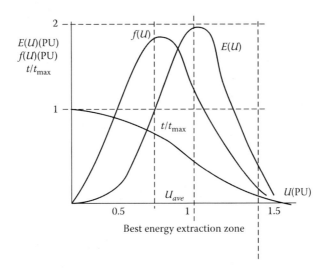

FIGURE 3.50 Sample time/speed frequency/speed (*f*) and energy E/speed.

The energy content of the wind E, during t_{max} (1 year) is then obtained from the integral:

$$\frac{\text{Energy during } t_{max}}{\rho \cdot (\text{disc area})} = \int_0^\infty u^3 f(u) du \tag{3.63}$$

where $E(u) = u^3 f(u)$, the energy available at speed *u*. Figure 3.50 illustrates this line of thinking.

The time average speed falls below the frequency/speed maximum f_{max}, which in turn is smaller than the maximum energy per unit speed range E_{max}.

In addition, the adequate speed zone for efficient energy extraction is apparent in Figure 3.50.

It should be kept in mind that these curves, or their approximations, depend heavily on location. In general, inland sites are characterized by large variations of speed over the day, month, while winds from the sea tend to have smaller variations in time.

Good extraction of energy over a rather large speed span as in Figure 3.50 implies operation of the wind turbine over a pertinent speed range. The electric generator has to be capable to operate at variable speed in such locations.

There are constant speed and variable speed wind turbines.

3.9.1 Principles and Efficiency of Wind Turbines

For centuries, wind mills have been operated in countries like Holland, Denmark, Greece, Portugal, and so forth. The best locations are situated either in the mountains or by the sea or by the ocean shore (or offshore).

Wind turbines are characterized by the following:

- Mechanical power P (W)
- Shaft torque (N m)
- Rotor speed *n* (rpm) or ω_r (rad/s)
- Rated wind speed U_R
- Tip speed ratio:

$$\lambda = \frac{\omega_r \cdot D_r/2}{U} = \frac{\text{rotor blade tip speed}}{\text{wind speed}} \tag{3.64}$$

The tip speed ratio $\lambda < 1$ for slow-speed wind turbines and $\lambda > 1$ for high-speed wind turbines.

- The power efficiency coefficient C_p:

$$C_p = \frac{8P}{\rho \pi D_r^2 \cdot U^3} <> 1 \tag{3.65}$$

In general, C_p is a single maximum function of λ that strongly depends on the type of the turbine. A classification of wind turbines is thus in order:

- Axial (with horizontal shaft)
- Tangential (with vertical shaft)

The axial wind turbines may be slow (Figure 3.51a) and rapid (Figure 3.51b).

The shape of the rotor blades and their number are quite different for the two configurations.

The slow axial wind turbines have a good starting torque and the optimum tip speed ratio $\lambda_{opt} \approx 1$, but their maximum power coefficient $C_{pmax}(\lambda_{opt})$ is moderate ($C_{pmax} \approx 0.3$). In contrast, high speed axial wind turbines self-start at higher speed (above 5 m/s wind speed) but, for an optimum tip speed ratio $\lambda_{opt} \geq 7$, they have maximum power coefficient $C_{pmax} \approx 0.4$. That is, a higher energy conversion ratio (efficiency).

For each location, the average wind speed U_{ave} is known. The design wind speed U_R is in general around 1.5 U_{ave}.

In general, the optimum tip speed ratio λ_{opt} increases as the number of rotor blades Z_1 decreases:

$$(\lambda_{opt}, Z_1) = (1, 8-24; 2, 6-12; 3, 3-6; 4, 2-4; 5, 2-3; >5, 2) \tag{3.66}$$

Three or two blades are typical for rapid axial wind turbines.

The rotor diameter D_r may be, to a first approximation, calculated from Equation 3.65 for rated (design conditions): P_{rated}, λ_{opt}, C_{popt}, U_R with the turbine speed from Equation 3.64.

Tangential—vertical shaft—wind turbines have been built in quite a few configurations. They are of two subtypes: drag type and lift type. The axial (horizontal axis) wind turbines are all of lift subtype.

Some of the tangential wind turbine configurations are shown in Figure 3.52.

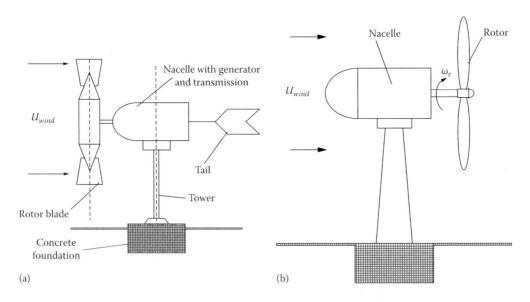

(a)   (b)

FIGURE 3.51 Axial wind turbines: (a) slow (multiblade) and (b) rapid (propeller).

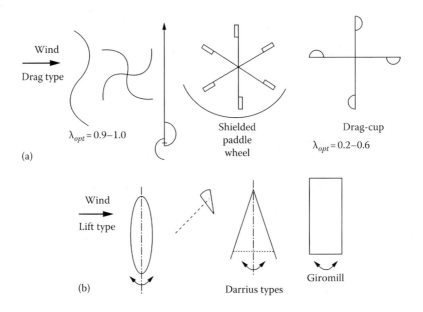

FIGURE 3.52 Tangential—vertical axis—wind turbines: (a) drag subtype and (b) lift subtype.

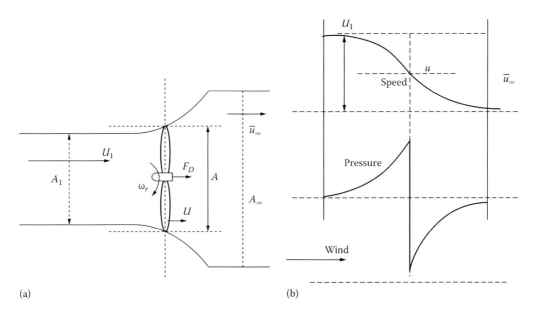

FIGURE 3.53 Basic wind turbine speed and pressure variation.

While drag subtype works at slow speeds ($\lambda_{opt} < 1$) the lift subtype works at high speeds ($\lambda_{opt} > 1$). Slow wind turbines have a higher self-starting torque but a lower power efficiency coefficient C_{pmax}.

The efficiency limit (Betz limit) may be calculated by portraying the ideal wind speed and pressure profile before and after the turbine (Figure 3.53).

The wind speed decreases immediately before and after the turbine disk plane while also a pressure differential takes place.

The continuity principle shows that

$$u_1 A_1 = u_\infty A_\infty \tag{3.67}$$

If the speed decreases along the direction of the wind speed, $u_1 > u_\infty$ and thus $A_1 < A_\infty$.

The wind power P_{wind} in front of the wind turbine is the product of mass flow to speed squared per 2:

$$P_{wind} = \rho U_1 A \cdot \frac{1}{2} \cdot U_1^2 = \frac{1}{2} \rho A U_1^3 \tag{3.68}$$

The power extracted from the wind, $P_{turbine}$ is

$$P_{turbine} = \rho U A \left(\frac{U_1^2}{2} - \frac{U_\infty^2}{2} \right) \tag{3.69}$$

Let us assume:

$$U \approx U_1^{-\Delta U_\infty /2}; \quad U_\infty = U_1 - \Delta U_\infty; \quad \Psi = \frac{\Delta U_\infty}{U_1} \tag{3.70}$$

The efficiency limit, η_{ideal}, is

$$\eta_{ideal} = \frac{P_{turbine}}{P_{wind}} = \frac{(1/2)\rho U A \left(U_1^2 - U_\infty^2 \right)}{(1/2)\rho A U_1^3} \tag{3.71}$$

With Equation 3.61, η_{ideal} becomes:

$$\eta_i = \left(1 - \frac{\Psi}{2} \right) \left[1 - \left(1 - \Psi \right)^2 \right] \tag{3.72}$$

The maximum ideal efficiency is obtained for $\partial \eta_i / \partial \Psi = 0$ at $\Psi_{opt} = 2/3$ ($U_\infty / U_1 = 1/3$) with $\eta_{imax} = 0.593$.

This ideal maximum efficiency is known as the Betz limit [26].

3.9.2 Steady-State Model of Wind Turbines

The steady-state behavior of wind turbines is carried out usually through the blade element momentum (BEM) model. The blade is divided into a number of sections whose geometrical, mechanical, and aerodynamic properties are given as functions of local radius from the hub.

At the local radius the cross-sectional airfoil element of the blade is shown in Figure 3.54.

The local relative velocity $U_{rel}(r)$ is obtained by superimposing the axial velocity $U(1 - a)$ and the rotation velocity $r\omega_r(1 + a')$ at the rotor plane.

The induced velocities ($-aU$ and $a'r\omega_r$) are produced by the vortex system of the machine.

The local attack angle α is

$$\alpha = \phi - \theta \tag{3.73}$$

with

$$\tan \phi = \frac{U(1-a)}{r\omega_r (1+a')} \tag{3.74}$$

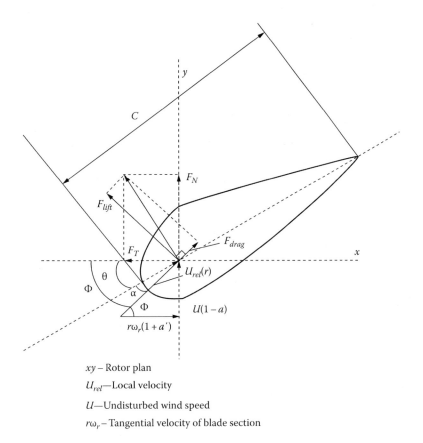

xy – Rotor plan
U_{rel}—Local velocity
U—Undisturbed wind speed
$r\omega_r$ – Tangential velocity of blade section

FIGURE 3.54 Blade element with pertinent speed and forces.

The local blade pitch θ is

$$\theta = \tau + \beta \tag{3.75}$$

where
   τ is the local blade twist angle
   β is the global pitch angle

The lift force F_{lift} is rectangular to U_{rel}, while the drag force F_{drag} is parallel to it.
The lift and drag forces F_{lift} and F_{drag} may be written as follows:

$$F_{lift} = \frac{1}{2}\rho U_{rel} \cdot C \cdot C_L; \quad \rho = 1.225\ \text{kg/m}^3 \tag{3.76}$$

$$F_{drag} = \frac{1}{2}\rho U_{rel} \cdot C \cdot C_D \tag{3.77}$$

where
   C is the local chord of the blade section
   C_L and C_D are lift and drag coefficients, respectively, known for a given blade section [26,27]

From lift and drag forces, the normal force (thrust), F_N, and tangential force F_T (along X,Y on the blade section plane) are simply

$$F_N(r) = F_{lift}\cos\Phi + F_{drag}\sin\Phi \qquad (3.78)$$

$$F_T(r) = F_{lift}\sin\Phi - F_{drag}\cos\Phi \qquad (3.79)$$

Various additional corrections are needed to account for the finite number of blades (B), especially for large values of a (axial induction factor).

Now the total thrust F_N per turbine is

$$F_T = B \int_0^{Dr/2} F_N(r)\,dr \qquad (3.80)$$

Similarly, the mechanical power P_T is as follows:

$$P_T = B \cdot \omega_r \cdot \int_0^{Dr/2} rF_T(r)\,dr \qquad (3.81)$$

Now, with the earlier definition of Equation 3.65, the power efficiency C_p may be calculated.

This may be done using a set of airfoil data for the given wind turbine, when a, a', C_L, and C_D are determined first.

A family of curves C_p–λ–β is thus obtained. This in turn may be used to investigate the steady-state performance of the wind turbine for various wind speeds U and wind turbine speeds ω_r.

As the influence of blade global pitch angle β is smaller than the influence of tip speed ratio λ in the power efficiency coefficient C_p, we may first keep $\beta = ct.$ and vary λ for a given turbine. Typical $C_p(\lambda)$ curves are shown on Figure 3.55 for three values of β.

For the time being let $\beta = ct.$ and rewrite formula Equation 3.65 by using λ (tip speed ratio):

$$P_M = \frac{1}{2}\rho C_p \pi D_r^2 U^3 = \frac{1}{2}\rho\pi\left(\frac{D_r}{2}\right)^5 \cdot \frac{C_p}{\lambda^3}\omega_r^3 \qquad (3.82)$$

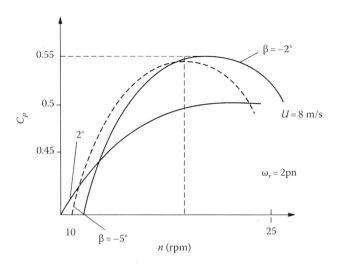

FIGURE 3.55 Typical C_p–$n(\lambda)$–β curves.

Adjusting turbine speed ω_r, that is λ, the optimum value of λ corresponds to the case when C_p is maximum, C_{pmax} (Figure 3.55).

Consequently, from Equation 3.82, we obtain the following:

$$P_M^{opt} = \frac{1}{2}\rho\pi\left(\frac{D_r}{2}\right)^5 \cdot \frac{C_{pmax}}{\lambda_{opt}^3}\omega_r^3 = K_W\omega_r^3 \tag{3.83}$$

Therefore, basically the optimal turbine power is proportional to the third power of its angular speed.

Within the optimal power range, the turbine speed ω_r should be proportional to wind speed U as follows:

$$\omega_r = U \cdot \frac{2}{D_r} \cdot \lambda_{opt} \tag{3.84}$$

Above the maximum allowable turbine speed, obtained from mechanical or thermal constraints in the turbine and electric generator, the turbine speed remains constant. As expected, in turbines with constant speed—imposed by the generator necessity to produce constant frequency and voltage power output, the power efficiency constant C_p varies with wind speed ($\omega_r = ct$) and thus less-efficient wind energy extraction is performed (Figure 3.56).

Typical turbine power versus turbine speed curves are shown in Figure 3.57.

Variable speed operation—which needs power electronics on the generator side—produces considerably more energy only if the wind speed varies considerably in time (inland sites). Not so in on or offshore sites, where wind speed variations are smaller.

However, the flexibility brought by variable speed in terms of electric power control of the generator and its power quality, with a reduction in mechanical stress in general (especially the thrust and torque reduction) is in favor of variable speed wind turbines.

There are two methods (Figure 3.58) to limit the power during strong winds ($U > U_{rated}$):

- Stall control
- Pitch control

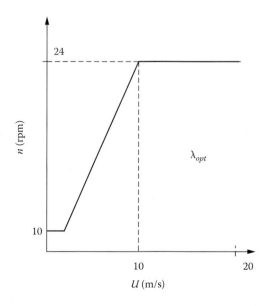

FIGURE 3.56 Typical optimum turbine/wind speed correlation.

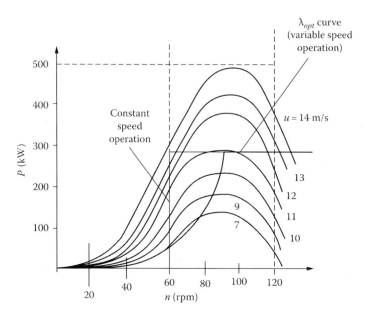

FIGURE 3.57 Turbine power versus turbine speed for various wind speed u values.

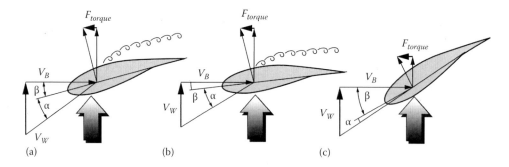

FIGURE 3.58 Stall and pitch control above rated wind speed U_{rated}: (a) passive stall, (b) active stall, and (c) pitch control.

Stalled blades act as a "wall in the wind." Stall occurs when the angle α between air flow and the blade chord is increased so much that the air flow separates from the airfoil in the suction side to limit the torque-producing force to its rated value.

For passive stall angle β stays constant as no mechanism to turn the blades is provided.

With a mechanism to turn the blades in place above rated wind speed U_{rated}, to enforce stall, the angle β is decreased by a small amount. This is the active stall method that may be used at low speeds also to increase power extraction by increasing power efficiency factor C_p (Figure 3.55).

With the pitch control (Figure 3.58) the blades are turned by notably increasing the angle β. The turbine turns to the position of the "flag in the wind" so that aerodynamic forces are reduced. As expected, the servodrive—for pitch control—to change β has to be designed for higher rating than for active stall.

3.9.3 Wind Turbine Models for Control

Besides wind slow variation with day or season time there are also under 1 Hz and over 1 Hz random wind speed variations (Figure 3.59) due to turbulent and wind gusts. Axial turbines (with 2,3 blades) experience 2,3 speed pulsations per revolution when the blades pass in front of the tower.

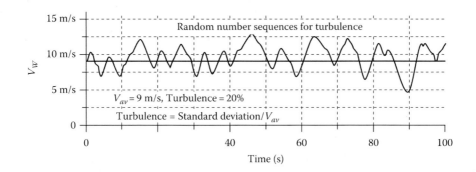

FIGURE 3.59 Wind speed typical variation with time.

Sideways tower oscillations also induce shaft speed pulsations.

Mechanical transmission and (or) the elasticity of blades, blades-fixtures, couplings produce additional oscillations. The pitch-servo dynamics has to be considered also.

The wind speed spectrum of wind turbine located in the wake of a neighboring one in a wind park, may also change. Care must be exercised in placing the components of a wind park [29].

Finally, electric load transients or faults are producing again speed variations.

All of the above clearly indicate the intricacy of wind turbine modeling for transients and control.

3.9.3.1 Unsteady Inflow Phenomena in Wind Turbines

The BEM model is based in steady state. It presupposes that an instant change of wind profile can take place (Figure 3.59).

Transition from state (1) to state (2) in Figure 3.59 corresponds to an increase of global pitch angle β by the pitch-servo.

Experiments have shown that in reality there are at least two time constants that delay the transition: one related to D_r/U and the other related to $2C/(D_r\omega_r)$ [30].

Time lags are related to the axial/and tangential induced velocities ($-aU$ and $+a'D_r\omega_r/2$).

The inclusion of a lead-lag filter to simulate the inflow phenomena seems insufficient due to considerable uncertainty in the modeling.

3.9.3.2 Pitch-Servo and Turbine Model

The pitch-servo is implemented as a mechanical hydraulic or electrohydraulic governor. A first-order (Figure 3.60) or a second-order model could be adopted.

In Figure 3.60, the pitch-servo is modeled as a simple delay T_{servo}, while the variation slope is limited between $d\beta_{min}/dt$ to $d\beta_{max}/dt$ (to take care of inflow phenomena). In addition, the global attack angle β span is limited from $\beta_{optimum}$ to $\beta_{maximum}$. β_{opt} is obtained from C_p–λ–β curve family for C_{pmax} with respect to β (Figure 3.61) [28].

Now from angle β to output power the steady-state model of the wind turbine is used (Figure 3.62) for a constant-speed active-stall wind turbine.

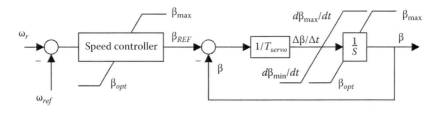

FIGURE 3.60 Optimum $\beta(U)$ for variable wind speed.

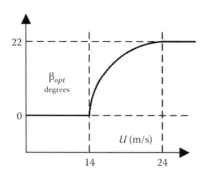

FIGURE 3.61 Wind profile transition from state to state.

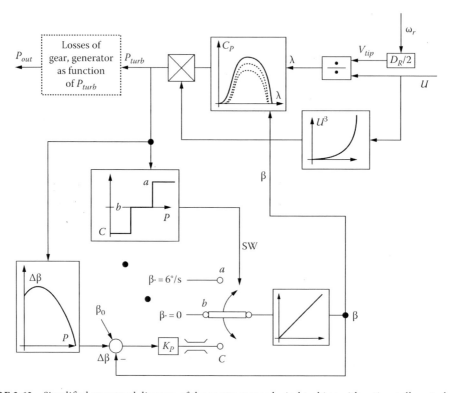

FIGURE 3.62 Simplified structural diagram of the constant speed wind turbine with active stall control.

When the turbine produces more than rated power, the switch is in position *a* and the angle β is increased at the rate of 6°/s to move the blades toward the "flag-in-the-wind" position. When the power is around rated value, the pitch drive stays idle with β = 0, and thus β = constant (position *b*).

Below rated power, the switch goes to position "C" and a proportional controller (K_p) produces the desired β. The reference value β* corresponds to its optimum value as function of mechanical power, that is maximum power.

This is only a sample of the constant speed turbine model with pitch-servo control for active stall above rated power and $β_{optimization}$ control below rated power.

As can be seen from Figure 3.62, the model is highly nonlinear. Still, the delays due to inflow phenomena, elasticity of various elements of the turbine are not yet included. Also the model of the pitch-servo is not included. Usually, there is a transmission between the wind turbine and the electric generator. A six-order drive train is shown in Figure 3.63.

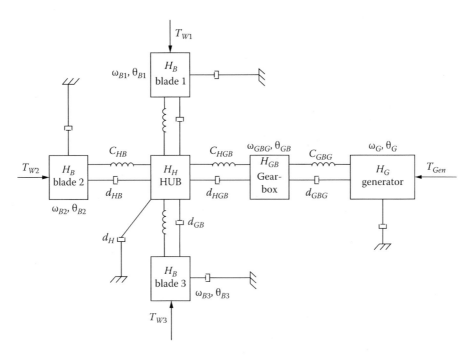

FIGURE 3.63 Six inertia drive train.

Inertias of hub, blades, gearbox, and of generator are denoted by H_i. Each part has also a spring and a dashpot element. The matrix dynamic equation of the drive train is of the following form [31]:

$$\frac{d}{dt}\begin{bmatrix}|\theta| \\ |\omega|\end{bmatrix} = \begin{bmatrix} 0 & I \\ -[2H^{-1}][C] & -[2H^{-1}]D \end{bmatrix} \cdot \begin{bmatrix}|\theta| \\ |\omega|\end{bmatrix} + \begin{bmatrix} 0 \\ 2H^{-1}\end{bmatrix}[T] \tag{3.85}$$

Quite a few "real-world" pulsations in speed or electric power may be detected by such models; resonance conditions may be avoided through design or control measures.

3.10 Summary

- Prime movers are mechanical machines that convert primary energy of a fuel (or fluid) into mechanical energy.
- Prime movers drive electric generators connected to the power grid or operating in isolation.
- Steam and gas turbines and internal combustion engines (spark-ignited or diesel) are burning fossil fuels to produce mechanical work.
- Steam turbines contain stationary and rotating blades grouped into (HP, IP, and LP sections on same shaft in tandem-compound and on two shafts in cross-compound configurations.
- Between stages, steam engines use reheaters; single reheat and double reheat, at most.
- Typically in steam turbines, the power is divided as 30% in the HP, 40% in the IP and 30% in the LP stage.
- Governor valves and IV are used to control the HP and, respectively, LP stages of the steam flow.
- The steam vessel may be modeled by a first-order delay, while the steam turbine torque is proportional to its steam flow rate.
- Three more delays related to inlet and steam chest, to the reheater and to the crossover piping may be identified.

- Speed governor for steam turbines include a speed relay with a first-order delay and a hydraulic servomotor characterized by a further delay.
- Gas turbines burn natural gas in combination with air that is compressed in a compressor driven by the gas turbine itself.
- The 500°C gas exhaust is used to produce steam that drives a steam turbine placed on the same shaft. These combined cycle unishaft gas turbines are credited with a total efficiency above 55%. Combined cycle gas turbines at large powers seem the way of the future. They are also introduced for cogeneration in high-speed small- and medium-power applications.
- Diesel engines are used from the kilowatt range to megawatt range power per unit for cogeneration or for standby (emergency) power sets.
- The fuel injection control in diesel engines is performed by a speed-governing system.
- The diesel engine model contains a nonconstant gain. The gain depends on the eer, which in turn is governed by the fuel/air ratio; a dead time constant dependent on engine speed is added to complete the diesel engine model.
- Diesel engines are provided with a turbocharger that has a turbine "driven" by the fuel exhaust that drives a compressor that provides the hot HP air for the air mix of the main engine. The turbocharger runs freely at high speed but is coupled to the engine at low speed.
- Stirling engines are "old" thermal piston engines with continuous heat supply. Their thermal cycles contains two isotherms. It contains, in a basic configuration, two opposed pistons and a regenerator in between. The efficiency of Stirling engine is temperature limited.
- Stirling engines are independent fuel type; use air, methane, He, or H_2 as working fluids. They did not reach commercial success in kinetic type due to problems with the regenerator and stabilization.
- Stirling engines with free piston-displacer mover and linear motion have reached recently the markets in units in the 50 W to a few kilowatts.
- The main merits of Stirling engines are related to their quietness and reduced noxious emissions, but they tend to be expensive and difficult to stabilize.
- Hydraulic turbines convert water energy of rivers into mechanical work. They are the oldest prime movers.
- Hydraulic turbines are of impulse type for heads above 300–400 m and of reaction type (below 300 m). In a more detailed classification, they are tangential (Pelton), radial–axial (Francis) and axial (Kaplan, Bulb, Straflo).
- High head (impulse) turbines use a nozzle with a needle-controller where water is accelerated and then it impacts the bowl-shaped buckets on the water wheel of the turbine. A jet deflector deflects water from runner to limit turbine speed when electric load decreases.
- Reaction turbines—at medium and low head—use wicket gates and rotor blade servomotors to control water flow in the turbine.
- Hydraulic turbines may be modeled by a first-order model if water hammer (wave) and surge effects are neglected. Such a rough approximation does not hold above 0.1 Hz.
- Second-order models for hydraulic turbines with water hammer effect in the penstock considered are valid up to 1 Hz. Higher orders are required above 1 Hz as the nonlinear model has a gain whose amplitude varies periodically. Second- or third-order models may be identified from tests through adequate curve fitting methods.
- Hydraulic turbine governors have one or two power levels. The lower power level may be electric, while the larger (upper) power level is a hydraulic servomotor. The speed controller of the governor has traditionally a permanent droop and a transient droop.
- Modern nonlinear control systems may now be used to control simultaneously the guide vane runner and the blade runner.
- Reversible hydraulic machines are used for pump storage power plants or for tidal-wave power plants. The optimal pumping speed is about 12%–20% above the optimal turbining speed. Variable speed operation is required. Therefore, power electronics on the electric side is mandatory.

- Wind turbines use the wind air energy. Nonuniformity and strength vary with location height and terrain irregularities. Wind speed duration versus speed, speed versus frequency and mean (average) speed using Raleigh or Weibull distribution are used to characterize wind on a location in time. The wind turbine rated wind speed is, in general, 150% of mean wind speed.
- Wind turbines are of two main types: axial (with horizontal shaft) and tangential (with vertical shaft).
- Wind turbines' main steady-state parameter is the power efficiency coefficient C_p, which is dependent on blade tip speed $R\omega_r$ to wind speed U (ratio λ). C_p depends on λ and on blade absolute attack angle β.
- The maximum C_p (0.3–0.4) with respect to β is obtained for $\lambda_{opt} \leq 1$ for low-speed axial turbines and for $\lambda_{opt} \geq 1$ for high-speed turbines.
- The ideal maximum efficiency limit of wind turbines is about 0.6 (Betz limit).
- Wind impacts on the turbine a thrust force and a torque. Only torque is useful. The thrust force and C_p depend on blade absolute attack angle β.
- The optimal power $P_T (\lambda_{opt})$ is proportional to u^3 (u—wind speed).
- Variable speed turbines will collect notably more power from a location if the speed varies significantly with time and season, such that λ may be kept optimum. Above rated wind speed (and power), the power is limited by passive stall, active stall, or pitch-servo control.
- Wind turbine steady-state models are highly nonlinear. Unsteady inflow phenomena show up in fast transients and have to be accounted for by more than lead-lag elements.
- Pitch-servo control is becoming more and more frequent even with variable speed operation, to allow speed limitation during load transients or power grid faults.
- First- or second-order models may be adopted for speed governors. Elastic transmission multimass models have to be added to complete the controlled wind turbine models for transients and control.
- R&D efforts on prime movers' modeling and control seem rather dynamic [32–34].
- Prime mover models will be used in following chapters where electric generators control will be treated in detail.

References

1. R. Decker, *Energy Conversion*, OUP, Oxford, UK, 1994.
2. P. Kundur, *Power System Stability and Control*, McGraw Hill, New York, 1994.
3. J. Machowski, J.W. Bialek, J.R. Bumby, *Power System Dynamics and Stability*, John Wiley & Sons, New York, 1997.
4. IEEE Working Group Report, Steam models for fossil fueled steam units in power-system studies, *IEEE Transactions*, PWRS-6(2), 753–761, 1991.
5. S. Yokokawa, Y. Ueki, H. Tanaka, Hi Doi, K. Ueda, N. Taniguchi, Multivariable adaptive control for a thermal generator, *IEEE Transactions*, EC-3(3), 479–486, 1988.
6. G.K. Venayagomoorthy, R.G. Harley, A continually on line trained microcontroller for excitation and turbine control of a turbogenerator, *IEEE Transactions*, EC-16(3), 261–269, 2001.
7. C.F. Taylor, *The Internal-Combustion Engine in Theory and Practice*, Vol. II: *Combustion, Fuels, Materials, Design*, The MIT Press, Cambridge, MA, 1968.
8. P.M. Anderson, M. Mirheydar, Analysis of a diesel-engine driven generating unit and the possibility for voltage flicker, *IEEE Transactions*, EC-10(1), 37–47, 1995.
9. N. Watson, M.S. Janota, *Turbocharging the Internal Combustion Engine*, McMillan Press, Englewoods, NJ, 1982.
10. S. Roy, O.P. Malik, G.S. Hope, Adaptive control of speed and equivalence ratio dynamics of a diesel driven power plant, *IEEE Transactions*, EC-8(1), 13–19, 1993.
11. A. Kusko, *Emergency Stand by Power Systems*, McGraw Hill, New York, 1989.

12. K.E. Yoager, J.R. Willis, Modelling of emergency diesel generators in an 800 MW nuclear power plant, *IEEE Transactions*, EC-8(3), 433–441, 1993.

13. U. Kieneke, L. Nielsen, *Automotive Control Systems*, Springer Verlag, Berlin, Germany, 2000.

14. G. Walker, O.R Fauvel, G. Reader, E.R. Birgham, *The Stirling Alternative*, Gordon and Breach Science Publishers, 1994.

15. R.W. Redlich, D.W. Berchowitz, Linear dynamics of free piston Stirling engine, *Proceedings of the IMechE*, 203–213, March 1985.

16. M. Barglazan, *Hydraulic Turbines and Hydrodynamic Transmissions Book*, University Politehnica of Timisoara, Timisoara, Romania, 1999.

17. IEEE Working Group on Prime Mover and Energy Supply, Hydraulic turbine and turbine control models for system stability studies, *IEEE Transactions*, PS-7(1), 167–179, 1992.

18. C.D. Vournas, Second order hydraulic turbine models for multimachine stability studies, *IEEE Transactions*, EC-5(2), 239–244, 1990.

19. C.K. Sanathanan, Accurate low order model for hydraulic turbine-penstock, *IEEE Transactions*, EC-2(2), 196–200, 1987.

20. D.D. Konidaris, N.A. Tegopoulos, Investigation of oscillatory problems of hydraulic generating units equipped with Francis turbines, *IEEE Transactions*, EC-12(4), 419–425, 1997.

21. D.J. Trudnowski, J.C. Agee, Identifying a hydraulic turbine model from measured field data, *IEEE Transactions*, EC-10(4), 768–773, 1995.

22. J.E. Landsberry, L. Wozniak, Adaptive hydrogenerator governor tuning with a genetic algorithm, *IEEE Transactions*, EC-9(1), 179–185, 1994.

23. Y. Zhang, O.P. Malik, G.S. Hope, G.P. Chen, Application of inverse input/output mapped ANN as a power system stabilizer, *IEEE Transactions*, EC-9(3), 433–441, 1994.

24. M. Djukanovic, M. Novicevic, D. Dobrijovic, B. Babic, D.J. Sobajic, Y.H. Pao, Neural-net based coordinated stabilizing control of the exciter and governor loops of low head hydropower plants, *IEEE Transactions*, EC-10(4), 760–767, 1995.

25. T. Kuwabara, A. Shibuya, M. Furuta, E. Kita, K. Mitsuhashi, Design and dynamics response characteristics of 400 MW adjustable speed pump storage unit for Obkawachi power station, *IEEE Transactions*, EC-11(2), 376–384, 1996.

26. L.L. Freris, *Wind Energy Conversion Systems*, Prentice Hall, New York, 1998.

27. V.H. Riziotis, P.K. Chaviaropoulos, S.G. Voutsinas, Development of the state of the art aerolastic simulator for horizontal axis wind turbines, Part 2: Aerodynamic aspects and application, *Wind Engineering*, 20(6), 223–440, 1996.

28. V. Akhmatov, Modelling of variable speed turbines with doubly-fed induction generators in short-term stability investigations, *Third International Workshop on Transmission Networks for Off-Shore Wind Farms*, Stockholm, Sweden, April 11–12, 2002.

29. T. Thiringer, J.A. Dahlberg, Periodic pulsations from a three-bladed wind turbine, *IEEE Transactions* EC-16(2), 128–133, 2001.

30. H. Suel, J.G. Schepers, Engineering models for dynamic inflow phenomena, *Journal of Wind Engineering and Industrial Aerodynamics*, 39, 267–281, 1992.

31. S.A. Papathanassiou, M.P. Papadopoulos, Mechanical stress in fixed-speed wind turbines due to network disturbances, *IEEE Transactions*, EC-16(4), 361–367, 2001.

32. S.H. Jangamshatti, V.G. Rau, Normalized power curves as a tool for identification of optimum wind turbine generator parameters, *IEEE Transactions*, EC-16(3), 283–288, 2001.

33. K. Natarajan, Robust PID controller design of hydro-turbines, *IEEE Transactions*, EC-20(3), 661–667, 2005.

34. D. Borkovski, T. Klegiel, Small hydropower plant with integrated turbine-generators working at variable speed, *IEEE Transactions*, EC-20(2), 453–459, 2013.

4

Large- and Medium-Power Synchronous Generators: Topologies and Steady State

4.1 Introduction

By large powers, we mean powers above 1 MW per unit, where in general the rotor magnetic field is produced with electromagnetic excitation. There are a few megawatt power permanent magnet (PM)-rotor synchronous generators (SGs), but they will be treated in a dedicated chapter, in Part 2 of EGH: variable speed generators.

Almost all electric energy generation is performed through SGs with power per unit up to 2000 MVA in thermal power plants and up to 700 MW (1000 MW in R&D) per unit in hydropower plants. SGs in the megawatt and tenth of megawatt range are used in diesel engine power groups for cogeneration or on locomotives and on ships.

We will start with a description of basic configurations, their main components, and principles of operation and then describe the steady-state operation in some detail.

4.2 Construction Elements

The basic parts of an SG are the stator, rotor, the framing (with cooling system), and the excitation system.

The stator is provided with a magnetic core made of silicon steel sheets (0.55 mm thick in general) in which uniform slots are stamped. A single, standard, magnetic sheet steel is produced up to 1 m diameter as a complete circle (Figure 4.1). Large turbo generators and most hydrogenerators have stator outer diameters well in excess of 1 m (up to 18 m), and thus the cores are made of 6–42 segments per circle (Figure 4.2).

The stator may also be split radially into two or more sections to allow handling and permit transport with windings in slots. The windings are inserted in slots section by section, and their connection together is performed at the power plant site.

When the stator with N_s slots is divided and the number of slot pitches per segment is m_p, the number of segments m_s is such that:

$$N_s = m_s \cdot m_p \tag{4.1}$$

Each segment is attached to the frame through two keybars or dovetail wedges that are uniformly distributed along the periphery (Figure 4.2).

In two successive layers (laminations), the segments are offset by half a segment.

The distance between wedges b is

$$b = \frac{m_p}{2} = 2a \tag{4.2}$$

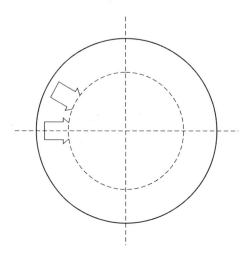

FIGURE 4.1 Single-piece stator core.

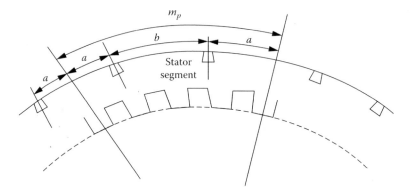

FIGURE 4.2 Divided stator core made of segments.

This is so to allow for offsetting the segments in subsequent layers by half a segment. Also, only one tool for stamping is required because all segments are identical. To avoid winding damage due to vibration, each segment should start and end in the middle of a tooth and span over an even number of slot pitches.

For the divided stator into S sectors, two types of segments are generally used. One type with m_p slot pitches and the other with n_p slot pitches such that

$$\frac{N_s}{S} = Km_p + n_p; \quad n_p < m_p; \quad m_p = 6 - 13 \tag{4.3}$$

With $n_p = 0$, the first case is obtained and, in fact, the number of segments per stator sector is an integer. This is not always possible and thus two types of segments are required.

The offset of segments in subsequent layers is $m_p/2$ if m_p is even, $(m_p \pm 1)/2$ for m_p odd, and $m_p/3$ if m_p is divisible by 3. In the particular case that $n_p = m_p/2$, we may just cut the main segment in two to obtain the second one, and thus again only one stamping tool is required. For more details, see Reference [1].

The slots of large- and medium-power SGs are rectangular and open (Figure 4.3).

The double layer winding, made usually of magnetic wires with rectangular cross-section, is "kept" inside the open slot by a wedge made of insulator material or from a magnetic material with a low equivalent tangential permeability that is μ_r times larger than that of air. The magnetic wedge may be made of magnetic powders or of laminations, with a rectangular prolonged hole (Figure 4.3b), "glued together" with a thermally and mechanically resilient resin.

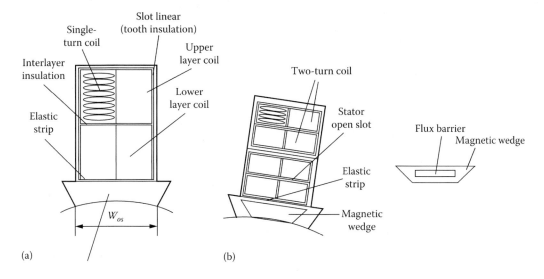

FIGURE 4.3 (a) Stator slotting and (b) magnetic wedge.

4.2.1 Stator Windings

The stator slots are provided with coils connected to form a three-phase winding.

The winding of each phase produces an air gap fixed magnetic field with $2p_1$ half-periods per revolution. With D_{is} as the internal stator diameter, the pole pitch τ—that is, the half period of winding mmf, is

$$\tau = \frac{\pi D_{is}}{2p_1} \tag{4.4}$$

The phase windings are phase shifted by $(2/3)\tau$ along the stator periphery and are symmetric. The average number of slots per pole per phase q is

$$q = \frac{N_s}{2p_1 \cdot 3} \tag{4.5}$$

The number q may be integer—with low number of poles ($2p < 8$–10)—or it may be a fractionary number:

$$q = a + b/c \tag{4.6}$$

Fractionary q windings are used mainly in SGs with a large number of poles, where a necessarily low integer q ($q \leq 3$) would produce too a high harmonics content in the generator electromagnetic force (emf).

Large- and medium-power SGs make use of typical lap (multiturn coil) windings (Figure 4.4) or of bar-wave (single-turn coil) windings (Figure 4.5).

The coils of phase A in Figures 4.4 and 4.5 are all in series. A single current path is thus available ($a = 1$). It is feasible to have "a" current paths in parallel, especially in large-power machines (line voltage is in general below 24 kV). With W_{ph} turns in series (per current path), we do have the relationship:

$$N_s = 3\frac{W_{ph} \cdot a}{n_c} \tag{4.7}$$

n_c turns per coil.

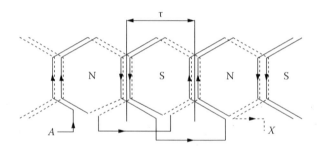

FIGURE 4.4 Lap winding (4 poles)—$q = 2$: phase A only.

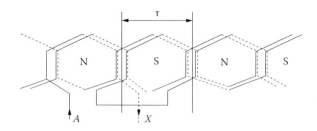

FIGURE 4.5 Basic bar-wave winding—$q = 2$: phase A only.

The coils may be multiturn lap coils, in general, or uniturn (bar) type—in wave coils.

A general comparison between the two types of windings (both with integer and fractionary q) reveals the following:

- The multiturn coils ($n_c > 1$) allow for greater flexibility in choosing the number of slots N_s, for given number of current paths a.
- Multiturn coils are however manufacturing-wise limited to 0.3 m long lamination stacks and pole pitches $\tau < 0.8$–1 m.
- Multiturn coils need bending as they are placed with one side in the bottom layer and with the other one in the top layer; bending needs to be done without damaging the electric insulation, which, in turn, has to be flexible enough for the intended purpose.
- Bar coils are used for heavy currents (above 1500 A). Wave bar coils imply a smaller number of connectors (Figure 4.5) and thus are less costly. The lap-bar coils allow for short-pitching—to reduce emf harmonics, while wave bar coils imply 100% average pitch coils.
- To avoid excessive eddy current (skin) effects in deep coil sides, transposition of individual strands is required. In multiturn coils ($n_c \geq 2$), one semi-Roebel transposition is enough, while in single-bar coils full Roebel transposition is required.
- Switching or lightning strokes along the transmission lines to SG produce steep-fronted voltage impulses between neighboring turns in the multiturn coil and thus additional insulation is required. Not so for the bar (single turn) coils where only interlayer and slot insulation is provided.
- Accidental short circuit in multiturn coil windings with a ≥ 2 current path in parallel produces a circulating current between current paths. This unbalance in current paths may be sufficient to trip off the pertinent circuit balance relay. Not so far the bar coils where the unbalance is less pronounced.
- Though slightly more expensive, the technical advantages of bar (single-turn) coils should make them the favorite solution in most cases.

AC windings for SGs may be built not only in two layers but also in one layer. In the latter case, there is a necessity to use 100% pitch coils that have longer end connections, unless bar coils are used.

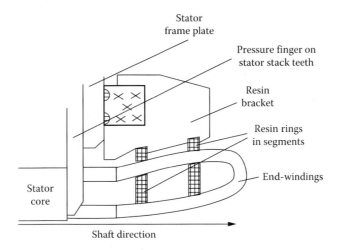

FIGURE 4.6 Typical support system for stator end-windings.

Stator end-windings have to be mechanically supported to avoid mechanical deformation during severe transients due to electrodynamic large forces between them and between them as a whole and the rotor excitation end-windings. As in general such forces are radial, the support for end-windings typically looks as in Figure 4.6.

Note: More on AC winding specifics in the chapter dedicated on SG design. Here we only derive the fundamental magneto-motive-force, mmf, wave of three-phase stator windings.

The mmf of a single-phase 4 pole winding with 100% pitch coils may be approximated with a step-like periodic function if the slot openings are neglected (Figure 4.7).

For the case in Figure 4.7. with $q = 2$ and 100% pitch coils, the mmf distribution is rectangular with only one step per half-period.

With chorded coils or $q > 2$, more steps would be visible in the mmf. That is, the distribution approximates then better a sinusoid waveform.

In general, the phase mmf fundamental distribution for steady state may be written as follows:

$$F_{1A}(x,t) = F_{1m} \cdot \cos\frac{\pi}{\tau}x \cdot \cos\omega_1 t \tag{4.8}$$

$$F_{1m} = 2\sqrt{2}\frac{W_1 K_{W1} I}{\pi p_1} \tag{4.9}$$

where
   W_1 turns per phase in series
   I is phase current (RMS)
   p_1 is pole pairs
   K_{W1} is the winding factor

$$K_{W1} = \frac{\sin \pi/6}{q \cdot \sin \pi/6q} \cdot \sin\left(\frac{y}{\tau}\frac{\pi}{2}\right) \tag{4.10}$$

with y/τ = coil pitch/pole pitch ($y/\tau > 2/3$)

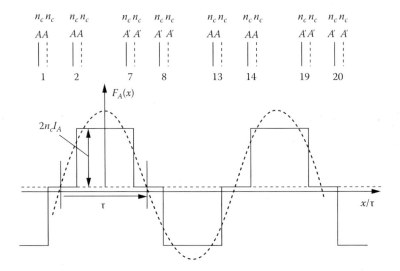

FIGURE 4.7 Stator phase mmf distribution ($2p = 4$, $q = 2$).

Equation 4.8 is strictly valid for integer q.

An equation similar to Equation 4.8 may be written for the νth space harmonic as follows:

$$F_{\nu A}(x,t) = F_{\nu m}\cos\nu\frac{\pi}{\tau}x\cos(\omega_1 t) \tag{4.11}$$

$$F_{\nu m} = \frac{2\sqrt{2}W_1 K_{W\nu}I}{\pi p_1 \nu} \tag{4.12}$$

$$K_{W\nu} = \frac{\sin\nu\pi/6}{q\cdot\sin(\nu\pi/6q)}\cdot\sin\frac{y}{\tau}\frac{\nu\pi}{2}$$

Phase B and phase C mmfs expressions are similar to Equation 4.8 but with $2\pi/3$ space and time lags. Finally, the total mmf (with space harmonics) produced by a three-phase winding is [2]

$$F_\nu(x,t) = \frac{3W_1 I\sqrt{2}K_{W\nu}}{\pi p_1 \nu}\left[K_{BI}\cos\left(\frac{\nu\pi}{\tau}-\omega_{1t}-(\nu-1)\frac{2\pi}{3}\right) - K_{BII}\cos\left(\frac{\nu\pi}{\tau}+\omega_{1t}-(\nu+1)\frac{2\pi}{3}\right)\right] \tag{4.13}$$

where

$$K_{BI} = \frac{\sin(\nu-1)\pi}{3\cdot\sin(\nu-1)\pi/3}$$

$$K_{BII} = \frac{\sin(\nu+1)\pi}{3\cdot\sin(\nu+1)\pi/3} \tag{4.14}$$

Equation 4.13 is valid for integer q.

For $\nu = 1$, the fundamental is obtained.

Due to full symmetry—with q integer—only odd harmonics exist. For $\nu = 1$, $K_{BI} = 1$, $K_{BII} = 0$, therefore the mmf fundamental represents a forward-traveling wave with the peripheral linear speed:

$$\frac{dx}{dt} = \frac{\tau\omega_1}{\pi} = 2\tau f_1 \tag{4.15}$$

The harmonic orders are $\nu = 3K \pm 1$. For $\nu = 7, 13, 19,\ldots$, $dx/dt = 2\tau f_1/\nu$ and for $\nu = 5, 11, 17,\ldots$, $dx/dt = -2\tau f_1/\nu$. That is, the first ones are direct traveling waves, while the second ones are backward-traveling waves. Coil chording ($y/\tau < 1$) and increased q may reduce harmonics amplitude (reduced $K_{w\nu}$), but the price is a reduction in the mmf fundamental too (K_{W1} decreases).

The rotor of large SGs may be built with salient poles (for $2p_1 > 4$) or with nonsalient poles ($2p_1 = 2, 4$).

The solid iron core of the nonsalient pole rotor is made of 12–20 cm thick (axially) rolled steel discs spigoted to each other to form a solid ring by using axial through bolts. Shaft ends are added (Figure 4.8).

Salient poles (Figure 4.9b) may be made of laminations packs tightened axially by through bolts and end plates and fixed to the rotor pole wheel by hammer-tail key bars.

In general, peripheral speeds around 110 m/s are feasible only with solid rotors made by forged steel. The field coils in slots (Figure 4.9) are protected from centrifugal forces by slot wedges that are made either of strong resins or of conducting material (copper), and the end-windings need bandages.

The interpole area in salient pole rotors (Figure 4.9b) is used to mechanically fix the field coil sides so that they do not move or vibrate, while the rotor rotates at its maximum allowable speed.

Nonsalient poles (high-speed) rotors show small magnetic anisotropy. That is, the magnetic reluctance of air gap along pole (longitudinal) axis d, and along interpole (transverse) axis q, is about the same, except for the case of severe magnetic saturation conditions.

In contrast, salient pole rotor experience a rather large (1.5–1 and more) magnetic saliency ratio between axis d and axis q. The damper cage bars placed in special rotor poles slots may be connected together through end rings (Figure 4.10a). Such a complete damper cage may be decomposed in two fictitious cages, one with the magnetic axis along d axis and the other along q axis (Figure 4.10), both with partial end rings (Figure 4.10).

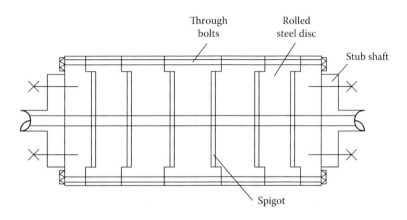

FIGURE 4.8 Solid rotor.

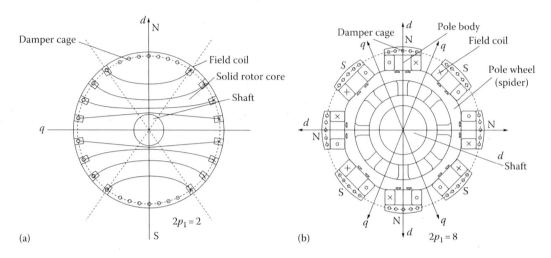

FIGURE 4.9   Rotor configurations: (a) with nonsalient poles and (b) with salient poles.

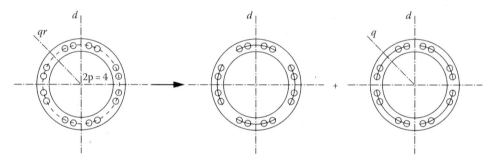

FIGURE 4.10   The damper cage and its *d* and *q* axis fictitious components.

4.3 Excitation Magnetic Field

The air gap magnetic field produced by the DC field (excitation) coils has a circumferential distribution that depends on the type of the rotor—with salient or nonsalient poles—and on the air gap variation along rotor pole span. For now let us consider that the air gap is constant under the rotor pole and the presence of stator slot openings is considered through the Carter coefficient K_{C1} which increases the air gap [2]:

$$K_{C1} \approx \frac{\tau_s}{\tau_s - \gamma_1 g} > 1, \quad \tau_s - \text{stator slot pitch} \tag{4.16}$$

$$\gamma_1 = \frac{4}{\pi}\left[\frac{(W_{os}/g)}{\tan(W_{os}/g)} - \ln\sqrt{1 + \left(\frac{W_{os}}{g}\right)^2} \right] \tag{4.17}$$

with

W_{os} is stator slot opening
g is air gap

The flux lines produced by the field coils (Figure 4.11) resemble the field coils mmf $F_F(x)$ as the air gap under the pole is considered constant (Figure 4.12).

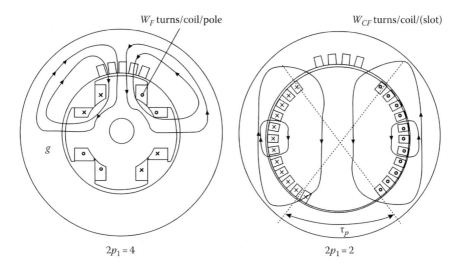

FIGURE 4.11 Basic field winding flux lines through air gap and stator.

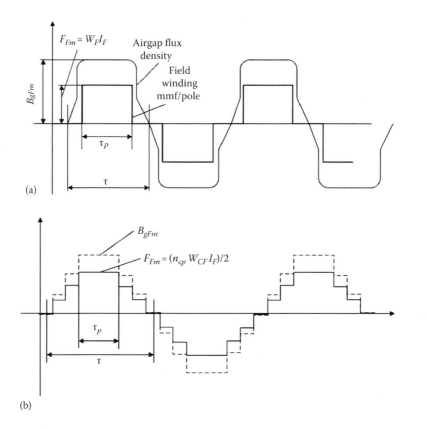

FIGURE 4.12 Field winding mmf and air gap flux density: (a) salient pole rotor and (b) nonsalient pole rotor.

The approximate distribution of no-load or field-winding-produced air gap flux density in Figure 4.12 has been obtained through Ampere's law.

For salient poles,

$$B_{gFm} = \frac{\mu_0 W_f I_F}{K_c g (1 + K_{S0})}, \quad \text{for} : |x| < \frac{\tau_p}{2} \tag{4.18}$$

and $B_{gFm} = 0$ otherwise.

In practice, $B_{gFm} = 0.6\text{–}0.9\,\text{T}$. Fourier decomposition of this rectangular distribution yields the following:

$$B_{gF\upsilon}(x) = K_{F\upsilon} \cdot B_{gFm} \cos \upsilon \frac{\pi}{\tau} x; \quad \upsilon = 1,3,5,\ldots \tag{4.19}$$

$$K_{F\upsilon} = \frac{4}{\pi} \sin \upsilon \frac{\tau_p}{\tau} \frac{\pi}{2} \tag{4.20}$$

Only the fundamental is useful. Both the fundamental distribution ($\nu = 1$) and the space harmonics depend on the ratio τ_p/τ (pole span/pole pitch). In general, $\tau_p/\tau \approx 0.6\text{–}0.72$. Also, to reduce the harmonics content, the air gap may be modified (increased), from the pole middle toward pole ends, as an inverse function of $\cos \pi x/\tau$:

$$g(x) = \frac{g}{\cos(\pi/\tau)x}, \quad \text{for} : \frac{-\tau_p}{2} < x < \frac{\tau_p}{2} \tag{4.21}$$

In practice, function of Equation 4.21 is not easy to generate, but approximations of it, easy to manufacture, are adopted.

Reducing the no-load air gap flux density harmonics causes a reduction of time harmonics in the stator emf (or no-load stator phase voltage).

For the nonsalient pole rotor,

$$B_{gFm} = \frac{\mu_0 (n_p/2) W_{CF} I_F}{K_C g (1 + K_{S0})}, \quad \text{for} : |x| < \frac{\tau_p}{2} \tag{4.22}$$

and stepwise varying otherwise (Figure 4.12b).

K_{S0} is the magnetic saturation factor that accounts for stator and rotor iron magnetic reluctance of the field paths.

$$B_{gF\upsilon}(x) = K_{F\upsilon} \cdot B_{gFM} \cdot \cos \frac{\upsilon \pi}{\tau} x \tag{4.23}$$

$$K_{F\upsilon} \approx \frac{8}{\upsilon^2 \pi^2} \frac{\cos \upsilon \dfrac{\tau_p}{\tau} \dfrac{\pi}{2}}{\left(1 - \upsilon \dfrac{\tau_p}{\tau}\right)} \tag{4.24}$$

It is obvious that in this case, the flux density harmonics are lower and thus constant air gap (cylindrical rotor) is feasible in all practical cases.

Let us consider only the fundamental of the no-load flux density in the air gap:

$$B_{gF1}(x_r) = B_{gFm1} \cos \frac{\pi}{\tau} x_r \qquad (4.25)$$

For constant rotor speed, the rotor coordinate x_r is related to stator coordinate x_s as follows:

$$\frac{\pi}{\tau} x_r = \frac{\pi}{\tau} x_s - \omega_r t - \theta_0 \qquad (4.26)$$

The rotor rotates at angular speed ω_r (in electrical terms: $\omega_r = p_1\Omega_r - \Omega_r$ mechanical angular velocity). θ_0 is an arbitrary initial angle; let $\theta_0 = 0$.

With Equation 4.26, Equation 4.25 becomes thus:

$$b_{gF1}(x_s,t) = B_{gFm1} \cos\left(\frac{\pi}{\tau} x_s - \omega_r t\right) \qquad (4.27)$$

Therefore, the excitation air gap flux density represents a forward-traveling wave at rotor speed. This traveling wave moves in front of the stator coils at the tangential velocity u_s:

$$u_s = \frac{dx_s}{dt} = \frac{\tau\omega_r}{\pi} \qquad (4.28)$$

It is now evident that with the rotor driven by a prime mover at speed ω_r and with the stator phases kept open, the excitation air gap magnetic field induces an emf in the stator windings:

$$E_{A1}(t) = -\frac{d}{dt} W_1 K_{W1} \int_{-(\tau/2)}^{+(\tau/2)} B_{gF1}(x_s,t)\,dx_s \qquad (4.29)$$

Finally,

$$E_{A1}(t) = E_1 \sqrt{2} \cos \omega_r t \qquad (4.30)$$

$$E_{1m} = \pi\sqrt{2}\left(\frac{\omega_r}{2\pi}\right) B_{gFm1} l_{stack} W_1 K_{W1} \frac{2\tau}{\pi} \qquad (4.31)$$

With l_{stack} is the stator stack length.

As the three phases are fully symmetric, the emfs in the three of them are as follows:

$$E_{A,B,C,1}(t) = E_{1m}\sqrt{2} \cos\left[\omega_r t - (i-1)\frac{2\pi}{3}\right] \qquad (4.32)$$

$$i = 1,2,3$$

Therefore, we notice that the excitation coil currents in the rotor are producing at no-load (open-stator phases) three symmetric emfs whose frequency ω_r is given by the rotor speed $\Omega_r = \omega_r/p_1$.

4.4 Two-Reaction Principle of Synchronous Generators

Let us now suppose that an excited SG is driven on no load at speed ω_r. When a balanced three-phase load is connected to the stator, the presence of emfs at frequency ω_r will naturally produce currents of same frequency. The phase shift between the emfs and the phase current ψ is dependent on load nature (power factor) and on machine parameters—not mentioned yet (Figure 4.13).

The sinusoidal emfs and currents are represented as simple phasors in Figure 4.13b.

The magnetic anisotropy of the rotor along axes d and q helps decompose each phase current in two components: one in phase with the emf and the other at 90° with respect to the former: I_{Aq}, I_{Bq}, and I_{Cq}, and I_{Ad}, I_{Bd}, and I_{Cd}, respectively.

As already proven in the paragraph on windings, three-phase symmetric windings flowed by balanced currents of frequency ω_r will produce traveling mmfs Equation 4.13:

$$F_d(x,t) = -F_{dm} \cos\left(\frac{\pi}{\tau}x_s - \omega_r t\right) \tag{4.33}$$

$$F_{dm} = \frac{3\sqrt{2}I_d W_1 K_{W1}}{\pi p_1}; \quad I_d = |I_{Ad}| = |I_{Bd}| = |I_{Cd}| \tag{4.34}$$

$$F_q(x,t) = F_{qm}\left(\frac{\pi}{\tau}x_s - \omega_r t - \frac{\pi}{2}\right) \tag{4.35}$$

$$F_{qm} = \frac{3\sqrt{2}I_q W_1 K_{W1}}{\pi p_1}; \quad I_q = |I_{Aq}| = |I_{Bq}| = |I_{Cq}| \tag{4.36}$$

In essence, the d-axis stator currents produce an mmf aligned to the excitation air gap flux density wave of Equation 4.26 but opposite in sign (for the situation in Figure 4.13b). This means that the d-axis mmf component produces a magnetic field "fixed" to the rotor and flowing along axis d as the excitation field does.

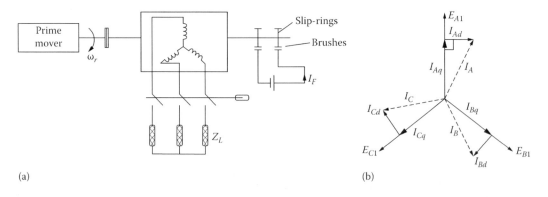

(a)   (b)

FIGURE 4.13 Illustration of SG principle: (a) the SG on load and (b) the emf and current phasors.

In contrast, the q-axis stator current components produces an mmf whose magnetic field is again "fixed" the rotor but flowing along axis q.

The emfs produced by motion in the stator windings might be viewed as produced by a fictitious three-phase AC winding flowed by symmetric currents I_{FA}, I_{FB}, and I_{FC} of frequency ω_r:

$$E_{A,B,C} = -j\omega_r M_{FA} I_{FA,B,C} \tag{4.37}$$

From what we have already discussed in this paragraph,

$$E_A(t) = 2\pi \frac{\omega_r}{2\pi} K_{W1} \frac{2}{\pi} l_{stack} \tau \cdot \frac{\mu_0 W_F I_f K_{F1}}{K_c g (1 + K_{S0})} \cos \omega_1 t \tag{4.38}$$

$$W_F = \frac{n_p}{2} W_{CF}; \quad \text{for nonsalient pole rotor (see Equation 4.22)}$$

The fictitious currents I_{FA}, I_{FB}, and I_{FC} are considered having the RMS value of I_F in the real field winding. From Equations 4.37 and 4.38, we obtain the following:

$$M_{FA} = \mu_0 \frac{\sqrt{2}}{\pi} \frac{W_1 W_F K_{W1} \tau \cdot l_{stack}}{K_C g (1 + K_{S0})} K_{F1} \tag{4.39}$$

M_{FA} is called the mutual rotational inductance between the field and armature (stator) phase windings.

The positioning of the fictitious I_F (per phase) in the phasor diagram (according to Equation 4.37) and that the stator phase current phasor I (in the first or second quadrant for generator and in the third or fourth quadrant for motor operation) are shown in Figure 4.14.

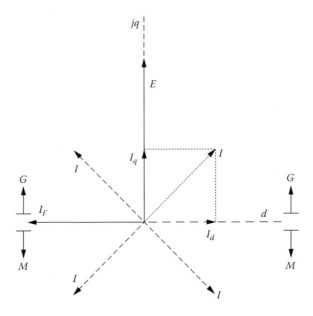

FIGURE 4.14 Generator—motor operation modes.

The generator—motor divide is determined solely by the electromagnetic (active) power:

$$P_{elm} = 3\,\mathrm{Re}\left(E \cdot I^*\right) > 0 \text{ generator, } < 0 \text{ motor} \tag{4.40}$$

The reactive power, Q_{elm} is

$$Q_{elm} = 3\,\mathrm{Imag}\left(E \cdot I^*\right) <> 0 \text{ (generator / motor)} \tag{4.41}$$

The *reactive power* may be either *positive* (delivered) or *negative* (drawn) both for *motor* and *generator* operation.

For reactive power "production," I_d should be opposite to I_F, that is, the longitudinal armature reaction air gap field will oppose the excitation air gap field. It is said that only with demagnetizing longitudinal armature reaction—machine overexcitation—the generator (motor) can "produce" reactive power. Therefore, for constant active power load, the reactive power "produced" by the synchronous machine may be increased by increasing the field current I_F. On the contrary, with underexcitation, the reactive power becomes negative; it is "absorbed." This extraordinary feature of the synchronous machine makes it suitable for voltage control, in power systems, through reactive power control via I_F control. On the other hand, the frequency ω_r, tied to speed, $\Omega_r = \omega_r/p_1$, is controlled through the prime mover governor, as discussed in Chapter 3. For constant frequency power output, speed has to be constant.

This is so because the two traveling fields—that of excitation and respectively that of armature windings—interact to produce constant (non-zero-average) electromagnetic torque only at standstill with each other.

This is expressed in Equation 4.40 by the condition that the frequency of $E_1 - \omega_r$—be equal to frequency of stator current $I_1 - \omega_1 = \omega_r$—to produce nonzero active power.

In fact, Equation 4.40 is valid only when $\omega_r = \omega_1$, but in essence the average instantaneous electromagnetic power is nonzero only in such conditions.

4.5 Armature Reaction Field and Synchronous Reactances

As during steady state, rotor (excitation) and stator (armature) produced magnetic field waves in the air gap are relatively at standstill, it follows that the stator currents do not induce voltages (currents) in the field coils on the rotor. The armature reaction—stator—field wave travels at rotor speed; the longitudinal I_{aA}, I_{aB}, I_{aC} and transverse I_{qA}, I_{qB}, I_{qC} armature currents (reaction) fields are fixed to the rotor, one along axis d and the other along axis q.

Thus, for these currents, the machine reacts with the magnetization reluctances of the air gap and of stator and rotor iron with no rotor-induced currents.

The trajectory of armature reaction d, q fields, and their distributions are shown in Figures 4.15 and 4.16.

The armature reaction mmfs F_{d1} and F_{q1} have a sinusoidal space distribution (only fundamental is considered), but their air gap flux densities does not have a sinusoidal space distribution. For constant air gap zones—such as it is that under the constant air gap salient pole rotors—the air gap flux density is sinusoidal. In the interpole zone of salient pole machine, the equivalent air gap is large and thus the flux density decreases quickly (Figures 4.15 and 4.16).

Only with the finite-element method (FEM) can the correct flux density distribution of armature (or excitation, or combined) mmfs be computed. For the time being, let us consider that, for the d axis mmf, the interpolar air gap is infinite and for the q axis mmf it is $g_q = 6g$. In axis q, the transverse armature mmf is maximum, and it is not practical to consider that the air gap in that zone is infinite as it would lead to large errors. Not so for d axis mmf, which is small toward axis q and thus the infinite air gap approximation is tolerable.

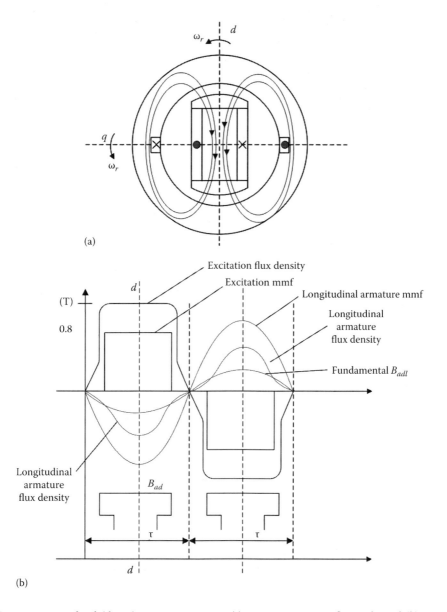

(a)

(b)

FIGURE 4.15 Longitudinal (*d* axis) armature reaction: (a) armature reaction flux paths and (b) air gap flux density and mmfs.

It should be noted that the *q*-axis armature reaction field is far from a sinusoid. This is so only for salient pole rotor SGs. For steady-state, we however operate only with fundamentals, and with respect to them we define the reactances, etc.

Therefore, we now extract the fundamentals of B_{ad} and B_{aq} to find the B_{ad1} and B_{aq1}:

$$B_{ad1} = \frac{1}{\tau} \int_0^\tau B_{ad}(x) \sin\left(\frac{\pi}{\tau} x_r\right) dx \qquad (4.42)$$

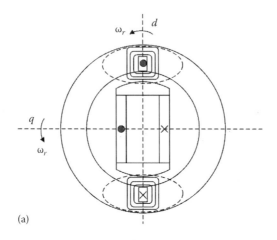

(a)

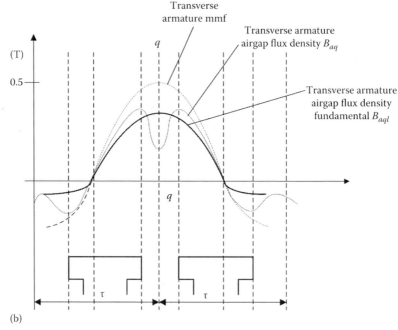

(b)

FIGURE 4.16 Transverse (*q* axis) armature reaction: (a) armature reaction flux paths and (b) air gap flux density and mmf.

with

$$B_{ad} = 0, \quad \text{for}: 0 < x_r < \frac{\tau - \tau_p}{2} \text{ and } \frac{\tau + \tau_p}{2} < x < \tau$$

$$B_{ad} = \frac{\mu_0 F_{dm} \sin\frac{\pi}{\tau} x_r}{K_c g (1 + K_{sd})} \quad \text{for} \quad \frac{\tau - \tau_p}{2} < x_r < \frac{\tau + \tau_p}{2}$$

(4.43)

Finally,

$$B_{ad1} = \frac{\mu_0 F_{dm} K_{d1}}{K_c g (1 + K_{sd})}; \quad K_{d1} \approx \frac{\tau_p}{\tau} + \frac{1}{\pi} \sin\frac{\tau_p}{\tau} \pi$$

(4.44)

In a similar way,

$$B_{aq} = \frac{\mu_0 F_{qm} \sin \frac{\pi}{\tau} x_r}{K_c g \left(1 + K_{sq}\right)}, \quad \text{for} : 0 \leq x_r \leq \frac{\tau_p}{2} \text{ and } \frac{\tau + \tau_p}{2} < x < \tau$$

(4.45)

$$B_{aq} = \frac{\mu_0 F_{qm} \sin \frac{\pi}{\tau} x_r}{K_c g_q \left(1 + K_{sq}\right)} \quad \text{for } \frac{\tau_p}{2} < x_r < \frac{\tau + \tau_p}{2}$$

$$B_{aq1} = \frac{\mu_0 F_{qm} K_{q1}}{K_c g};$$

(4.46)

$$K_{q1} = \frac{\tau_p}{\tau} - \frac{1}{\pi} \sin \frac{\tau_p}{\tau} \pi + \frac{2}{3\pi} \cos \left(\frac{\tau_p}{\tau} \frac{\pi}{2} \right)$$

Note that the integration variable was x_r referred to as rotor coordinates.

Equation 4.44 and 4.46 warrant the following remarks:

- The fundamental armature reaction flux density in axes d and q are proportional to the respective stator mmfs and are inversely proportional to air gap and magnetic saturation equivalent factors K_{sd} and K_{sq} ($K_{sd} \neq K_{sq}$ in general).
- B_{ad1} and B_{aq1} are also proportional to equivalent armature reaction coefficients K_{d1} and K_{q1}. Both being smaller than unity ($K_{d1} < 1, K_{q1} < 1$) account for air gap nonuniformity (slotting is considered only by Carter coefficient). Other than that B_{ad1} and B_{aq1} formulae are very similar to the air gap flux density fundamental B_{a1} in a uniform air gap machine with same stator, B_{a1}:

$$B_{a1} = \frac{\mu_0 F_1}{K_c g (1 + K_S)}; \quad F_1 = \frac{3\sqrt{2} W_1 K_{W1} I_1}{\pi p_1}$$

(4.47)

The cyclic magnetization inductance X_m of uniform air gap machine with a three-phase winding is straightforward as the self-emf in such a winding, E_{a1}, is [3]

$$E_{a1} = \omega_r W_1 K_{W1} \Phi_{a1}; \quad \Phi_{a1} = \frac{2}{\pi} B_{a1} \tau \cdot l_{stack}$$

(4.48)

From Equations 4.47 and 4.48, X_m is

$$X_m = \frac{E_{a1}}{I_1 \sqrt{2}} = \frac{6 \mu_0 \omega_r}{\pi^2} \frac{\left(W_1 K_{W1}\right)^2 \tau \cdot l_{stack}}{K_C g (1 + K_S) p_1}$$

(4.49)

It follows logically that the so-called cyclic magnetization reactances of synchronous machines X_{dm} and X_{qm} are proportional to their flux density fundamentals:

$$X_{dm} = X_m \frac{B_{ad1}}{B_{a1}} = X_m \cdot K_{d1}$$

(4.50)

$$X_{qm} = X_m \frac{B_{aq1}}{B_{a1}} = X_m K_{q1} \tag{4.51}$$

$K_{sd} = K_{sq} = K_s$ was implied.

The term "cyclic" comes from the fact that these reactances manifest themselves only with balanced stator currents and symmetric windings and only for steady state.

During steady state with balanced load, the stator currents manifest themselves by two distinct magnetization reactances, one for axis d and the other for axis q, acted upon by the d and q phase current components. We should add to these the leakage reactance typical to any winding, X_{1l}, to compose the so-called synchronous reactances of the synchronous machine (X_d and X_q):

$$X_d = X_{1\sigma} + X_{dm} \tag{4.52}$$

$$X_q = X_{1\sigma} + X_{qm} \tag{4.53}$$

The damper cage currents are zero during steady state with balanced load, as the armature reaction field fundamental components are at standstill with the rotor and have constant amplitudes (due to constant stator current amplitude).

We are thus ready to proceed with SG equations for steady state under balanced load.

4.6 Equations for Steady State with Balanced Load

We already introduced stator fictitious AC three-phase field currents $I_{F,A,B,C}$ to emulate the field winding motion-produced emfs in the stator phases $E_{A,B,C}$.

The decomposition of each stator phase currents $I_{qA,B,C}$, $I_{dA,B,Cl}$, which then produces the armature reaction field waves at standstill with respect to excitation field wave, has lead to the definition of cyclic synchronous reactances X_d and X_q. Consequently, as our fictitious machine is under steady state with zero rotor currents, the per-phase equations in complex (phasors) are simply as follows:

$$I_1 R_1 + V_1 = E_1 - jX_d I_d - jX_q I_q$$

$$E = -jX_{Fm} \times I_F; X_{Fm} = \omega_r M_{FA} \tag{4.54}$$

$$I_1 = I_d + I_q \tag{4.55}$$

RMS values all over in Equations 4.54 and 4.55.

To secure the correct phasing of currents, let us consider I_F along axis d (real). Then, according to Figure 4.13,

$$I_q = I_q \times \left(-j \frac{I_F}{I_F}\right); \quad I_d = I_d \frac{I_F}{I_F}; \quad I_1 = \sqrt{I_d^2 + I_q^2} \tag{4.56}$$

With $I_F > 0$, I_d is positive for underexcitation ($E_1 < V_1$) and negative for overexcitation ($E_1 > V_1$). Also, I_q in Equation 4.56 is positive for generating and negative for motoring.

The terminal phase voltage V_1 may represent the power system voltage or an independent load Z_L:

$$Z_L = \frac{V_1}{I_1} \tag{4.57}$$

A power system may be defined by an equivalent internal emf—E_{PS} and an interior impedance Z_{PS}:

$$V_1 = E_{PS} + Z_{PS}I_1 \tag{4.58}$$

For an infinite power system, $P_1 + jQ_1 = 3V_1I_1^* = 3E_1I_1^* - 3(I_1)^2 R_1 - 3jX_{1l}(I_1)^2 - 3j(X_{dm}I_d + X_{qm}I_q)I_1^*$ and E_{PS} is constant.

For a limited power system either only $|Z_{PS}| \neq 0$, or also E_{PS} varies in amplitude, phase, or frequency.

The power system impedance Z_{PS} includes the impedance of multiple generators in parallel, of transformers and of power transmission lines.

The power balance applied to Equation 4.54, after multiplication by $3I_1^*$, yields the following:

$$P_1 + jQ_1 = 3V_1I_1^* = 3E_1I_1^* - 3(I_1)^2 R_1 - 3jX_{1l}(I_1)^2 - 3j(X_{dm}I_d + X_{qm}I_q)I_1^* \tag{4.59}$$

The real part represents the active output power P_1 and the imaginary part is the reactive power, both positive if delivered by the SG:

$$P_1 = 3E_1I_q - 3I_1^2 R_1 + 3(X_{dm} - X_{qm})I_dI_q = 3V_1I_1 \cos\phi_1 \tag{4.60}$$

$$Q_1 = -3E_1I_d - 3I_1^2 X_{sl} - 3(X_{dm}I_d^2 + X_{qm}I_q^2) = 3V_1I_1 \sin\phi_1 \tag{4.61}$$

As can be seen from Equations 4.60 and 4.61, the active power is positive (generating) only with $I_q > 0$. Also with $X_{dm} \geq X_{qm}$, the anisotropy active power is positive (generating) only with positive I_d (magnetization armature reaction along axis d). But positive I_d in Equation 4.61 means definitely negative (absorbed) reactive power and the SG is underexcited.

In general, $X_{dm}/X_{qm} = (1.0–1.7)$ for most SGs with electromagnetic excitation. Consequently, the anisotropy electromagnetic power is notably smaller than the interaction electromagnetic power. In nonsalient pole machines $X_{dm} \approx (1.01–1.05)X_{qm}$ due to the presence of rotor slots in axis q that increase the equivalent air gap (K_C increases due to double slotting). Also, when the SG saturates (magnetically) the level of saturation under load may be in some regimes larger than in axis d. In other regimes when magnetic saturation is larger in axis d, nonsalient pole rotor may have a slight inverse magnetic saliency ($X_{dm} < X_{qm}$). As only the stator winding losses have been considered ($3R_1I_1^2$), the total electromagnetic power P_{elm} is

$$P_{elm} = 3E_1I_q + 3(X_{dm} - X_{qm})I_dI_q \tag{4.62}$$

Now the electromagnetic torque T_e is

$$T_e = \frac{P_{elm}}{(\omega_r/p_1)} = 3p_1\left[M_{FA}I_FI_q + (L_{dm} - L_{qm})I_dI_q\right] \tag{4.63}$$

where

$$L_{dm} = \frac{X_{dm}}{\omega_r}; \quad L_{qm} = \frac{X_{qm}}{\omega_r} \tag{4.64}$$

And, from Equation 4.37,

$$E_1 = \omega_r M_{FA} I_F \tag{4.65}$$

We may also separate in the stator phase flux linkage Ψ_1, the two components Ψ_d and Ψ_q:

$$\Psi_d = M_{FA} + L_d I_d$$
$$\Psi_q = L_q I_q \tag{4.66}$$

$$L_d = \frac{X_d}{\omega_r}; \quad L_q = \frac{X_q}{\omega_r} \tag{4.67}$$

The total stator phase flux linkage Ψ_1 is

$$\Psi_1 = \sqrt{\Psi_d^2 + \Psi_q^2}; \quad I_1 = \sqrt{I_d^2 + I_q^2} \tag{4.68}$$

As expected, from Equation 4.63, the electromagnetic torque does not depend on frequency (speed) ω_r, but only on field current and stator current components, besides the machine inductances: the mutual one, M_{FA}, and magnetization ones L_{dm} and L_{qm}. The currents I_F, I_d, I_q influence the level of magnetic saturation in stator and rotor cores and thus M_{FA}, L_{dm} and L_{qm} are functions of all of them.

Magnetic saturation is an involved phenomenon to be treated in Chapter 5.

The shaft torque T_a differs from electromagnetic torque T_e by the mechanical power loss (p_{mec}) braking torque:

$$T_a = T_e + \frac{p_{mec}}{(\omega_r / p_1)} \tag{4.69}$$

For generator operation mode T_e is positive and thus $T_a > T_e$. Still missing are the core losses located mainly in the stator.

4.7 Phasor Diagram

Equations 4.54, 4.55, and 4.66 through 4.68 lead to a new voltage equation:

$$I_1 R_1 + V_1 = -j\omega_r \Psi_1 = E_t; \quad \Psi_1 = \Psi_d + j\Psi_q \tag{4.70}$$

where E_t is total flux phase emf in the SG. Now two phasor diagrams, one suggested by Equation 4.54 and other by Equation 4.70 are presented in Figure 4.17a and b.

The time phase angle δ_V between the emf E_1 and the phase voltage V_1 is traditionally called the *internal (power) angle* of the SG.

As we have written Equations 4.54 and 4.70 for the generator association of signs, $\delta_V > 0$ for generating ($I_q > 0$) and $\delta_V < 0$ for motoring ($I_q < 0$).

For large SGs, even the stator resistance may be neglected for more clarity in the phasor diagrams but at the cost of "loosing" the copper loss consideration.

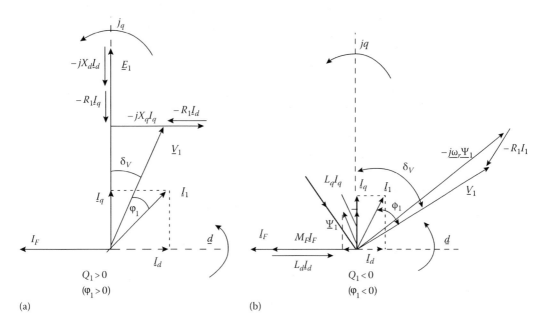

FIGURE 4.17 Phasor diagrams: (a) standard and (b) modified but equivalent.

4.8 Inclusion of Core Losses in the Steady-State Model

The core loss due to the fundamental component of magnetic field wave produced together by both excitation and armature mmf occurs only in the stator. This is so because the two fundamental field waves travel at rotor speed. We may consider, to a first approximation, that the core losses are related directly to the main (air gap) magnetic flux linkage Ψ_{1m}:

$$\Psi_{1m} = M_{FA}I_F + L_{dm}I_d + L_{qm}I_q = \Psi_{dm} + j\Psi_{qm} \tag{4.71}$$

$$\Psi_{dm} = M_{FA}I_F + L_{dm}I_d; \quad \Psi_{qm} = L_{qm}I_q \tag{4.72}$$

The leakage flux linkage components $L_{sl}I_d$ and $L_{sl}I_q$ do not produce significant core losses as $L_{sl}/L_{dm} < 0.15$ in general and most of the leakage flux lines flow within air zones (slot, end-windings, air gap).

Now, we may consider a three-phase stator fictitious short circuited resistive-only winding—R_{Fe}—which accounts for the core loss.

Neglecting the reaction field of core loss currents I_{Fe}, we have

$$-\frac{d\Psi_{1m0}}{dt} = R_{Fe}I_{Fe} = -j\omega_r\Psi_{1m0} \tag{4.73}$$

R_{Fe} is thus "connected" in parallel to the main flux emf $(-j\omega_r\Psi_{1m})$. The voltage equation then becomes thus:

$$I_{1t}\left(R_1 + jX_{sl}\right) + V_1 = -j\omega_1\Psi_{1m} \tag{4.74}$$

$$\text{with } I_{1t} = I_d + I_q + I_{Fe} = I_1 + I_{Fe} \tag{4.75}$$

The new phasor diagram of Equation 4.74 is shown in Figure 4.18:

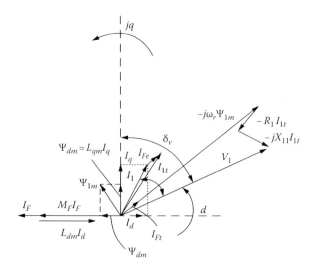

FIGURE 4.18 Phasor diagram with core loss included.

Though core losses are small in large SGs, and thus they do not change the phasor diagram notably, their inclusion allows for a correct calculation of efficiency (at least at low loads) and of stator currents as the power balance yields the following:

$$P_1 = 3V_1 I_{1t} \cos\phi_1 = 3\omega_r M_{FA} I_F I_q + 3\omega_r \left(L_{dm} - L_{qm} \right) I_d I_q - 3R_1 I_{1t}^2 - 3\frac{\omega_r^2 \Psi_{1m}^2}{R_{Fe}} \qquad (4.76)$$

$$\Psi_{1m} = M_{FA} I_F + L_{dm} I_d - j L_{qm} I_q \qquad (4.77)$$

$$I_{Fe} = \frac{-j\omega_r \Psi_{1m}}{R_{Fe}}; \quad I_{1t} = I_d + I_q + I_{Fe} \qquad (4.78)$$

Once the SG parameters R_1, R_{Fe}, L_{dm}, L_{qm}, M_{FA}, excitation current I_F, speed (frequency)—$\omega_r/p_1 = 2\pi n$ (rps)—are known, the phasor diagram in Figure 4.17 allows for the computation of I_d, I_q provided the power angle δ_v and the phase voltage V_1 are also given. After that, the active and reactive power delivered by the SG may be computed. Finally, the efficiency η_{SG} is as follows:

$$\eta_{SG} = \frac{P_1}{P_1 + p_{Fe} + p_{copper} + p_{mec} + p_{add}} = \frac{P_{elm} - p_{copper} - p_{Fe} - p_{add}}{P_{elm} + p_{mec}} \qquad (4.79)$$

with p_{add} is additional losses on load.

Alternatively, with I_F as parameter, I_d and I_q can be modified (given) such that $\sqrt{I_d^2 + I_q^2} = I_1$ be given as a fraction of full load current.

Note: While decades ago the phasor diagrams were used for graphical computation of performance, nowadays they are used only to illustrate performance and derive equations for a pertinent computer program to calculate the same performance faster and with increased precision.

Example 4.1

The following data are obtained from a salient pole rotor synchronous hydrogenerator: S_N = 72 MVA, V_{lline} = 13 kV/star connection, $2p_1$ = 90, f_1 = 50 Hz, q_1 = 3 slots/pole/phase, I_{1r} = 3000 A, R_1 = 0.0125 Ω, $(\eta_r)_{\cos \varphi1 = 1}$ = 0.9926, $p_{Fen} = p_{mecn}$. Additional data include stator interior diameter D_{is} = 13 m, stator active stack length l_{stack} = 1.4 m, constant air gap under the poles g = 0.020 m, Carter coefficient K_C = 1.15, τ_p/τ = 0.72. The equivalent unique saturation factor K_s = 0.2.

The number of turns in series per phase is $W_1 = p_1 q_1 \times 1$ turn/coil = 45 × 3 × 1 = 115 turns/phase. Let us calculate the following:

a. The stator winding factor K_{W1};
b. The d–q magnetization reactances X_{dm}, X_{qm};
c. With $X_{1l} = 0.2X_{dm}$, determine X_d, X_q;
d. Rated core and mechanical losses $P_{Fen} = p_{mecn}$;
e. x_d, x_q, r_1 in P.U. with $Z_n = V_{1ph}/I_{1r}$
f. Neglecting all losses at $\cos \varphi_1 = 1$ and $\delta_v = 30°$ calculate E_1, I_d, I_q, I_1, E_1, P_1, Q_1
g. The no load air gap flux density ($K_s = 0.2$) and the corresponding rotor-pole mmf $W_F I_F$

Solution:

1. The winding factor K_{W1} (Equation 4.10) is as follows:

$$K_{W1} = \frac{\sin \pi/6}{3\sin(\pi/6 \cdot 3)} \sin\left(\frac{1}{1} \cdot \frac{\pi}{2}\right) = 0.9598$$

Full-pitch coils are required ($y/\tau = 1$) as the single layer case is considered.
2. The expressions of X_{dm} and X_{qm} are shown in Equations 4.49 through 4.51:

$$X_{dm} = X_m \cdot K_{d1}$$

$$X_{qm} = X_m \cdot K_{q1}$$

From Equation 4.44,

$$K_{d1} = \frac{\tau_p}{\tau} + \frac{1}{\pi}\sin\frac{\tau_p}{\tau}\pi = 0.72 + \frac{1}{\pi}\sin 0.72 \cdot \pi = 0.96538$$

$$K_{q1} = \frac{\tau_p}{\tau} - \frac{1}{\pi}\sin\frac{\tau_p}{\tau}\pi + \frac{2}{3\pi}\cos\frac{\tau_p}{\tau}\frac{\pi}{2} = 0.4776 + 0.0904 = 0.565$$

$$X_m = \frac{6\mu_0}{\pi^2}\omega_r\frac{(W_1 K_{W1})^2 \cdot \tau \cdot l_{stack}}{K_C g(1+K_s)p_1}$$

with : $\tau = \pi D_{is}/2p = \pi \cdot 13/90 = 0.45355$ m

$$X_m = \frac{6 \cdot 4\pi \times 10^{-7} \cdot 2\pi \cdot 50 \cdot (115 \cdot 0.9598)^2 \times 0.45355 \times 1.4}{\pi^2 \times 1.15 \times 0.020(1+0.2) \times 45} = 1.4948$$

$$X_{dm} = 1.4948 \times 0.96538 = 1.443 \ \Omega$$

$$X_{qm} = 1.4948 \times 0.565 = 0.8445 \ \Omega$$

3. With $X_{1l} = 0.2 \times 1.4948 = 0.2989\,\Omega$, the synchronous reactances X_d and X_q are

$$X_d = X_{1l} + X_{dm} = 0.2989 + 1.443 \approx 1.742\ \Omega$$

$$X_q = X_{1l} + X_{qm} = 0.2989 + 0.8445 = 1.1434\ \Omega$$

4. As the rated efficiency at $\cos \varphi_1 = 1$ is $\eta_r = 0.9926$ and using Equation 4.79,

$$\sum p = p_{copper} + p_{Fen} + p_{mec} = S_n \left(\frac{1}{\eta_r} - 1 \right) = 72 \cdot 10^6 \left(\frac{1}{0.9926} - 1 \right) = 536 \cdot 772\ \text{kW}$$

The stator winding losses p_{copper} are

$$p_{copper} = 3 R_1 I_{1r}^2 = 3 \cdot 0.0125 \cdot 3000^2 = 337.500\ \text{kW}$$

Therefore,

$$p_{Fe} = p_{mec} = \frac{\sum p - p_{copper}}{2} = \frac{536.772 - 337.500}{2} = 99.636\ \text{kW}$$

5. The normalized impedance Z_n is

$$Z_n = \frac{V_{1ph}}{I_{1r}} = \frac{13 \cdot 10^3}{\sqrt{3} \cdot 3000} = 2.5048\ \Omega$$

$$x_d = \frac{X_d}{Z_n} = \frac{1.742}{2.5048} = 0.695$$

$$x_q = \frac{X_q}{Z_n} = \frac{1.1434}{2.5048} = 0.45648$$

$$r_1 = \frac{R_1}{Z_n} = \frac{0.0125}{2.5048} = 4.99 \times 10^{-3}$$

6. After neglecting all losses, the phasor diagram in Figure 4.16a, for $\cos \varphi_1 = 1$, can be shown.

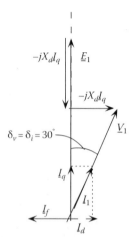

Phasor diagram for cos $\varphi_1 = 1$ and zero losses.

The phasor diagram uses phase quantities in RMS values.

From the adjacent phasor diagram,

$$I_q = \frac{V_1 \sin \delta_V}{X_q} = \frac{13{,}000}{\sqrt{3}} \frac{0.5}{1.1434} = 3286 \text{ A}$$

$$I_d = -I_q \tan 30^0 = -3286 \frac{1}{\sqrt{3}} = -1899.42 \text{ A}$$

$$I_1 = \sqrt{I_d^2 + I_q^2} = 3796 \text{ A}$$

And, the emf per phase E_1 is

$$E_1 = V_1 \cos \delta_V + X_d |I_d| = \frac{13{,}000}{\sqrt{3}} \cdot \frac{\sqrt{3}}{2} + 1.742 \cdot 1899 = 9.808 \text{ kV}$$

$$P_1 = 3 V_1 I_1 \cos \phi_1 = 3 \cdot \frac{13{,}000}{\sqrt{3}} \cdot 3796 = 85.372 \text{ MW}$$

$$Q_1 = 3 V_1 I_1 \sin \phi_1 = 0$$

It could be inferred that the rated power angle δ_{Vr} is smaller than 30° in this practical example.

7. We may use Equation 4.48 to calculate E_1 at no load as follows:

$$E_1 = \frac{\omega_r}{\sqrt{2}} W_1 K_{W1} \Phi_{pole1}$$

$$\Phi_{pole1} = \frac{2}{\pi} B_{g1} \cdot \tau \cdot l_{stack}$$

Then, from Equation 4.20, we obtain

$$B_{g1} = B_{gFM} K_{F1}; \quad K_{F1} = \frac{4}{\pi} \sin \frac{\tau_p}{\tau} \frac{\pi}{2}$$

Also, from Equation 4.78, we obtain

$$B_{gFM} = \frac{\mu_0 W_F I_F}{K_C g (1 + K_S)}$$

Thus, gradually,

$$\Phi_{pole1} = \frac{9808 \times \sqrt{2}}{2\pi 50 \times 115 \times 0.9596} = 0.3991 \text{ Wb}$$

$$B_{g1} = \frac{0.3991 \times 3.14}{2 \times 0.45355 \times 1.4} = 0.9868$$

$$B_{gFM} = \frac{0.9868}{\frac{4}{\pi}\sin 0.72 \cdot \frac{\pi}{2}} = 0.8561\,\text{T}$$

$$W_F I_F = \frac{0.8561 \times 1.15 (1+0.2) \times 2 \times 10^{-2}}{1.256 \times 10^{-6}} = 18{,}812\,\text{A turns/pole}$$

Note that the large air gap ($g = 2 \times 10^{-2}$ m) justifies the rather moderate saturation (iron reluctance) factor $K_S = 0.2$.

The field-winding losses have not been considered in the efficiency as they are covered from a separate power source.

4.9 Autonomous Operation of Synchronous Generators

Autonomous operation of SGs is required by numerous applications. Also, some SG characteristics in autonomous operation, obtained through special tests or by computation, may be used to characterize the SG rather comprehensively.

Typical characteristics at constant speed are as follows [4]

    a. No-load saturation curve $E_1(I_F)$
    b. Short-circuit saturation curve: $I_{1sc}(I_F)$ for $V_1 = 0$ and $\cos\varphi_1 = ct$
    c. Zero power factor saturation curve: $V_1(I_1)$; $I_F = ct \cos\varphi_1 = ct$

These curves may be computed or obtained from standard tests.

4.9.1 No-Load Saturation Curve: $E_1(I_F)$; $n = ct$, $I_1 = 0$

At zero load (stator) current, the excited machine is driven at the speed $n_1 = f_1/p_1$ by a smaller power rating motor. The stator no-load voltage, in fact the emf (per phase or line) E_1, and the field current are measured. The field current is monotonously raised from zero to a positive value I_{Fmax} corresponding to 120%–150% of rated voltage V_{1r} at rated frequency f_{1r} ($n_{1r} = f_{1r}/p_1$). The experimental arrangement is shown in Figure 4.19.

At zero field current, the remnant magnetization of rotor poles iron produces a small emf E_{1r} (2%–8% of V_{1r}), and thus the experiments start at point A or A'. The field current is then increased in small

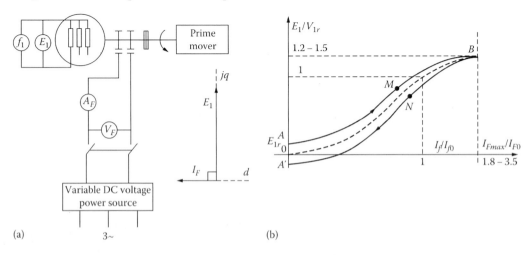

FIGURE 4.19 No-load saturation curve test: (a) the experimental arrangement and (b) the characteristic.

increments until the no-load voltage E_1 reaches 120%–150% of rated voltage (point B, along the trajectory AMB). Then the field current is decreased steadily to zero in very small steps and the characteristic evolves along BNA' trajectory. It may be that the starting point is A', and this is confirmed when I_F increases from zero and the emf decreases first and then increases. In the latter case, the characteristic travels along the way A'NBMA. The hysteresis phenomenon in the stator and the rotor cores is the cause for the difference between the rising and falling sides of the curve. The average curve represents the no-load saturation curve.

The increase in emf well above rated voltage is required to check the required field current for the lowest design power factor at full load (I_{Fmax}/I_{F0}). This ratio is, in general, $I_{Fmax}/I_{F0} = 1.8$–3.5. The lower the lowest power factor at full-load and rated voltage, the larger the ratio I_{Fmax}/I_{F0}. The ratio also varies with the air gap-to-pole pitch ratio (g/τ) and with the number of poles pairs p_1. It is important to know the corresponding I_{Fmax}/I_{F0} ratio for a proper thermal design of the SG.

The no-load saturation curve may also be computed: either analytically or through FEM. As FEM analysis will be dealt with later, here we dwell on the analytical approach. To do so, we draw two typical flux line pairs corresponding to the no-load operation of SG (Figure 4.20).

There are two basic analytical approaches of practical interest. Let us call them here the flux-line method and, respectively, the multiple magnetic circuit method.

The rather simplified flux-line method considers the Ampere's law along a basic flux line and applies the flux conservation in the rotor yoke, rotor pole body and rotor pole shoe and, respectively, in the stator teeth and yoke.

The magnetic saturation in these regions is considered through a unique (average) flux density and also an average flux-line length. It is thus an approximate method as the level of magnetic saturation varies tangentially along rotor-pole body, shoe, in the salient rotor pole and, respectively, in the rotor teeth of the nonsalient pole.

The leakage flux lost between the salient rotor pole bodies and their shoes is also very approximately considered.

However, if a certain average air gap flux density value B_{gFm} is assigned for start, the rotor pole mmf $W_F I_F$ required to produce it, accounting for magnetic saturation, though approximately, may be computed without any iteration. If the air gap under the rotor salient poles increases from center to pole ends

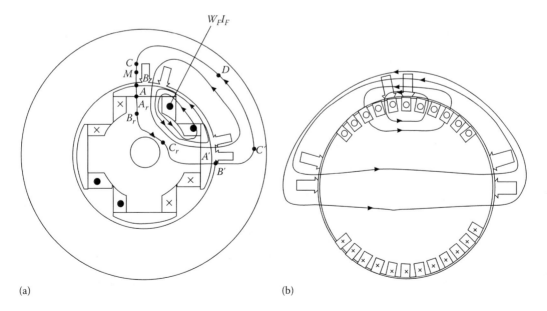

(a) (b)

FIGURE 4.20 Flux lines at no load: (a) the salient pole rotor and (b) the nonsalient pole rotor.

(to produce a more sinusoidal air gap flux density), an average value again is to be considered to simplify the computation. Once the $B_{gFm}(I_F)$ curve is calculated, the $E_1(I_F)$ curve is straightforward (based on Equation 4.30):

$$E_1(I_F) = \frac{\omega_r}{\sqrt{2}} \times \frac{2}{\pi} \tau B_{gFm}\left(I_f\right) K_{F1} \cdot l_{stack} \cdot W_1 K_{W1} [V(\text{RMS})] \tag{4.80}$$

The analytical flux-line method is illustrated here through a case study (Example 4.2).

Example 4.2

A three-phase salient pole rotor SG with $S_n = 50$ MVA, $V_l = 10{,}500$ V, $n_1 = 428$ rpm, $f_1 = 50$ Hz has the following geometrical data: internal stator diameter $D_r = 3.85$ m, $2p_1 = 14$ poles, $l_{stack} \approx 1.39$ m, pole pitch $\tau = \pi D_r/2p_1 = 0.864$ m, air gap g (constant) = 0.021 m, $q_1 = 6$ slots/pole/phase, open stator slots with $h_s = 0.130$ m (total slot height, with 0.006 m reserved for the wedge); $W_s = 0.020$ m (slot width), stator yoke $h_{ys} = 0.24$ m, and rotor geometry is as in Figure 4.21.

Let us consider only the rated flux density condition, with $B_{gFm1} = 0.850$ T. The stator lamination magnetization curve is given in Table 4.1.

The Ampere's law along the contour ABCDC′B′A′ relates the mmf drop from rotor-to-rotor pole $F_{AA'}$:

$$F_{AA'} = 2\left(H_{gFm1}gK_c + \frac{1}{6}\left(H_B + 4H_M + H_C\right)h_{st} + H_{YS}l_{YS} \right) \tag{4.81}$$

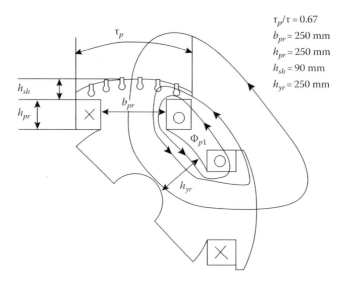

$\tau_p/\tau = 0.67$
$b_{pr} = 250$ mm
$h_{pr} = 250$ mm
$h_{sh} = 90$ mm
$h_{yr} = 250$ mm

FIGURE 4.21   Rotor geometry and rotor pole leakage flux Φ_{pl}.

TABLE 4.1   The Magnetization Curve $B(H)$ for the Iron Cores

B (T)	0.1	0.2	0.3	0.4	0.5	0.6	0.7	0.8	0.9		
H (A/m)	35	49	65	76	90	106	124	148	177		
B (T)	1	1.1	1.2	1.3	1.4	1.5	1.6	1.7	1.8	1.9	2.0
H (A/m)	220	273	356	482	760	1340	2460	4800	8240	10,200	34,000

The air gap magnetic field H_{gFm1} is

$$H_{gFm1} = \frac{B_{gFm1}}{\mu_0} = \frac{0.85}{1.256 \times 10^{-6}} = 0.676 \times 10^6 \text{ A/m} \tag{4.82}$$

The magnetic fields at the stator tooth top, middle, and bottom (H_B, H_M, H_C) are related to Figure 4.22 which shows that the stator tooth is trapezoidal as the slot is rectangular.

The flux density in the three tooth cross-sections is as follows:

$$B_B = B_{gFm} \cdot \frac{\tau_s}{\tau_s - W_s}; \quad W_S = 0.02 \text{ m}; \quad \tau_S = \frac{\tau}{qm} = \frac{0.8635}{6 \cdot 3} = 0.048 \text{ m}$$

$$B_M = B_B \cdot \frac{\tau - W_S}{W_{tm}}; \quad W_{tm} = \frac{\pi(D_{is} + h_{st})}{2pqm} - W_S$$

$$B_C = B_B \cdot \frac{(\tau_s - W_S)}{W_{tB}}; \quad W_{tB} = \frac{\pi(D_{is} + 2h_{st})}{2pqm} - W_S$$

Finally,

$$B_B = 0.85 \frac{48}{48 \cdot 20} = 1.457 \text{ T}$$

$$W_{tm} = \frac{\pi(3.85 + 0.130)}{14 \times 6 \times 3} - 0.02 = 0.0296 \text{ m}$$

$$B_M = 1.457 \cdot \frac{48 - 20}{29.6} = 1.378 \text{ T}$$

$$W_{tB} = \frac{\pi(3.85 + 2 \times 0.130)}{14 \times 6 \times 3} - 0.02 = 0.0312 \text{ m}$$

$$B_C = 1.457 \cdot \frac{(48 - 20)}{31.2} = 1.307 \text{ T}$$

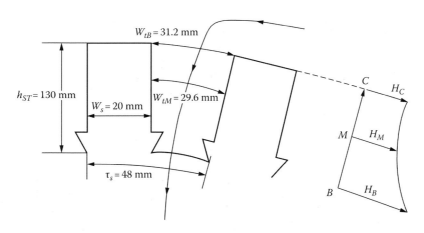

FIGURE 4.22 Stator slot geometry and the no-load magnetic field.

From the magnetization curve (Table 4.1) through linear interpolation,

$$H_B = 1090.6 \text{ A/m}$$

$$H_M = 698.84 \text{ A/m}$$

$$H_C = 501.46 \text{ A/m}$$

The maximum flux density in stator yoke B_{YS} is as follows:

$$B_{ys} = \frac{\tau}{\pi} \cdot \frac{B_{gFm1}}{h_{ys}} = \frac{0.8635}{\pi} \cdot \frac{0.85}{0.24} = 0.974 \text{ T}$$

From Table 4.1,

$$H_{ys} = 208.82 \text{ A/m}$$

Now the average length of the flux line in the stator yoke "reduced" to the peak yoke flux density B_{ys} is approximately

$$l_{ys} \approx \frac{\pi \left(D_{is} + 2h_{st} + h_{ys} \right)}{4p} \cdot K_{ys}; \quad 0.5 < K_{ys} < 1$$

The value of K_{ys} depends on the level of saturation, etc. FEM digital simulations may be used to find the value of the fudge factor K_{ys}. A reasonable value would be $K_{ys} \approx 2/3$.
Therefore,

$$l_{ys} = \frac{\pi \left(3.85 + 2.013 + 0.24 \right)}{4 \times 7} \times \frac{2}{3} = 0.3252 \text{ m}$$

We may now calculate $F_{AA''}$ from Equation 4.82:

$$F_{AA'} = 2 \left(0.676 \cdot 10^6 \times 2 \cdot 10^{-2} + \frac{1}{6} \left(1090.6 + 4 \times 698.84 + 501.46 \right) \right) \cdot 0.130$$

$$+ 208.82 \cdot 0.3252 = 27365.93 \text{ A turns}$$

Now the leakage flux Φ_{pl} in the rotor—between rotor poles (Figure 4.20)—is proportional to $F_{AA''}$. Alternatively, Φ_{pl} may be considered as a fraction of pole flux Φ_p:

$$\Phi_p = \frac{2}{\pi} B_{gFm1} \cdot \tau \cdot l_{stack} = \frac{2}{\pi} 0.85 \cdot 0.8635 \times 0.39 = 0.649 \text{ Wb}$$

$$\Phi_{pl} = K_{sl} \Phi_{pl}; \quad K_{sl} \approx 0.15 - 0.25$$

$$\Phi_{pl} = 0.15 \cdot 0.649 = 0.09747 \text{ T}$$

Therefore, the total flux in the rotor pole Φ_{pr} is

$$\Phi_{pr} = \Phi_p + \Phi_{pl} = 0.649 + 0.09747 = 0.7465 \text{ Wb}$$

The rotor pole shoe is not saturated in the no load despite of the presence of rotor damper bars, but the pole body and rotor yoke may be rather saturated. The mmf required to magnetize the rotor F_{rotor} is as follows:

$$F_{rotor} = \left(F_{ABr} + F_{BrCr}\right) \times 2 = 2\left(H_{pr} \cdot h_{pr} + H_{yr} \cdot l_{yr}\right) \tag{4.83}$$

With the rotor pole body width $b_{pr} = 0.25$ m, the flux density in the pole body B_{pr} is as follows:

$$B_{pr} \approx \frac{\Phi_{pr}}{l_{stack} \cdot b_{pr}} = \frac{0.7465}{1.39 \times 0.25} = 2.148 \text{ T!!}$$

This very large flux density level does not occur along the entire rotor height h_{pr}. At the top of pole body, approximately, $\Phi_{pr} \approx \Phi_p = 0.649$ Wb.
Therefore,

$$\left(B_{pr}\right)_{Ar} \approx \frac{\Phi_p}{l_{stack} b_{pr}} = \frac{0.649}{1.39 \times 0.25} = 1.8676 \text{ T!!}$$

Let us consider an average $B_{prav} = \dfrac{2.148 + 1.8676}{2} = 2 \text{ T!!}$. For this value, Table 4.1 gives $H_{pr} = 34{,}000$ A/m.
In the rotor yoke B_{yr} is as follows:

$$B_{yr} = \frac{\Phi_{pr}}{2h_{yr} \cdot l_{stack}} = \frac{0.7465}{2 \times 0.25 \times 1.39} = 1.074 \text{ T!!}$$

Therefore, $H_{yr} = 257$ A/m.
The average length of field path in the rotor yoke l_{yr} is

$$l_{yr} \approx \frac{\pi\left(D_{is} - 2g - 2\left(h_{sh} + h_{rp}\right) - h_{yr}\right)}{4p_1}$$

$$= \frac{\pi\left(3.85 - 2 \cdot 0.02 - 2\left(0.09 + 0.25\right) - 0.25\right)}{4 \cdot 7} = 0.32185 \text{ m}$$

Thus, from Equation 4.83, F_{rotor} is

$$F_{rotor} = 2\left[34{,}000 \cdot \left(0.09 + 0.25\right) + 257 \cdot 0.32185\right] = 23{,}285 \text{ A turns}$$

Now the total mmf per two neighboring poles (corresponding to a complete flux line) $2W_F I_F$ is

$$2 \times W_F I_F = F_{AA'} + F_{rotor} = 27{,}365 + 23{,}285 = 50{,}650 \text{ A turns}$$

The air gap mmf requirements are as follows:

$$F_g = 2H_{gFm1} \cdot g = 2 \times 0.676 \times 10^6 + 2 \times 10^{-2} = 27{,}040 \text{ A turns}$$

Therefore, the contribution of the iron in the mmf requirement, defined as a saturation factor K_s, is

$$1 + K_S = \frac{2W_F I_F}{F_g} = \frac{50,650}{27,040} = 1.8731$$

Therefore, $K_S = 0.8731$.

For the case in point, the main contribution is placed in the rotor pole. This is natural as the pole body width b_{pr} has to leave room to place the field windings. Therefore, in general, b_{pr} is around $\tau/3$, at most. The above routine illustrates the computational procedure for one point of the no load magnetization curve $B_{gFm1}(I_F)$. Other points may be calculated is a similar way. A more precise solution, at the price of larger computation time, may be obtained through the multiple magnetic circuit method [5], but real precision results require FEM as shown in Chapter 5.

4.9.2 Short-Circuit Saturation Curve $I_1 = f(I_F)$; $V_1 = 0$, $n_1 = n_r = ct$

The short-circuit saturation curve is obtained by driving the excited SG at rated speed n_r with short-circuited stator terminals. The field DC current I_F is varied downward gradually and both I_F and stator current I_{sc} are measured. In general, measurements for 100%, 75%, 50%, and 25% of rated current is necessary to reduce the winding temperature during that test. The results are plotted on Figure 4.23b.

From the voltage equation (Equation 4.54) with $V_1 = 0$ and $I_1 = I_{3sc}$, one obtains

$$E_1(I_f) = R_1 I_{3sc} + jX_d I_{dsc} + jX_q I_{qsc} \tag{4.84}$$

$$E_1(I_f) = -jX_F I_F \tag{4.85}$$

Neglecting stator resistance and observing that, with zero losses, $I_{sc3} = I_{dsc}$, as $I_{qsc} = 0$ (zero torque), we obtain

$$E_1(I_F) = jX_d I_{3sc} \tag{4.86}$$

The magnetic circuit is characterized by very low flux density. This is so because the armature reaction strongly reduces the resultant emf E_{1res} to

$$E_{1res} = -jX_{1l} \cdot I_{3sc} = E_1 - jX_{dm} I_{3sc} \tag{4.87}$$

which represents a low value on the no-load saturation curve, corresponding to an equivalent small field current:

$$I_{F0} = I_F - I_{3sc} \cdot \frac{X_{dm}}{X_{FA}} = OA - AC \tag{4.88}$$

Adding the no-load saturation curve, the short-circuit triangle may be portrayed (Figure 4.24). Its sides are all quasiproportional to the short-circuit current.

In general, by making use of the no-load and short-circuit saturation curves, saturated values of d axis synchronous reactance may be obtained:

$$X_{ds} = \frac{E_1(I_F)}{I_{sc3}(I_F)} = \frac{\overline{AA''}}{\overline{AA'}} \tag{4.89}$$

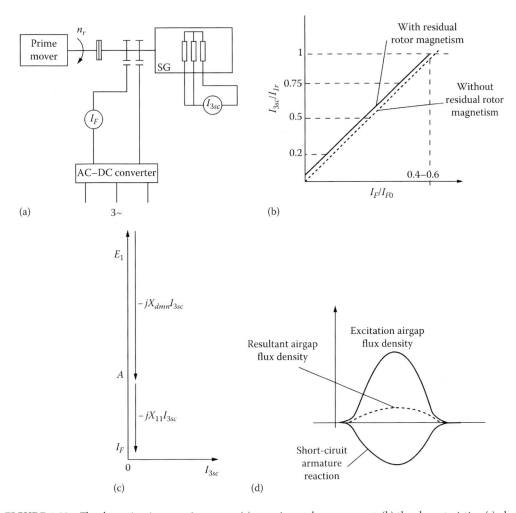

FIGURE 4.23 The short-circuit saturation curve: (a) experimental arrangement, (b) the characteristics, (c) phasor diagram with $R_1 = 0$, and (d) air gap flux density (slotting neglected).

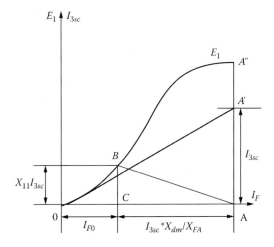

FIGURE 4.24 The short-circuit triangle.

Under load, the magnetization state differs from that of no-load situation and the value of X_{ds} from Equation 4.89 is of somewhat limited practical utilization.

4.9.3 Zero Power Factor Saturation Curve $V_1(I_F)$; I_1 = ct, cos φ_1 = 0, $n_1 = n_r$

Under zero power factor and zero losses, the voltage equation of Equation 4.54 becomes

$$V_1 = E_1 - jX_dI_d; \quad I_1 = I_d; \quad I_q = 0 \tag{4.90}$$

Again for pure reactive load and zero losses, the electromagnetic torque is zero; and so is I_q.

An underexcited synchronous machine acting as a motor on no load is, in general, used to represent the reactive load for the SG under zero power factor operation with constant stator current I_d.

The field current of the SG is reduced simultaneously with the increase in field current of the underexcited no-load synchronous motor (SM), to keep the stator current I_d constant (at rated value), while the terminal voltage decreases. This way $V_1(I_F)$ for constant I_d is obtained (Figure 4.25).

The abscissa of the short-circuit triangle OCA is moved at the level of rated voltage, then a parallel $0B'$ to $0B$ is drawn that intersects the no-load curve at B'. The vertical segment $0B'$ is defined as follows:

$$X_pI_1 = \overline{B'C} \tag{4.91}$$

Though we started with the short-circuit triangle in our geometrical construction $BC < B'C$ because magnetic saturation conditions are different. Therefore, in fact, X_p (Potier's reactance)> X_{1l}, in general, especially in salient pole rotor SG.

The main practical purpose of zero power factor saturation curve today would be to determine the leakage reactance and for temperature tests. One way to reduce the value of X_p and thus fall closer to X_{1l} is to raise the terminal voltage above rated one in the $V_1(I_F)$ curve and thus obtain the triangle ACB″ whose CB″ $\approx X_{1l}I_1$. It is claimed, however, that (115%–120%) voltage is required, which might not be allowed by some manufacturers. Alternative methods to measure the stator leakage reactance X_{1l} are to be presented in Chapter 8. The zero-power factor load testing may be used for temperature on load estimation without requiring active power full load.

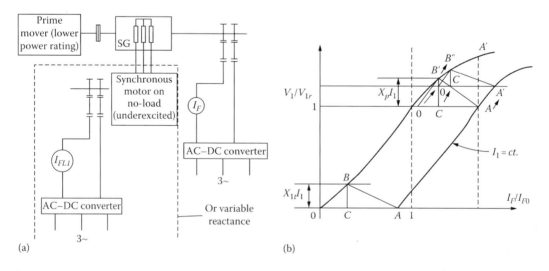

(a)   (b)

FIGURE 4.25 The zero-power factor saturation curve: (a) the experimental arrangement and (b) the "extraction" of Potier reactance X_p.

4.9.4 V_1–I_1 Characteristic, $I_F = ct$, $\cos \varphi_1 = ct$, $n_1 = n_r$

The V_1–I_1 characteristic refers to terminal voltage versus load current I_1, for balanced load at constant field current, load power factor and speed.

Obtaining the V_1–I_1 curve needs full real load, therefore in fact it is feasible either only on small and medium power autonomous SGs at manufacturer's site or the testing may be performed after the commissioning at user's site.

The voltage equation, phasor diagram, and the no-load saturation curve should provide information such that, with magnetic saturation coarsely accounted for, to calculate $V_1(I_1)$ for given load impedance per phase $Z_L(Z_L, \varphi_1)$ (Figure 4.26).

$$I_1 R_1 + Z_L I_1 = E_1 - jX_d I_d - jX_q I_q$$

$$I_d(R_1 + R_L) + I_q(R_1 + R_L) + j(X_d + X_L)I_d + j(X_q + X_L)I_q = E_1 \tag{4.92}$$

$$\cos\phi_1 = \frac{R_s}{Z_s}$$

As Figure 4.26b suggests, Equation 4.92 may be divided into two equations:

$$E_1 = I_q(R_1 + R_L) + (X_d + X_L)(I_d); \quad I_d \lessgtr 0$$

$$0 = (R_1 + R_L)(-I_d) - (X_q + X_L)I_q; \quad I_q > 0 \tag{4.93}$$

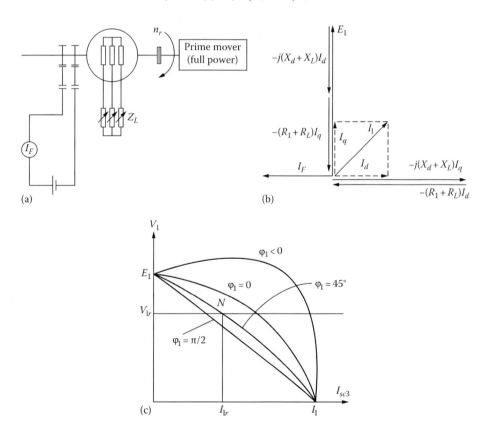

FIGURE 4.26 V_1–I_1 curve: (a) the experimental arrangement and (b) the phasor diagram for load Z_L.

With I_F given, E_1 is extracted from the no-load saturation curve. Then, with cos φ_1 given, we may choose to modify R_L (load resistance) only as X_L is

$$X_L = R_L \tan \varphi_1 \tag{4.94}$$

Then, Equation 4.93 can be simply solved to calculate I_d and I_q. The phase current I_1 is

$$I_1 = \sqrt{I_d^2 + I_q^2} \tag{4.95}$$

Finally, the corresponding terminal voltage V_1 is

$$V_1 = \frac{R_L}{\cos \varphi_1} \cdot I_1 \tag{4.96}$$

Typical $V_1(I_1)$ curves are shown in Figure 4.26c. The voltage decreases with load (I_1) for resistive ($\varphi_1 = 0$) and resistive—inductive ($\varphi_1 > 0$) load, and it increases and then decreases for resistive—capacitive load ($\varphi_1 < 0$).

Such characteristics may be used to calculate the voltage regulation ΔV_1:

$$\left(\Delta V_1\right)_{I_1/\cos\phi_1/I_F} = \frac{E_1 - V_1}{E_1} = \frac{\text{no-load voltage load voltage}}{\text{no-load voltage}} \tag{4.97}$$

Autonomous SGs are designed to provide operation at rated load current and rated voltage and a minimum (lagging) power factor cos $\varphi_{1\min} = 0.6$–0.8 (point N on Figure 4.25c). It should be evident that I_{1r} should be notably smaller than I_{3sc}.

Consequently,

$$\frac{X_{dsat}}{Z_n} = x_{dsat} < 1; \quad Z_n = \frac{V_N}{I_N} \tag{4.98}$$

The air gap in SGs for autonomous operation has to be large to secure such a condition. Consequently, notable field current mmf is required. And thus the power loss in the field winding increases. This is one reason to consider permanent magnet rotor SGs for autonomous operation for low-medium power units; though full power electronics is needed.

Note that for calculations with errors below 1%–2%, when using Equation 4.93, careful consideration of magnetic saturation level that depends simultaneously on I_F, I_d, I_q must be observed. This subject will be treated in more detail in Chapter 5.

4.10 SG Operation at Power Grid (in Parallel)

SGs in parallel constitute the basis of a regional, national, or continental electric power system (grid). SGs have to connected to the power grid one by one.

For the time being, we suppose that the power grid is of infinite power, that is, of fixed-voltage amplitude, frequency, and phase.

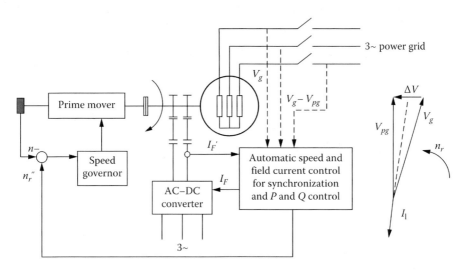

FIGURE 4.27 SG connection to the power grid.

In order to connect the SGs to the power grid without large current and power transients, the amplitude, frequency, sequence, and phase of the SG no-load voltages have to coincide with the same parameters of the power grid. As the power switch does not react instantaneously, some transients will always occur. However, they have to be limited. Automatic synchronization of the SG to the power grid is today performed through coordinated speed (frequency and phase) and field current control (Figure 4.27).

The active power transients during the connection to the power grid may be positive (generating) or negative (motoring) (see Figure 4.26).

4.10.1 Power/Angle Characteristic: $P_e (\delta_V)$

The power (internal) angle δ_V is the angle between the terminal voltage V_1 and the field current produced emf E_1. This angle may be calculated both for the autonomous and for the power-grid-connected generator. Traditionally, the power/angle characteristic is calculated and widely used for power-grid-connected generators mainly because of stability computation opportunities. For a large power grid, the voltage phasors in the phasor diagram are fixed in amplitude and phase. For more clarity, we neglect here the losses in the SG. We repeat here the phasor diagram in Figure 4.17a, but with $R_1 = 0$ (Figure 4.28).

The active and reactive powers P_1, Q_1 from Equations 4.60 and 4.61 with $R_1 = 0$ become

$$P_1 = 3E_1 I_q + 3(X_{dm} - X_q) I_d I_q \tag{4.99}$$

$$Q_1 = -3E_1 I_d - 3X_d I_d^2 - 3X_q I_q^2 \tag{4.100}$$

From Figure 4.28,

$$I_d = \frac{V_1 \cos \delta_V - E_1}{X_d}; \quad I_q = \frac{V_1 \sin \delta_V}{X_q} \tag{4.101}$$

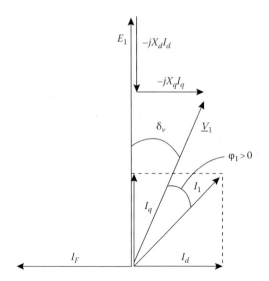

FIGURE 4.28 SG phasor diagram (zero losses).

With Equation 4.101, Equations 4.99 and 4.100 become

$$P_1 = \frac{3E_1 V_1 \sin \delta_V}{X_d} + \frac{3}{2} V_1^2 \left(\frac{1}{X_q} - \frac{1}{X_d} \right) \sin 2\delta_V \qquad (4.102)$$

$$Q_1 = \frac{3E_1 V_1 \cos \delta_V}{X_d} - 3V_1^2 \left(\frac{\cos^2 \delta_V}{X_d} + \frac{\sin^2 \delta_V}{X_q} \right) \qquad (4.103)$$

Unity power factor is obtained with $Q_1 = 0$, that is,

$$(E_1)_{Q_1=0} = V_1 \left(\cos \delta_V + \frac{X_d}{X_q} \frac{\sin^2 \delta_V}{\cos^2 \delta_V} \right) \qquad (4.104)$$

For same power angle δ_V and V_1, E_1 should be larger for the salient pole rotor SG as $X_d > X_q$. The active power has two components: one due to the interaction of stator and rotor fields and the second one due the rotor magnetic saliency ($X_d > X_q$).

As in standard salient pole rotor SGs $X_d/X_q < 1.7$, the second term in P_e, called here saliency active power, is relatively small unless the SG is severely under excited: $E_1 \ll V_1$. For given E_1, V_1, the SG reactive and active power delivery depend on the power (internal) angle δ_V (Figure 4.29).

The graphs in Figure 4.29 warrant the following remarks:

- The generating and motoring modes are characterized (for zero losses) by positive and, respectively, negative power angles
- As δ_V increases up to the critical value δ_{VK}, which corresponds to maximum active power delivery P_{1K}, the reactive power goes from leading to lagging for given emf E_1, V_1 frequency (speed) ω_1.
- The reactive power is independent of the sign of power angle δ_V.

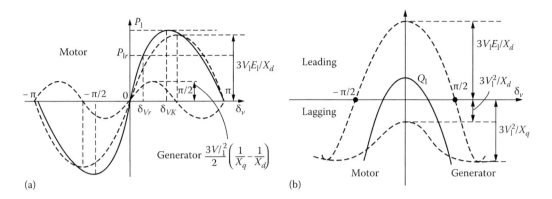

FIGURE 4.29 Active P_1 (a) and reactive Q_1 (b) powers versus power angle.

- In salient pole rotor SGs, the maximum power P_{1K} for given V_1, E_1, and speed, is obtained for a power angle $\delta_{VK} < 90°$, while for nonsalient pole rotor SGs ($X_d = (1 - 1.05)\, X_q$), $\delta_{VK} \approx 90°$.
- The rated power angle δ_{Vr} is chosen in general around 22°–30° for nonsalient pole rotor SGs and around 30°–40° for salient pole rotor SGs. The lower speed, higher relative inertia, stronger damper cage of the latter might secure better stability, which justifies the lower power reserve (or ratio P_{1K}/P_{1r}).

4.10.2 V-Shaped Curves: $I_1(I_F)$, P_1 = ct, V_1 = ct, n = ct

The V curves represent a family of $I_1(I_F)$ curves, drawn at constant V_1, speed (ω_1), with active power P_1 as parameter.

The computation of V shape curve is rather straightforward once $E_1(I_F)$—no-load saturation curve—and X_d and X_q are known. Unfortunately, when I_F varies from low to large values so does I_1 (i.e., I_d and I_q); local magnetic saturation varies, despite of the fact that basically the total flux linkage $\Psi_s \approx V_1/\omega_1$ stays constant. This is due to rotor magnetic saliency ($X_d \neq X_q$) where local saturation conditions vary rather notably. However, to a first approximation, for constant V_1 and ω_1 (i.e., Ψ_1), with E_1 calculated at a first fixed total flux, the value of M_{FA} stays constant and thus $E_1 \approx M_{FA} \cdot I_F \approx C_{FA} \cdot I_F$; $C_{FA} \approx (V_{1r}/(\omega_r I_{F0}))$. I_{F0} is the field current value that produces $E_1 = V_{1r}$ at no load.

For given I_F, $E_1 = C_{FA} \times I_F$ and P_1 assigned a value from Equation 4.102, we may compute δ_V, then from Equation 4.105, the corresponding stator current I_1 is found:

$$I_1 = \sqrt{I_d^2 + I_q^2} = \sqrt{\left(\frac{-E_1 + V_1 \cos\delta_V}{X_d}\right)^2 + \left(\frac{V_1 \sin\delta_V}{X_q}\right)^2} \tag{4.105}$$

As expected, for given active delivered power, the minimum value of stator current is obtained for a field current I_F corresponding to unity power factor ($Q_1 = 0$). That is, $(E_1)_{I_{1min}} = (E_1)_{Q1 = 0}$ may be determined from Equation 4.105 with δ_V already known from Equation 4.102. Then, $I_{FK} = E_1/C_{FA}$. The maximum power angle admitted for a given power P_1 limits the lowest field current admissible for steady state.

Finally, graphs as in Figure 4.30 are obtained.

Knowing the field current lower limit, for given active power, is paramount in avoiding an increase on the power angle above δ_{VK}. In fact δ_{VK} decreases with P_1 raising.

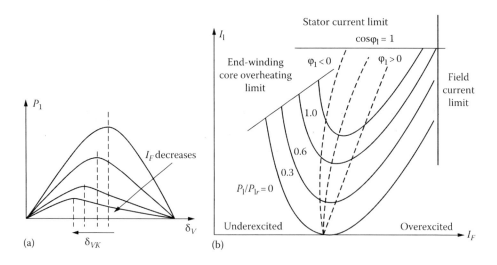

FIGURE 4.30   V shape curves: (a) P_1/δ_V assisting curves with I_F as parameter and (b) the $I_1(I_F)$ curves for constant P_1.

4.10.3 Reactive Power Capability Curves

The maximum limitation of I_F is due to thermal reasons. However, the SG heating depends on both I_1 and I_F as both winding losses are very important. In addition, I_1, I_F, and δ_V determine the core losses in the machine at given speed.

When reactive power request is increased, the increase in I_F raises the field winding losses and thus the stator winding losses, that is the active power P_1, have to be limited.

The rationale for V-shape curves [6,7] may be continued to find the reactive power Q_1 for the given P_1 and I_F. As shown in Figure 4.29, there are three distinct thermal limits: I_F limit (vertical), I_1 limit (horizontal), and the end-winding overheating (inclined) limit at low values of field current.

To explain this latest, rather obscure, limitation, please refer to Figure 4.31.

For the underexcited SG the field current and armature-current-produced fields are having an angle smaller than 90° (the angle between I_F and I_1 in the phasor diagram). Consequently, their end windings fields more or less add to each other. This resultant end-region field enters at 90° the end-region stator laminations and produces severe eddy current losses, which limit thermally the SG reactive power

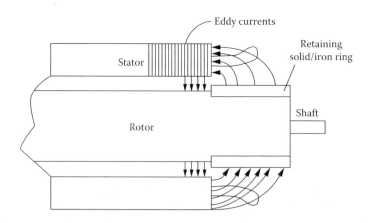

FIGURE 4.31   End-region field path for the underexcited SG.

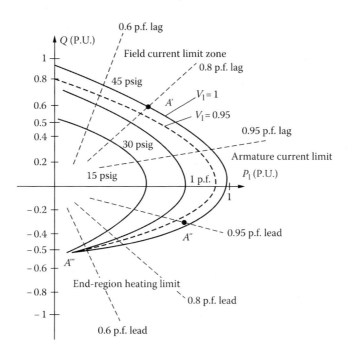

FIGURE 4.32 Reactive power capability curves for a hydrogen cooled SG.

absorption ($Q_1 < 0$). This phenomenon is so strong because the retaining ring solid iron eddy currents (produced solely by the stator end-windings currents) are small and thus incapable of attentuating severely the end-region resultant field. This is because the solid iron retaining ring is not saturated magnetically as the field current is small. When the SG is overexcited, this phenomenon is not important because the stator and rotor fields are opposite (I_F and I_1 phase angle shift is above 90°), and the retaining magnetic ring is saturated by the large field current. Consequently, the stator end windings-current-produced field in the stator penetrates first deeply into the retaining rings producing large eddy currents that further attenuate this resultant field in the end-region zone (the known short-circuit transformer effect on inductance).

Finally, the $Q_1(P_1)$ curves are shown in Figure 4.32.

The reduction of hydrogen pressure leads to a reduction of reactive and active power capability of the machine.

As expected, the machine reactive power absorption capability ($Q_1 < 0$) is notably smaller than reactive power delivery capability. Both, the end-region lamination loss limitation and the rise of power angle closer to its maximum limitation, seem responsible for such an asymmetric behavior (Figure 4.32).

4.10.4 Defining Static and Dynamic Stability of SGs

The fact that SGs require constant speed to deliver electric power at constant frequency introduces special restrictions and precautionary measures to preserve SG stability, when tied to an electric power system (grid). The problem of stability is very complex. To preserve and extend it, active speed and voltage (active and reactive power) close loop controls are provided. We will deal in some detail with stability and control in a dedicated chapter. Here, we only introduce the problem in a more phenomenological manner. Two main concepts are standard in defining stability: static stability and dynamic stability.

The *static stability* is the property of SG to remain in synchronism to the power grid in presence of slow variations in the shaft power (output active power, when losses are neglected).

According to the rising side of $P_1(\delta_V)$ curve (Figure 4.28), when the mechanical (shaft) power increases so does the power angle δ_V as the rotor slowly advances ahead the phase of E_1, with the phase of V_1 as fixed. When δ_V increases, the active power delivered electrically, by the SG, increases.

This way, the energy balance is kept and no important energy increment is accumulated in the SG's inertia. The speed stays constant but when P_1 increases, so does δ_V. The SG is statically stable if $\partial P_1/\partial \delta_V > 0$.

We denote by P_{1S}, this power derivative with angle and call it synchronization power:

$$P_{1S} = \frac{\partial P_1}{\partial \delta_V} = 3E_1V_1\cos\delta_V + 3V_1^2\left(\frac{1}{X_d} - \frac{1}{X_q}\right)\cos 2\delta_V \qquad (4.106)$$

P_{1S} is maximum at $\delta_V = 0$ and decreases to zero when δ_V increases toward δ_{VK}, where $P_{1S} = 0$.

At the extent that the field current decreases so does δ_{VK} and thus the static stability region diminishes. In reality, the SG is allowed to operate at values of δ_V, notably below δ_{VK} to preserve dynamic stability.

The *dynamic* stability is the property of the SG to remain in synchronism (with the power grid) in the presence of quick variations of shaft power or of electric load short circuit, etc.

As the combined inertia of SGs and their prime movers is relatively large, the speed and power angle transients are much slower than electrical (current and voltage) transients. Therefore, for example, we can still consider the SG under electromagnetic steady state when the shaft power (water admission in a hydraulic turbine) varies to produce slow speed and power angle transients. The electromagnetic torque T_e is thus, still, approximately,

$$T_e \approx \frac{P_1 \cdot p_1}{\omega_r} = \frac{3p_1}{\omega_r}\left[\frac{E_1V_1\sin\delta_V}{X_d} + \frac{V_1^2}{2}\left(\frac{1}{X_q} - \frac{1}{X_d}\right)\sin 2\delta_V\right] \qquad (4.107)$$

Let us consider a step variation of shaft power from P_{sh1} to P_{sh2} (Figure 4.33a) in a lossless SG.

The SG power angle power angle should vary slowly from δ_{V1} to δ_{V2}. In reality, the power angle δ_V will overshoot δ_{V2} and, after a few attenuated oscillations, will settle at δ_{V2} if the machine remains in synchronism.

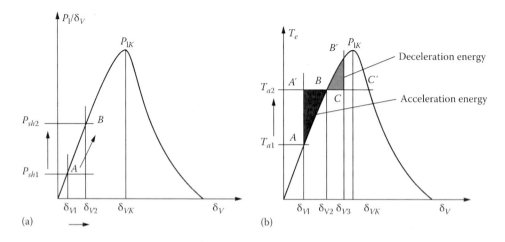

FIGURE 4.33 Dynamic stability: graphical illustration.

Neglect the rotor damper cage effects that occur during transients, and the motion equation is then written as follows:

$$\frac{J}{p_1}\frac{d\omega_r}{dt} = T_{shaft} - T_e; \quad \omega_r - \omega_{r0} = \frac{d\delta_V}{dt} \tag{4.108}$$

with ω_{r0} is the synchronous speed.

Multiplying Equation 4.108 by $d\delta_V/dt$, one obtains

$$d\left(\frac{J}{2p_1}\left(\frac{d\delta_V}{dt}\right)^2\right) = \left(T_{shaft} - T_e\right)d\delta_V = \Delta T \cdot d\delta_V = dW \tag{4.109}$$

Equation 4.109 illustrates the variation of kinetic energy of the prime-mover generator set translated in an acceleration area $AA'B$ and a deceleration area $BB'C$.

$$W_{AB} = \text{area of } AA'B \text{ triangle} = \int_{\delta_{V1}}^{\delta_{V2}} \left(T_{shaft} - T_e\right)d\delta_V \tag{4.110}$$

$$W_{AB'} = \text{area of } BB'C \text{ triangle} = \int_{\delta_{V2}}^{\delta_{V3}} \left(T_{shaft} - T_e\right)d\delta_V \tag{4.111}$$

Only when the two areas are equal to each other, there is the possibility that the SG will come back from B' to B after a few attenuated oscillations. Attenuation comes from the asynchronous torque of damper cage currents, neglected so far.

This is the so-called criterion of areas.

The maximum shaft torque or electric power step variation that can be accepted with the machine still remaining in synchronism is shown in Figure 4.34a and b and corresponds to the case when point C coincides with C'.

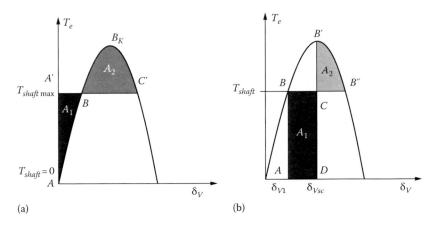

(a)   (b)

FIGURE 4.34 Dynamic stability ideal limits: (a) maximum shaft torque step variation from zero and (b) maximum short-circuit clearing time (angle: $\delta_{Vsc} - \delta_{V1}$) from load.

Let us illustrate the dynamic stability by the situation of a loaded SG at power angle δ_{V1}. A three-phase short-circuit occurs at δ_{V1} with its transients attenuated very quickly such that the electromagnetic torque is zero ($V_1 = 0$, zero losses also).

Therefore, the SG starts accelerating until the short circuit is cleared at δV_{sc}, which corresponds to a few tens of a second at most. Then the electromagnetic torque T_e becomes larger than the shaft torque and the SG decelerates. Only if the

$$\text{Area of } \overline{ABCD} \geq \text{Area of } CB'B'' \tag{4.112}$$

there are chances for the SG to remain in synchronism; that, is to be dynamically stable.

4.11 Unbalanced Load Steady-State Operation

SGs connected to the power grid, but especially those in autonomous applications, operate often on unbalanced three-phase loads.

That is, the stator currents in the three phases have different amplitudes and their phasing differs from 120°:

$$I_A(t) = I_1 \cos\left(\omega_1 t - \gamma_1\right)$$

$$I_B(t) = I_2 \cos\left(\omega_1 t - \frac{2\pi}{3} - \gamma_2\right) \tag{4.113}$$

$$I_C(t) = I_3 \cos\left(\omega_1 t + \frac{2\pi}{3} - \gamma_3\right)$$

For balanced load, $I_1 = I_2 = I_3$ and $\gamma_1 = \gamma_2 = \gamma_3$. These phase currents may be decomposed in direct, inverse, and homopolar sets according to Fortesque's transform (Figure 4.35).

$$I_{A+} = \frac{1}{3}\left(I_A + aI_B + a^2 I_C\right); \quad a = e^{j\frac{2\pi}{3}}$$

$$I_{A-} = \frac{1}{3}\left(I_A + a^2 I_B + aI_C\right); \quad a^2 = e^{-j\frac{2\pi}{3}}$$

$$I_{A0} = \frac{1}{3}\left(I_A + I_B + I_C\right)$$

$$I_{B+} = a^2 I_{A+}; \tag{4.114}$$

$$I_{C+} = aI_{A+};$$

$$I_{B-} = aI_{A-};$$

$$I_{C-} = a^2 I_{A-}$$

Unfortunately, the superposition of the current sets flux linkages is admissible only in the absence of magnetic saturation. Let us suppose that the SG is indeed unsaturated and lossless ($R_1 = 0$).

For the direct components I_{A+}, I_{B+}, I_{C+}, which produce a forward-traveling mmf at rotor speed, the theory unfolded so far still holds. Therefore, let us write the voltage equation for phase A and direct component of current I_{A+}:

$$V_{A+} = E_{A+} - jX_d I_{dA+} - jX_q I_{qA+} \tag{4.115}$$

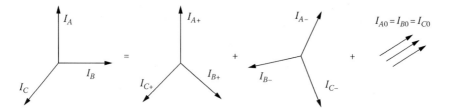

FIGURE 4.35 The symmetrical component sets.

The inverse (negative) components of stator currents I_{A-}, I_{B-}, I_{C-} produce an mmfs that travels at opposite rotor speed $-\omega_r$.

The relative angular speed of the inverse mmf with respect to rotor speed is thus $2\omega_r$. Consequently, voltages and currents are induced in the rotor damper windings and in the field winding at $2\omega_r$ frequency, in general. The behavior is similar to an induction machine at slip $S = 2$, but which has nonsymmetrical windings on the rotor and nonuniform air gap. We may approximate the SG behavior with respect to the inverse component as follows:

$$I_{A-} \cdot Z_- + U_{A-} = E_{A-}$$
$$Z_- = R_- + jX_-$$

(4.116)

Unless the stator windings are not symmetric or some of the field coils have short circuited turns $E_{A-} = 0$. Z_- is the inverse impedance of the machine and represents a kind of multiple winding rotor induction machine impedance at $2\omega_r$ frequency.

The homopolar components of currents produce mmfs in the three phases that are phase-shifted spatially by 120° and have the same amplitude and time phasing. They produce a zero-traveling field in the air gap and thus do not interact with the rotor in terms of the fundamental component. The corresponding homopolar impedance is $Z_0 \approx R_1 + jX_0$ and

$$X_0 < X_{1l}$$

(4.117)

Therefore,

$$X_d > X_q > X_- > X_{1l} > X_0$$

(4.118)

The stator equation for the homopolar set is

$$jI_{A0}X_0 + V_{A0} = 0$$

(4.119)

Finally,

$$V_A = V_{A+} + V_{A-} + V_{A0}$$

(4.120)

Similar equations are valid for the other two phases. We have assimilated here the homopolar with the stator leakage reactance ($X_{1l} \approx X_0$). Truth is that this assertion is not valid if chorded coils are used, when $X_0 < X_{1l}$. It seems that due to the placement of stator winding in slots, the stator homopolar mmf has a step-like distribution with $\tau/3$ as half-period and does not rotate; it is an AC field.

This third space-harmonic-like mmf may be decomposed in a forward and backward wave, and both move with respect to rotor and induce eddy currents at least in the damper cage. Additional losses occur in the rotor occur. As we are not prepared by now theoretically to calculate Z_- and X_0, we refer to some experiments to measure them so that we get some confidence in using the aforementioned theory of symmetrical components.

4.12 Measuring X_d, X_q, Z_-, Z_0

We will treat here some basic measurements procedures for SG reactances: X_d, X_q, Z_-, Z_0. For example, to measure X_d and X_q, the open field winding SG, supplied with symmetric forward voltages (ω_{r0}, frequency) through a variable-ratio transformer, is driven at speed ω_r very close but different from the stator frequency ω_{r0} (Figure 4.36):

$$\omega_r = \omega_{r0} \cdot \left(1.01 - 1.02\right) \tag{4.121}$$

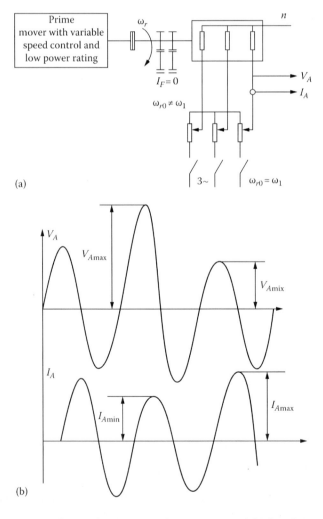

FIGURE 4.36   Measuring X_d and X_q: (a) the experimental arrangement and (b) the voltage and current waveforms.

We need not measure this speed precisely, but we need to notice the slow pulsation in the stator current with frequency $\omega_r - \omega_{r0} \approx (0.01-0.02)\omega_{r0}$.

Identifying the maxima and minima in the stator voltage $V_A(t)$ and current $I_A(t)$ leads to approximate values of X_d and X_q:

$$X_d \approx \frac{V_{A\max}}{I_{A\min}}; \quad X_q = \frac{V_{A\min}}{I_{A\max}} \tag{4.122}$$

The slip $S = (\omega_r - \omega_{r0})/\omega_{r0}$ has to be very small so that the currents induced in the rotor damper cage may be neglected. If they are not negligible, X_d and X_q are smaller than in reality due to the damper eddy current screening effect.

The saturation level will be medium if currents around or above rated value are used.

Identifying the voltage and current maxima, even if the voltage and current are digitally acquired and off-line processed in a computer, is doable with practical precision.

The inverse- (negative-) sequence impedance Z_- may be measured by driving the rotor, with the field winding short circuited, at synchronous speed ω_r, while feeding the stator with a purely negative sequence of low level voltages (Figure 4.37).

The power analyzer is used to produce

$$|Z_-| = \frac{V_{A-}}{I_{A-}}; \quad R_- = \frac{(P_-)_{phase}}{I_{A-}^2} \tag{4.123}$$

$$X_- = \sqrt{|Z_-| - (R_-)^2} \tag{4.124}$$

Again the frequency of currents induced in the rotor damper and field windings is $2\omega_{r0} = 2\omega_1$ and the corresponding slip is $S = 2.0$. Alternatively, it is possible to AC supply at standstill the stator between two phases only:

$$Z_- \approx \frac{U_{AB}}{2I_A} \tag{4.125}$$

This time the torque is zero and thus the SG stays at standstill, but the frequency of currents in the rotor is only $\omega_{r0} = \omega_1$ (the negative sequence impedance will be addressed in detail in Chapter 8). The homopolar

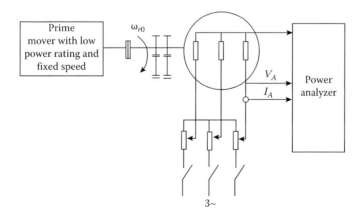

FIGURE 4.37 Negative sequence testing for Z_-.

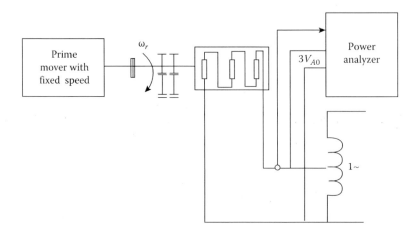

FIGURE 4.38   Measuring homopolar Z_0.

impedance Z_0 may be measured by supplying the stator phases connected in series from a single-phase AC source. The test may be made at zero speed or at rated speed ω_{r0} (Figure 4.38). For the rated speed test, the SG has to be driven at shaft.

The power analyzer yields the following:

$$|Z_0| = \frac{3V_{A0}}{3I_{A0}}; \quad R_0 = \frac{P_0}{3I_{A0}^2}; \quad X_0 = \sqrt{|Z_0|^2 - R_0^2} \tag{4.126}$$

A good portion of R_0 is the stator resistance R_1 so $R_0 \approx R_1$.

The voltage in measurements for Z_- and Z_0 should be made low to avoid high currents.

4.13 Phase-to-Phase Short Circuit

The three-phase—balanced—short circuit has already been investigated in a previous paragraph with the current I_{sc3}:

$$I_{3sc} = \frac{E_1}{X_d} \tag{4.127}$$

The phase-to-phase short circuit is a severe case of unbalanced load. When a short circuit between two phases occurs, with the third phase open, the currents are related to each other as (Figure 4.39a):

$$I_B = -I_C = I_{sc2}; \quad V_B = V_C; \quad I_A = 0 \tag{4.128}$$

From Equation 4.111, the symmetrical components of I_A are

$$I_{A+} = \frac{1}{2}\left(aI_B + a^2I_C\right) = \frac{1}{3}\left(a - a^2\right)I_{sc2} = \frac{+j}{\sqrt{3}}I_{sc2}$$

$$I_{A-} = -I_{A+}; \quad I_{A0} = 0 \tag{4.129}$$

The star connection leads to the absence of zero-sequence current.

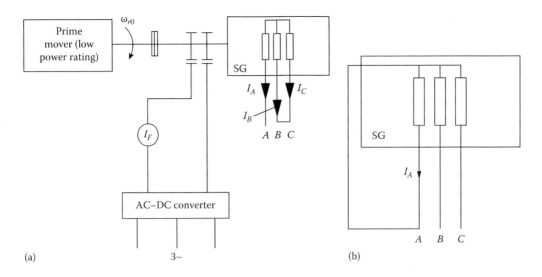

FIGURE 4.39 Unbalanced short circuit: (a) phase-to-phase and (b) single phase.

The terminal voltage of phase A, V_A, for a nonsalient pole machine ($X_d = X_q = X_+$) is obtained from Equation 4.117 with Equations 4.114 and 4.115:

$$V_A = V_{A+} + V_{A-} + V_{A0} = E_{A+} - jX_+ \cdot I_{A+} - Z_- I_{A-}$$

$$= E_{A+} - \frac{j}{\sqrt{3}} I_{sc2} \left(jX_+ - Z_- \right) \tag{4.130}$$

In a similar way,

$$V_B = a^2 V_{A+} + a V_{A-} = a^2 E_{A+} - \frac{jI_{sc2}}{\sqrt{3}} \left(ja^2 X_+ - aZ_- \right)$$

$$V_C = a V_{A+} + a^2 V_{A-} = a E_{A+} - \frac{jI_{sc2}}{\sqrt{3}} \left(ajX_+ - a^2 Z_- \right) \tag{4.131}$$

But $V_B = V_C$, and thus

$$E_{A+} = -\frac{jI_{sc2}}{\sqrt{3}} \left(jX_+ + Z_- \right) \tag{4.132}$$

$$\text{and} \quad V_A = \frac{2j}{\sqrt{3}} I_{sc2} Z_- = -2V_B \tag{4.133}$$

Finally,

$$jX_+ + Z_- = -j\sqrt{3} \frac{E_{A+}(I_F)}{I_{sc2}} \tag{4.134}$$

A few remarks are in order:

- Equation 4.134 with the known no-load magnetization curve, and the measured short-circuit current I_{sc2}, apparently allows for the computation of negative impedance if the positive one $jX_+ = jX_d$, for nonsalient pole rotor SG is given. Unfortunately, the phase shift between E_{A1} and I_{sc2} is hard to measure. Thus, if we only let:

$$Z_- \approx -jX_-$$ (4.135)

Equation 4.134 becomes usable as

$$X_+ + X_- = \frac{E_{A+} \cdot \sqrt{3}}{I_{sc2}}$$ (4.136)

RMS values enter Equation 4.136.
- Apparently, Equation 4.133 provides a good way of computing the negative impedance Z_- directly, with V_A and I_{sc2} measured. Their phase shift can be measured if the SG null point is used as common point for V_A and I_{2sc} measurements.
- During short circuit, be it phase to phase, the air gap magnetic flux density is small, but also distorted. Therefore, it is not easy to verify Equation 4.131 unless the voltage V_A and current I_{sc2} are first filtered to extract the fundamental.
- Thus, only Equation 4.134 can be used directly to approximate X_-, with X_+ unsaturated known. As $X_+ \gg X_-$ for strong damper cage rotors, the precision of computing X_- from the sum $(X_+ + X_-)$ is not so good.
- In a similar way as above for the single-phase short circuit (Figure 4.39b):

$$X_+ + X_- + X_0 \approx \frac{3E_{A+}(I_F)}{I_{sc1}}$$ (4.137)

with $X_+ > X_- > X_0$
- To a first approximation,

$$I_{sc3} : I_{sc2} : I_{sc1} \approx 3 : \sqrt{3} : 1$$ (4.138)

for an SG with a strong damper cage rotor.
- E_{A+} should be calculated for the real-field current I_F, but, as during short circuit the real saturation level is low, the unsaturated value of X_{FA} should be used: $E_{A+} = I_f(X_{FA})_{unsaturated}$.
- Small autonomous SGs may have the null available for single-phase loads, and thus the homopolar component shows up.
- The negative sequence currents in the stator produces double frequency-induced currents in the rotor damper cage and in the field windings. If the field winding is supplied from a static power converter, the latter may prevent the occurrence of AC currents in the field winding. Consequently notable overvoltages may occur in the latter. They should be considered when designing the field winding power electronics supply. Also, the double-frequency currents in the damper cage, produced by the negative component set, have to be limited as they affect the rotor overtemperature. Therefore, the ratio I_-/I_+, that is the level of current unbalance, is limited by standards below 10%–12%.
- A similar phenomenon occurs in autonomous SGs where the acceptable level of current unbalance I_-/I_f is given as a specification item and then considered in the thermal design. Finally, experiments are needed to make sure that the SG can really stand the predicted current unbalance.

- The phase-to-phase or single-phase short circuits are extreme cases of unbalanced load. The symmetrical components method presented here can be used for actual load unbalance situations where the +, −, 0 current components sets may be calculated first. A numerical example follows:

Example 4.3

A three-phase lossless two pole SG with S_n = 100 kVA, at V_{1l} = 440 V and f_1 = 50 Hz has the following parameters: $x_+ = x_d = x_q = 0.6$ pu, $x_- = 0.2$ pu, $x_0 = 0.12$ pu and supplies a single-phase-resistive load at rated current. Calculate the load resistance and the phase voltages V_A, V_B, V_C if the no-load voltage E_{1l} = 500 V.

Solution

We start with the computation of symmetrical current components sets (with $I_B = I_C = 0$):

$$I_{A+} = I_{A-} = I_{A0} = I_r/3$$

The rated current for star connection I_r is

$$I_r = \frac{S_n}{\sqrt{3}V_{1l}} = \frac{100,000}{\sqrt{3} \cdot 440} \approx 131 \text{ A}$$

The nominal impedance Z_n is

$$Z_n = \frac{U_{1l}}{\sqrt{3}I_r} = \frac{440}{\sqrt{3} \cdot 131} = 1.936 \, \Omega$$

Therefore,

$$X_+ = Z_n x_+ = 1.936 \times 0.6 = 1.1616 \, \Omega$$

$$X_- = Z_n x_- = 1.936 \times 0.2 = 0.5872 \, \Omega$$

$$X_0 = Z_n x_0 = 1.936 \times 0.12 = 0.23232 \, \Omega$$

From Equation 4.114, the positive sequence voltage equation is

$$V_{A+} = E_{A+} - jX_+ \frac{I_r}{3}$$

From Equation 4.115,

$$V_{A-} = -jX_- \frac{I_r}{3}$$

Also from Equation 4.118,

$$V_{A0} = -jX_0 \frac{I_r}{3}$$

The phase of voltage V_A is the summation of its components

$$V_A = E_{A+} - j(X_+ + X_- + X_0)\frac{I_r}{3}$$

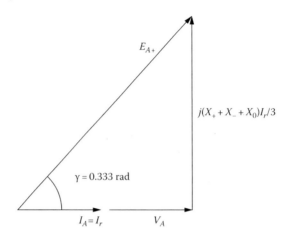

FIGURE 4.40 The phasor diagram.

As the single-phase load has been declared as resistive, I_r is in phase with V_A and thus

$$E_{A+} = V_{A+} + j\left(X_+ + X_- + X_0\right)\frac{I_r}{3}$$

A phasor diagram could be built as shown in Figure 4.40.

With $E_{A+} = 500\ \text{V/sqrt(3)} = 280\ \text{V}$ and $I_r = 131\ \text{A}$ known, we may calculate the phase voltage of loaded phase, V_A:

$$V_A = \sqrt{E_{A+}^2 - \left[\left(X_+ + X_- + X_0\right)I_r/3\right]^2} = \sqrt{(289)^2 - \left[(1.1616 + 0.3872 + 0.23232)\cdot 131/3\right]^2}$$

$$= 278.348\ \text{V}$$

The voltages along phases B and C are as follows:

$$V_B = E_{A+}e^{-j\frac{2\pi}{3}} - jX_+\frac{I_r}{3}e^{-j\frac{2\pi}{3}} - jX_-\frac{I_r}{3}e^{+j\frac{2\pi}{3}} - jX_0\frac{I_r}{3}$$

$$V_C = E_{A+}e^{+j\frac{2\pi}{3}} - jX_+\frac{I_r}{3}e^{j\frac{2\pi}{3}} - jX_-\frac{I_r}{3}e^{-j\frac{2\pi}{3}} - jX_e\frac{I_r}{3}$$

$$E_{A+} = E_{A+}\cdot e^{j\gamma_0}; \quad \gamma_0 = 0.333\,\text{rad}$$

The real axis falls along V_A and I_A, in the horizontal direction:

$$V_B = -83.87 - j\times 270\,[\text{V}]$$

$$V_C = -188.65 + j213.85\,[\text{V}]$$

The phase voltages are not symmetric anymore ($V_A = 278$ V, $V_B = 282.67$ V, $V_C = 285$ V). The voltage regulation is not very large as $x_+ = 0.6$, and the phase voltage unbalance is not large either since the homopolar reactance is usually small, $x_0 = 0.12$. And so is X_- due to a strong damper cage on the rotor. A small x_+ presupposes a notably large air gap and thus the field winding mmf should be rather large to produce acceptable values of flux density in the air gap on no load ($B_{gFm} = 0.7–0.9$ T), to secure a reasonable volume SG.

4.14 Synchronous Condenser

As already pointed out earlier in this chapter, the reactive power capability of a synchronous machine is basically the same for motor or generating mode (Figure 4.28b). It is thus feasible to use a synchronous machine as a motor without any mechanical load, connected to the local power grid (system), to "deliver" or "drain" reactive power and thus contribute to the overall power factor correction or (and) local voltage control.

The reactive power flow is controlled through field current control (Figure 4.41).

The phasor diagram (with zero losses) springs from voltage equation (Equation 4.54) with $I_q = 0$ and $R_1 = 0$ (Figure 4.42)

$$V_1 = E_1 - jX_dI_d; \quad I_1 = I_d \tag{4.139}$$

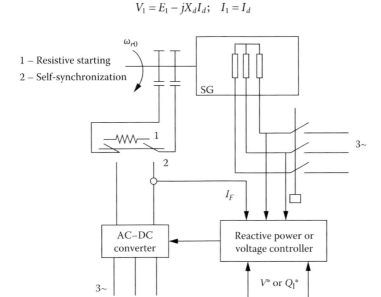

FIGURE 4.41 Synchronous condenser.

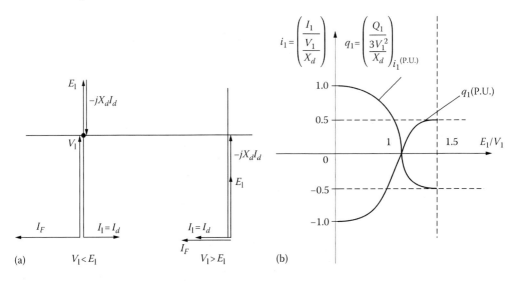

FIGURE 4.42 Phasor diagrams and reactive power of synchronous condenser: (a) phasor diagrams and (b) p.u. current (i_1) and reactive power (q_1).

The reactive power Q_1 of Equation 4.104, with $\delta_V = 0$ is

$$Q_1 = 3V_1 \frac{(E_1 - V_1)}{X_d} = -3V_1 I_d \tag{4.140}$$

$$Q_1 = \frac{3V_1^2}{X}; \quad X = \frac{X_d}{E_1/V_1 - 1} \tag{4.141}$$

As expected Q_1 changes sign at $E_1 = V_1$, and so does the current:

$$I_d = \frac{(V_1 - E_1)}{X_d}; \quad X = \begin{cases} >0 & \text{for } E_1/V_1 > 1 \\ <0 & \text{for } E_1/V_1 < 1 \end{cases} \tag{4.142}$$

Negative I_d means demagnetizing I_d or $E_1 > V_1$. As magnetic saturation depends on the resultant magnetic field, for constant voltage V_1, the saturation level stays about the same irrespective of field current I_F. Therefore,

$$E_1 \approx \omega_r (M_{FA})_{V_1} \cdot I_F \tag{4.143}$$

Also, X_d should not vary notably for constant voltage V_1. The maximum delivered reactive power depends on I_d, but the thermal design should account for both stator and rotor field winding losses, together with core losses located in the stator core.

Therefore, it seems that the synchronous condenser should be designed at maximum delivered (positive) reactive power $Q_{1\max}$:

$$Q_{1\max} = 3 \frac{V_1}{X_d} \left[E_{1\max}(I_{F\max}) - V_1 \right]; \quad I_1 = \frac{E_{1\max} - V_1}{X_d} \tag{4.144}$$

To reduce the size of such a machine acting as a no-load motor, two pole rotor configurations seem appropriate.

The synchronous condenser is in fact a positive/negative reactance with continuous control through field current via a low power rating AC–DC converter.

It does not introduce significant voltage or current harmonics in the power systems. However, it makes noise, has a sizeable volume, experience losses and needs maintenance. These are a few reasons for the increase in use of pulse-width modulator (PWM) converter controlled capacitors in parallel with inductors to control voltage in power systems.

Existing synchronous motors are also used whenever possible, to control reactive power and voltage locally while driving their loads, or on no-load, in synchronous condenser operation mode.

4.15 PM-Assisted DC-Excited Salient Pole Synchronous Generators

Small and medium (tens of kW to MW/unit) SGs with salient-pole vectors can benefit from the assistance of permanent magnets placed between rotor poles (Figure 4.43) to [8]:

- Reduce DC excitation leakage flux between rotor poles
- Reduce the magnetic saturation level in the rotor under load

The main advantage is that increase in air gap flux produced by rotor for smaller field current and thus 20%–25% more power at better efficiency is obtained.

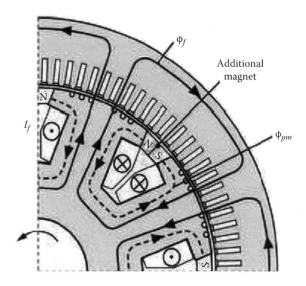

FIGURE 4.43 PM-assisted DC-excited salient pole SG.

Care must be exercised in designing and placing the PMs between rotor poles to avoid their local demagnetization in the most critical operation modes (sudden short circuits, etc.). Adding damper bars above the PMs (between rotor poles) screens the PMs from high armature transient reaction fields.

In a case study [8] referring to a 2.3 MVA, 3 phase, 8 pole, 50 Hz, 650 V, 2021 A, SG, with a stator diameter of 1240 mm and a stack length of 507 mm, additional NdFeB ($B_r = 1.19$ T) produce a notable increase in the no load line voltage (Figure 4.44).

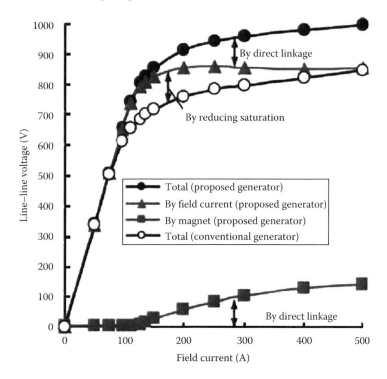

FIGURE 4.44 FEM calculated no load line voltage [8].

The no load voltage increase is produced two ways:

1. By the reduction of magnetic saturation in the rotor
2. By the reduction of excitation flux leakage between adjacent rotor poles, but without strong PM

While the power factor is affected [8], the field current is notably reduced (for given armature current), Figure 4.45a, which is reflected into a larger efficiency (Figure 4.45b), as also proved in laboratory work at small power [8].

It should be noted the total amount of PMs is significantly smaller than in permanent magnet SGs of equivalent power and speed; which makes the solution very tempting/practical.

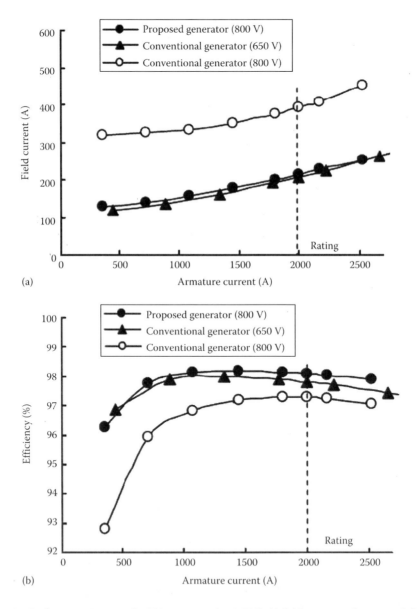

FIGURE 4.45 Performance increases by PM assistance (2.3 MVA): (a) field current reduction and (b) efficiency increasing at 800 V [8].

4.16 Multiphase Synchronous Machine Inductances via Winding Function Method

Three-phase SGs are standard by now. But, especially when used with static power converters for variable speed, more than three-phase stator windings are introduced to:

- Reduce the switching frequency in the static power converters, required by either power increase or fundamental frequency (or speed) increase
- Reduce the power per phase, to reduce power electronics costs and increase redundancy (reliability) by increased fault tolerance

While phases 6, 9, 12, and 15 are typical, phases 5, 7, 11, and 13 may also be used. The current in the phases is not necessarily sinusoidal; and thus in an inverter/rectifier with a null point, the phase current summation is not zero anymore. It is not zero even with sinusoidal current in a six-phase winding with a 30° electrical shift between phases. Therefore, an additional inverter leg in the converter is provided and controlled to symmetrize the phase currents. For an odd number of phases, this null current is about equal to the phase current, while it is doubled for even number of phases (e.g., six phases).

Let us consider here, for simplicity, an "m" phase SG whose phases are shifted by $\alpha = \pi/m$ (asymmetric winding: for $m = 6$, $\alpha = \pi/6 = 30°$), Figure 4.46.

According to Reference [9], a rather comprehensive modified winding function method can be used to develop an analytical circuit model to calculate the machine inductance matrix that compares favorably with the FEM results, as long as magnetic saturation is small and there is no damper cage on the rotor.

In essence, each machine inductance l_{ij} is made of a leakage, l_{lij} and an air gap component, l_{mij}; the latter is dependent on rotor position θ_{er}. While the leakage inductances have rather standard analytical formulae, the air gap (magnetization) one may be written as follows:

$$l_{mij}\left(\theta_{er}\right) = Rl_{stack}u_0 \int_0^{2\pi} \frac{p_s\left(\xi\right) \cdot p_r\left(\xi - \theta_{er}\right)}{g_0} \omega_i\left(\xi\right)\omega_j\left(\xi\right)d\xi \qquad (4.145)$$

where
l_{stack} is the stack length
R_r is rotor radius
g_0 is minimum air gap
ω_i, ω_j are winding functions
$p_s(\xi), p_r(\xi)$ are stator and rotor relative magnetic permeances (between 0 and 1)

For the winding arrangement in Figure 4.46:

$$\omega_i\left(\xi\right) = \omega_0\left(\xi - i\alpha\right); \quad \alpha = \frac{\pi}{m} \qquad (4.146)$$

Consequently,

$$l_{mij}\left(\theta_{er}\right) = Rl_{stack}u_0 \int_0^{2\pi} \frac{p_s\left(\xi\right) \cdot p_r\left(\xi\right)}{g_0} \omega_0\left(\xi - i\alpha\right)\omega_0\left(\xi - j\alpha\right)d\xi \qquad (4.147)$$

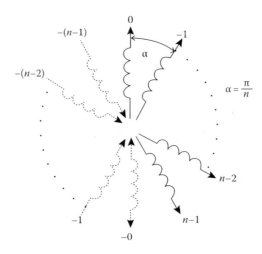

FIGURE 4.46 Two pole m-phase asymmetric stator AC winding.

Therefore, only the winding factor $\omega_0(\xi)$ has to be calculated. For q slots/pole/phase, coil pitch γ and N_c turns per coil and a current path (and slot pitch $\alpha_c = \alpha/q$):

$$\omega_0(\xi) = \frac{4}{\pi}\frac{N_c}{a} \sum_{\nu=1,3,5,7} \frac{\sin(\nu(\gamma/2))\sin(\alpha_s q(\nu/2))}{\nu \sin(\alpha_s(\nu/2))}\cos(\nu\xi) \tag{4.148}$$

$$\text{and} \quad f(\xi) = p_s(\xi)\omega_0(\xi) \tag{4.149}$$

The only left unknowns are $p_s(\xi)$ and $p_r(\xi - \theta_{er})$. Reference 9 uses FEM in three simplified topology situations models to calculate $p_s(\xi)$ and $p_r(\xi)$. Results as in Figure 4.47 are obtained.

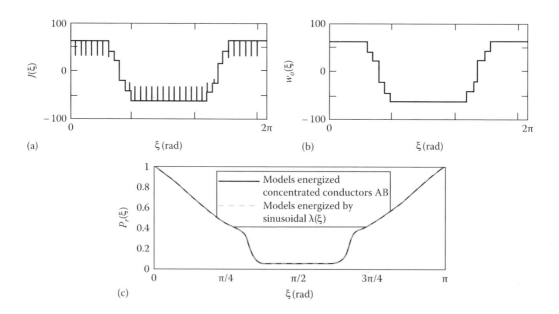

FIGURE 4.47 Functions: (a) $f(\xi)$, (b) $\omega_0(\xi)$, and (c) $p_r(\xi)$.

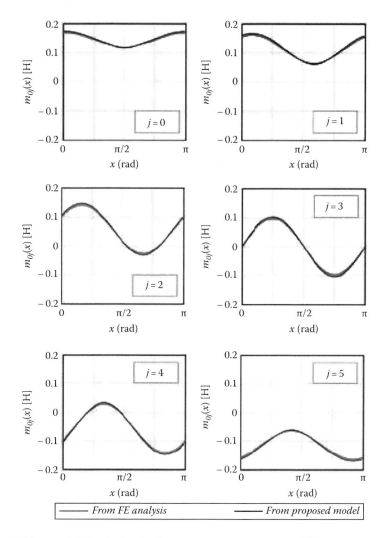

FIGURE 4.48 FEM versus MWF calculated inductances versus rotor position [9].

The FEM calculated functions: $f(\xi)$ and $p_r(\xi - \theta_{er})$ may be decomposed in harmonics $h(1, 3, 5,...)$ and $e_{ij}(\theta_{er})$ may be detailed further; in a similar way, the neutral inductances between stator and excitation winding can be derived [9].

Finally, in the absence of magnetic saturation, the FEM calculated and MWF calculated inductances show remarkable agreement in their dependence on rotor position $\theta_r = x$ in radians (Figure 4.48). The same good agreement was proven by the no-load voltage waveform with validation through test results, also [9].

4.17 Summary

- Large and medium power SGs are built with DC excitation windings on the rotor with either salient or nonsalient poles.
- Salient rotor poles are built for $2p_1 > 4$ poles and nonsalient rotor poles for $2p_1 = 2,4$.
- The stator core of SGs is made of silicon-steel laminations (0.5 mm thick in general) with uniform slotting. The slots house the three-phase windings.

- The stator core is made of one piece only if the outer diameter is below 1 m, otherwise it is made of segments. Sectionable cores are wound section by section and the wound sections are mounted together at the user's site.
- The slots in SGs are in general open, provided with nonmagnetic or magnetic wedges (to reduce emf harmonic content).
- Stator windings are of single- and double-layered type and are made of lap (multiturn) coils or the bar-wave (single-turn) coils (to reduce the lengthy connections between coils).
- Stator windings are built in general with integer slots/pole/phase q; only for large number of poles $2p_1 > 16–20$, q may be fractionary: 3.5, 4.5 (to reduce emf harmonics content).
- The symmetric AC currents of stator windings produce a positive mmf wave that travels with the ω_1/p_1 angular speed (with respect to the stator) $\omega_1 = 2\pi f_1$, f_1 is the frequency of currents.
- The core of salient pole rotors is made of a solid iron pole wheel spider on top of which $2p_1$ salient poles made in general of laminations (1 mm thick in general) are placed. The poles are attached to the pole wheel spider through hammer or dove-tail keybars or bolts and screws with end plates.
- Nonsalient pole rotors are made of solid iron with machined radial slots over two-thirds of periphery to house distributed field winding coils. Costs constrained and higher peripheral speeds have led to solid cores for nonsalient poles rotors with $2p_1 = 2,4$ poles.
- The rotor poles are provided with additional (smaller) slots filled with copper or brass bars short circuited by partial or total end rings. This is the damper winding. A complete damper winding (with a short-circuit end ring) is required to reduce the open phase peak voltage during line to line short circuit.
- The air gap flux density produced by the rotor field windings has a fundamental and space harmonics. They are to be limited in order to reduce the stator emf (no-load voltage) harmonics. Larger air gap under the salient poles is used for the scope. Uniform air gap is used for nonsalient poles because their distributed field coils produce lower harmonics in the air gap flux density. The design air gap flux density flat top value is about 0.7–0.8 T in large and medium power SGs. The emf harmonics may be further reduced by the type of the stator winding (larger or fractionary q, chorded coils).
- The air gap flux density of the rotor field winding currents is a traveling wave at rotor speed $\Omega_r = \omega_1/p_1$.
- When $\omega_r = \omega_1$, the stator AC currents and rotor DC currents air gap fields are at standstill with each other. These conditions lead to an interaction between the two fields with nonzero average electromagnetic torque. This is the speed of synchronism or the synchronous speed.
- When an SG is driven at the speed ω_r (electrical rotor angular speed; $\Omega_r = \omega_r/p_1$ is the mechanical rotor speed), the field rotor DC currents produce emfs in the stator windings whose frequency ω_1 is $\omega_1 = \omega_r$. If a balanced three-phase load is connected to the stator terminals, the occurring stator currents will naturally have the same frequency $\omega_1 = \omega_r$; their mmf will, consequently, produce an air gap traveling field at the speed $\omega_1 = \omega_r$. Their phase shift with respect to phase emfs depends on load character (inductive-resistive or capacitive-resistive) and on SG reactances (not discussed yet). This is the principle of SG.
- The air gap field of stator AC currents is called *armature reaction*.
- The phase stator currents may be decomposed in two components (I_d, I_q), one in phase with the emf and the other at 90°. Thus two mmfs are obtained, whose air gap fields are at standstill with respect to the moving rotor. One along d (rotor pole) axis—called longitudinal—and the other one along q axis, called transverse. This decomposition is the core of two-reaction theory of SGs.
- The two stator mmfs field are tied to rotor d and q axes and thus their cyclic magnetization reactances X_{dm} and X_{qm} may be easily calculated. Leakage reactances are added to get X_d, X_q the synchronous reactances. With zero damper currents and DC field currents on the rotor, the steady-state voltage equation is straightforward:

$$I_1 R_1 + V_1 = E_1 - jX_d I_d - jX_q I_q; \quad I_1 = I_d + I_q$$

- The SG "delivers" both active and reactive power P_1 and Q_1. They both depend on X_d, X_q, R_1 and on the power angle δ_V—the phase angle between the emf and terminal voltage (phase variables).
- Core losses may be included in the SG equations at steady state as pure resistive short circuited stator fictitious windings whose currents are produced by the resultant air gap or stator phase linkage.
- The SG loss components are stator winding losses, stator core losses, rotor field winding losses, additional losses (mainly in the rotor damper cage), and mechanical losses. The efficiency of large SGs is very good (above 98%, total, including field winding losses).
- The SGs may operate in standalone applications or connected to the local (or regional) power system. No load, short circuit, zero power factor saturation curves, together with output $V_1(I_1)$ curve fully characterize standalone operation with balanced load. Voltage regulation tends to be large with SGs as the synchronous reactances in PU (x_d or x_q) are larger than 0.5–0.6, to limit the rotor field winding losses.
- Operation of SGs at power system is characterized by the angular curve $P_1(\delta_V)$, V-shape curves $I_1(I_F)$ for $P_1 = ct$ and the reactive power capability $Q_1(P_1)$.
- The $P_1(\delta_V)$ curve shows a single maximum value at $\delta_{VK} \leq 90°$; this critical angle decreases when the field current I_F decreases for constant stator terminal voltage V_1 and speed.
- Static stability is defined as the property of SG to remain at synchronism for slow shaft torque variations. Basically up to $\delta_V = \delta_{VK}$, the SG is statically stable.
- The dynamic stability is defined as the property of the SG to remain in synchronism for fast shaft torque or electric power (short circuiting until clearing) transients. The area criterion is used to forecast the reserve of dynamic stability for each transient. Dynamic stability limits the rated power angle to 22°–40°, much less than its maximum value $\delta_{VK} \leq 90°$.
- The stand-alone SG may encounter unbalanced loads. The symmetrical components (Fortesque) method may be applied to describe SG operation under such conditions, provided saturation level does not change (or is absent). Impedances for the negative and zero components of stator currents, Z_- and Z_0, are defined and basic methods to measure them are described. In general $|Z_+| > |Z_-| > |Z_0|$, and thus the stator phase voltage unbalance under unbalanced loads is not very large. However, the negative-sequence stator currents induce voltages and thus produce currents of double stator frequency in the rotor damper cage and field winding. Additional losses are thus present. They have to be limited, to keep rotor temperature within reasonable limits. The maximum I_-/I_+ ratio is standardized (for power system SGs) or specified (for stand-alone SGs).
- The synchronous machine acting as a motor with no shaft load is used for reactive power absorption (I_F small) or delivery (I_F large). This regime is called "synchronous condenser" since the machine is seen by the local power system either as a capacitor (I_F large, overexcited $E_1 > V_1$) or as an inductor (I_F small, under-excited machine $E_1/V_1 < 1$). Its role is to raise or control local power factor or voltage in the power system.
- Inter-rotor-pole PMs in SG have been shown to increase output by more than 20% for slightly better efficiency.
- Multiphase (5, 7, 11, 6, 12) SG may be used when full power electronics control of output is used, to reduce system losses and costs; modified winding function method has been shown instrumental in calculating their inductance matrix in absence of magnetic saturation; when FEM usage is mandatory.

References

1. R. Richter, Electrical machines, vol. 2, *Synchronous Machines*, Verlag Birkhauser, Basel, Switzerland, 1954, (in German).
2. J.H. Walker, *Large Synchronous Machines*, Clarendon Press, Oxford, UK, 1981.
3. I. Boldea, S.A. Nasar, *Induction Machine Handbook*, CRC Press, Boca Raton, FL, 2001, p. 101.

4. IEEE Std. 115—1995: Test Procedures for Synchronous Machines.

5. V. Ostovic, *Dynamics of Saturated Electric Machines*, Springer Verlag, New York, 1989.

6. M. Kostenko, L. Piotrovski, *Electrical Machines*, vol. 2, MIR Publishers, Moscow, Russia, 1974.

7. C. Concordia, *Synchronous Machines*, John Wiley and Sons, Hoboken, NJ, 1951.

8. K. Yamazaki, K. Nishioka, K. Shima, T. Fukami, K. Shirai, Estimation of assist effects by additional PMs in salient pole SGs, *IEEE Trans.*, IE-59(6), 2515–2523, 2012.

9. A. Tessarolo, Accurate computation of multiphase SM inductances based on winding function theory, IEEE Trans., EC-27(4), 895–904, 2012.

5

Synchronous Generators: Modeling for Transients

5.1 Introduction

Chapter 4 dealt with the synchronous generator (SG) principles and steady state based on the two reaction theories. In essence, the concept of traveling field (rotor) and stator (stator) mmfs and air gap fields at standstill with each other has been presented.

Each stator phase current can be decomposed in two components under the steady state: one in phase with the emf and the other phase-shifted by 90°.

Two stator mmfs both traveling at rotor speed have been identified. One of these produces an airgap field whose maximum is aligned to the rotor poles (d axis), whereas the other is aligned to the q axis (between poles).

The d and q axis magnetization inductances X_{dm}, X_{qm} are thus defined. The voltage equations with balanced three-phase stator currents, under steady state, are then obtained.

Further on, this equation is exploited to derive all performance aspects for steady state when no currents are induced into the rotor damper winding, and the field-winding current is direct (unipolar). Though unbalanced load steady state was also investigated, the negative-sequence impedance Z_- could not be explained theoretically, and thus a basic experiment to measure it has been described in Chapter 4. Further on, during transients, when the stator current amplitude and frequency, rotor damper, and field currents and speed vary, a more general (advanced) model is required to handle the machine behavior properly.

Advanced models for transients include the following

- The phase-variable model
- The orthogonal axis (dq) model
- Finite element (FE)/circuit model

The first two models are essentially lumped circuit models, while the third model is a coupled, field (distributed parameter) and circuit, model.

Also, the first two are analytical models while the third is a numerical model. The presence of solid iron rotor core, damper windings, and distributed field coils on the rotor of nonsalient rotor pole SGs (turbo-generators, $2p_1 = 2,4$) further complicates the FE/circuit model to account for the eddy currents in the solid iron rotor—so influenced by local magnetic saturation level.

In view of such a complex problem, in this chapter we begin with the phase coordinate model whose inductances (some of them) are dependent on rotor position, that is, time. To get rid of rotor position dependence of self- and mutual (stator/rotor) inductances, the *dq* model is used. Its derivation is straight-forward through the Park matrix transform. The *dq* model is then exploited to describe the steady state. Further on, the operational parameters are presented and used to portray electromagnetic (constant speed) transients such as the three-phase sudden short circuit, etc.

A rather extended discussion on magnetic saturation inclusion into the *dq* model is then housed and illustrated for steady state and transients.

The electromechanical transients (speed variation too) are presented both for small perturbations (through linearization) and large perturbations, respectively. For the latter case, numerical solutions of state space equations are required, and are therefore illustrated.

Mechanical (or slow) transients, such as SG free or forced "oscillations," are presented for electromagnetic steady state.

Simplified *dq* models, adequate for power system stability studies, are introduced and justified in some detail. Illustrative examples are worked out. The asynchronous running is also presented as it is the regime that shows up the asynchronous (damping) torque, so critical to SG stability and control. Though the operational parameters with $s = j\omega$ lead to various SG parameters and time constants, their analytical expressions are given in Chapter 7 (on design) and their measurement is presented as part of Chapter 8, on testing.

The chapter ends with some FE/coupled circuit models related to SG steady state and transients.

5.2 Phase-Variable Model

The phase-variable model is a circuit model. Consequently, the SG is described by a set of three stator circuits coupled through motion with two (or a multiple of two) orthogonally placed (*d* and *q*) damper windings and a field winding (along axis *d*: of largest magnetic permeance)—Figure 5.1.

The stator and rotor circuits are magnetically coupled with each other.

It should be noted that the convention of voltage–current signs (directions) is based on the respective circuit nature: source on the stator and sink on the rotor. This is in agreement with Poynting vector direction, toward the circuit for the sink and outward for the source (Figure 5.1).

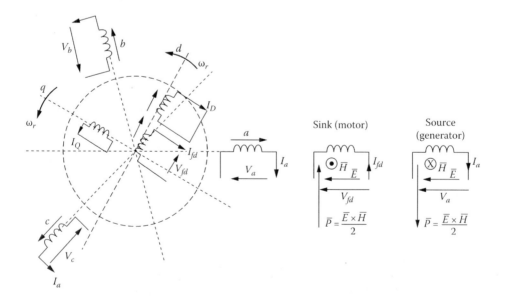

FIGURE 5.1   Phase-variable circuit model with single damper cage.

The phase voltage equations, in stator coordinates for the stator, and rotor coordinates for the rotor, are simply missing any "apparent" motion induced voltages:

$$i_A R_s + v_a = -\frac{d\Psi_A}{dt}$$

$$i_B R_S + v_b = -\frac{d\Psi_B}{dt}$$

$$i_C R_S + v_c = -\frac{d\Psi_C}{dt} \tag{5.1}$$

$$i_D R_D = -\frac{d\Psi_D}{dt}$$

$$i_Q R_Q = -\frac{d\Psi_Q}{dt}$$

$$I_f R_f - V_f = -\frac{d\Psi_f}{dt}$$

The rotor quantities are not yet reduced to the stator. The essential parts missing in Equation 5.1 are the flux linkage to current relationships, that is, self- and mutual inductances between the six coupled circuits in Figure 5.1. For example,

$$\Psi_A = L_{AA}I_a + L_{AB}I_b + L_{AC}I_c + L_{Af}I_f + L_{AD}I_D + L_{AQ}I_Q \tag{5.2}$$

Let us now define the stator phase self- and mutual inductances L_{AA}, L_{BB}, L_{CC}, L_{AB}, L_{BC}, L_{CA} for a salient-pole rotor SG. For now, let us consider the stator and rotor magnetic cores of having infinite magnetic permeability.

As already demonstrated in Chapter 4, the magnetic permeance of air gap along axes d and q differ (Figure 5.2).

The phase A mmf has a sinusoidal space distribution, because all space harmonics are neglected.

The magnetic permeance of the air gap is maximum in axis d, P_d, and minimum in axis q and may be approximated to the following:

$$P(\theta_{er}) = P_0 + P_2 \cos 2\theta_{er} = \frac{P_d + P_q}{2} + \left(\frac{P_d - P_q}{2}\right)\cos 2\theta_{er} \tag{5.3}$$

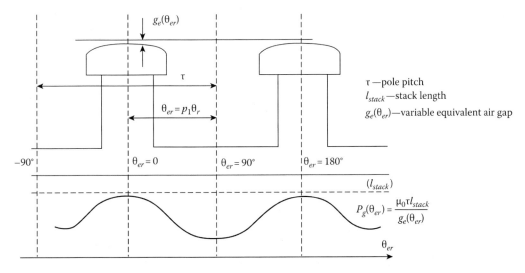

FIGURE 5.2 The air gap permeance per pole versus rotor position.

Therefore, the air gap self-inductance of phase A depends on that of a uniform air gap machine (single-phase fed) and on the ratio of the permeance $P(\theta_{er})/(P_0 + P_2)$ (see Chapter 4):

$$L_{AAg} = \frac{4}{\pi^2 p_1}\left(W_1 K_{W1}\right)^2 \left(P_0 + P_2 \cos 2\theta_{er}\right) \tag{5.4}$$

$$P_0 + P_2 = \frac{\mu_0 \tau l_{stack}}{g_{ed}}; \quad P_0 - P_2 = \frac{\mu_0 \tau l_{stack}}{g_{eq}}; \quad g_{ed} < g_{eq} \tag{5.5}$$

In addition,

$$L_{AAg} = L_0 + L_2 \cos 2\theta_{er} \tag{5.6}$$

To complete the definition of the self-inductance of phase A, the phase leakage inductance L_{sl} has to be added (the same for all three phases if they are fully symmetric):

$$L_{AA} = L_{sl} + L_0 + L_2 \cos 2\theta_{er} \tag{5.7}$$

Ideally, for a nonsalient-pole rotor SG, $L_2 = 0$; but in reality, a small saliency still exists due to a more accentuated magnetic saturation level along axis q where the distributed field coil slots are located.

In a similar way,

$$L_{BB} = L_{sl} + L_0 + L_2 \cos\left(2\theta_{er} + \frac{2\pi}{3}\right) \tag{5.8}$$

$$L_{CC} = L_{sl} + L_0 + L_2 \cos\left(2\theta_{er} - \frac{2\pi}{3}\right) \tag{5.9}$$

The mutual inductance between phases is considered to be only in relation to air gap permeances. It is evident that with ideally (sinusoidally) distributed windings, $L_{AB}(\theta_{er})$ varies with θ_{er} as L_{CC} and again has two components (to a first approximation):

$$L_{AB} = L_{BA} = L_{AB0} + L_{AB2} \cos\left(2\theta_{er} - \frac{2\pi}{3}\right) \tag{5.10}$$

Now as phases A and B are 120° phase shifted, it follows that

$$L_{AB0} \approx L_0 \cos\frac{2\pi}{3} = -\frac{L_0}{2} \tag{5.11}$$

The variable part of L_{AB} is quite similar to that of Equations 5.9, and thus

$$L_{AB2} = L_2 \tag{5.12}$$

Relationships of Equations 5.11 and 5.12 are valid for ideal conditions. In reality, there are some small differences even for symmetric windings.

Further on,

$$L_{AC} = L_{CA} = -\frac{L_0}{2} + L_2 \cos\left(2\theta_{er} + \frac{2\pi}{3}\right) \qquad (5.13)$$

$$L_{BC} = L_{CB} = -\frac{L_0}{2} + L_2 \cos 2\theta_{er} \qquad (5.14)$$

FE analysis of field distribution with only one phase supplied with DC current could provide ground for more exact approximations of self- and mutual stator inductance dependence on θ_{er}. Based on this, additional terms in $\cos(4\theta_{er})$, even $6\theta_{er}$ may be added. For fractionary q windings, more intricate θ_{er} dependences may be developed.

The mutual inductances between stator phases and rotor circuits are straightforward as they vary with $\cos(\theta_{er})$ and $\sin(\theta_{er})$.

$$L_{Af} = M_f \cos\theta_{er}$$

$$L_{Bf} = M_f \cos\left(\theta_{er} - \frac{2\pi}{3}\right)$$

$$L_{Cf} = M_f \cos\left(\theta_{er} + \frac{2\pi}{3}\right)$$

$$L_{AD} = M_D \cos\theta_{er}$$

$$L_{BD} = M_D \cos\left(\theta_{er} - \frac{2\pi}{3}\right) \qquad (5.15)$$

$$L_{CD} = M_D \cos\left(\theta_{er} + \frac{2\pi}{3}\right)$$

$$L_{AQ} = -M_Q \sin\theta_{er}$$

$$L_{BQ} = -M_Q \sin\left(\theta_{er} - \frac{2\pi}{3}\right)$$

$$L_{CQ} = -M_Q \sin\left(\theta_{er} + \frac{2\pi}{3}\right)$$

Note that

$$L_0 = \frac{\left(L_{dm} + L_{qm}\right)}{2}$$

$$L_2 = \frac{\left(L_{dm} - L_{qm}\right)}{2} \qquad (5.16)$$

where
  L_{dm} and L_{qm} have been defined in Chapter 4 with all stator phases on
  M_f is the maximum of field/armature inductance derived also in Chapter 4

We may now define the SG phase-variable 6×6 matrix $\left|L_{ABCfDQ}\left(\theta_{er}\right)\right|$:

$\left|L_{ABCfDQ}(\theta_{er})\right| =$

	A	B	C	f	D	Q
A	$L_{sl}+L_0+L_2\cos 2\theta_{er}$	$-\frac{L_0}{2}+L_2\cdot\cos\left(2\theta_{er}-\frac{2\pi}{3}\right)$	$-\frac{L_0}{2}+L_2\cdot\cos\left(2\theta_{er}+\frac{2\pi}{3}\right)$	$M_f\cos\theta_{er}$	$M_D\cos\theta_{er}$	$-M_Q\cos\theta_{er}$
B		$L_{sl}+L_0+L_2\cos\left(2\theta_{er}+\frac{2\pi}{3}\right)$	$-\frac{L_0}{2}+L_2\cdot\cos 2\theta_{er}$	$M_f\cos\left(\theta_{er}-\frac{2\pi}{3}\right)$	$M_D\cos\left(\theta_{er}-\frac{2\pi}{3}\right)$	$-M_Q\sin\left(\theta_{er}-\frac{2\pi}{3}\right)$
C			$L_{sl}+L_0+L_2\cos\left(2\theta_{er}+\frac{2\pi}{3}\right)$	$M_f\cos\left(\theta_{er}+\frac{2\pi}{3}\right)$	$M_D\cos\left(\theta_{er}+\frac{2\pi}{3}\right)$	$-M_Q\sin\left(\theta_{er}-\frac{2\pi}{3}\right)$
f				$L_{fl}^r+L_{fm}^r$	M_D^r	0
D				$L_{Dl}^r+L_{Dm}^r$		0
Q				0	0	$L_{Ql}^r+L_{Qm}^r$

$$(5.17)$$

A mutual coupling leakage inductance L_{fDl} also occurs between the field winding f and the d-axis cage winding D in salient-pole rotors. The zeroes in Equations 5.17 reflect the zero coupling between orthogonal windings in the absence of magnetic saturation. L_{fm}^r, L_{Dm}^r, and L_{Qm}^r are typical main (air gap permeance) self-inductances of rotor circuits. L_{fl}^r, L_{Dl}^r, and L_{Ql}^r are the leakage inductances of rotor circuits in axes d and q.

The resistance matrix is of diagonal type:

$$R_{ABCfdq} = \text{Diag}\left[R_s, R_r, R_s, R_f^r, R_D^r, R_Q^r\right] \tag{5.18}$$

Provided core losses, space harmonics, magnetic saturation, and frequency (skin effects) in the rotor core are all neglected, the voltage/current matrix equation fully represents the SG at constant speed:

$$\left[I_{ABCfDQ}\right]\left[R_{ABCfDQ}\right]+\left[V_{ABCfDQ}\right] = \frac{-d\Psi_{ABCfDQ}}{dt} = -\left[L_{ABCfDQ}\left(\theta_{er}\right)\right]\frac{d}{dt}\left[I_{ABCfDQ}\right]-\frac{\partial\left[L_{ABCfDQ}\right]}{\partial\theta_{er}}\frac{d\theta_{er}}{dt}\left[I_{ABCfDQ}\right] \tag{5.19}$$

with

$$V_{ABCfDQ} = \left[V_A, V_B, V_C, -V_f, 0, 0\right]^T; \quad \frac{d\theta_{er}}{dt} = \omega_r \tag{5.20}$$

$$\Psi_{ABCfDQ} = \left[\Psi_A, \Psi_B, \Psi_C, \Psi_f^r, \Psi_D^r, \Psi_Q^r\right]^T \tag{5.21}$$

The minus sign for V_f arises from the motor association of signs convention for rotor.

The first term on the right-hand side of Equation 5.19 represents the transformer induced voltages, and the second term refers to the motion-induced voltages.

Multiplying Equation 5.19 by $[I_{ABCfDQ}]^T$ yields the following:

$$\left[I_{ABCfDQ}\right]^T\left[V_{ABCfDQ}\right] = -\frac{1}{2}\left[I_{ABCfDQ}\right]^T\frac{\partial\left[L_{ABCfDQ}\left(\theta_{er}\right)\right]}{\partial\theta_{er}}\left[I_{ABCfDQ}\right]\cdot\omega_r$$

$$-\frac{d}{dt}\left[\frac{1}{2}\left[I_{ABCfDQ}\right]^T\cdot L_{ABCfDQ}\left(\theta_{er}\right)\cdot\left[I_{ABCfDQ}\right]\right]-\left[I_{ABCfDQ}\right]^T\left[I_{ABCfDQ}\right]\left[R_{ABCfDQ}\right] \quad (5.22)$$

The instantaneous power balance Equation 5.22 serves to identify the electromagnetic power that is related to the motion-induced voltages:

$$P_{elm} = -\frac{1}{2}\left[I_{ABCfDQ}\right]^T\cdot\frac{\partial}{\partial\theta_{er}}\left[L_{ABCfDQ}\left(\theta_{er}\right)\right]\left[I_{ABCfDQ}\right]\omega_r \quad (5.23)$$

P_{elm} should be positive for the generator regime.

The electromagnetic torque T_e opposes motion when positive (generator model) and is:

$$T_e = \frac{+P_{elm}}{\left(\omega_r/p_1\right)} = -\frac{p_1}{2}\left[I_{ABCfDQ}\right]^T\frac{\partial\left[L_{ABCfDQ}\left(\theta_{er}\right)\right]}{\delta\theta_{er}}\left[I_{ABCfDQ}\right] \quad (5.24)$$

The equation of motion is

$$\frac{J}{p_1}\frac{d\omega_r}{dt} = T_{shaft}-T_e; \quad \frac{d\theta_{er}}{dt} = \omega_r \quad (5.25)$$

The phase-variable equations constitute an eighth-order model with time-variable coefficients (inductances).

Such a system may be solved as it is either with flux linkages vector as the variable or with the current vector as the variable, together with speed ω_r and rotor position θ_{er} as motion variables.

Numerical methods such as Runge–Kutta–Gill or predictor–corrector may be used to solve the system for various transient or steady-state regimes, once the initial values of all variables are given. Also, the time variation of voltages and of shaft torque must be known. Inverting the matrix of time-dependent inductances at every time integration step is however a tedious job. Moreover, as it is, the phase-variable model does offer little in terms of interpreting the various phenomena and operation modes in an intuitive manner.

This is how the *dq* model has been developed out of the necessity to solve quickly various transient operation modes of SGs connected to the power grid (or in parallel).

5.3 *dq* Model

The main aim of the *dq* model is to eliminate the dependence of inductances on rotor position. To do so, the system of coordinates should be attached to the machine part that has magnetic saliency—the rotor, for SGs.

The *dq* model should express both stator and rotor equations in rotor coordinates, aligned to rotor *d* and *q* axis because, at least in the absence of magnetic saturation, there is no coupling between the two orthogonal axes. The rotor windings *f*, *D*, and *Q* are already aligned along *d* and *q* axes. The rotor circuits voltage equations have been written in rotor coordinates in Equation 5.1.

It is only the stator voltages, V_A, V_B, V_C, currents I_A, I_B, I_C, flux linkages Ψ_A, Ψ_B, Ψ_C that have to be transformed to rotor orthogonal coordinates.

The transformation of coordinates ABC—dq_0, known also as the Park transform, valid for voltages, currents and flux linkages as well is

$$\left[P(\theta_{er})\right] = \frac{2}{3} \begin{vmatrix} \cos(-\theta_{er}) & \cos\left(-\theta_{er} + \frac{2\pi}{3}\right) & \cos\left(-\theta_{er} - \frac{2\pi}{3}\right) \\ \sin(-\theta_{er}) & \sin\left(-\theta_{er} + \frac{2\pi}{3}\right) & \sin\left(-\theta_{er} - \frac{2\pi}{3}\right) \\ \frac{1}{2} & \frac{1}{2} & \frac{1}{2} \end{vmatrix} \tag{5.26}$$

Therefore,

$$\begin{vmatrix} V_d \\ V_q \\ V_0 \end{vmatrix} = \left|P(\theta_{er})\right| \cdot \begin{vmatrix} V_A \\ V_B \\ V_C \end{vmatrix} \tag{5.27}$$

$$\begin{vmatrix} I_d \\ I_q \\ I_0 \end{vmatrix} = \left|P(\theta_{er})\right| \cdot \begin{vmatrix} I_A \\ I_B \\ I_C \end{vmatrix} \tag{5.28}$$

$$\begin{vmatrix} \Psi_d \\ \Psi_q \\ \Psi_0 \end{vmatrix} = \left|P(\theta_{er})\right| \cdot \begin{vmatrix} \Psi_A \\ \Psi_B \\ \Psi_C \end{vmatrix} \tag{5.29}$$

The inverse transformation—that conserves powers—is

$$\left[P(\theta_{er})\right]^{-1} = \frac{3}{2}\left[P(\theta_{er})\right]^{T} \tag{5.30}$$

The expressions of Ψ_A, Ψ_B, Ψ_C from the flux/current matrix are

$$\left|\Psi_{ABCfDQ}\right| = \left|L_{ABCfDQ}(\theta_{er})\right|\left|I_{ABCfDQ}\right| \tag{5.31}$$

The phase currents I_A, I_B, I_C are recovered from I_d, I_q, I_0 by

$$\begin{vmatrix} I_A \\ I_B \\ I_C \end{vmatrix} = \frac{3}{2}\left[P(\theta_{er})\right]^{T} \cdot \begin{vmatrix} I_d \\ I_q \\ I_0 \end{vmatrix} \tag{5.32}$$

An alternative Park transform uses sqrt(2/3) instead of 2/3 for direct and inverse transform. This one is fully orthogonal (power direct conservation).

The rather short and elegant expressions of Ψ_d, Ψ_q, Ψ_0 are obtained as follows:

$$\Psi_d = \left(L_{sl} + L_0 - L_{AB0} + \frac{3}{2} L_2 \right) I_d + M_f I_f^r + M_D I_D^r$$

$$\Psi_q = \left(L_{sl} + L_0 - L_{AB0} - \frac{3}{2} L_2 \right) I_q + M_Q I_q^r \qquad (5.33)$$

$$\Psi_0 = \left(L_{sl} + L_0 + 2 L_{AB0} \right) I_0; \quad L_{AB0} \approx -L_0/2$$

From Equation 5.16,

$$L_{dm} = \frac{3}{2} \left(L_0 + L_2 \right);$$

$$\qquad (5.34)$$

$$L_{qm} = \frac{3}{2} \left(L_0 - L_2 \right)$$

are exactly the "cyclic" magnetization inductances along axes d and q as defined in Chapter 4.

Therefore, Equation 5.33 becomes

$$\Psi_d = L_d I_d + M_f I_f^r + M_D I_D^r;$$

$$\qquad (5.35)$$

$$L_d = L_{sl} + L_{dm}$$

$$\Psi_q = L_q I_q + M_Q I_Q^r;$$

$$\qquad (5.36)$$

$$L_q = L_{sl} + L_{qm}$$

$$\Psi_0 \approx L_{sl} I_0 \qquad (5.37)$$

In a similar way, for the rotor

$$\Psi_f^r = \left(L_{fl}^r + L_{fm} \right) I_f^r + \frac{3}{2} M_f I_d + \frac{3}{2} M_{fD} I_D^r$$

$$\Psi_D^r = \left(L_{Dl}^r + L_{Dm} \right) I_D^r + \frac{3}{2} M_D I_d + \frac{3}{2} M_{fD} I_f^r \qquad (5.38)$$

$$\Psi_Q^r = \left(L_{Ql}^r + L_{Qm} \right) I_Q^r + \frac{3}{2} M_Q I_q$$

As can be seen in Equation 5.37, the zero components of stator flux and current Ψ_0, I_0 are related simply by the stator phase leakage inductance L_{sl}, and thus they do not participate in the energy conversion through the fundamental components of mmfs and fields in the SGs.

It is thus acceptable to consider it separately. Consequently, the dq transformation may be visualized as representing a fictitious SG with orthogonal stator axes fixed magnetically to the rotor d–q axes. The magnetic field axes of the respective stator windings are fixed to the rotor d–q axes, but their conductors

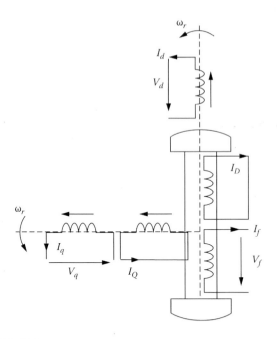

FIGURE 5.3 The *dq* model of SGs.

(coils) are at standstill (Figure 5.3)—fixed to the stator. The *dq* model equations may also be derived directly through the equivalent fictitious orthogonal axis machine (Figure 5.3).

$$I_d R_s + V_d = -\frac{d\Psi_d}{dt} + \omega_r \Psi_q$$

$$I_q R_s + V_q = -\frac{d\Psi_q}{dt} - \omega_r \Psi_d$$

(5.39)

The rotor equations are added:

$$I_f R_f - V_f = -\frac{d\Psi_f}{dt}$$

$$i_D R_D = -\frac{d\Psi_D}{dt}$$

$$i_Q R_Q = -\frac{d\Psi_Q}{dt}$$

(5.40)

From Equation 5.39, it is assumed that

$$\frac{d\Psi_d}{d\theta_{er}} = -\Psi_q$$

$$\frac{d\Psi_q}{d\theta_{er}} = \Psi_d$$

(5.41)

The assumptions are true if the windings dq are sinusoidally distributed and the air gap is constant but with a radial flux barrier along axis d. Such a hypothesis is valid for distributed stator windings to a good approximation only if the air gap flux density fundamental is considered. The null (zero) component equation is simply as follows:

$$I_0 R_s + V_0 = -L_{sl}\frac{di_0}{dt} = -\frac{d\Psi_0}{dt};$$

$$I_0 = \frac{I_A + I_B + I_C}{3}$$

(5.42)

The equivalence between the real three-phase SG and its dq model in terms of instantaneous power, losses, and torque is marked by the 2/3 coefficient in Park's transformation.

$$V_A I_A + V_B I_B + V_C I_C = \frac{3}{2}\left(V_d I_d + V_q I_q + 2V_0 I_0\right)$$

$$T_e = -\frac{3}{2}p_1\left(\Psi_d I_q - \Psi_q I_d\right)$$

(5.43)

$$T_e = -\frac{P_e}{\omega_r / p_1}$$

$$R_s\left(I_A^2 + I_B^2 + I_C^2\right) = \frac{3}{2}R_s\left(I_d^2 + I_q^2 + 2I_0^2\right)$$

(5.44)

The electromagnetic torque, T_e, calculated in Equation 5.43, is considered positive when opposite to motion.

Note: For the Park transform with $\sqrt{2/3}$ coefficients, the power, torque, and loss equivalence in Equation 5.43 and Equation 5.44 lack the 3/2 factor. Also in this case, Equations 5.38 have $\sqrt{3/2}$ instead of 3/2 coefficients.

The motion equation is

$$\frac{J}{p_1}\frac{d\omega_r}{dt} = T_{shaft} + \frac{3}{2}p_1\left(\Psi_d I_q - \Psi_d I_q\right)$$

(5.45)

Reducing the rotor variables to stator variables is common in order to reduce the number of inductances.
   But first the dq model flux/current relations derived directly from Figure 5.4 with rotor variables reduced to stator would be

$$\Psi_d = L_{sl}I_d + L_{dm}\left(I_d + I_D + I_f\right)$$

$$\Psi_q = L_{sl}I_q + L_{qm}\left(I_q + I_Q\right)$$

$$\Psi_f = L_{fl}I_q + L_{dm}\left(I_d + I_D + I_f\right)$$

(5.46)

$$\Psi_D = L_{Dl}I_D + L_{dm}\left(I_d + I_D + I_f\right)$$

$$\Psi_Q = L_{Ql}I_Q + L_{qm}\left(I_q + I_Q\right)$$

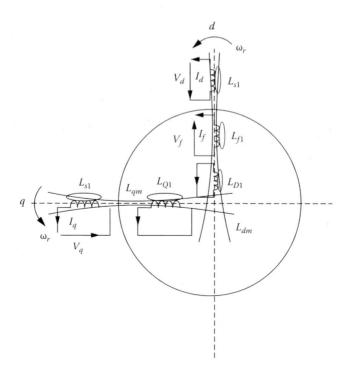

FIGURE 5.4 Inductances of *dq* model.

The mutual and self-inductances of air gap (main) flux linkage are identical to L_{dm} and L_{qm} after rotor to stator reduction. Comparing Equations 5.38 with 5.46, the following definitions of current reduction coefficients are considered valid:

$$I_f = I_f^r \cdot K_f$$

$$I_D = I_D^r \cdot K_D$$

$$I_Q = I_Q^r \cdot K_Q$$

$$K_f = \frac{M_f}{L_{dm}}$$

$$K_D = \frac{M_D}{L_{dm}}$$ (5.47)

$$K_Q = \frac{M_Q}{L_{Qm}}$$

We may now use coefficients in Equation 5.38 to obtain the following:

$$\Psi_f^r \cdot \frac{2}{3} \frac{L_{dm}}{M_f} = \Psi_f = L_{fl} I_f + L_{dm} \left(I_f + I_D + I_d \right)$$ (5.48)

where

$$L_{fl} = \frac{2}{3} L_{fl}^r \cdot \frac{L_{dm}^2}{M_f^2} = L_{fl}^r \frac{2}{3K_f^2}$$

$$L_{fm} \frac{2}{3} \frac{L_{dm}}{M_f^2} \approx 1 \qquad (5.49)$$

$$\frac{2}{3} \frac{L_{dm}}{M_f M_D} \approx 1$$

$$\Psi_D^r \cdot \frac{2}{3} \frac{L_{dm}}{M_D} = \Psi_D = L_{Dl} I_D + L_{dm} \left(I_f + I_D + I_d \right) \qquad (5.50)$$

where

$$L_{Dl} = \frac{2}{3} L_{Dl}^r \frac{L_{dm}^2}{M_D^2} = L_{Dl}^r \cdot \frac{2}{3} \cdot \frac{1}{K_D^2}$$

$$L_{Dm} \cdot \frac{2}{3} \frac{L_{dm}}{M_D^2} \approx 1 \qquad (5.51)$$

$$\Psi_Q^r \frac{2}{3} \frac{L_{qm}}{M_Q} = \Psi_Q = L_{Ql} I_Q + L_{qm} \left(I_q + I_Q \right) \qquad (5.52)$$

where

$$L_{Ql} = \frac{2}{3} L_{Ql}^r \cdot \left(\frac{L_{qm}}{M_Q} \right)^2 = L_{Ql}^r \frac{2}{3} \frac{1}{K_Q^2}$$

$$L_{Qm} \cdot \frac{2}{3} \frac{L_{qm}}{M_Q^2} \approx 1 \qquad (5.53)$$

We still need to reduce the rotor circuit resistances R_f^r, R_D^r, R_Q^r and the field-winding voltage to stator quantities. This may be done by power equivalence as follows:

$$\frac{3}{2} R_f \left(I_f^2 \right) = R_f^r I_f^{r2}$$

$$\frac{3}{2} R_D \left(I_D^2 \right) = R_D^r I_D^{r2} \qquad (5.54)$$

$$\frac{3}{2} R_Q \left(I_Q^2 \right) = R_Q^r I_Q^{r2}$$

$$\frac{3}{2} V_f I_f = V_f^r I_f^r \qquad (5.55)$$

Finally

$$R_f = R_f^r \frac{2}{3} \frac{1}{K_f^2}$$

$$R_D = R_D^r \frac{2}{3} \frac{1}{K_D^2}$$

$$R_Q = R_Q^r \frac{2}{3} \frac{1}{K_Q^2}$$ (5.56)

$$V_f = V_f^r \frac{2}{3} \frac{1}{K_f}$$

Note that resistances and leakage inductances are reduced by same coefficients, as expected for power balance.

A few remarks are in order:

- The "physical" *dq* model in Figure 5.4. presupposes that there is a single common (main) flux linkage along each of the two orthogonal axes that embrace all windings along those axes.
- The flux/current relationships of Equation 5.46 for the rotor uses stator-reduced rotor current, inductances, and flux linkage variables. In order to be valid, the approximations:

$$L_{fm}L_{dm} \approx \frac{3}{2}M_f^2$$

$$M_{fD}L_{dm} \approx \frac{3}{2}M_f M_D$$ (5.57)

$$L_{Dm}L_{dm} \approx \frac{3}{2}M_D^2$$

$$L_{Qm}L_{qm} \approx \frac{3}{2}M_Q^2$$

have to be accepted.

- The validity of approximations of Equation 5.57 is related to the condition that air gap field distribution produced by stator, and, respectively, rotor currents is the same. As far as the space fundamental is concerned, this condition holds. Once heavy local magnetic saturation conditions occur, Equation 5.57 depart from reality.
- No leakage flux coupling between the *d* axis damper cage and the field winding ($L_{fDl} = 0$) has been considered thus far, though in salient-pole rotors $L_{fDl} \neq 0$ may be needed to properly assess the SG transients, especially in the field winding.
- The coefficients K_f, K_D, and K_Q used in the reduction of rotor voltage (V_f^r), currents I_f^r, I_D^r, and I_Q^r, leakage inductances L_{fl}^r, L_{Dl}^r, and L_{Ql}^r, and resistances R_f^r, R_D^r, and R_Q^r, to the stator may be calculated through analytical or numerical (field distribution) methods, but they may be also measured. Care must be exercised as K_f, K_D, and K_Q depend slightly on the saturation level in the machine.
- The reduced number of inductances in of Equation 5.57 should be instrumental in their estimation (through experiments).

Note: When $\sqrt{2/3}$ is used in the Park transform (matrix) K_P, K_D, and K_Q in Equation 5.47 have to be all multiplied by $\sqrt{3/2}$, but the factor 2/3 (or 3/2) disappears completely from Equations 5.48 through 5.57 (see also Reference [1]).

5.4 Per Unit (P.U.) *dq* Model

Once the rotor variables $(V_f^r, I_f^r, I_D^r, I_Q^r, R_f^r, R_D^r, R_Q^r, L_{fl}^r, L_{Dl}^r, L_{Ql}^r)$ have been reduced to the stator, according to relationships Equations 5.47, 5.54, 5.55, and 5.56 the P.U. *dq* model requires base quantities only for the stator.

Though the selection of base quantities leaves room for choice, the following set is rather widely accepted:

$$V_b = V_n \sqrt{2} \text{ —peak stator phase nominal voltage} \tag{5.58}$$

$$I_b = I_n \sqrt{2} \text{ —peak stator phase nominal current} \tag{5.59}$$

$$S_b = 3 V_n I_n \text{ —nominal apparent power} \tag{5.60}$$

$$\omega_b = \omega_{rn} \text{ —rated electrical angular speed } (\omega_{rn} = p_1 \Omega_{rn}) \tag{5.61}$$

Based on this restricted set, additional base variables are derived:

$$T_{eb} = \frac{S_b \cdot p_1}{\omega_b} \text{ —base torque} \tag{5.62}$$

$$\Psi_b = \frac{V_b}{\omega_b} \text{ —base flux linkage} \tag{5.63}$$

$$Z_b = \frac{V_b}{I_b} = \frac{V_n}{I_n} \text{ —base impedance (valid also for resistances and reactances)} \tag{5.64}$$

$$L_b = \frac{Z_b}{\omega_b} \text{ —base inductance} \tag{5.65}$$

Inductances and reactances are the same in P.U. Though in some instances time is also provided with a base quantity $t_b = 1/\omega_b$, we choose here to leave time in seconds as it seems more intuitive.

The inertia H_b (in seconds) is consequently

$$H_b = \frac{1}{2} J \left(\frac{\omega_b}{p_1} \right)^2 \cdot \frac{1}{S_b} \tag{5.66}$$

It follows that the time derivative in P.U. becomes thus:

$$\frac{d}{dt} \to \frac{1}{\omega_b} \frac{d}{dt} ; s \to \frac{s}{\omega_b} \text{ (Laplace operator)} \tag{5.67}$$

The P.U. variables and coefficients (inductances, reactances, and resistances) are in general denoted by lower-case letters.

Consequently, the P.U. *dq* model equations—extracted from Equations 5.39 through 5.41, 5.43, and 5.46 become

$$\frac{1}{\omega_b}\frac{d}{dt}\psi_d = \omega_r\psi_q - i_d r_s - v_d; \quad \psi_d = l_{sl}i_d + l_{dm}\left(i_d + i_D + i_f\right)$$

$$\frac{1}{\omega_b}\frac{d}{dt}\psi_q = -\omega_r\psi_d - i_q r_s - v_q; \quad \psi_q = l_{sl}i_d + l_{qm}\left(i_q + i_Q\right)$$

$$\frac{1}{\omega_b}\frac{d}{dt}\psi_0 = -i_0 r_0 - v_0$$

$$\frac{1}{\omega_b}\frac{d}{dt}\psi_f = -i_f r_f + v_f; \quad \psi_f = l_{fl}i_f + l_{dm}\left(i_Q + i_D + i_F\right)$$

$$\frac{1}{\omega_b}\frac{d}{dt}\psi_D = -i_D r_D; \quad \psi_D = l_{Dl}i_D + l_{dm}\left(i_d + i_D + i_F\right) \tag{5.68}$$

$$\frac{1}{\omega_b}\frac{d}{dt}\psi_Q = -i_Q r_Q; \quad \psi_Q = l_{Ql}i_Q + l_{qm}\left(i_q + i_Q\right)$$

$$2H\frac{d}{dt}\omega_r = t_{shaft} - t_e; \quad t_{shaft} = \frac{T_{shaft}}{T_{eb}}; \quad t_e = \frac{T_e}{T_{eb}}$$

$$t_e = -\left(\psi_d i_q - \psi_q i_d\right); \quad \frac{1}{\omega_b}\frac{d\theta_{er}}{dt} = \omega_r; \quad \theta_{er} - \text{in radians}$$

t_e is the P.U. torque is positive when opposite to the direction of motion (generator mode).

The Park transformation (matrix) in P.U. basically retains its original form. Its usage is essential to make transition between the real machine and *d–q* model voltages (in general). $v_d(t)$, $v_q(t)$, $v_f(t)$, and $t_{shaft}(t)$ are needed to investigate any transient or steady-state regime of the machine. Finally, the stator currents of the *d–q* model (i_d, i_q) are transformed back into i_A, i_B, i_C to find the real machine stator currents, behavior for the case in point.

The field-winding current I_f and the damper cage currents I_D, I_Q are the same for the *dq* model and for the original machine. Note that all the quantities in Equation 5.68 are reduced to stator and thus directly related in P.U. to stator base quantities.

In Equation 5.68, all quantities except for time *t* and *H* are in P.U. (time *t* and inertia *H* are given in seconds and ω_b in rad/s).

Equations 5.68 represent the dq_0 model of a three-phase SG with single damper circuits along rotor orthogonal axes *d* and *q*.

Also, the coupling of windings along axis *d*, and, respectively, *q* takes place only through the main (air gap) flux linkage.

Magnetic saturation is not yet included, and only the fundamental of air gap flux distribution is considered.

Instead of P.U. inductances l_{dm}, l_{qm}, l_{fl}, l_{Dl}, and l_{Ql}, the corresponding reactances may be used: x_{dm}, x_{qm}, x_{fl}, x_{Dl}, and x_{Ql}, as the two sets are identical (in numbers, in P.U.). Also, $l_d = l_{sl} + l_{dm}$, $x_d = x_{sl} + x_{dm}$, $l_q = l_{sl} + l_{dm}$, $x_q = x_{sl} + x_{qm}$.

5.5 Steady State via the *dq* Model

During steady state, the stator voltages and currents are sinusoidal and the stator frequency ω_1 is equal to rotor electrical speed $\omega_r = \omega_1 = $ const.:

$$V_{A,B,C}(t) = V\sqrt{2}\cos\left[\omega_1 - (i-1)\frac{2\pi}{3}\right]$$

$$I_{A,B,C}(t) = I\sqrt{2}\cos\left[\omega_1 - \phi_1 - (i-1)\frac{2\pi}{3}\right]$$

(5.69)

Using the Park transformation with $\theta_{er} = \omega_1 t + \theta_0$, the *dq* voltages are obtained:

$$V_{d0} = \frac{2}{3}\left(V_A(t)\cos(-\theta_{er}) + V_B(t)\cos\left(-\theta_{er} + \frac{2\pi}{3}\right) + V_C(t)\cos\left(-\theta_{er} - \frac{2\pi}{3}\right)\right)$$

$$V_{q0} = \frac{2}{3}\left(V_A(t)\sin(-\theta_{er}) + V_B(t)\sin\left(-\theta_{er} + \frac{2\pi}{3}\right) + V_C(t)\cos\left(-\theta_{er} - \frac{2\pi}{3}\right)\right)$$

(5.70)

Substituting Equation 5.69 into Equation 5.70 yields

$$V_{d0} = V\sqrt{2}\cos\theta_0$$

$$V_{q0} = -V\sqrt{2}\sin\theta_0$$

(5.71)

In a similar way, we obtain the currents I_{d0} and I_{q0}:

$$I_{d0} = I\sqrt{2}\cos(\theta_0 + \phi_1)$$

$$I_{q0} = -I\sqrt{2}\sin(\theta_0 + \phi_1)$$

(5.72)

Under steady state, the *dq* model stator voltages and currents are DC quantities. Consequently, for steady state, we should consider $d/dt = 0$ in Equations 5.68:

$$V_{d0} = \omega_r \Psi_{q0} - I_{d0}r_s; \quad \Psi_{q0} = l_{sl}I_{q0} + l_{qm}I_{q0}$$

$$V_{q0} = -\omega_r \Psi_{d0} - I_{q0}r_s; \quad \Psi_{d0} = l_{sl}I_{d0} + l_{dm}(I_{d0} + I_{f0})$$

$$V_{f0} = r_f I_{f0}; \quad \Psi_{f0} = l_{fl}I_{f0} + l_{dm}(I_{d0} + I_{f0})$$

$$I_{D0} = I_{Q0} = 0; \quad \Psi_{D0} = l_{dm}(I_{d0} + I_{f0})I_{D0}; \quad l_d = l_{dm} + l_{sl}$$

$$t_e = -(\Psi_{d0}I_{q0} - \Psi_{q0}I_{d0}); \quad \Psi_{Q0} = l_{qm}I_{q0}; \quad l_q = l_{qm} + l_{sl}$$

(5.73)

We may now introduce space phasors for the stator quantities:

$$\bar{\Psi}_{s0} = \Psi_{d0} + j\Psi_{q0}$$

$$\bar{I}_{s0} = I_{d0} + jI_{q0} \tag{5.74}$$

$$\bar{V}_{s0} = V_{d0} + jV_{q0}$$

The stator equations in Equation 5.73 become thus:

$$\bar{V}_{s0} = -r_s\bar{I}_{s0} - j\omega_r\bar{\Psi}_{s0} \tag{5.75}$$

The space-phasor (or vector) diagram corresponding to Equation 5.74 is shown in Figure 5.5.

With $\varphi_1 > 0$, both the active and reactive power delivered by the SG are positive. This condition implies that I_{d0} goes against I_{f0} in the vector diagram; also, for generating, I_{q0} falls along the negative direction of axis q. Note that axis q is ahead of axis d in the direction of motion and, for $\varphi_1 > 0$, $\bar{I}_{s0}$ and $\bar{V}_{s0}$ are contained in the third quadrant. Also, the positive direction of motion is trigonometric. The voltage vector $\bar{V}_{s0}$ will stay in the third quadrant (for generating), while $\bar{I}_{s0}$ may be placed either in the third or fourth quadrant. We may use Equation 5.72 to calculate the stator currents I_{d0}, I_{q0} provided V_{d0} and V_{q0} are known.

The initial angle θ_0 of Park transformation represents, in fact, the angle between the rotor pole (d axis) axis and the voltage vector angle. Axis d is behind $\bar{V}_{s0}$, and this is explained in Figure 5.5:

$$\theta_0 = -\left(\frac{3\pi}{2} - \delta_{V0}\right) \tag{5.76}$$

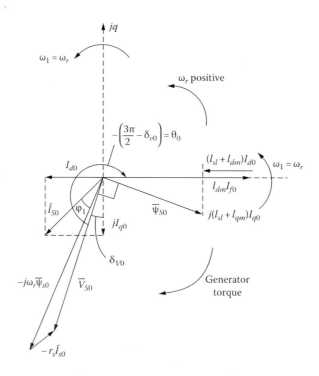

FIGURE 5.5   The space phasor (vector) diagram of SGs.

Substituting Equation 5.75 into Equation 5.71, we obtain the following:

$$V_{d0} = -V\sqrt{2}\sin\delta_{V0} < 0$$

$$V_{q0} = -V\sqrt{2}\cos\delta_{V0} < 0$$

$$I_{d0} = -I\sqrt{2}\sin(\delta_{V0}+\phi_1) <> 0 \qquad (5.77)$$

$$I_{q0} = -I\sqrt{2}\cos(\delta_{V0}+\phi_1) < 0 \text{ for generating}$$

The active and reactive powers P_1 and Q_1 are as expected

$$P_1 = \frac{3}{2}\left(V_{d0}I_{d0} + V_{q0}I_{q0}\right) = 3VI\cos\phi_1$$

$$Q_1 = \frac{3}{2}\left(V_{d0}I_{q0} - V_{q0}I_{d0}\right) = 3VI\sin\phi_1 \qquad (5.78)$$

In P.U.: $v_{d0} = -v \times \sin\delta_{v0}$, $v_{q0} = -v\cos\delta_{v0}$, $i_{d0} = -i\sin(\delta_{v0}+\phi_1)$, $i_{q0} = -i\cos(\delta_{v0} + \phi_1)$.
The no-load regime is obtained with $I_{d0} = I_{q0} = 0$, and thus

$$V_{d0} = 0$$

$$V_{q0} = -\omega_r \Psi_{d0} = -\omega_r l_{dm} I_{f0} = \frac{-V}{\sqrt{2}} \qquad (5.79)$$

For no load in Equation 5.75, $\delta_v = 0$ and $I = 0$. V_0 is the no-load phase voltage (RMS value).
For the steady-state short circuit $V_{d0} = V_{q0} = 0$ in Equation 5.73. If, in addition, $r_s \approx 0$, then $I_{qs} = 0$ and

$$I_{d0sc} = \frac{-l_{dm}I_{f0}}{l_d};$$

$$I_{sc3} = \frac{I_{d0sc}}{\sqrt{2}} \qquad (5.80)$$

where I_{sc3} is the phase short-circuit current (RMS value).

Example 5.1

A hydrogenerator with 200 MVA, 24 kV (star connection), 60 Hz, unity power factor, at 90 rpm has the following P.U. parameters: $l_{dm} = 0.6$, $l_{qm} = 0.4$, $l_{sl} = 0.15$, $r_s = 0.003$, $l_{fl} = 0.165$, and $r_f = 0.006$. The field circuit is supplied at 800 V DC. ($V_f^r = 800$ V).

When the generator works at rated MVA, $\cos\phi_1 = 1$ and rated terminal voltage, calculate the following:

(i) The internal angle δ_{V0}
(ii) P.U. values of V_{d0}, V_{q0}, I_{d0}, I_{q0}
(iii) Air gap torque in P.U. and in Nm
(iv) The P.U. field current I_{f0} and its actual value in Amperes

Solution

(i) The vector diagram simplifies as $\cos\varphi_1 = 1$ ($\varphi_1 = 0$), but it is worth deriving a formula to directly calculate the power angle δ_{v0}.

Substituting Equations 5.71 and 5.72 into Equation 5.73 yields finally

$$\delta_{V0} = \tan^{-1}\left(\frac{\omega_1 l_q I\cos\phi_1 - r_s I\sin\phi_1}{V + r_s I\cos\phi_1 + \omega_1 l_q I\sin\phi_1}\right)$$

with $\varphi_1 = 0$ and $\omega_1 = 1$, $I = 1$ P.U (rated current), $v = 1$ P.U. (rated voltage):

$$\delta_{V0} = \tan^{-1}\frac{1\times0.45\times1\times1-0.0}{1+0.003\times1\times1\times0} = 24.16°$$

(ii) The field current may be calculated from Equation 5.73:

$$i_{f0} = \frac{-V_{q0} - I_{q0}r_s - \omega_r l_d I_{d0}}{\omega_r l_{dm}} = \frac{0.912 + 0.912\times0.003 + 1\cdot(0.6+0.15)0.4093}{1.0\cdot0.6} = 2.036 \text{ P.U.}$$

The base current

$$I_0 = I_n\sqrt{2} = \frac{S_n \cdot \sqrt{2}}{\sqrt{3}V_{nl}} = \frac{200\cdot10^6 \cdot \sqrt{2}}{\sqrt{3}\cdot24,000} = 6792 \text{ A}$$

(iii) The field circuit P.U. resistance $r_f = 0.006$, and thus the P.U. field circuit voltage, reduced to the stator is:

$$V'_{f0} = r_f \cdot I_{f0} = 2.036\times0.006 = 12.216\times10^{-3} \text{ P.U.}$$

Now with $V'_f = 800$ V, the reduction to stator coefficient K_f for field current is as follows:

$$K_f = \frac{2}{3}\frac{V^r_f}{v_{f0}\cdot V_b} = \frac{2}{3}\frac{800}{12.216\times10^{-3}\cdot24,000/\sqrt{3}\cdot\sqrt{2}} \approx 2.224$$

Consequently, the field current (in Amperes) I^r_{f0} is

$$I^r_{fo} = \frac{i_{f0}\cdot I_b}{K_f} = \frac{2.036\cdot6792}{2.224} = 6218 \text{ A}$$

Therefore, the excitation power

$$P_{exc} = V^r_{f0}I^r_{f0} = 800\times6218 = 4.9744 \text{ MW}$$

(iv) The P.U. electromagnetic torque is

$$t_e \approx p_e + r_s I^2 = 1.0 + 0.003\cdot1^2 = 1.003$$

The torque in Nm is ($2p_1 = 80$ poles)

$$T_e = t_e \cdot T_{eb} = 1.003 \times \frac{200 \times 10^6}{2\pi \cdot 60 / 40} = 21.295 \times 10^6 \text{ Nm}(!)$$

5.6 General Equivalent Circuits

Let us replace d/dt in the P.U. dq model (Equation 5.68) by the Laplace operator s/ω_b, which means that the initial conditions are implicitly zero. If they are not, their initial values should be added.

The general equivalent circuits illustrate Equation 5.68 with d/dt replaced by s/ω_b after separating the main flux linkage components Ψ_{dm}, Ψ_{qm}:

$$\Psi_d = l_{sl}I_d + \Psi_{dm}; \quad \Psi_q = l_{sl}I_q + \Psi_{qm}$$

$$\Psi_{dm} = l_{dm}\left(I_d + I_D + I_f\right); \quad \Psi_{qm} = l_{qm}\left(I_q + I_Q\right)$$

$$\Psi_f = l_{fl}I_f + \Psi_{dm}; \quad \Psi_D = l_{sl}I_D + \Psi_{dm} \tag{5.81}$$

$$\left(r_0 + \frac{s}{\omega_b}l_0\right)i_0 = -V_0$$

with

$$\left(r_f + \frac{s}{\omega_b}l_{fl}\right)I_f - V_f = -\frac{s}{\omega_b}\Psi_{dm}$$

$$\left(r_D + \frac{s}{\omega_b}l_{Dl}\right)I_D = -\frac{s}{\omega_b}\Psi_{dm}$$

$$\left(r_Q + \frac{s}{\omega_b}l_{Ql}\right)I_Q = -\frac{s}{\omega_b}\Psi_{qm} \tag{5.82}$$

$$\left(r_S + \frac{s}{\omega_b}l_{sl}\right)I_d + V_d - \omega_r\Psi_q = -\frac{s}{\omega_b}\Psi_{dm}$$

$$\left(r_S + \frac{s}{\omega_b}l_{sl}\right)I_q + V_q + \omega_r\Psi_d = -\frac{s}{\omega_b}\Psi_{qm}$$

Equation 5.82 contains three circuits in parallel along axis d and two equivalent circuits along axis q. It is also implicit that the coupling of the circuits along axis d and q is performed only through the main flux components Ψ_{dm} and Ψ_{qm}.

Magnetic saturation and frequency effects are not yet considered.

Based on Equations 5.82, the general equivalent circuits of SG are shown in Figure 5.6.

A few remarks on Figure 5.6 are as follows:

- The magnetization current components I_{dm} and I_{qm} are defined as the sum of the dq model currents:

$$I_{dm} = I_d + I_D + I_f;$$

$$I_{qm} = I_q + I_Q \tag{5.83}$$

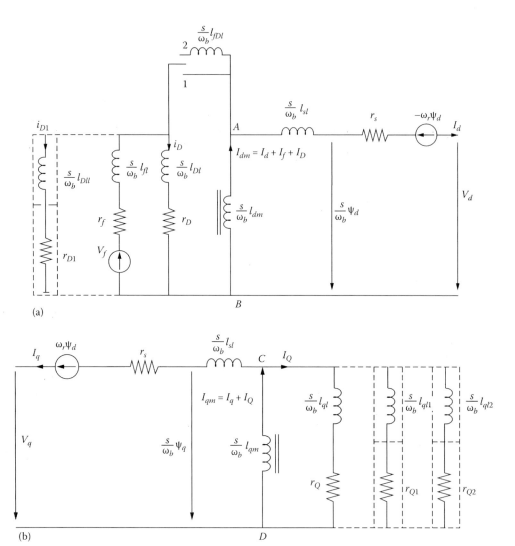

FIGURE 5.6 General equivalent circuits of SGs: (a) along axis d and (b) along axis q.

- There is no magnetic coupling between the orthogonal axes d and q because magnetic saturation is either ignored or considered separately along each axis as follows:

$$l_{sl}(i_s); \quad l_{dm}(I_{dm}); \quad l_{sl}(I_s), l_{qm}(I_{qm}); \quad I_s = \sqrt{I_d^2 + I_q^2}$$

- Should the frequency (skin) effect be present in the rotor damper cage (or in the rotor pole solid iron), additional rotor circuits are added in parallel. In general, one additional circuit along axis d and two along axis q are sufficient even for solid-rotor pole SGs (Figure 5.6). In these cases, additional equations have to be added to Equation 5.82, but their composure is straightforward.
- Figure 5.6a also exhibits the possibility to consider the additional, leakage type, flux linkage (inductance, l_{fDl}) between the field and damper cage windings, in salient-pole rotors. This inductance is considered instrumental when the field-winding parameter identification is checked after the stator parameters have been estimated in tests with measured stator variables. Sometimes, l_{fDl} is estimated as negative.

- For steady state $s = 0$ in the equivalent circuits, and thus the voltages V_{AB} and V_{CD} are zero. Consequently, $I_{D0} = I_{Q0} = 0$, $V_{f0} = -r_f I_{f0}$ and the steady-state dq model equations may be "read" from Figure 5.6.
- The null component voltage equation in Equation 5.81, $V_0 + (r_0 + (s/\omega_b)l_0)i_0 = 0$, does not appear, as expected, in the general equivalent circuit because it does not interfere with the main flux fundamental. In reality, the null component may produce some eddy currents in the rotor cage through its third space harmonic mmf.

5.7 Magnetic Saturation Inclusion in the *dq* Model

The magnetic saturation level is, in general, different in various regions of SG cross-section. Also the distribution of the flux density in the air gap is not quite sinusoidal.

However, in the dq model, only the flux–density fundamental is considered.

Further on, the leakage flux paths saturation is influenced by the main flux paths saturation. A fully realistic model of saturation would mean that all leakage and main inductances depend on all currents in the dq model.

However, such a model would be too cumbersome to be practical.

Consequently, we will present here only two main approximations of magnetic saturation inclusion in the dq model from the many proposed so far [2–7]. These two appear to us to be rather representative. Both of them include cross-coupling between the two orthogonal axes due to main flux path saturation. While the first one presupposes the existence of an unique magnetization curve along axis d, and, respectively q, in relation to total mmf ($I_m = \sqrt{I_{dm}^2 + I_{qm}^2}$), the second one curve-fits the family of curves $\Psi_{dm}^*(I_{dm}, I_{qm})$, $\Psi_{qm}^*(I_{dm}, I_{qm})$, keeping the dependence on both I_{dm} and I_{qm}.

In both the models, the leakage flux path saturation is considered separately by defining transient leakage inductances l_{sl}^t, l_{Dl}^t, l_{fl}^t.

$$l_{sl}^t = l_{sl} + \frac{\partial l_{sl}}{\partial i_s} i_s \le l_{sl}; \quad I_s = \sqrt{I_d^2 + I_q^2}$$

$$l_{Dl}^t = l_{Dl} + \frac{\partial l_{Dl}}{\partial i_D} i_D < l_{Dl}$$

$$l_{fl}^t = l_{fl} + \frac{\partial l_{fl}}{\partial i_f} i_f < l_{fl} \tag{5.84}$$

$$l_{Ql}^t = l_{Ql} + \frac{\partial l_{Ql}}{\partial i_Q} i_Q < l_{Ql}$$

Each of transient inductances in Equation 5.84 is considered as dependent on the respective current.

5.7.1 The Single *dq* Magnetization Curve Model

According to this model of main flux paths saturation, the distinct magnetization curves along axis d and q depend only on the total magnetization current I_m [2,3].

$$\Psi_{dm}^*(I_m) \ne \Psi_{qm}^*(I_m); \quad I_m = \sqrt{I_{dm}^2 + I_{qm}^2}$$

$$I_{dm} = I_d + I_D + I_f \tag{5.85}$$

$$I_{qm} = I_q + I_Q$$

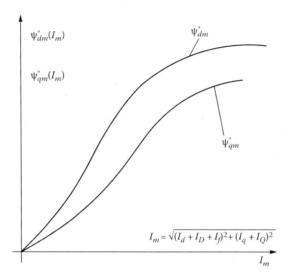

FIGURE 5.7 The unique d, q magnetization curves.

Note: The two distinct, but unique, d and q axes magnetization curves, shown in Figure 5.7, represent a disputable approximation. It is only recently that FEM investigations have shown that the concept of unique magnetization curves does not hold with the SG for underexcited (draining reactive power) conditions [4]: $I_m < 0.7$ P.U.

For $I_m > 0.7$, the model apparently works well for a wide range of active and reactive power load conditions.

The magnetization inductances l_{dm}, l_{qm} are also functions of I_m, only:

$$\Psi_{dm} = l_{dm}\left(I_m\right) \cdot I_{dm}$$
$$\Psi_{qm} = l_{qm}\left(I_m\right) \cdot I_{qm} \tag{5.86}$$

with

$$l_{dm}\left(I_m\right) = \frac{\Psi_{dm}^*\left(I_m\right)}{I_m}$$
$$l_{qm}\left(I_m\right) = \frac{\Psi_{qm}^*\left(I_m\right)}{I_m} \tag{5.87}$$

We should note that the $\Psi_{dm}^*\left(I_m\right), \Psi_{qm}^*\left(I_m\right)$ may be obtained through tests where either only one or both components (I_{dm}, I_{qm}) of magnetization current I_m are present. This detail should not be missed if coherent results are to be expected.

It is advisable to use a few combinations of I_{dm}, I_{qm} for each axis and use curve fitting methods to derive the unique magnetization curves $\Psi_{dm}^*\left(I_m\right), \Psi_{qm}^*\left(I_m\right)$.

Based on Equations 5.86 and 5.87, the main flux time derivatives are obtained:

$$\frac{d\Psi_{dm}}{dt} = \frac{d\Psi_{qm}^*}{dI_m}\frac{dI_m}{dt} \cdot \frac{I_{dm}}{I_m} + \frac{\Psi_{dm}^*}{I_m^2}\left(I_m\frac{dI_{dm}}{dt} - I_{dm}\frac{dI_m}{dt}\right)$$
$$\frac{d\Psi_{qm}}{dt} = \frac{d\Psi_{qm}^*}{dI_m}\frac{dI_m}{dt} \cdot \frac{I_{qm}}{I_m} + \frac{\Psi_{qm}^*}{I_m^2}\left(I_m\frac{dI_{qm}}{dt} - I_{qm}\frac{dI_m}{dt}\right) \tag{5.88}$$

where

$$\frac{dI_m}{dt} = \frac{I_{qm}}{I_m}\frac{dI_{qm}}{dt} + \frac{I_{dm}}{I_m}\frac{dI_{dm}}{dt} \tag{5.89}$$

Finally,

$$\frac{d\Psi_{dm}}{dt} = l_{ddm}\frac{dI_{dm}}{dt} + l_{qdm}\frac{dI_{qm}}{dt}$$

$$\frac{d\Psi_{qm}}{dt} = l_{dqm}\frac{dI_{dm}}{dt} + l_{qqm}\frac{dI_{qm}}{dt} \tag{5.90}$$

$$l_{ddm} = l_{dmt}\frac{I_{dm}^2}{I_m^2} + l_{dm}\frac{I_{qm}^2}{I_m^2}$$

$$l_{qqm} = l_{qmt}\frac{I_{qm}^2}{I_m^2} + l_{qm}\frac{I_{dm}^2}{I_m^2} \tag{5.91}$$

$$l_{dqm} = l_{qdm} = \left(l_{dmt} - l_{dm}\right)I_{dm}\frac{I_{qm}}{I_m^2}$$

$$l_{dmt} - l_{dm} = l_{qmt} - l_{qm} \tag{5.92}$$

$$l_{dmt} = \frac{d\Psi_{dm}^*}{dI_m}$$

$$l_{qmt} = \frac{d\Psi_{qm}^*}{dI_m} \tag{5.93}$$

The equality of coupling transient inductances $l_{dqm} = l_{qdm}$ between the two axes is based on the reciprocity theorem. l_{dmt} and l_{qmt} are the so-called differential d and q axes magnetization inductances, while l_{ddm} and l_{qqm} are the transient magnetization self-inductances with saturation included.

All these four inductances depend, in fact, on both I_{dm} and I_{qm}, while l_{dm}, l_{dmt}, l_{qm}, and l_{qmt} depend only on I_m.

For the situation when DC premagnetization occurs, the differential magnetization inductances l_{dmt} and l_{qmt} should be replaced by the so-called incremental inductances l_{dm}^i and l_{qm}^i:

$$l_{dm}^i = \frac{\Delta\Psi_{dm}^*}{\Delta I_m}$$

$$l_{qm}^i = \frac{\Delta\Psi_{qm}^*}{\Delta I_m} \tag{5.94}$$

l_{dm}^i and l_{qm}^i are related to the incremental permeability in the iron core when a superposition of DC and AC magnetization occurs (Figure 5.8.)

The normal permeability of iron $\mu_n = B_m/H_m$ is used when calculating the magnetization inductances l_{dm} and l_{qm}, $\mu_d = dB_m/dH_m$ for l_{dm}^t and l_{qm}^t and $\mu_i = \Delta B_m/\Delta H_m$ (Figure 5.8) for the incremental magnetization inductances l_{md}^i, l_{mq}^i.

For the incremental inductances, the permeability μ_i corresponds to a local small hysteresis cycle (in Figure 5.8) and thus $\mu_i < \mu_d < \mu_n$. For zero DC premagnetization and small AC voltages (currents) at

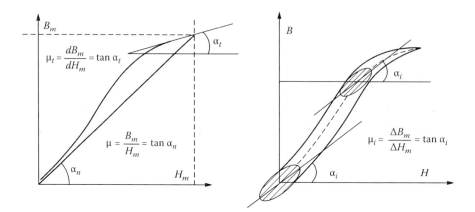

FIGURE 5.8   Iron permeabilities.

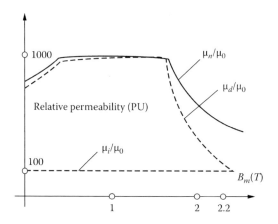

FIGURE 5.9   Typical P.U. normal, differential, and incremental permeabilities of silicon laminations.

standstill, for example, $\mu_i \approx 100\text{--}150 \, \mu_0$ which explains why the magnetization inductances correspond to l_{md}^i and l_{mq}^i rather than to l_{dm}, l_{qm} and are much smaller than the latter (Figure 5.9).

Once $l_{dm}(I_m), l_{qm}(I_m), l_{dm}^t(I_m), l_{qm}^t(I_m), l_{dm}^i(I_m), l_{qm}^i(I_m)$ are determined, either through field analysis or through experiments, then $l_{ddm}(I_{dm}, I_{qm}), l_{qqm}(I_{dm}, I_{qm})$ may be calculated with I_{dm} and I_{qm} given.

Interpolation through tables or analytical curve fitting may be applied to produce easy-to-use expressions for digital simulations.

The single unique $d\text{--}q$ magnetization curves model include the cross-coupling implicitly in the expressions of Ψ_{dm}, Ψ_{qm}, but it considers it explicitly in the $d\Psi_{dm}/dt$ and $d\Psi_{qm}/dt$ expressions, that is, in the transients. Either with currents $I_{dm}, I_{qm}, I_f, I_D, I_Q, \omega_r, \theta_{er}$ or with flux linkages $\Psi_{dm}, \Psi_{qm}, \Psi_f, \Psi_D, \Psi_Q, \omega_r, \theta_{er}$ (or with quite a few intermediary current, flux–linkage combinations) as variables, models based on same concepts may be developed and used rather handily both for the study of steady states and for transients [5]. The computation of $\Psi_{dm}^*(I_m), \Psi_{qm}^*(I_m)$ functions or their measurement from standstill tests is rather straightforward.

This tempting simplicity is payed for by the limitation that the unique $d\text{--}q$ magnetization curves concept does not seem to hold when the machine is notably underexcited, with the emf lower than the terminal voltage, because the saturation level is smaller despite the fact that I_m is about the same as for lagging power factor at constant voltage [4].

This limitation justifies the search for a more general model valid for the whole range of active (reactive) power capability envelope of the SG.

We call this the "multiple magnetization curve model."

5.7.2 Multiple *dq* Magnetization Curve Model

This kind of models presupposes that the d and q axis flux linkages Ψ_d and Ψ_q are explicit functions of I_d, I_q, I_{dm}, I_{qm}.

$$\Psi_d = l_{sl}I_d + l_{dq}\left(I_q + I_Q\right) + l_{dm}\left(I_f + I_D + I_d\right) = l_{sl}I_d + l_{dq}I_{qm} + l_{dms}I_{dm} = l_{sl}I_d + \Psi_{dms}$$

$$\Psi_q = l_{sl}I_d + l_{qm}\left(I_q + I_Q\right) + l_{dq}\left(I_d + I_D + I_f\right) = l_{sl}I_q + l_{qm}I_{qm} + l_{dq}I_{dm} = l_{sl}I_q + \Psi_{qms}$$

$$\Psi_f = l_{fl}I_f + l_{dm}\left(I_f + I_D + I_d\right) + l_{dq}\left(I_q + I_Q\right) = l_{fl}I_f + l_{dms}I_{dm} + l_{dq}I_{qm} = l_{fl}I_f + \Psi_{dms} \qquad (5.95)$$

$$\Psi_D = l_{Dl}I_D + l_{dm}\left(I_f + I_D + I_d\right) + l_{dq}\left(I_q + I_Q\right) = l_{Dl}I_D + l_{dms}I_{dm} + l_{dq}I_{qm} = l_{Dl}I_D + \Psi_{dms}.$$

$$\Psi_Q = l_{Ql}I_f + l_{dq}\left(I_f + I_D + I_d\right) + l_{qm}\left(I_q + I_Q\right) = l_{Ql}I_Q + l_{dq}I_{dm} + l_{qms}I_{qm} = l_{Ql}I_Q + \Psi_{qms}$$

Now l_{dm}, l_{qm} are functions of both I_{dm} and I_{qm}. Conversely $\Psi_{dms}(I_{dm}, I_{qm})$ and $\Psi_{qms}(I_{dm}, I_{qm})$ are two families of magnetization curves that have to be found either by computation or by experiments.

For steady state $I_D = I_Q = 0$, but otherwise Equations 5.95 hold as they are. Basically, the Ψ_{dms} and Ψ_{qms} curves look like the ones shown Figure 5.10.

$$\Psi_{dms} = l_{dms}(I_{dm}, I_{qm})I_{dm}$$

$$\Psi_{qms} = l_{qms}(I_{dm}, I_{qm})I_{qm} \qquad (5.96)$$

Once this family of curves is acquired (by FE analysis or by experiments), various analytical approximations may be used to curve-fit them adequately.

Then, with $\Psi_d, \Psi_q, \Psi_f, \Psi_D, \Psi_Q, \omega_r, \theta_{er}$ as variables and I_f, I_D, I_Q, I_d, I_q as dummy variables, the $\Psi_{dms}(I_{dm}, I_{qm})$, $\Psi_{qms}(I_{dm}, I_{qm})$ functions are used in Equation 5.95 to calculate iteratively each time step, the dummy variables.

When using flux linkages as variables no additional inductances, responsible for cross-coupling magnetic saturation, need consideration. As they are not constant, their introduction does not seem very practical. However, such attempts keep reoccurring [6,7] as the problem seems far from a definitive solution.

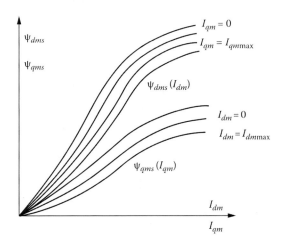

FIGURE 5.10 Family of magnetization curves.

Considering cross-coupling due to magnetic saturation seems necessary when calculating the field current, stator current, and power angle, under steady state for given active, reactive power and voltage, with an error less than 2% for the currents and around an 1° error for the power angle [4,7]. Also, during large disturbance transients, where the main flux varies notably, the cross-coupling saturation effect is to be considered.

Though magnetic saturation is very important for refined steady state and for transients investigations, most of the theory of transients for the control SGs is developed for constant parameter conditions; operational parameters is such a case.

5.8 Operational Parameters

In the absence of magnetic saturation variation, the general equivalent circuits of SG—Figure 5.6—lead to the following generic expressions of operational parameters in the Ψ_d and Ψ_q operational expressions:

$$\Psi_d(s) = l_d(s) \cdot I_d(s) + g(s)v_{ex}(s) \text{ [P.U.]}$$

$$\Psi_q(s) = l_q(s) \cdot I_q(s)$$

(5.97)

with

$$l_d(s) = \frac{(1+sT_d')(1+sT_d'')}{(1+sT_{d0}')(1+sT_{d0}'')} \cdot l_d \text{ [P.U.]}$$

$$l_q(s) = \frac{(1+sT_q'')}{(1+sT_{q0}'')} \cdot l_q \text{ [P.U.]}$$

$$g(s) = \frac{(1+sT_D)}{(1+sT_{d0}')(1+sT_{d0}'')} \cdot \frac{l_{dm}}{r_f} \text{ [P.U.]}$$

(5.98)

Sparing the analytical derivations, the time constants T_d'', T_d'', T_{d0}', T_{d0}'', T_q', T_{q0}'' have the expressions

$$T_d' \approx \frac{1}{\omega_b r_f}\left(l_{fl} + l_{fDl} + \frac{l_{dm}l_{sl}}{l_{dm}+l_{sl}}\right); [s]$$

$$T_d'' \approx \frac{1}{\omega_b r_D}\left(l_{Dl} + \frac{l_{dm}l_{fDl}l_{fl} + l_{dm}l_{sl}l_{fl} + l_{sl}l_{fDl}l_{fl}}{l_{dm}l_{fl} + l_{fl}l_{sl} + l_{dm}l_{fDl} + l_{sl}l_{fDl} + l_{fDl}l_{fl} + l_{dm}l_{sl}}\right); [s]$$

$$T_{d0}' \approx \frac{1}{\omega_b r_f}\left(l_{dm} + l_{fDl} + l_{fl}\right); [s]$$

$$T_{d0}'' \approx \frac{1}{\omega_b r_D}\left(l_{Dl} + \frac{l_{fl}(l_{dm}+l_{fDl})}{l_{fl}+l_{dm}+l_{fDl}}\right); [s]$$

(5.99)

$$T_q'' \approx \frac{1}{\omega_b r_Q}\left(l_{Ql} + \frac{l_{qm}l_{sl}}{l_{qm}+l_{sl}}\right); [s]$$

$$T_{q0}'' \approx \frac{1}{\omega_b r_Q}\left(l_{Ql} + l_{qm}\right); [s]$$

$$T_D \approx \frac{l_{Dl}}{\omega_b r_D}; [s]$$

As already mentioned, with ω_b in rad/s, the time constants are all in seconds, while all resistances and inductances are in P.U. values.

The time constants differ between each other up to more 100–1 ratios. T'_{d0} is of the order of seconds in large SGs, while T'_d, T''_{d0}, and T''_{q0} are in the order of a few tenth of a second, T''_d, T''_q in the order of a few tens of milliseconds and T_D is a few milliseconds.

Such a broad spectrum of time constants indicates that the SG equations for transients Equation 5.82 represent a stiff system. Consequently, the solution through numerical methods needs time integration steps smaller than the lowest time constant, in order to portray correctly all occurring transients. The time constants are catalog data for SGs:

- T'_{d0}—d axis open circuit field-winding (transient) time constant ($I_d = 0$, $I_D = 0$)
- T''_{d0}—d axis open circuit damper-winding (subtransient) time constant ($I_d = 0$)
- T'_d—d axis transient time constant ($I_D = 0$)—field-winding time constant with short-circuited stator but with open damper winding
- T''_d—d axis subtransient time constant—damper-winding time constant with short-circuited field winding and stator
- T''_{q0}—q axis open circuit damper-winding (subtransient) time constant ($I_q = 0$)
- T''_q—q axis subtransient time constant (q axis damper-winding time constant with short-circuited stator)
- T_D—d axis damper-winding self-leakage time constant

In the industrial practice of SGs the limit—initial and final—values of operational inductances have become catalog data:

$$l''_d = \lim_{\substack{s \to \infty \\ (t \to 0)}} l_d(s) = l_d \cdot \frac{T'_d T''_d}{T'_{d0} T''_{d0}}$$

$$l'_d = \lim_{\substack{s \to \infty \\ T''_d = T''_{d0} = 0}} l_d(s) = l_d \cdot \frac{T'_d}{T'_{d0}}$$

$$l_d = \lim_{\substack{s \to 0 \\ t \to \infty}} l_d(s) = l_d \qquad (5.100)$$

$$l''_q = \lim_{\substack{s \to \infty \\ t \to 0}} l_q(s) = l_q \frac{T''_q}{T''_{q0}}$$

$$l_q = \lim_{\substack{s \to 0 \\ t \to \infty}} l_q(s) = l_q$$

where

l''_d, l'_d, and l_d are d-axis subtransient, transient, and synchronous inductances
l''_q and l_q are q-axis subtransient, transient, and synchronous inductances

Typical values of the time constants (in seconds) and subtransient and transient and synchronous inductances (in P.U.) are shown in Table 5.1.

As Table 5.1 suggests, various inductances and time constants that characterize the SG are constants. In reality, they depend on magnetic saturation and skin effects (in solid rotors) as suggested in previous paragraphs. There are, however, transient regimes where the magnetic saturation stays practically the same as it corresponds to small disturbance transients.

On the other hand, in high-frequency transients, the l_d, l_q variation with magnetic saturation level is less important, while the leakage flux paths saturation becomes notable for large values of stator and rotor current (the beginning of a sudden short-circuit transient).

TABLE 5.1 Typical SG Parameter Values

Parameter	Two-Pole Turbo Generator	Hydro Generators
l_d (P.U.)	0.9–1.5	0.6–1.5
l_q (P.U.)	0.85–1.45	0.4–1.0
l_d' (P.U.)	0.12–0.2	0.2–0.5
l_d'' (P.U.)	0.07–0.14	0.13–0.35
l_{fDl} (P.U.)	−0.05–+0.05	−0.05–+0.05
l_0 (P.U.)	0.02–0.08	0.02–0.2
l_p (P.U.)	0.07–0.14	0.15–0.2
r_s (P.U.)	0.0015–0.005	0.002–0.02
T_{d0}' (s)	2.8–6.2	1.5–9.5
T_d' (s)	0.35–0.9	0.5–3.3
T_d'' (s)	0.02–0.05	0.01–0.05
T_{d0}'' (s)	0.02–0.15	0.01–0.15
T_q'' (s)	0.015–0.04	0.02–0.06
T_{q0}'' (s)	0.04–0.08	0.05–0.09
l_q'' (P.U.)		0.2–0.45

l_p—Potier inductance in P.U. ($lp \geq l_d$).

To make the treatment of transients easier to approach, we distinguish here a few types of transients:

- Fast (electromagnetic) transients: speed is constant
- Electromechanical transients: electromagnetic + mechanical transients (speed also varies)
- Slow (mechanical) transients: electromagnetic steady state; speed varies

In what follows, we will treat each of these transients in some detail.

5.8.1 Electromagnetic Transients

In fast (electromagnetic) transients, the speed may be considered constant and thus the equation of motion is ignored.

The stator voltage equations of Equation 5.82 in Laplace form with Equation 5.97 become thus:

$$-v_d(s) = r_s I_d + \frac{s}{\omega_b} l_d(s) I_d - \omega_r l_q(s) I_q + g(s) \frac{s}{\omega_b} v_f(s)$$

$$-v_q(s) = r_s I_q + \frac{s}{\omega_b} l_q(s) I_q + \omega_r \left[l_d(s) I_d + g(s) v_f(s) \right]$$

(5.101)

Note that ω_r is in relative units and, for rated rotor speed, $\omega_r = 1$.

If the initial values I_{d0}, I_{q0} of variables I_d and I_q and the time variation of $v_d(t)$, $v_q(t)$ and $v_f(t)$ may be translated into Laplace forms of $v_d(s)$, $v_q(s)$, $v_f(s)$, then Equations 5.101 may be solved to obtain the $i_d(s)$ and $i_q(s)$:

$$\begin{vmatrix} -v_d(s) & -g(s)\dfrac{s}{\omega_b}v_f(s) \\ -v_q(s) & -\omega_{r0}g(s)v_f(s) \end{vmatrix} = \begin{vmatrix} r_s + \dfrac{s}{\omega_b l_d(s)} & -\omega_{r0} l_q(s) \\ \omega_{r0} l_d(s) & r_s + \dfrac{s}{\omega_b} l_q(s) \end{vmatrix} \begin{vmatrix} i_d(s) \\ i_q(s) \end{vmatrix}$$

(5.102)

Though $I_d(s)$ and $I_q(s)$ may be directly derived from Equation 5.102, their expressions are hardly practical in the general case.

However, there are a few particular operation modes where their pursuit is important. The sudden three-phase short circuit from no-load and the step voltage or AC operation at standstill are considered here. For start, the voltage build-up at no-load, in the absence of a damper winding, is treated.

Example 5.2: The Voltage Build Up at No Load

Let us apply Equation 5.102 for the stator voltage build-up at no load in an SG without a damper cage on the rotor when the 100% step DC voltage is applied to the field winding.

Solution

With I_d and I_q being zero what remains from Equation 5.102 is the following:

$$-v_d(s) = g(s)\frac{s}{\omega_b}v_f(s)$$

$$-v_q(s) = \omega_r \cdot g(s) \cdot v_f(s)$$

(5.103)

The Laplace transform of a step function is applied to the field-winding terminals:

$$v_f(s) = \frac{v_f}{s}\omega_b$$

(5.104)

The transfer function $g(s)$ from Equation 5.98 with $l_{Dl} = l_{fDl} = 0$, $v_D = 0$ and zero stator currents is as follows:

$$g(s) = \frac{l_{dm}}{r_f} \times \frac{1}{1 + sT'_{d0}}$$

(5.105)

$$T'_{d0} = \frac{l_{dm} + l_{fl}}{r_f \cdot \omega_b}$$

(5.106)

Finally,

$$v_d(t) = -v_f \frac{l_{dm}}{\left(l_{dm} + l_{fl}\right)} e^{-\frac{t}{T'_{d0}}}$$

(5.107)

$$v_q(t) = -v_f \frac{\omega_r l_{dm}}{r_f} e^{-\frac{t}{T'_{d0}}}$$

(5.108)

With $\omega_r = 1$ P.U., $l_{dm} = 1.2$ P.U., $l_{fl} = 0.2$ P.U., $r_f = 0.003$ P.U., $v_{f0} = 0.003$ P.U., and $\omega_b = 2\pi \times 60 = 377$ rad/s.

$$T'_{d0} = \frac{1.2 + 0.2}{0.003 \cdot 377} = 1.2378 \text{ s}$$

$$v_d(t) = -0.003\frac{1.2}{1.2 + 0.2}e^{-\frac{t}{1.2378}} = -2.5714 \times 10^{-3} \cdot e^{-t/1.2378} \text{ [P.U.]}$$

$$v_q(t) = -0.003 \times 1 \times \frac{1.2}{0.003}\left(1 - e^{-\frac{t}{1.2378}}\right) = -1.2\left(1 - e^{-t/1.2378}\right) \text{[P.U.]}$$

The phase voltage of phase A is Equation 5.30:

$$v_A(t) = v_d(t)\cos\omega_b t - v_q(t)\sin\omega_b t$$

$$= -2.5714 \times 10^{-3} e^{-t/1.2378} \cos(\omega_b t + \theta_0) + 1.2\left(1 - e^{-t/1.2378}\right)\sin(\omega_b t + \theta_0)\left[\text{P.U.}\right]$$

For no load, from Equations 5.76, with zero power angle ($\delta_{v0} = 0$):

$$\theta_0 = -\frac{3\pi}{2}$$

In a similar way, $v_B(t)$ and $v_C(t)$ are obtained using Park inverse transformation.

The stator symmetrical phase voltages may be expressed simply in volts by multiplying the voltages in P.U. to the base voltage $V_b = V_n \times \text{sqrt}(2)$; V_n is the base RMS phase voltage.

5.8.2 Sudden Three-Phase Short Circuit from No Load

The initial no-load conditions are characterized by $I_{d0} = I_{q0} = 0$. Also if the field-winding terminal voltage is constant:

$$v_{f0} = I_{f0} \cdot r_f \tag{5.109}$$

From Equation 5.102, this time with $s = 0$ and $I_{d0} = I_{q0} = 0$, it follows that

$$\left(v_{d0}\right)_{s=0} = 0 \tag{5.110}$$

$$\left(v_{q0}\right)_{s=0} = -\omega_{r0}\frac{l_{dm}}{r_f}v_{f0} \tag{5.111}$$

Therefore, already for the initial conditions, the voltage along axis d, v_{d0}, is zero under no load. For axis q, the no-load voltage occurs. To short-circuit the machine, we simply have to apply along axis q the opposite voltage $-v_{q0}$.

It should be noted that, as $v_f = v_{f0}$, $v_f(s) = 0$, Equation 5.102 becomes

$$\begin{vmatrix} 0 \\ -\dfrac{v_{q0}\cdot\omega_b}{s} \end{vmatrix} = \begin{vmatrix} r_s + \dfrac{s}{\omega_b}l_d(s) & -\omega_{r0}l_q(s) \\ \omega_{r0}l_d(s) & r_0 + \dfrac{s}{\omega_b}l_q(s) \end{vmatrix} \cdot \begin{vmatrix} I_d(s) \\ I_q(s) \end{vmatrix} \tag{5.112}$$

The solution of Equation 5.112 is rather straightforward, with

$$I_d^*(s) = \frac{-v_{q0}\omega_b^3\omega_r}{sl_d(s)\left[\omega_b^2\omega_r^2 + s^2 + sr_s\omega_b\left(\dfrac{1}{l_d(s)} + \dfrac{1}{l_q(s)}\right) + \dfrac{r_s^2\omega_b^2}{l_d(s)\cdot l_q(s)}\right]} \tag{5.113}$$

As it is, $I_d(s)$ would be difficult to handle and thus two approximations are made: the terms in r_s^2 are neglected, and therefore

$$\frac{r_s\omega_b}{2}\left(\frac{1}{l_d(s)}+\frac{1}{l_q(s)}\right)\approx\frac{1}{T_a}=\text{const.}\tag{5.114}$$

with

$$\frac{1}{T_a}\approx\frac{r_s\omega_b}{2}\left(\frac{1}{l_d''}+\frac{1}{l_q''}\right)\tag{5.115}$$

With Equations 5.114 and 5.115, Equation 5.113 becomes

$$I_d(s)\approx\frac{-V_{q0}\omega_b^3\omega_r}{s\left(s^2+\dfrac{2}{T_a}s+\omega_b^2\omega_r^2\right)}\cdot\frac{1}{l_d(s)}\tag{5.116}$$

$$I_q(s)\approx\frac{-V_{q0}\omega_b^2\omega_r}{\left(s^2+\dfrac{2}{T_a}s+\omega_b^2\omega_r^2\right)}\cdot\frac{1}{l_q(s)}\tag{5.117}$$

Using Equations 5.97 and 5.98, $1/l_d(s)$ and $1/l_q(s)$ may be expressed as follows:

$$\frac{1}{l_d(s)}=\frac{1}{l_d}+\left(\frac{1}{l_d'}-\frac{1}{l_d}\right)\frac{s}{s+1/T_d'}+\left(\frac{1}{l_d''}-\frac{1}{l_d'}\right)\frac{s}{s+1/T_d''}\tag{5.118}$$

$$\frac{1}{l_q(s)}=\frac{1}{l_q}+\left(\frac{1}{l_q''}-\frac{1}{l_q'}\right)\frac{s}{s+1/T_q''}\tag{5.119}$$

With T_d', T_q'' larger than $1/\omega_b$ and $\omega_r=1.0$, after some analytical derivations with approximations, the inverse Laplace transforms of $I_d(s)$ and $I_q(s)$ are obtained:

$$I_d(t)\approx-v_{q0}\left[\frac{1}{l_d}+\left(\frac{1}{l_d'}-\frac{1}{l_d}\right)e^{-t/T_d'}+\left(\frac{1}{l_d''}-\frac{1}{l_d'}\right)e^{-t/T_d''}-\frac{1}{l_d''}e^{-t/T_a}\cos\omega_b t\right]$$

$$I_q(t)\approx-\frac{v_{q0}}{l_q''}e^{-1/T_a}\sin\omega_b t;\tag{5.120}$$

$$I_d(t)=I_{d0}+I_d(t)$$

$$I_q(t)=I_{q0}+I_q(t)$$

The phase sudden short-circuit current from no load ($I_{d0} = I_{q0} = 0$) is obtained using the following:

$$I_A(t) = \left[I_d(t)\cos(\omega_b t + \gamma_0) - I_q(t)\sin(\omega_b t + \gamma_0) \right]$$

$$= -v_{q0}\left\{ \left[\frac{1}{l_d} + \left(\frac{1}{l'_d} - \frac{1}{l_d} \right)e^{-t/T'_d} + \left(\frac{1}{l''_d} - \frac{1}{l'_d} \right)e^{-t/T''_d} \right]\cos(\omega_b t + \gamma_0) \right.$$

$$\left. - \frac{1}{2}\left(\frac{1}{l''_d} + \frac{1}{l''_q} \right)e^{-t/T_a} - \frac{1}{2}\left(\frac{1}{l''_d} - \frac{1}{l''_q} \right)e^{-t/T_a}\cos(2\omega_b t + \gamma_0) \right\} \tag{5.121}$$

The relationship between $I_f(s)$ and $I_d(s)$ for the case in point ($v_f(s) = 0$) is

$$I_f(s) = -g(s)\frac{s}{\omega_b}I_d(s) \tag{5.122}$$

Finally,

$$I_f(t) = I_{f0} + I_{f0} \cdot \frac{(l_d - l'_d)}{l_d}\left[e^{-t/T'_d} - \left(1 - T_D/T''_D\right)e^{-t/T''_d} - \frac{T_D}{T''_d}e^{-t/T_a}\cos\omega_b t \right] \tag{5.123}$$

Typical sudden short-circuit currents are shown in Figure 5.11a–d.

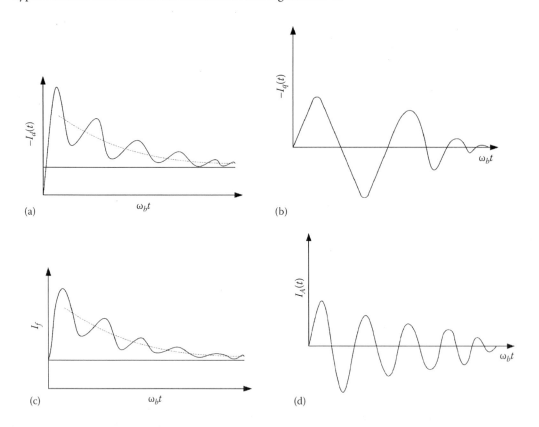

FIGURE 5.11 Sudden short-circuit currents: (a) $I_d(t)$, (b) $I_q(t)$, (c) $I_f(t)$, and (d) $I_A(t)$.

Further on the flux linkages $\Psi_d(s)$, $\Psi_q(s)$ (with $v_f(s) = 0$) are as follows:

$$\Psi_d(s) = l_d(s)I_d(s) = \frac{-\omega_b^3\omega_r^2}{\left(s^2 + \dfrac{2s}{T_a} + \omega_b^2\omega_r^2\right)} \cdot \frac{v_{q0}}{s}$$

$$\Psi_q(s) = l_q(s)I_q(s) = \frac{-\omega_b^2 v_{q0}\omega_r^2}{\left(s^2 + \dfrac{2s}{T_a} + \omega_b^2\omega_r^2\right)}$$

(5.124)

With $T_a \gg 1/\omega_b$, the total flux linkage components are approximately

$$\Psi_d(t) = \Psi_{d0} + \Psi_d(t) \approx v_{q0} \times e^{-t/T_a} \times \cos\omega_b t$$

$$\Psi_q(t) = 0 + \Psi_q(t) \approx -v_{q0} \times e^{-t/T_a} \times \sin\omega_b t$$

(5.125)

Note: Due to various approximations, the final flux–linkage in axis d and q are zero. In reality (with $r_s \neq 0$), none of them are quite zero.

The electromagnetic torque t_e (P.U.) is

$$t_e(t) = -\left(\Psi_d(t)I_q(t) - \Psi_q(t)I_d(t)\right)$$

(5.126)

The above approximations ($r_a^2 \approx 0$, $T_a \gg 1/\omega_b$, $\omega_b = ct$. T_d', $T_{d0}' \gg T_d''$, T_{d0}'', T_s, $r_a < sl_q(s)/\omega_b$) have been proven to yield correct results in stator current waveform during *unsaturated* short circuit—with given unsaturated values of inductance and time constant terms—within an error below 10%.

It may be argued that this is a notable error; but at the same time, it should be noted that instrumentation errors are within this range.

The use of expressions Equations 5.121 and 5.123 to estimate the various inductances and time constants, based on measured stator and field currents transients during a provoked short circuit at lower than rated no-load voltage, is included in both ANSI and IEC standards. Traditionally, graphoanalytical methods of curve fitting have been used to estimate the parameters from the sudden three-phase short circuit.

With today's available computing power, various nonlinear programming approaches to the SG parameter estimation from short-circuit current versus time curve have been proposed [8,9].

Despite notable progress along this path, there are still uncertainties and notable errors as the magnetic saturation of both leakage and main flux paths is present and varies during the short-circuit transients. The speed also varies in many cases during the short circuit, while the model assumes it constant. To avoid the zero-sequence currents due to nonsimultaneous phase short circuit, ungrounded three-phase short circuit should be performed.

Moreover, additional damper cage circuits are to be added for solid-rotor pole SGs (turbogenerators) to account for frequency (skin) effects.

Sub-subtransient circuits and parameters are introduced to model these effects.

For most large SGs, the sudden three-phase short-circuit test, be it at lower no-load voltage and (or) speed, may be performed only at the user's site, during commissioning, using the turbine as controlled speed prime mover. This way, the speed during the short circuit may be kept constant.

5.9 Standstill Time-Domain Response Provoked Transients

Flux (current) raise or decay tests may be performed at standstill, with the rotor aligned to axis d and q or for any given rotor position, in order to extract, by curve fitting, the stator-current and field currents time response for the appropriate SG model [10]. Any voltage versus time signal may be applied, but the frequency response standstill tests have become recently worldwide accepted. All these standstill tests are purely electromagnetic tests as the speed is kept constant (zero in this case).

The situation in Figure 5.12a corresponds to axis d, while Figure 5.12b refers to axis q.

For axis d,

$$I_A + I_B + I_C = 0; \quad V_B = V_C; \quad V_A + V_B + V_C = 0$$

$$I_d(t) = \frac{2}{3}\left[I_A + I_B \cos\frac{2\pi}{3} + I_C \cos\left(-\frac{2\pi}{3}\right)\right] = I_A(t)$$

$$I_q(t) = 0$$

$$V_d(t) = \frac{2}{3}\left[V_A + V_B \cos\frac{2\pi}{3} + V_C \cos\left(-\frac{2\pi}{3}\right)\right] = V_A(t)$$

(5.127)

$$V_q(t) = 0$$

$$V_A - V_B = V_{D1D2} = V_A - \left(-\frac{V_A}{2}\right) = \frac{3}{2}V_A = \frac{3}{2}V_d(t)$$

$$V_{ABC}/I_A = \frac{3}{2}V_d/I_d$$

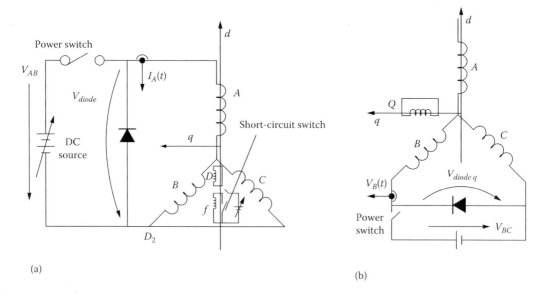

(a)   (b)

FIGURE 5.12 Arrangement for standstill voltage response transients: (a) axis d and (b) axis q.

For axis q,

$$I_A = 0, \quad I_B = -I_C$$

$$V_q = -\frac{2}{3}\left[V_A \sin(0) + V_B \sin\frac{2\pi}{3} + V_C \sin\left(-\frac{2\pi}{3}\right) \right]$$

$$= -\frac{2}{3}\left(V_B - V_C\right) \cdot \frac{\sqrt{3}}{2} = -\left(V_B - V_C\right)/\sqrt{3}$$

$$I_q = -\frac{2}{3}\left[I_A \sin(0) + I_B \sin\frac{2\pi}{3} + I_C \sin\left(-\frac{2\pi}{3}\right) \right] = -\frac{2}{3} I_B \sqrt{3} = -2I_B/\sqrt{3}$$

$$V_q/I_q = \left(V_B - V_C\right)/2I_B = \frac{V_{BC}}{2I_B}; \quad I_d = 0$$

(5.128)

Now, for axis d, we simply apply Equations 5.102 with $V_f(s) = 0$, if the field winding is short circuited, and with $I_q = 0$, $\omega_{r0} = 0$.

Also

$$V_d = +\frac{2}{3} V_{ABC}$$

$$\frac{2}{3} V_{ABC} \cdot \frac{\omega_b}{s} = \left(r_s + \frac{s}{\omega_b} l_d(s) \right) I_d(s); \quad I_d = I_A$$

(5.129)

$$I_f(s) = -\frac{s}{\omega_b} g(s) I_d(s)$$

For axis q, $I_d = 0$, $\omega_{r0} = 0$,

$$V_q = -\frac{V_{BC}}{\sqrt{3}}$$

$$-\frac{1}{\sqrt{3}} V_{BC} \cdot \frac{\omega_b}{s} = \left(r_s + \frac{s}{\omega_b} l_q(s) \right) I_q(s);$$

(5.130)

$$I_q = \frac{2I_B}{\sqrt{3}}$$

The standstill time-domain transients may be explored by investigating both the current rise for step voltage application or current decay when the stator is short circuited through the freewheeling diode, after the stator was disconnected from the power source.

For current decay, the left-hand side of Equations 5.129 and 5.130 should express only $-2/3\ V_{Diode}$ in axis d and $-1/\text{sqrt}(3)\ V_{diode}$ in axis q. The diode voltage $V_{diode}(+)$ should be acquired through proper instrumentation.

For the standard equivalent circuits with $l_d(s)$, $l_q(s)$ having expressions of Equation 5.97, from Equations 5.129 and 5.130:

$$I_d(s) = \frac{+\dfrac{2}{3}V_{AB0}(s)\omega_b}{s\left[r_s + \dfrac{s}{\omega_b}l_d\dfrac{(1+sT_d')(1+sT_d'')}{(1+sT_{d0}')(1+sT_{d0}'')}\right]} = I_A(s)$$

$$I_f(s) = -\frac{s}{\omega_b}\frac{l_{dm}}{r_f}\frac{1+sT_D}{(1+sT_{d0}')(1+sT_{d0}'')}\cdot I_d(s) \tag{5.131}$$

$$I_q(s) = \frac{-\dfrac{V_{BC}(s)}{\sqrt{3}}\omega_b}{s\left[r_s + \dfrac{s}{\omega_b}l_q\dfrac{(1+sT_q'')}{(1+sT_{q0}'')}\right]} = \frac{2I_B(s)}{\sqrt{3}}$$

With approximations similar to the case of sudden short-circuit transients expressions of $I_d(t)$, $I_f(t)$ are obtained. They are simpler as no interference from axis q occurs.

In a more general case, where additional damper circuits are included to account for skin effects in solid rotor SG, the identification process of parameters from step voltage responses becomes more involved. Nonlinear programming methods such as the least squared error, maximum likelihood, and the more recent evolutionary methods such as genetic algorithms could be used to identify separately the d and q inductances and time constants from standstill time-domain responses.

The starting point of all such "curve fitting" methods is the fact that the stator current response contains a constant component and a few aperiodic components with time constant close to T_d'', T_d', and T_{ad}:

$$\frac{1}{T_{ad}} \approx \frac{r_s}{l_d''}\omega_b \tag{5.132}$$

for axis d, and T_q'' and T_{aq} for axis q

$$\frac{1}{T_{aq}} \approx \frac{r_s\omega_b}{l_q''} \tag{5.133}$$

When an additional damper circuit is added in axis d, a new time constants T_d''' occurs. Transient and sub-subtransient time constants in axis $q(T_q''', T_q')$ appear when three circuits are considered in axis q:

$$I_d(t) = I_{d0} + I_d'e^{-t/T_d'} + I_d''e^{-t/T_d''} + I_{da}e^{-t/T_{ad}} + I_d'''e^{-t/T_d'''}$$

$$I_q(t) = I_{q0} + I_q'e^{-t/T_q'} + I_q''e^{-t/T_q''} + I_{qa}e^{-t/T_{aq}} + I_q'''e^{-t/T_q'''} \tag{5.134}$$

$$I_f(t) = I_{f0} + I_f'e^{-t/T_d'} + I_f''e^{-t/T_d''} + I_f'''e^{-t/T_d'''} + I_{f0}e^{-t/T_{ad}}$$

Curve fitting the measured $I_d(t)$ and $I_q(t)$, respectively, with the calculated ones based on Equations 5.134 yields the time constants and l_d''', l_d'', l_d', l_q''', l_q'', l_q'. The main problem with step voltage standstill tests is that they do not excite properly all "frequencies"—time constants—of the machine.

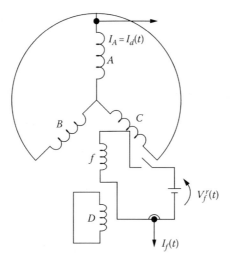

FIGURE 5.13 Field-winding SSTR arrangement.

A random cyclic PWM voltage excitation at standstill seems more adequate for parameter identification [11–13]. Also, for more coherency, care must be exercised to set initial (unique) values for stator leakage inductance l_{sl} and for the rotor to stator reduction ratio K_f:

$$K_f I_f^r = I_f \tag{5.135}$$

In reality, $I_f^r(t)$ is acquired and processed and not $I_f(t)$. Finally, it is more and more often suggested, for practicality, to excite (with a random cyclic PWM voltage) the field winding with short circuited and opened stator windings rather than the stator, at standstill, because higher saturation levels (up to 25%) may be obtained without machine overheating.

Evidently, such tests are feasible only in axis d (Figure 5.13).

Again $I_d = I_A$, $V_d = V_q = 0$, $I_q = 0$. Therefore, from Equation 5.102 with $I_q = 0$ and $V_d(s) = 0$, $V_q(s) = 0$:

$$I_d(s) = \frac{-g(s)\dfrac{s}{\omega_b}V_f(s)}{r_s + \dfrac{s}{\omega_b}l_d(s)}$$

$$I_f(s) = \frac{v_f(s)\left[1 - g(s)\dfrac{s^2}{\omega_b^2}\left(1 - \dfrac{l_{sl}}{r_s + sl_d(s)/\omega_b}\right)\right]}{r_f + \dfrac{s}{\omega_b}l_{fl}} \tag{5.136}$$

Again the voltage and current rotor/stator reduction ratios, which are rather constant, are needed:

$$I_f = K_f I_f^r; \quad V_f = \frac{2}{3K_f}V_f^r; \quad r_f = \frac{2}{3K_f^2}r_f^r \tag{5.137}$$

To eliminate hysteresis effects (with $V_f(t)$ made from constant height pulses with randomly large timings and zeroes) and reach pertinent frequencies, $V_f(t)$ may change polarity cyclically.

The main advantage of standstill time response tests (SSTRs) consists in the fact that testing time is rather short.

5.10 Standstill Frequency Response

Another provoked electromagnetic phenomenon at zero speed that is being used to identify the SG parameters (inductances and time constants) is the standstill frequency response (SSFR). The SG is supplied in axis d and q respectively, through a single-phase AC voltage applied to the stator (with the field winding short circuited), or to the field winding (with the stator short circuited). The frequency of the applied voltage is varied in general from 0.001 Hz to more than 100 Hz, while the voltage is adapted to keep the AC current small enough (below 5% of rated value) to avoid winding overheating. The whole process of raising frequency level may be mechanized, but still considerable testing time is required. The arrangement is identical to that in Figures 5.12 and 5.13, but now $s = j\omega$, with ω in rad/s.

Equation 5.131 becomes

$$\frac{V_{ABC}}{I_A} = \frac{3}{2}\left[r_s + j\frac{\omega}{\omega_b}l_d(j\omega) \right] = \frac{3}{2}Z_d(j\omega)$$

$$I_f(j\omega) = -j\frac{\omega}{\omega_b}g(j\omega) \cdot I_d(j\omega)$$

(5.138)

for axis d, and

$$\frac{V_{BC}}{I_B} = 2\left[r_s + j\frac{\omega}{\omega_b}l_q(j\omega) \right] = 2Z_q(j\omega)$$

for axis q.

Complex number definitions have been used as a single-frequency voltage is applied at any time.

The frequency range is large enough to encompass whole spectrum of electrical time constants that spreads from a few milliseconds to a few seconds.

When the SSFR tests are performed on the field winding (with the stator short circuited—Figure 5.13), the response in I_A and I_f is adapted from Equation 5.136 with $s = j\omega$:

$$I_d(j\omega) = I_A(j\omega) = \frac{-g(j\omega)j\dfrac{\omega}{\omega_b}V_f(j\omega)}{r_s + j\dfrac{\omega}{\omega_b}l_d(j\omega)}$$

(5.139)

$$I_f(j\omega) = -v_f(j\omega)\frac{\left[1 + g(j\omega)\dfrac{\omega^2}{\omega_b^2}\left(1 - \dfrac{l_{sl}}{\left(rs + j\omega l_d(j\omega)/\omega_b\right)}\right)\right]}{rf + j\dfrac{\omega}{\omega_b}l_{sl}}$$

The general equivalent circuits emanate $l_d(j\omega)$, $l_q(j\omega)$, $g(j\omega)$. For the standard case, equivalent circuits of Figure 5.6 and Equations 5.97 and 5.98 are used.

For a better representation of frequency (skin) effects, more damper circuits are added along axis d (one in general) and along axis q (two in general) (Figure 5.14).

The leakage coupling inductance l_{fDl} between the field winding and the damper windings also called Canay's inductance [14], though small in general (less than stator leakage inductance), proved to be necessary to simultaneously fit the stator current and the field current frequency responses in axis d. Adding even one more such a leakage coupling inductance (say between the two damper circuits in d axis) failed so far to produce improved results but hampered the convergence of the nonlinear programming estimation method used to identify the SG parameters [15].

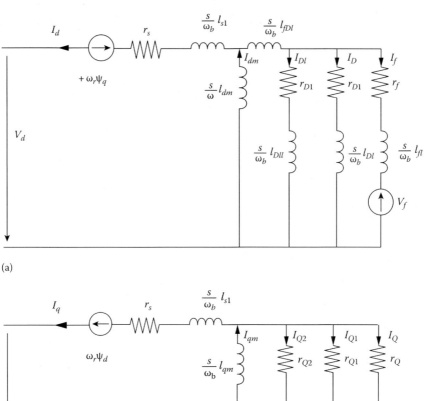

(a)

(b)

FIGURE 5.14 SG equivalent circuits with three rotor circuits: (a) axis d and (b) axis q.

The main argument in favor of $l_{fDl} \neq 0$ should be its real physical meaning (Figure 5.15).

A myriad of mathematical methods have been proposed recently to identify the SG parameters from SSFR with mean squared error [16] and maximum likelihood [17] as some of the most frequent. A rather detailed description of such methods is presented in Chapter 8.

5.10.1 Asynchronous Running

When the speed $\omega_r \neq \omega_{r0} = \omega_1$, the stator mmf induces currents in the rotor windings, mainly at slip frequency:

$$S = \frac{\omega_1 - \omega_r}{\omega_1} \tag{5.140}$$

S is the slip.

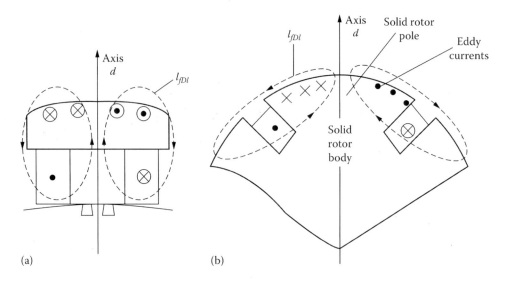

FIGURE 5.15 The leakage coupling inductance l_{fDl}: (a) salient-pole and (b) nonsalient-pole solid rotor.

These currents interact with the stator field to produce an asynchronous torque t_{as} as in an induction machine. For $S > 0$, the torque is motoring while for $S < 0$ it is generating.

As the rotor magnetic and electric circuits are not fully symmetric, there will be also asynchronous torque pulsations even with $\omega_r \neq \omega_1 = $ const. Also, the average synchronous torque t_{eav} is zero as long as $\omega_r \neq \omega_1$.

The d–q model can be used directly to handle transients at a speed $\omega_r \neq \omega_1$. Here, we use the dq model to calculate the average asynchronous torque and currents considering that the power source that supplies the field winding has a zero internal impedance. That is, the field winding is short circuited for asynchronous (AC) currents.

The Park transform may be applied to the symmetrical stator voltages:

$$V_{A,B,C}(t) = V \cos\left(\omega_1 \omega_b t - (i-1)\frac{2\pi}{3} \right); [\text{P.U.}]$$

$$V_d = -V \cos\left(\omega_1 \omega_b t - \theta_e \right) \tag{5.141}$$

$$V_q = +V \sin\left(\omega_1 \omega_b t - \theta_e \right)$$

where

$$\frac{1}{\omega_b} \frac{d\theta_e}{dt} = \omega_r = ct.$$

$$\theta_e = \int \omega_r \omega_b dt + \theta_0$$

Also, ω_1 in PU, ω_b is in rad/s, t is in seconds and V is in P.U. are given.

We may now introduce complex number symbols:

$$\underline{V}_d = -V$$

$$\underline{V}_q = jV \tag{5.142}$$

As the speed is constant d/dt in the dq model is replaced by

$$\frac{1}{\omega_b}\frac{d}{dt} \to j\left(\omega_1 - \omega_r\right) = jS\omega_1$$

$$\omega_r = \omega_1\left(1 - S\right)$$

(5.143)

As the field-winding circuit is short circuited for the AC current, $v_f(jS\omega_1) = 0$.

We may use once more Equation 5.102 in complex numbers and $\omega_r = \omega_1(1 - S)$:

$$\begin{vmatrix} -\underline{V}_d\left(jS\omega_1\omega_b\right) \\ -\underline{V}_q\left(jS\omega_1\omega_b\right) \end{vmatrix} = \begin{vmatrix} r_s + jS\omega_1 l_d\left(jS\omega_1\omega_b\right) & -\omega_1\left(1-S\right)l_q\left(jS\omega_1\omega_b\right) \\ \omega_1\left(1-S\right)l_d\left(jS\omega_1\omega_b\right) & r_s + jS\omega_1 l_q\left(jS\omega_1\omega_b\right) \end{vmatrix} \cdot \begin{vmatrix} \underline{I}_d \\ \underline{I}_q \end{vmatrix}$$

(5.144)

System of Equation 5.144 may be solved for I_d and I_q with $\underline{V}_d$ and $\underline{V}_q$ from Equation 5.142. The average torque t_{asav} is

$$t_{asav} = -\text{Re}\left[\underline{\Psi}_d\left(jS\omega_1\omega_b\right)\underline{I}_q^*\left(jS\omega_1\omega_b\right) - \underline{\Psi}_q\left(jS\omega_1\omega_b\right)\underline{I}_d^*\left(jS\omega_1\omega_b\right)\right]$$

(5.145)

where

$$\underline{\Psi}_d\left(jS\omega_1\omega_b\right) = \underline{I}_d l_d\left(jS\omega_1\omega_b\right);$$

$$\underline{\Psi}_q\left(jS\omega_1\omega_b\right) = \underline{I}_q l_q\left(jS\omega_1\omega_b\right)$$

(5.146)

The field-winding AC current I_f is obtained from Equation 5.138:

$$\underline{I}_f\left(jS\omega_1\omega_b\right) = -jS\omega_1 g\left(jS\omega_1\omega_b\right)\cdot\underline{I}_d\left(jS\omega_1\omega_b\right)$$

(5.147)

If the stator resistance is neglected in Equation 5.144,

$$\underline{I}_d = \frac{+V}{j\omega_1 l_d\left(jS\omega_1\omega_b\right)}$$

$$\underline{I}_q = \frac{-jV}{j\omega_1 l_q\left(jS\omega_1\omega_b\right)}$$

(5.148)

In such conditions, the average asynchronous torque is

$$t_{asav} \approx -\frac{v^2}{\omega_1^2}\text{Re}\left[\frac{1}{jl_d^*\left(jS\omega_1\omega_b\right)} + \frac{1}{jl_q^*\left(jS\omega_1\omega_b\right)}\right], \left[\text{P.U.}\right]$$

(5.149)

The torque is positive when generating (opposite to the direction of motion). This happens only for $S < 0$ ($\omega_r > \omega_1$).

There is a pulsation in the asynchronous torque due to magnetic anisotropy and rotor circuit asymmetry.

Its frequency is ($2S\omega_1\omega_b$) in rad/s.

The torque pulsations may be emphasized by switching back from I_d, I_q, $\underline{\Psi}_d$, $\underline{\Psi}_q$, complex number form to their instantaneous values:

$$I_d(t) = \mathrm{Re}\left(\underline{I}_d e^{j\left[(S\omega_1\omega_b)t - \theta_0 \right]} \right)$$

$$I_q(t) = \mathrm{Re}\left(\underline{I}_q e^{j\left[(S\omega_1\omega_b)t - \theta_0 \right]} \right)$$

$$\Psi_d(t) = \mathrm{Re}\left(\underline{\Psi}_d e^{j\left[(S\omega_1\omega_b)t - \theta_0 \right]} \right) \tag{5.150}$$

$$\Psi_q(t) = \mathrm{Re}\left(\underline{\Psi}_q e^{j\left[(S\omega_1\omega_b)t - \theta_0 \right]} \right)$$

$$I_A(t) = I_d(t)\cos\left(\omega_1\omega_b (1-S)t + \theta_0 \right) - I_q(t)\sin\left(\omega_1\omega_b (1-S)t + \theta_0 \right)$$

$I_d(t)$, $I_q(t)$, $\Psi_d(t)$, $\Psi_q(t)$ will exhibit solely components at slip frequency, while I_A will show the fundamental frequency ω_1 (P.U.) and the $\omega_1 (1 - 2S)$ component when $r_s \neq 0$.

The instantaneous torque at constant speed in asynchronous running is

$$t_{as}(t) = -\left(\Psi_d(t)I_q(t) - \Psi_q(t)I_d(t) \right); \left[\mathrm{P.U.} \right] \tag{5.151}$$

The $2S\omega_1$ P.U. component (pulsation) in t_{as} is thus physically evident from Equation 5.151. This pulsation may run as high as 50% in P.U. The average torque t_{asav} (P.U.) for an SG with the data: $V = 1$, $l_{sl} = 0.15$, $l_{dm} = 1.0$, $l_{fl} = 0.3$, $l_{Dl} = 0.2$, $l_{qm} = 0.6$, $l_{Ql} = 0.12$, $r_s = 0.012$, $r_D = 0.03$, $r_Q = 0.04$, $r_f = 0.03$, $V_f = 0$ is shown in Figure 5.16.

A few remarks are in order:

- The average asynchronous torque may equal or surpass the base torque. As the currents are very large, the machine should not be allowed to work asynchronously for more than 2 min, in general, in order to avoid severe overheating. Also, the SG draws reactive power from the power system while it delivers active power.

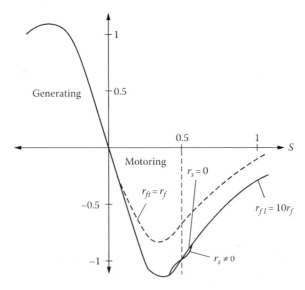

FIGURE 5.16 Asynchronous running of an SG.

- The torque shows a small inflexion around $S = 1/2$ when $r_s \neq 0$. This is not present for $r_s \approx 0$.
- With no additional resistance in the field winding ($r_{fl} = r_f$), no inflexion in torque—speed curve around $S = 1/2$ occurs. The average torque is smaller also in comparison with the case of $r_{fl} = 10\, r_f$ (additional resistance in the field winding is included).

Example 5.3: Asynchronous Torque Pulsations

For the above data, but with $r_s = 0$ and $\omega_1 = 1$ (rated frequency) derive the formula of instantaneous asynchronous torque.

Solution

From Equation 5.148,

$$\underline{I}_d = \frac{V}{jl_d\left(js\omega_b\right)}$$

$$\underline{I}_q = \frac{-jV}{jl_q\left(jS\omega_b\right)} \tag{5.152}$$

$$\underline{\Psi}_d = l_d\underline{I}_d = -jV$$

$$\underline{\Psi}_q = l_q\underline{I}_q = -V$$

Also let us define $l_d(jS\omega_b)$ and $l_q(jS\omega_b)$ as follows:

$$jl_d\left(jS\omega_b\right) = r_d + jl_{dr} = \left|l_d\right|e^{j\varphi_d}$$

$$jl_q\left(jS\omega_b\right) = r_q + jl_{qr} = \left|l_q\right|e^{j\varphi_q} \tag{5.153}$$

According to Equation 5.150 with $\theta_0 = 0$ (for simplicity), from Equation 5.152:

$$I_d = \frac{V}{\left|l_d\right|}\cos\left(S\omega_b t - \phi_d\right)$$

$$I_q = +\frac{V}{\left|l_q\right|}\sin\left(S\omega_b t - \phi_q\right) \tag{5.154}$$

$$\Psi_d = V\sin\left(S\omega_b t\right)$$

$$\Psi_q = -V\cos\left(S\omega_b t\right)$$

$$t_{as}(t) = -\left(\Psi_d I_q - \Psi_q I_d\right) = \left[-\frac{V^2}{2}\left(\frac{\cos\phi_d}{\left|l_d\right|} + \frac{\cos\phi_q}{\left|l_q\right|}\right) + \frac{V^2}{2\left|l_q\right|}\cos\left(2S\omega_b t - \phi_q\right)\right.$$

$$\left. -\frac{V^2}{2\left|l_d\right|}\cos\left(2S\omega_b t - \phi_d\right)\right] \tag{5.155}$$

The first term in Equation 5.155 represents the average torque, while the last two terms refer to the pulsating torque. As expected, for a completely symmetric rotor (as that of an induction machine) the pulsating asynchronous torque is zero because $\left|l_d\right| = \left|l_q\right|$, $\left|\Psi_d\right| = \left|\Psi_q\right|$ for all slip (speed values).

Form motoring ($S > 0$), φ_d, $\varphi_q < 90°$ and for generating ($S < 0$), $90° < \varphi_d$, $\varphi_q < 180°$. For small values of slip, the asynchronous torque versus slip may be approximated to a straight line (see Figure 5.16):

$$t_{asav} \approx -K_{as}S\omega_1 = +K_{as}\left(\omega_r - \omega_1\right) \tag{5.156}$$

Equation 5.155 may provide a good basis to calculate K_{as} for high power SGs ($r_s \approx 0$).

Example 5.4: DC Field—Current-Produced Asynchronous Stator Losses

The DC current in the field winding, if any, produces additional losses in the stator windings through currents at a frequency equal to speed $\omega' = \omega_1(1-S)$(P.U.) Calculate these losses and their torque.

Solution

In rotor coordinates, these stator dq currents I'_d, I'_q are DC, at constant speed; to calculate them separately, only the motion-induced voltages are considered, with $V'_d = V'_q = 0$ (the stator is short circuited, a sign that the power system has a zero internal impedance).
Also, $I'_D = I'_Q = 0$, $I_f = I_{f0}$:

$$0 = -\left(1-S\right)\omega_1 l_q I'_q + r_s I'_d$$
$$0 = +\left(1-S\right)\omega_1 \left(l_d I_d + l_{dm} I_{f0}\right) + r_s I'_q \tag{5.157}$$

From Equation 5.157 and for $\omega_1 = 1$,

$$I'_d = \frac{-l_{dm} I_{F0} \left(1-S\right)^2 l_q}{r_s^2 + \left(1-S\right)^2 l_d l_q}$$
$$I'_q = \frac{-l_{dm} I_{F0} \left(1-S\right) r_s}{r_s^2 + \left(1-S\right)^2 l_d l_q} \tag{5.158}$$

The stator losses P'_{CO} are

$$P'_{CO} = \frac{3}{2} r_s \left(I'^2_d + I'^2_q\right) \tag{5.159}$$

The corresponding braking torque t'_{as} is

$$t'_{as} = \frac{P'_{CO}}{\omega_1 \left(1-S\right)} > 0 \tag{5.160}$$

The maximum value of t'_{as} occurs at a rather large slip S'_K:

$$S'_k \approx 1 - \sqrt{\frac{2 l_d l_q - r_s^2}{2 l_q^2 + l_d l_q}} \tag{5.161}$$

The maximum torque $t_{as}^{\prime k}$ is

$$t_{as}^{\prime k} \approx t_{as}^{\prime}\left(S_K^{\prime}\right) \tag{5.162}$$

Close to the synchronous speed ($S = 0$), this torque becomes negligible.

5.11 Simplified Models for Power System Studies

The P.U. system of SG Equation 5.68 describes rather completely the standard machine for any transients. The complexity of such a model makes it less practical for power system stability studies where tens or hundreds of SGs and consumers are involved and have to be modeled. Simplifications in the SG model are required for such a purpose. Some of them are discussed later, while much more is available in the literature on power systems stability and control [1,17].

5.11.1 Neglecting the Stator Flux Transients

When neglecting the stator transients in the dq model, it means to make $(\partial \Psi d)/\partial t = (\partial \Psi q)/\partial t = 0$.

It has been demonstrated that it is necessary also to simultaneously consider—only in the stator voltage equations—constant (synchronous) speed:

$$
\begin{aligned}
&V_d = -I_d r_s + \Psi_q \omega_{r0} \\[4pt]
&V_q = -I_q r_s - \Psi_d \omega_{r0} \\[4pt]
&\frac{1}{\omega_b}\frac{d\Psi_f}{dt} = -I_f r_f + V_f \\[4pt]
&\frac{1}{\omega_b}\frac{d\Psi_D}{dt} = -I_D r_D \\[4pt]
&\frac{1}{\omega_b}\frac{d\Psi_Q}{dt} = -I_Q r_Q \\[4pt]
&\Psi_d = l_{sl}I_d + l_{dm}\left(I_d + I_f + I_D\right) \\[4pt]
&\Psi_q = l_{sl}I_q + l_{qm}\left(I_q + I_Q\right) \\[4pt]
&\Psi_f = l_{fl}I_f + l_{dm}\left(I_d + I_f + I_D\right) + l_{fDl}\left(I_f + I_D\right) \\[4pt]
&\Psi_D = l_{fl}I_D + l_{dm}\left(I_d + I_f + I_D\right) + l_{fDl}\left(I_f + I_D\right) \\[4pt]
&\Psi_Q = l_{Ql}I_q + l_{qm}\left(I_q + I_Q\right) \\[4pt]
&2H\frac{d\omega_r}{dt} = t_{shaft} - t_e; \quad t_e = \Psi_d I_q - \Psi_q I_d; \quad \omega_b\frac{d\delta v}{dt} = \omega_r - \omega_0 \\[4pt]
&\frac{1}{\omega_b}\frac{d\theta_{er}}{dt} = \omega_r; \quad
\begin{vmatrix} V_d \\ V_q \\ V_0 \end{vmatrix} = \left|P\left(\theta_{er}\right)\right| \begin{vmatrix} V_a \\ V_b \\ V_c \end{vmatrix}
\end{aligned}
\tag{5.163}
$$

The flux–current relationships are the same as in Equation 5.68. The state variables may be I_d, I_q, Ψ_f, Ψ_D, Ψ_Q, ω_r, θ_{er}. I_d and I_q are calculated from the, now algebraic, equations of stator. The system order has been reduced by two units.

As expected, fast 50 (60) Hz frequency transients, occurring in I_d, I_q, t_e, are eliminated. Only the "average" transient torque is "visible." Allowing for constant (synchronous) speed in the stator equations with $(\partial \Psi d)/\partial t = \partial \Psi q/\partial t = 0$ counteracts the effects of such an approximation at least for small signal transients, in terms of speed and angle response [17]. Neglecting of stator transients leads to steady-state stator voltage equations. Consequently, if the power network transients are neglected, the connection of the SG model to the power network model is rather simple with steady state all over.

A drastic computation time saving is thus obtained in power system stability studies.

5.11.2 Neglecting the Stator Transients and the Rotor Damper-Winding Effects

This time, in addition, the damper-winding currents are zero $I_D = I_Q = 0$ and thus:

$$V_d = -I_d r_s + \Psi_q \omega_{r0}$$

$$V_q = -I_q r_s - \Psi_d \omega_{r0}$$

$$\frac{1}{\omega_b} \frac{d\Psi_f}{dt} = -I_f r_f + V_f$$

$$\Psi_d = l_{sl} I_d + l_{dm} \left(I_d + I_f \right)$$

$$\Psi_q = l_{sl} I_q + l_{qm} I_q \tag{5.164}$$

$$\Psi_f = l_{fl} I_f + l_{dm} \left(I_d + I_f \right)$$

$$2H \frac{d\omega_r}{dt} = t_{shaft} - t_e; \quad t_e = \Psi_d I_q - \Psi_q I_d; \quad \omega_b \frac{d\delta v}{dt} = \omega_r - \omega_{r0}$$

$$\frac{1}{\omega_b} \frac{d\theta_{er}}{dt} = \omega_r; \quad \begin{vmatrix} V_d \\ V_q \\ V_0 \end{vmatrix} = \left| P\left(\theta_{er} \right) \right| \begin{vmatrix} V_A \\ V_B \\ V_C \end{vmatrix}$$

The order of the system has been further reduced by two units. Additional computation time saving is obtained, with only one electrical transient left—the one produced by the field winding. The model is adequate for slow transients (seconds and more).

5.11.3 Neglecting All Electrical Transients

The field current is now considered also constant. We are dealing with very slow (mechanical) transients:

$$V_d = -I_d r_s + \Psi_q \omega_{r0}$$

$$V_q = -I_q r_s - \Psi_d \omega_{r0}$$

$$I_f = \frac{V_f}{r_f} \tag{5.165}$$

$$2H \frac{d\omega_r}{dt} = t_{shaft} - t_e; \quad t_e = \Psi_d I_q - \Psi_q I_d; \quad \omega_b \frac{d\delta v}{dt} = \omega_r - \omega_{r0}$$

$$\omega_b \frac{d\theta_{er}}{dt} = \omega_r; \quad \begin{vmatrix} V_d \\ V_q \\ V_0 \end{vmatrix} = \left| P\left(\theta_{er} \right) \right| \begin{vmatrix} V_A \\ V_B \\ V_C \end{vmatrix}$$

This time we start again with initial values of variables: I_{d0}, I_{q0}, I_{f0}, $\omega_r = \omega_{r0}$, $\delta_{V0}(\theta_{er0})$ and of V_{d0}, V_{q0}, $t_{e0} = t_{shaft0}$.

Therefore, the machine is under steady state electromagnetically, while making use of motion equation to handle mechanical transients. In very slow transients (tens of seconds), such a model is appropriate.

Note: A plethora of constant flux approximate models (with or without rotor damper cage), in use for power system studies [1,17], are not followed here [17].

Amongst the simplified models, we do illustrate here only the "mechanical" model as it helps explaining SG self-synchronization, step shaft torque response and SG oscillations (free and forced).

5.12 Mechanical Transients

When the prime-mover torque varies in a nonperiodical or periodical fashion, the large inertia of the SG leads to a rather slow speed (power angle δ_V) response. To illustrate such a response, the electromagnetic transients may be altogether neglected as suggested by the "mechanical model" presented in the previous paragraph.

As the speed varies, an asynchronous torque t_{as} occurs, besides the synchronous torque. The motion equation becomes (in P.U.)

$$2H\frac{d\omega_r}{dt} = t_{shaft} - t_e - t_{as} \tag{5.166}$$

with ($r_s = 0$):

$$t_e = e_f \frac{V\sin\delta_V}{l_d} + \frac{V^2}{2}\left(\frac{1}{l_q} - \frac{1}{l_d}\right)\sin 2\delta_V \tag{5.167}$$

$$t_{as} = \frac{K_{as}}{\omega_b}\frac{d\delta_V}{dt}; \tag{5.168}$$

$$\frac{1}{\omega_b}\frac{d\delta_V}{dt} = \omega_r - \omega_1$$

$e_f = l_{dm}I_f$ represents the no-load voltage for given field current.

Only the average asynchronous torque is considered here. The model in Equations 5.166 through 5.168 may be solved numerically for ω_r and δ_V as variables once their initial values are given together with the prime mover torque t_{shaft} versus time or versus speed with or without a speed governor.

For small deviations, Equations 5.166 through 5.168 become:

$$\frac{2H}{\omega_b}\frac{d^2\Delta\delta_V}{dt^2} + \left(\frac{\partial t_e}{\partial \delta_V}\right)_{\delta_{V0}} \cdot \Delta\delta_V + \frac{K_{as}}{\omega_b}\frac{d\Delta\delta_V}{dt} = \Delta t_{shaft} \tag{5.169}$$

$$\left(t_e\right)_{\delta_{V0}} = t_{shaft0}$$

$$\left(\frac{\partial t_e}{\partial \delta_V}\right)_{\delta_{V0}} = t_{eso} \tag{5.170}$$

Equation 5.170 reflects the starting steady-state conditions at initial power angle δ_{V0}. t_{eso} is the so-called synchronizing torque (as long as $t_{eso} > 0$ static stability is secured).

5.12.1 Response to Step Shaft Torque Input

For step shaft torque input, Equation 5.169 allows for an analytical solution:

$$\Delta\delta_V(t) = \frac{\Delta t_{shaft}}{t_{es0}} + \mathrm{Re}\left[A_1 e^{\gamma_1 t} + B_1 e^{\gamma_2 t}\right] \tag{5.171}$$

$$\underline{\gamma}_{1,2} = \frac{-(K_{as}/\omega_b) \pm \sqrt{\left(\dfrac{K_{as}}{\omega_b}\right)^2 - \dfrac{8H \cdot t_{es0}}{\omega_b}}}{4(H/\omega_b)} = -\frac{1}{T_{as}} \pm j\omega' \tag{5.172}$$

$$\frac{1}{T_{as}} = \frac{K_{as}}{4H};$$

$$(\omega')^2 = -\left(\frac{1}{T_{as}}\right)^2 + \omega_0^2; \tag{5.173}$$

$$\omega_0 = \sqrt{\frac{t_{es0}\omega_b}{2H}}$$

Finally,

$$\Delta\delta_V(t) = \frac{\Delta t_{shaft}}{t_{es0}}\left[1 - e^{-\frac{t}{T_{as}}}\frac{\sin(\omega t + \Psi)}{\sin\Psi}\right]$$

The constant Ψ is obtained by assuming initial steady-state conditions:

$$(\Delta\delta_V)_{t=0} = 0,$$

$$\left(\frac{d(\Delta\delta)}{dt}\right)_{t=0} = 0 \tag{5.174}$$

Finally,

$$\tan\Psi = \omega' T_{as} \tag{5.175}$$

The power angle and speed response for step shaft torque is shown qualitatively in Figure 5.17.

Note that ω_0 (Equation 5.173) is traditionally known as the eigen mechanical frequency of the SG. Unfortunately, ω_0 varies with power angle δ_V, field current I_{f0} and with inertia. It decreases with increasing δ_V and increases with increasing I_{f0}.

In general, $f_0 = \omega_0/2\pi$ varies from less or about 1 Hz to a few Hertz for large and, respectively, medium-power SGs.

5.12.2 Forced Oscillations

Shaft torque oscillations may occur due to various reasons. The Diesel engine prime movers are a typical case as their torque varies with rotor position.

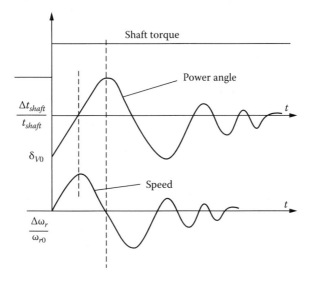

FIGURE 5.17 Power angle and speed responses to step shaft torque input.

The shaft torque oscillations may be written as follows:

$$\Delta t_{shaft} = \sum t_{sh\upsilon} \cos\left(\Omega_v t - \Psi_v\right) \tag{5.176}$$

with Ω_v in rad/s.

Let us consider first an autonomous SG without any asynchronous torque at all ($K_{as} = 0$). This is the ideal case of free oscillations.

From Equation 5.169,

$$\frac{2H}{\omega_b}\frac{d^2\Delta\delta_V}{dt^2} = \Sigma t_{sh\upsilon}\cos\left(\Omega_v t - \Psi_v\right) \tag{5.177}$$

The steady-state solution of this equation is straightforward:

$$\Delta\delta_{vv}^a = -\Delta\delta_{vvm}\cos(\Omega_v t - \Psi_v)$$

$$-\frac{\omega_b t_{shv}}{2H\Omega_v^2} = \Delta\delta_{vvm}^a \tag{5.178}$$

The amplitude of this free oscillation (for harmonic υ) is thus inversely proportional to inertia and to the frequency of oscillation squared.

For the SG with rotor damper cage and connected to the power system, both K_{as} and t_{es0} (synchronizing torque) are nonzero and thus Equation 5.169 has to be solved as it is:

$$\frac{2H}{\omega_b}\frac{d^2\Delta\delta_V}{dt^2} + \frac{K_{as}}{\omega_b}d\frac{\Delta\delta_V}{dt} + t_{es0}\Delta\delta_V = \Sigma\Delta t_{shv}\cos\left(\Omega_v t - \Psi_v\right) \tag{5.179}$$

Again the steady-state solution is sought:

$$\Delta\delta_{vv} = \Delta\delta_{vvm}\sin\left(\Omega_v t - \Psi_v - \phi_v\right) \tag{5.180}$$

where

$$\Delta\delta_{vvm} = \frac{\Delta t_{shv}}{\sqrt{\left(\dfrac{2H}{\omega_b}\Omega_v^2 - t_{es0}\right)^2 + \left(\dfrac{K_{as}}{\omega_b}\Omega_v\right)^2}};$$

$$\phi_v = \tan^{-1}\left(\frac{\left(-\dfrac{2H}{\omega_b}\Omega_v^2 + t_{es0}\right)}{K_{as}\dfrac{\Omega_v}{\omega_b}}\right)^{-1} \tag{5.181}$$

The ratio of the power angle amplitudes $\Delta\delta_{vvm}$ and $\Delta\delta a_{vvm}$ of forced, and, respectively, free oscillations is called the modulus of mechanical resonance K_{mv}:

$$K_{mv} = \frac{\Delta\delta_{vvm}}{\Delta\delta_{vvm}^a} = \frac{1}{\sqrt{\left(1 - \dfrac{\omega_0^2}{\Omega_v^2}\right)^2 + \left(\dfrac{K_{as}}{2H\Omega_v}\right)^2}} \tag{5.182}$$

The damping coefficient K_{dv} is

$$K_{dv} = \frac{K_{as}}{2H\Omega_v} \tag{5.183}$$

Typical variations of K_{mv} with ω_0/Ω_v ratio for various K_{dv} values are given in Figure 5.17.

The resonance conditions for $\omega_0 = \Omega_v$ are evident. To reduce the amplification effect, K_{dv} is increased, but this is feasible only up to a point by enforcing the damper cage (more copper). Therefore, in general, for all shaft torque frequencies Ω_v, it is appropriate to fall outside the hatched region in Figure 5.18:

$$1.25 > \frac{\omega_0}{\Omega_v} > 0.8 \tag{5.184}$$

As ω_0 is the eigen mechanical frequency of Equation 5.173—varies with load (δ_v) and with field current for a given machine, condition of Equation 5.184 is not so easy to fulfill for all shaft torque pulsations. The elasticity of shafts and of mechanical couplings between them in an SG set is a source of additional oscillations to be considered for the constraint in Equation 5.184. The case of the autonomous SG with damper cage rotor requires a separate treatment, which is beyond our scope here.

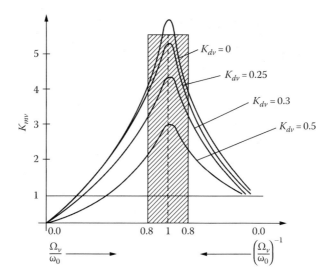

FIGURE 5.18 The modulus of mechanical resonance K_{mv}.

5.13 Small Disturbance Electromechanical Transients

After the investigation of fast (constant speed)—electromagnetic and, respectively, slow (mechanical) transients, we return to the general case when both electrical and mechanical transients are to be considered.

Electric power load variations are typically causing such complex transients.

For multiple SGs and loads, power systems and voltage and frequency control system design the investigation of transients is generally based on small disturbance theories to capitalize on linear control systems theoretical heritage and also reduce digital simulation time.

In essence, the complete (or approximate) dq model of Equation 5.68 of the SG is linearized about a chosen initial steady-state point by using only the first Taylor's series component. The linearized system is written in the state space form:

$$\Delta \dot{X} = A\Delta X + B\Delta V$$

$$\Delta Y = C\Delta X + D\Delta V$$

(5.185)

$\Delta \dot{X}$ is the state variables vector
ΔV is the input vector
ΔY is the output vector

The voltage or (and) speed controller systems may be included in Equation 5.185, and thus the small disturbance stability of the controlled generator is investigated by the eigenvalues method:

$$\det\left(A - \lambda I\right) = 0$$

(5.186)

I is unity diagonal matrix.

The eigenvalues λ may be real or complex numbers.

For system stability to small disturbances, all eigenvalues should have a NEGATIVE real part.

The unsaturated *dq* model of Equation 5.88 in P.U. may be linearized as follows:

$$\Delta V_d = -\Delta I_d r_s - \frac{1}{\omega_b} d\left(\frac{\Delta \Psi_d}{dt}\right) + \omega_{r0} \Delta \Psi_q + \Delta \omega_r \Psi_{q0}$$

$$\Delta V_q = -\Delta I_d r_s - \frac{1}{\omega_b} d\left(\frac{\Delta \Psi_q}{dt}\right) - \omega_{r0} \Delta \Psi_d - \Delta \omega_r \Psi_{d0}$$

$$\Delta V_f = \Delta I_f r_f + \frac{1}{\omega_b} d\left(\frac{\Delta \Psi_f}{dt}\right) \qquad (5.187)$$

$$0 = -\Delta I_D r_D - \frac{1}{\omega_b} d\left(\frac{\Delta \Psi_D}{dt}\right)$$

$$0 = -\Delta I_Q r_Q - \frac{1}{\omega_b} d\left(\frac{\Delta \Psi_Q}{dt}\right)$$

$$\Delta t_{shaft} = \Delta t_e + 2H \frac{d}{dt}\left(\Delta \omega_r\right)$$

$$\Delta t_e = -\left(\Psi_{d0}\Delta I_q + I_{q0}\Delta \Psi_{d0} - \Psi_{q0}\Delta I_d - I_{d0}\Delta \Psi_q\right) \qquad (5.188)$$

$$\frac{1}{\omega_b} \frac{d\Delta \delta_V}{dt} = \Delta \omega_r$$

For the initial (steady-state point),

$$V_{d0} = -I_{d0} r_s + \Psi_{q0} \omega_{r0}$$

$$V_{q0} = -I_{q0} r_s - \Psi_{d0} \omega_{r0}$$

$$\Psi_{d0} = l_{sl} I_{d0} + l_{dm}\left(I_{d0} + I_{f0}\right)$$

$$\Psi_{q0} = \left(l_{sl} + l_{qm}\right) I_{q0} \qquad (5.189)$$

$$V_{f0} = I_{f0} r_f; \quad I_{D0} = I_{Q0} = 0$$

$$t_{e0} = -\left(\Psi_{d0} I_{d0} - \Psi_{q0} I_{q0}\right) = t_{shaft}$$

$$\delta_V = \tan^{-1} \frac{V_{d0}}{V_{q0}}$$

$$V_{d0} = -V_0 \sin \delta_0$$

$$V_{q0} = -V_0 \cos \delta_0$$

$$\Delta V_d \approx -\Delta V \sin \delta_{V0} - V_0 \cos \delta_{V0} \Delta \delta_V$$

$$\Delta V_q \approx -\Delta V \cos \delta_{V0} + V_0 \sin \delta_{V0} \Delta \delta_V \qquad (5.190)$$

$$\Delta \Psi_d = l_{sl} \Delta I_d + l_{dm} \Delta I_{dm}; \quad \Delta I_{dm} = \Delta I_d + \Delta I_f + \Delta I_D$$

$$\Delta \Psi_q = l_{sl} \Delta I_q + l_{qm} \Delta I_{qm}; \quad \Delta I_{qm} = \Delta I_q + \Delta I_Q$$

$$\Delta \Psi_D = l_{Dl} \Delta I_D + l_{dm} \Delta I_{dm}$$

$$\Delta \Psi_Q = l_{Ql} \Delta I_Q + l_{qm} \Delta I_{qm}$$

Eliminating the flux linkage disturbances in Equations 5.187 and 5.188 by using Equation 5.190 leads to the definition of the following state space variable vector ΔX:

$$\left|\Delta X\right| = \left(\Delta I_{ds}, \Delta I_f, \Delta I_{dm}, \Delta I_{qs}, \Delta I_{qm}, \Delta \omega_r, \Delta \delta_V\right)^t \tag{5.191}$$

The input vector is

$$\left|\Delta V\right| = \left[-\Delta V \sin \delta_0, \Delta V_f, 0, -\Delta V \cos \delta_0, 0, \Delta t_{shaft}, 0\right]^t \tag{5.192}$$

We may now put Equations 5.187 and 5.188 with Equation 5.190 into matrix form as follows:

$$\left|\Delta V\right| = -\left|L\right|\frac{1}{\omega_b}d\frac{(\Delta X)}{dt} - \left|R\right|\Delta X \tag{5.193}$$

where

$$|L| = \begin{array}{c} \\ 1 \\ 2 \\ 3 \\ 4 \\ 5 \\ 6 \\ 7 \end{array} \begin{array}{ccccccc} 1 & 2 & 3 & 4 & 5 & 6 & 7 \\ l_{sl} & 0 & l_{dm} & 0 & 0 & 0 & 0 \\ 0 & l_{fl} & l_{dm} & 0 & 0 & 0 & 0 \\ -l_{sl} & -l_{Dl} & l_{dm}+l_{Dl} & 0 & 0 & 0 & 0 \\ 0 & 0 & 0 & l_{sl} & l_{qm} & 0 & 0 \\ 0 & 0 & 0 & -l_{Ql} & l_{qm}+l_{Ql} & 0 & 0 \\ 0 & 0 & 0 & 0 & 0 & -1 & 0 \\ 0 & 0 & 0 & 0 & 0 & 0 & -1 \end{array}$$

	1	2	3	4	5	6	7
1	r_s	0	0	$-\omega_{r0}l_{sl}$	$-\omega_{r0}l_{qm}$	$-\Psi_{q0}$	$V_0 \cos \delta_{V0}$
2	0	r_f	0	0	0	0	0
3	$-r_D$	$-r_D$	$+r_D$	0	0	0	0
$\lvert R\rvert=4$	$\omega_{r0}l_{sl}$	0	$\omega_{r0}l_{dm}$	r_s	0	Ψ_{d0}	$-V_0 \sin \delta_{V0}$
5	0	0	0	$-r_Q$	r_Q	0	0
6	$\dfrac{l_{sl}I_{q0}-\Psi_{q0}}{2H\omega_b}$	0	$\dfrac{l_{dm}I_{q0}}{2H\omega_b}$	$\dfrac{\Psi_{d0}-l_{sl}I_{d0}}{2H\omega_b}$	$\dfrac{-l_{dm}I_{d0}}{2H\omega_b}$	-1	0
7	0	0	0	0	0	1	0

Comparing Equation 5.185 with Equation 5.193,

$$A = -\left|L\right|^{-1} \cdot \left|R\right|;$$
$$B = -\left[L\right]^{-1}\left|1,1,1,1,1,1,1\right| \tag{5.194}$$

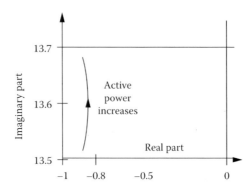

FIGURE 5.19 Critical mode eigenvalues of unsaturated (I_s, I_f, I_m) model when active power rises from 0.8 to 1.2 P.U.

All that remains to calculate the eigenvalues of matrix A and thus establish the small disturbance stability performance of an SG.

The choice of I_{dm} and I_{qm} as variables, instead of rotor damper cage currents I_D and I_Q, makes [L] more sparse and leaves way to somehow consider magnetic saturation. At least for steady state when the dependences of l_{dm} and l_{qm} on both I_{dm} and I_{qm} (Figure 5.10) may be in advance established by tests or by finite element calculations.

This is easy to apply as values of I_{dm0} and I_{qm0} are straightforward:

$$I_{dm0} = I_{d0} + I_{f0}$$

$$I_{qm0} = I_{q0}$$

(5.195)

and the level of saturation may be considered "frozen" at the initial conditions (uninfluenced by small perturbations). Typical critical eigenvalues change with active power for the I_s, I_f, I_m model above are shown in Figure 5.19 [18].

It is possible to choose the variable vector in different ways by combining various flux linkages and current as variables [18]. Not much to gain with such choices unless magnetic saturation is not rigorously considered. It has been shown that care must be exercised in representing magnetic saturation by single magnetization curves (dependent on $I_m = \sqrt{I_{dm}^2 + I_{qm}^2}$) in the underexcited regimes of SG when large errors may occur. Only using complete $I_{dm}(\Psi_{dm}, \Psi_{qm})$, $I_{qm}(\Psi_{dm}, \Psi_{qm})$ families of saturation curves will lead to good results throughout the whole active and reactive power capability region of the SG.

As during small perturbation transients the initial steady-state level is paramount, we leave out the transient saturation influence here, to consider it in the study of large disturbance transients when the latter really matters notably.

5.14 Large Disturbance Transients Modeling

Large disturbance transients modeling of single SG should take into account both magnetic saturation and frequency (skin rotor) effects.

Completing the *dq* model, with two *d* axis damper circuits and three *q*-axis damper circuits, satisfies such standards, if magnetic saturation is included, even separately, along each axis (Figure 5.20).

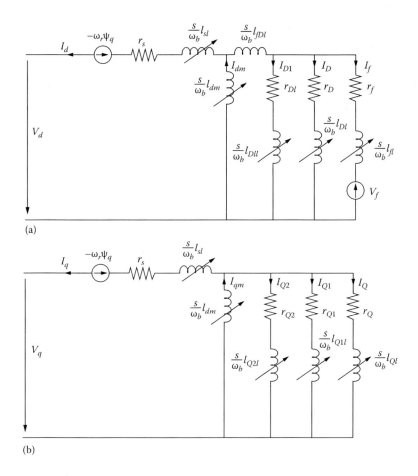

FIGURE 5.20 Three-circuit SG model with some variable inductances: (a) axis d and (b) axis q.

For the sake of completion, all equations of such a model are given in what follows:

$$\Psi_d = l_{sl}I_d + \Psi_{dm}$$

$$\Psi_q = l_{sl}I_q + \Psi_{qm}$$

$$\Psi_f = l_{fl}I_f + \Psi_{dm} + l_{fDl}\left(I_f + I_D + I_{D1}\right)$$

$$\Psi_D = l_{Dl}I_D + \Psi_{dm} + l_{fDl}\left(I_f + I_D + I_{D1}\right)$$

$$\Psi_{D1} = l_{D1l}I_{D1} + \Psi_{dm} + l_{fDl}\left(I_f + I_D + I_{D1}\right)$$

$$\Psi_Q = l_{Ql}I_Q + \Psi_{qm}$$

$$\Psi_{Q1} = l_{Q1l}I_{Q1} + \Psi_{qm}$$

$$\Psi_{Q2} = l_{Q2l}I_{Q2} + \Psi_{qm}$$

$$I_{dm} = I_d + I_D + I_{D1} + I_f$$

$$I_{qm} = I_q + I_Q + I_{Q1} + I_{Q2}$$

(5.196)

$$\Psi_{dm} = l_{dm}(I_{dm}, I_{qm})I_{dm}$$

$$\Psi_{qm} = l_{qm}(I_{dm}, I_{qm})I_{qm}$$

(5.197)

The magnetization curve families in Equation 5.197 have to be obtained either through SSTR or through FEM magnetostatic calculations at standstill. Once polynomial splines approximations of l_{dm} and l_{qm} functions are developed, the time derivatives of Ψ_{dm} and Ψ_{qm} may be obtained as follows:

$$\frac{s}{\omega_b}\frac{d\Psi_{dm}}{dt} = l_{ddm}\frac{s}{\omega_b}\frac{dI_{dm}}{dt} + l_{qdm}\frac{s}{\omega_b}\frac{dI_{qm}}{dt}$$

$$\frac{s}{\omega_b}\frac{d\Psi_{qm}}{dt} = l_{dqm}\frac{s}{\omega_b}\frac{dI_{dm}}{dt} + l_{qqm}\frac{s}{\omega_b}\frac{dI_{qm}}{dt}$$

(5.198)

$$l_{ddm} = l_{dm} + \frac{\partial l_{dm}I_{dm}}{\partial I_{dm}}; \quad l_{qdm} = \frac{\partial l_{dm}I_{dm}}{\partial I_{qm}}$$

$$l_{dqm} = \frac{\partial l_{qm}I_{qm}}{\partial I_{dm}}; \quad l_{qqm} = l_{qm} + \frac{\partial l_{qm}}{\partial I_{qm}}I_{qm}$$

The leakage saturation occurs only in the field winding and in the stator winding and is considered to depend solely on the respective currents:

$$\frac{s}{\omega_b}l_{sl}(I_s) = \left(l_{sl}(I_s) + \frac{\partial l_{sl}}{\partial I_s}I_s\right)\frac{s}{\omega_b} = l_{slt}(I_s)\frac{s}{\omega_b}$$

$$I(s) = \sqrt{I_d^2 + I_q^2}$$

(5.199)

$$\frac{s}{\omega_b}l_{fl}(I_f) = \left(l_{fl}(I_f) + \frac{\partial l_{fl}}{\partial I_f}I_f\right)\frac{s}{\omega_b}$$

Therefore, the leakage inductances of stator and field winding in the equivalent circuit are replaced by their transient value l_{slt}, l_{flt}. However, in the flux–current relationship their steady-state values l_{sl}, l_{fl} occur. They are all functions of their respective currents. It should be mentioned that only for currents in excess of 2 P.U., leakage saturation influence is worth considering. Sudden short circuit represents such a transient process.

Equation 5.198 suggest that a cross-coupling between the equivalent circuits of axes d and q is required. This is rather simple to operate (Figure 5.21).

From the reciprocity condition, $l_{dqm} = l_{qdm}$.

The rather general equivalent circuit may be identified:

- The magnetization curve family of Equation 5.197, Equation 5.198 and leakage inductance functions $l_{sl}(I_s)$, $l_{fl}(I_f)$ are first determined from time domain standstill tests (or FEM).
- From frequency response at standstill or through FEM, all the other components are calculated.

The dependence of l_{dm} and l_{qm} of both I_{dm} and I_{qm} should lead to model suitability in all magnetization conditions, including the disputed case of underexcited SG when the concept $l_{dm}(I_m)$, $l_{dm}(I_m)$ unique functions or of total magnetization current (mmf) fails [4].

The machine equations are straightforward from the equivalent scheme and thus are not repeated here. The choice of variables is as in the paragraph on small perturbations.

Tedious FEM, tests, and their processing are all required before such a rather complete circuit model of the SG is established in all rigor.

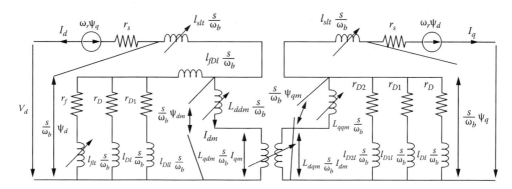

FIGURE 5.21 General three-circuit SG model with cross-coupling saturation.

That the *dq* model may be used to investigate various symmetrical transients is very clear. The same model may be used in asymmetrical stator connections also, as long as the time functions of V_d and V_q may be obtained. But $v_d(t)$ and $v_q(t)$ may be defined only if $V_A(t)$, $V_B(t)$, $V_C(t)$ functions are available. Alternatively, the load voltage-current relationships have to be amenable to state space form.

Let us illustrate this idea with a few examples.

5.14.1 Line-to-Line Fault

A typical line fault at machine terminals is shown in Figure 5.22.

The power source–generator voltage relationships after the short circuit are as follows:

$$V_B - V_C = E_B - E_C$$
$$V_A = V_C; \quad I_A + I_B + I_C = 0, \quad V_0 = 0$$

(5.200)

Consequently,

$$V_C(t) = \frac{1}{3}\left(E_C - E_B\right)$$
$$V_B(t) = -2V_C(t)$$
$$V_A(t) = V_C(t)$$
$$E_{A,B,C}(t) = V\sqrt{2}\cos\left(\omega_{1t} - (i-1)\frac{2\pi}{3}\right); \quad i = 1,2,3$$

Consequently, all stator voltages $V_A(t)$, $V_B(t)$, and $V_C(t)$ may be defined right after the short circuit.

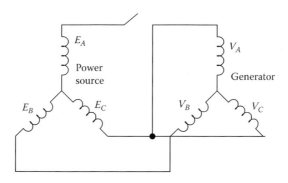

FIGURE 5.22 Line-to-line SG short circuit.

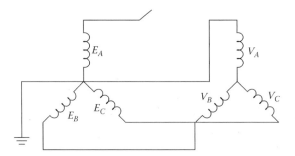

FIGURE 5.23   Line-to-neutral fault.

5.14.2 Line-to-Neutral Fault

In this case, a phase of the synchronous machine is connected to the neutral of the power system (Figure 5.23), which may be connected to the ground or not.

According to Figure 5.23:

$$V_B - V_C = E_B - E_C$$

$$V_C - V_A = E_C; \quad V_A + V_B + V_C = 0 \tag{5.201}$$

Consequently,

$$V_A = -\frac{(E_C + E_B)}{3}$$

$$V_B = V_A + E_B \tag{5.202}$$

$$V_C = V_A + E_C$$

Therefore, again, provided E_A, E_B, and E_C and time functions $V_A(t)$, $V_B(t)$, and $V_C(t)$ are also known.

5.15 Finite Element SG Modeling

The numerical methods for field distribution calculation in electric machines are by now an established field with a rather long history even before 1975 when finite difference methods prevailed.

Since then, the finite element (integral) methods took over to produce revolutionary results ever since.

For the basics of finite element method, see Reference [19]. In 1976, SG time-domain responses at standstill have been approached successfully by FEM, making use of the conductivity matrices concept [20]. In 1980, the SG sudden three-phase short circuit has been calculated [21,22] by FEM.

The relative motion between stator and rotor during balanced and unbalanced short-circuit transients has been reported in 1987 [23].

In 1990s, the time stepping and coupled-field and circuit concepts have been successfully introduced [24] to eliminate circuit simulation restrictions based in conductivity matrices representations. Typical results related to no-load and steady-state short-circuit curves obtained through FEM for a 150 MVA 13.8 kV SG are shown in Figure 5.24, after [4].

Also for steady state, FE have been proved to predict correctly (within 1%–2%) the field current required for various active and reactive power loads over the whole P–Q capability curve of same SG [4].

Finally, the rotor angle during steady state has been predicted within 2 degrees for the whole spectrum of active and reactive power loads (Figure 5.25) [4].

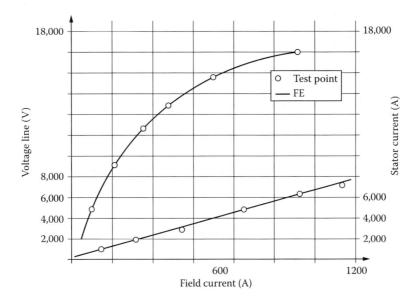

FIGURE 5.24 Open- and short-circuit curves of a 150 MVA, 13.8 kV, 2 pole SG.

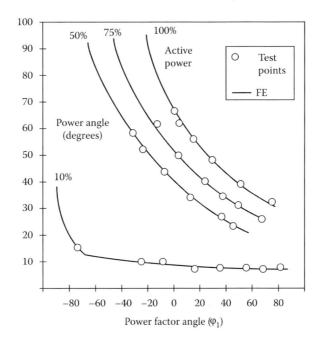

FIGURE 5.25 Power angle δ_V versus power factor angle φ_1 of a 150 MVA, 13.8 kV, 2 pole SG.

The FEM has also been successful in calculating SG response to standstill time domain and frequency responses, then used to identify the general equivalent circuit elements [25–27].

A complete picture of FE flux paths during an ongoing sudden short-circuit process is shown in Figure 5.26 [23].

The traveling field is visible. And so is the fact that quite a lot of flux paths cross the slots, as expected.

Finally, FE simulation of an 120 MW, 13.8 kV, 50 Hz SG on load during and after a three-phase fault has been successfully done [28] Figure 5.27.

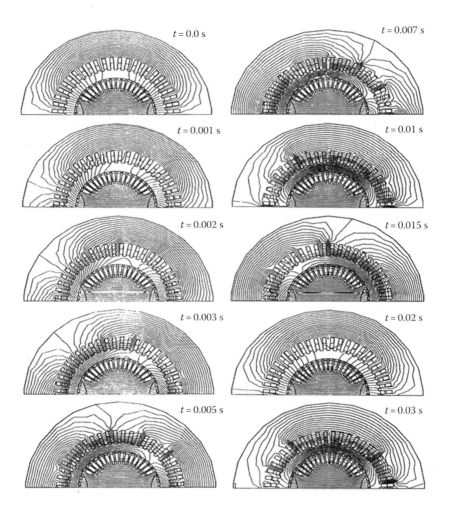

FIGURE 5.26 Flux distribution during a 0.5 P.U. balanced short circuit at a 660 MW SG terminals (contour intervals are 0.016 Wb/m).

This is a severe transient as the power angle reaches over 90° during the transients when notable active power is delivered as the field current is also large. The plateau in the line voltage recovery from 0.4 to 0.5 s (Figure 5.27d, e, and c) is this way explainable.

It is almost evident that FEM has reached the point of being able to simulate virtually any operation mode in an SG. The only question is the amount of computation time required, not to mention the extraordinary expertise involved in wisely using it.

The FEM is the way of the future in SG, but there may be two avenues to that future:

1. Use FEM to calculate and store SG lumped parameters for an ever wider saturation and frequency effects and then use the general equivalent circuits for the study of various transients with the machine alone or incorporated in a power system [30,31].

2. Use FEM directly to calculate the electromagnetic and mechanical transients through coupled-field circuit models or even directly through powers, torques, and motion equations [32–35].

While the first avenue looks more practical for power system studies, the second one may be used for the design refinements of a new SG.

The few existing dedicated FEM software packages are expected to improve further, at reasonably additional computation time and costs.

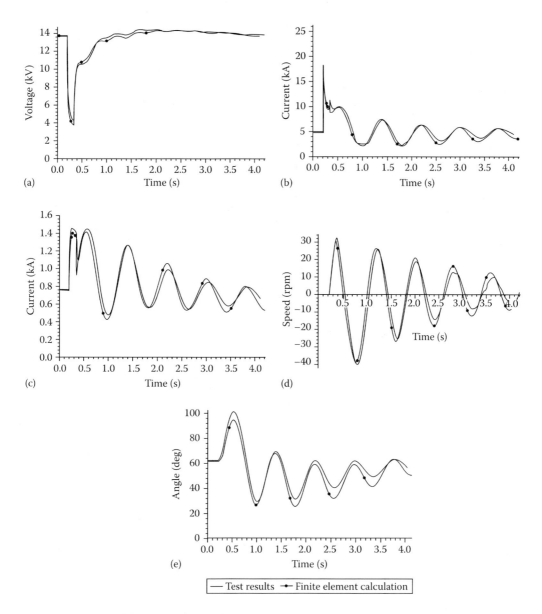

FIGURE 5.27 On load three-phase fault and recovery transients by FE and tests: (a) line voltage, (b) line current, (c) field current, (d) rotor speed, and (e) power angle.

5.16 SG Transient Modeling for Control Design

In Section 5.10, simplified models for power system studies have been described. One specific approximation has gained wide acceptance, especially for SG control design.

It is related to neglecting of stator transients with neglecting the damper-winding effects altogether (third-order model) or with considering one damper winding along axis *q* but no damper winding along axis *d* (fourth-order model) [29].

We start with Equation 5.164, the third-order model:

$$V_d = -I_d r_s + \Psi_q \omega_r$$

$$V_q = -I_q r_s - \Psi_d \omega_r$$

$$\begin{vmatrix} \Psi_d \\ \Psi_f \end{vmatrix} = \begin{vmatrix} l_d & l_{dm} \\ l_{dm} & l_f \end{vmatrix} \begin{vmatrix} I_d \\ I_f \end{vmatrix}; \quad \Psi_q = l_q I_q$$

$$\frac{1}{\omega_b} \dot{\Psi}_f = -I_f r_f + v_f \tag{5.203}$$

$$t_e = -\left(\Psi_d I_q - \Psi_q I_d \right)$$

$$2H \dot{\omega}_r = t_{\text{shaft}} - t_e;$$

$$\omega_b \dot{\delta} = \omega_r - \omega_1$$

Eliminating I_f from the flux–current relationships and deriving the transient emf e_q' as

$$e_q' = +\omega_r \frac{l_{dm}}{l_f} \Psi_f \tag{5.204}$$

the q-axis stator equation becomes

$$V_q = -r_s I_q - x_d' I_d - e_q' \tag{5.205}$$

where

$$x_d' = \omega_r \left(l_d - \frac{l_{dm}^2}{l_f} \right); \quad l_f = l_{fl} + l_{dm} \tag{5.206}$$

We may now rearrange the field circuit equation in (Equation 5.203) with e_q' and x_d' to obtain the following:

$$\dot{e}_q' = \frac{v_f - e_q' + i_d(x_d - x_d')}{T_{d0}'}; \quad T_{d0}' = \frac{l_f}{r_f} \tag{5.207}$$

Considering a damper winding Q along axis q, a similar equation to Equation 5.204 is obtained with e_d' as follows:

$$e_d' = \omega_r \frac{l_{qm}}{l_Q} \Psi_Q;$$

$$l_Q = l_{Ql} + l_{qm} \tag{5.208}$$

$$\dot{e}_d' = \frac{-i_q(x_q - x_q') - e_d'}{T_{q0}'}; \quad T_{q0}' = \frac{l_Q}{r_Q}; \quad x_q' = l_q - \frac{l_{qm}^2}{l_Q} \tag{5.209}$$

$$V_d = -r_s I_d - x_q' I_q + e_d'$$

As expected, in the absence of the rotor q axis damping winding, e'_d is taken as zero and the third order of the model is restored. x'_d and x'_q are the transient reactances (in P.U.) as defined earlier in this chapter. The two equations of motion in Equation 5.203, Equations 5.207 and 5.208, have to be added to form the fourth order (transient model) of the SG.

As the transient emf e'_q differential equation includes the field-winding voltage v_f, the application of this model to control is greatly facilitated as shown in Chapter 6, dedicated to the control of SG.

The initial values of transients emfs e'_d, e'_q are calculated from stator equations in Equations 5.205 and 5.208 for steady state:

$$\left(e'_q\right)_{t=0} = \omega_{r0} \frac{l_{dm}}{l_f} \cdot l_f \left(I_f\right)_{t=0} + l_{dm}\left(I_d\right)_{t=0}$$

$$\left(e'_d\right)_{t=0} = \omega_{r0} \frac{l^2_{qm}}{l_Q} \cdot \left(I_q\right)_{t=0}$$

(5.210)

In a similar way, a subtransient model may be defined for the first moments after a fast transient process [2]. Such a model is not suitable for control design purposes where the excitation current control is rather slow anyway.

Linearization of the transient model with the rotor speed ω_r as variable (even in the stator equations) proves to be very useful in automatic voltage regulator (AVR) design as shown in Chapter 6.

5.17 Summary

- SGs undergo transient operation modes when the currents, flux linkages, voltages, load, or speed vary with time [29]. Connection of the SGs to a power system, electrical load, or shaft torque variations produce transients. The steady-state, two reaction, model in Chapter 4, based on traveling fields at standstill to each other, is valid only for steady-state operation.

    The main SG models for transients are as follows:
- The phase-variable circuit model
- The orthogonal axis dq circuit model
- The coupled FEM field/circuit model
    - The phase-variable model is based on the SG structure as multiple electric and magnetic circuits coupled together electrically and (or) magnetically. The stator/rotor circuits mutual inductances always depend on rotor position. In salient-pole rotors, the stator phase self-inductances vary also with rotor position, that is with time. With one damper circuit along each rotor axis, a field winding and three stator phases, an eight-order nonlinear system with time-variable coefficients is obtained, when the two equations of motion are added. The basic variables are I_A, I_B, I_C, I_f, I_D, I_Q, ω_r, and θ_r.
    - Solving a variable coefficient state space system requires the inverting of a time-variable sixth-order matrix for each integration step time interval. Only numerical methods such as Runge–Kutta–Gill, predictor–corrector can handle the solving of such a stiff state—space high-order system. In addition, the computation time is prohibitive due to the time dependence of inductances. Finally, the complexity of the model leaves little room for intuitive interpretation of phenomena and trends for various transients. And it is unpractical for control design. In conclusion, the phase-variable model should be used only for special situations such as highly unbalanced transients or faults, etc.
    - Simpler models are needed to handle transients in a more practical manner.
    - The orthogonal axis (dq) model is characterized by inductances between windings that are independent from rotor position. The dq model may be derived from the phase-variable

model either through a mathematical change of variables (Park transform) or through a physical orthogonal axis model.

- The Park transform is an orthogonal change of variables such that to conserve powers, losses and torque:

$$\begin{vmatrix} V_d \\ V_q \\ V_0 \end{vmatrix} = \left| P(\theta_{er}) \right| \cdot \begin{vmatrix} V_A \\ V_B \\ V_C \end{vmatrix}$$

$$\left| P(\theta_{er}) \right| = \frac{2}{3} \begin{bmatrix} \cos(-\theta_{er}) & \cos\left(-\theta_{er} + \frac{2\pi}{3}\right) & \cos\left(-\theta_{er} - \frac{2\pi}{3}\right) \\ \sin(-\theta_{er}) & \sin\left(-\theta_{er} + \frac{2\pi}{3}\right) & \sin\left(-\theta_{er} - \frac{2\pi}{3}\right) \\ \frac{1}{2} & \frac{1}{2} & \frac{1}{2} \end{bmatrix} \tag{5.211}$$

$$\left| P(\theta_{er}) \right|^{-1} = \frac{3}{2} \left| P(\theta_{er}) \right|^T ; \quad \frac{d\theta_{er}}{dt} = \omega_r$$

The physical *dq* model consists of a fictitious machine with same rotor *f*, *D*, *Q* orthogonal rotor windings as in the actual machine and with two stator windings whose magnetic axes (mmfs) are always fixed to the rotor *d* and *q* orthogonal axes. The fact that the rotor *dq* axes move at rotor speed and are always aligned to axis *d* and *q* secure the independence of the *dq* model inductances of rotor position.

- Steady state means *d.c.* in the *dq* model of SG.
- For complete equivalence of the *dq* model with the real machine a null component is added. This component does not produce torque through the fundamental and its current is determined by the stator resistance and leakage inductance:

$$V_0 = -r_s I_0 - l_{sl} \frac{s}{\omega_b} \frac{dI_0}{dt} \tag{5.212}$$

- The *dq* model parameters dependence on the real machine parameters is rather simple.
- To reduce the number of inductances in the *dq* model, the rotor quantities are reduced to the stator under the assumption that the air gap main magnetic field couples all windings along axis *d* and, respectively, *q*. Thus the stator-rotor coupling inductances become equal to the stator magnetization inductances l_{dm}, l_{qm}.
- An additional leakage coupling inductance l_{fDl} between the field winding and the *d*-axis damper winding is introduced also, as it proves useful in providing for good results both in the stator and field current transient responses.
- The *dq* model is in general used in the per unit (P.U.) form to reduce the range of variables during transients from zero to say 20 for all transients and all power ranges. The base quantities, voltage, current, power are the rated ones, in general, but other choices are possible. In this chapter, all variables and parameters are in P.U., but the time *t* and inertia *H* are in seconds. In this case, $d/dt \rightarrow s/\omega_b$ in Laplace form (ω_b—the base (rated) angular frequency in rad/s).
- The rotor-to-stator reduction coefficients conserve losses and power, as expected.

- The *dq* model equations writing assumes source circuits in the stator and sink circuits in the rotor and the induced voltage: $e = -d\Psi/dt$. Also, implicitly, the Poynting vector $\overline{P} = \dfrac{\overline{E} \times \overline{H}}{2}$ enters the sink circuits and exits the source circuits. This way, all flux/current relations contain only + signs. This choice leads to the fact that while power and torque is positive for generating, the components V_d, V_q, and I_q are always negative for generating. But I_d is either negative or positive depending on the load power factor angle. This is valid for the trigonometric positive motion direction.
- The space-vector diagram at steady state evidentiates the power angle δ_V between the voltage vector and axis q in the third quadrant, with $\delta_V > 0$ for generating.
- Based on the *dq* model, state-space equations in P.U. of Eqaution 5.82, two distinct general equivalent circuits may be drawn (Figure 5.6). They are very intuitive in the sense that all *dq* model equations may be derived by inspection. The distinct d and q equivalent circuits for transients indicate that there is no magnetic coupling between the two orthogonal axes.
- In reality, in heavily saturated SGs, there is a cross-coupling due to magnetic saturation between the two orthogonal axes. Putting this phenomenon into the *dq* model has received a lot of attention lately but here only two representative solutions are described.
- One uses distinct but unique magnetization curves along axes d and q: $\Psi_{dm}^*\left(I_m\right), \Psi_{qm}^*\left(I_m\right)$ where I_m is the total magnetization current (or mmf): $I_m = \sqrt{I_{dm}^2 + I_{qm}^2}$. This approximation seems to fail when the SG is "deep" in the leading power factor mode (underexcited, with $I_m < 0.7$ P.U.). However when $I_m > 0.7$ P.U. it has not been proved wrong yet.
- The second model for saturation presupposes a family of magnetization curves along axis *dq*, respectively, q: $\Psi_{dm}^*\left(I_{dm}, I_{qm}\right), \Psi_{qm}^*\left(I_{dm}, I_{qm}\right)$. This model, after adequate analytical approximations of these functions, should not fail over the entire active/reactive power envelope of an SG. But it requires more computation efforts.
- The magnetization curves along axes d and q may be obtained either from experiments or through FEM field calculations.
- The cross-coupling magnetic saturation may be rather handily included in the *dq* general equivalent circuit for both axes (Figure 5.21).
- The general equivalent circuits are based on the flux/stator current relationships without rotor currents:

$$\Psi_d(s) = l_d(s)I_d(s) + g(s)v_f(s)$$
$$\Psi_q(s) = l_q(s)I_q(s)$$
(5.213)

- $l_d(s)$, $l_q(s)$, and $g(s)$ are known as the operational parameters of the SG (s—Laplace operator).
- Operational parameters include the main inductances and time constants that are catalogue data of the SG: subtransient, transient, and synchronous inductances l_d'', l_d', l_d, along d-axis and l_q'', l_q' along q axis. The corresponding time constants are T_d'', T_d', T_{d0}'', T_{d0}', T_q'', and T_{q0}'. Additional terms are added when frequency (skin) effect imposes: additional fictitious rotor circuits: l_d''', l_q''', l_q', T_d''', T_q''', T_{d0}''', and T_{q0}'''.
- Though transients may be handled directly via the complete d–q model through its solving by numerical methods, a few approximations have led to very practical solutions.
- Electromagnetic transients are those that occur at constant speed. The operational calculus may be applied with some elegant analytical approximated solutions as a result. Sudden three-phase short circuit falls into this category. It is used for unsaturated parameter identification by comparing the measured and calculated current waveforms during a sudden short circuit. Graphical models (20 years ago) and advanced nonlinear programming methods have been used for curve fitting the theory and tests with parameter estimation as the main goal.

- Electromagnetic transients may be provoked also at zero speed with the applied voltage vector along axis d or q with or without DC in the other axis. The applied voltage may be DC step voltage or PWM cyclical voltage rich in frequency content. Alternatively, single-frequency voltages may be applied one after the other and the impedance measured (amplitude and phase). Again, a way to estimate the general equivalent circuit parameters has been born through standstill electromagnetic transients processing. Alternatively, FEM calculations may replace or accompany the tests for same purpose: parameter estimation.
- For multimachine transients investigation, simpler lower-order dq models have become standard. One of them is based on neglecting the stator (fast) transients. In this model, the fast decaying AC components in stator currents and torque are missing. Further on, it seems better that in the case the rotor speed in the stator equations be kept equal to the synchronous speed $\omega_r = \omega_{r0} = \omega_1$. The speed in the model varies as the equation of motion remains in place.
- Gradually, the damper circuits transients may be also neglected ($I_D = I_Q = 0$), and thus only one electrical transient, determined by the field winding, remains. With the two motion equations, this one makes for a third-order system. Finally, all electric transients may be neglected to be left only with the motion equations, for very slow (mechanical transients).
- Asynchronous running at constant speed may also be tackled as an electromagnetic transient with $s \rightarrow jS\omega_1$; S—slip; $S = (\omega_1 - \omega_r)/\omega_1$. An inflexion in the asynchronous torque is detected around $S = 1/2$ ($\omega_r = \omega_1/2$). It tends to be small in large machines as r_s is very small in P.U.
- In close to synchronous speed, $|S| < 0.05$, the asynchronous torque is proportional to slip speed $S\omega_1$. Also, torque pulsations occur in the asynchronous torque, they have to be accounted for during transients, for better precision, as asynchronous torque t_{as} is

$$t_{as} = t_{asav} + t_{asp}\cos\left(2S\omega_1\omega_b + \Psi\right) \tag{5.214}$$

Its pulsations frequency is small because S gets smaller, as it is the case around synchronism ($|S| < 0.05$, $\omega_r \approx (0.95 - 1.05)\omega_1$).
- Mechanical (very slow) transients may be treated easily with the SG at steady state electromagnetically. Only speed ω_r and power angle δ_v vary in the so-called rotor swing equation. Through numerical methods, the nonlinear model may be solved in general but, for small perturbations, a simple analytical solution for $\omega_r(t)$ and $\delta_v(t)$ may be found for particular inputs such as shaft torque step or frequency response.
- For the SG in stand-alone operation without a damper cage, a proper mechanical frequency f_0 is defined. It varies with field current I_f, power angle δ_v, and inertia, but it is in general less than 2–3 Hz for large- and medium-power standard SGs. When an SG is connected to the power grid and the shaft torque presents pulsations at Ω_v, resonance conditions may occur. They are characterized by severe oscillations amplification (in δ_v), that are tempered by a large inertia or (and) a strong damper cage. Torsional shaft frequencies by the prime mover shaft and coupling to the SG may also produce resonance mechanical conditions that have to be avoided.
- Electromechanical transients are characterized by both electrical and mechanical transients. For small perturbations, the dq model provides a very good way to investigate the SG stability—without or with voltage and speed control—by the eigenvalue method, after linearization around a steady state given point.
- For large disturbance transients, the full dq model with magnetic saturation and frequency effects (Figure 5.21) is recommended. Numerical methods may solve the transients altogether, but direct stability methods typical to nonlinear systems may also be used.
- Finite element analysis is by now widely used to asses various SG steady state and transients through coupled field/circuit models. The computation time is still prohibitive for use in design

optimization or for controller design. With the computation power of microcomputers rising by the year, the FEM is becoming the norm in analyzing the FEM steady-state and transient performance: electromagnetic, thermal, or mechanical.

- Still the circuit models, with the parameters calculated through FEM and then curve fitted by analytical approximations will eventually remain the norm for preliminary and optimization design, particular transients, and SG control.
- The approximate (circuit) transient (fourth-order) model of SG is finally given as it will be used in Chapter 6 dedicated to the control of SGs.

References

1. J. Machowski, J.W. Bialek, J.R. Bumby, *Power System Dynamics and Stability*, Wiley and Sons, Chichester, U.K., 1997.
2. M. Namba, J. Hosoda, S. Dri, M. Udo, Development for measurement of operating parameters of synchronous generator and control system, *IEEE Trans.*, PAS-200(2), 618–628, 1981.
3. I. Boldea, S.A. Nasar, Unified treatment of core losses and saturation in orthogonal axis model of electrical machines, *IEE Proc*, 134(6), 355–363, 1987.
4. N.A. Arjona, D.C. Macdonald, A new lumped steady-state synchronous machine model derived from finite element analysis, *IEEE Trans*, EC-14 (1), 1–7, 1999.
5. E. Levi, Saturation modeling in *d–q* models of salient pole synchronous machines, *IEEE Trans.*, EC-14(1), 44–50, 1999.
6. A.M. El-Serafi, J. Wu, Determination of parameter representing cross-coupling effect in saturated synchronous machines, *IEEE Trans.*, EC-8(3), 333–342, 1993.
7. K. Ide, S. Wakmi, K. Shima, K. Miyakawa, Y. Yagi, Analysis of saturated synchronous reactances of large turbine generator by considering cross-magnetizing reactances using finite elements, *IEEE Trans.*, EC-14(1), 66–71, 1999.
8. I. Kamwa, P. Viarouge, R. Mahfoudi, Phenomenological models of large synchronous machines from short-circuit tests during commissioning—a classical modern approach, *IEEE Trans.*, EC-9(1), 85–97, 1994.
9. S.A. Soliman, M.E. El-Hawary, A.M. Al-Kandari, Synchronous machine optimal parameter estimation from digitized sudden short-circuit armature current, *Record of ICEM-2000*, Espoo, Finland.
10. A. Keyhani, H. Tsai, T. Leksan, Maximum likelihood estimation of synchronous machine parameters from standstill time response data, *IEEE Trans.*, EC-9(1), 98–114, 1994.
11. I. Kamwa, P. Viarouge, J. Dickinson, Identification of generalized models of synchronous machines from time domain tests, *IEEE Pro.*, 138(6), 485–491, 1991.
12. S. Horning, A. Keyhani, I. Kamwa, On line evaluation of a round rotor synchronous machine parameter set estimated from standstill time domain data, *IEEE Trans.*, EC-12(4), 289–296, 1997.
13. K. Beya, R. Pintelton, J. Schonkens, B. Mpanda-Maswe, P. Lataire, M. Dehhaye, P. Guillaume, Identification of synchronous machine parameter, using broadband excitation, *IEEE Trans.*, EC-9(2), 270–280, 1994.
14. I.M. Canay, Causes of discrepancies in calculation of rotor quantities and exact equivalent diagrams of the synchronous machine, *IEEE Trans.*, PAS-88, 1114–1120, 1969.
15. I. Kamwa, P. Viarouge, On equivalent circuit structures for empirical modeling of turbine-generators, *IEEE Trans.*, EC-9(3), 579–592, 1994.
16. P.L. Dandeno, A.T. Poray, Development of detailed equivalent circuits from standstill frequency response measurements, *IEEE Trans.*, PAS-100(4), 1646–1655, 1981.
17. A. Keyhani, S. Hao, R.P. Schultz, Maximum likelihood estimation of generator stability constants using SSFR test data, *IEEE Trans.*, EC-6(1), 140–154, 1991.

18. P. Kundur, *Power System Stability and Control*, Mc. Graw-Hill Inc., 1993, pp. 169–191.

19. J.V. Milanovic, F. Al-Jowder, E. Levi, Small disturbance stability of saturated anisotropic synchronous machine models, *Record of ICEM-2000*, Espoo, Finland, Vol. 2, pp. 898–902.

20. Y. Hannalla, D.C. Macdonald, Numerical analysis of transient field in electrical machines, *Proc. IEE.*, 123, 183–186, 1976.

21. A.Y. Hanalla, D.C. Macdonald, Sudden 3-phase shortcircuit characteristics of turbine generators from design data using electromagnetic field calculation, *Proc. IEE.*, 127, 213–220, 1980.

22. S.C. Tandon, A.F. Armor, M.V.K. Chari, Nonlinear transient FE field computation for electrical machines and devices, *IEEE Trans.*, PAS-102, 1089–1095, 1983.

23. P.J. Turner, FE simulation of turbine-generator terminal faults and application to machine parameter prediction, *IEEE Trans.*, EC-2(1), 122–131, 1987.

24. S.I. Nabita, A. Foggia, J.L. Coulomb, G. Reyne, FE simulation of unbalanced faults in a synchronous machine, *IEEE Trans.*, MAG-32, 1561–1564, 1996.

25. D.K. Sharma, D.H. Baker, J.W. Daugherty, M.D. Kankam, S.H. Miunich, R.P. Shultz, Generator simulation-model constants by FE comparison with test results, *IEEE Trans.*, PAS-104, 1812–1821, 1985.

26. M.A. Arjona, D.C. Macdonald, Characterizing the *D*-axis machine model of a turbogenerator using finite elements, *IEEE Trans.* EC-14(3), 340–346, 1999.

27. M.A. Arjona, D.C. Macdonald, Lumped modeling of open circuit turbogenerator operational parameters, *IEEE Trans.*, EC-14(3), 347–353, 1999.

28. J.P. Sturgess, M. Zhu, D.C. Macdonald, Finite element simulation of a generator on load during and after a three phase fault, *IEEE Trans.*, EC-7(4), 787–793, 1992.

29. T. Laible, *Theory of Synchronous Machines in Transient Regimes*, Springer Verlag, Berlin, Germany, 1952.

30. D.C. Aliprantis, S.D. Sudhoff, B.T. Kuhn, A synchronous machine model with saturation and arbitrary rotor representation, *IEEE Trans.*, EC-20(3), 584–594, 2005.

31. D.C. Aliprantis, S.D. Sudhoff, B.T. Kuhn, Experimental characterization procedure for a synchronous machine model with saturation and arbitrary rotor network representation, *IEEE Trans.*, EC-20(3), 595–603, 2005.

32. A. Tessarolo, C. Bassi, D. Giulivo, Time stepping FEA of a 14 MVA salient pole shipboard alternator for different damper winding design solutions, *IEEE Trans.*, IE-59(6), 2524–2535, 2012.

33. S.E. Dallas, A.N. Safacas, J.C. Kappatou, Interturn stator fault analysis of a 200 MVA Hydrogenerator during transient operation using FEM, *IEEE Trans.*, EC-26(4), 1151–1160, 2011.

34. L. Weili, G. Chunwei, Z. Ping, Calculation of a complex 3D model of a turbogenerator with end region regarding electrical losses, cooling and heating, *IEEE Trans.*, EC-26(4), 1073–1080, 2011.

35. D. Zarko, D. Ban, I. Vazdar, V. Jaric, Calculation of unbalanced magnetic pull in a salient pole SG using FEM and measured shaft orbit, *IEEE Trans.*, IE-59(6), 2536–2549, 2012.

6

Control of Synchronous Generators in Power Systems

6.1 Introduction

A satisfactory operation of an AC power system is achieved when frequency and voltage remain nearly constant or vary in a limited and controlled manner when active and reactive loads vary.

Active power flow is related to prime mover's energy input and thus to the speed of speed governors (SGs). On the other hand, reactive power control is related to terminal voltage. Too a large electric active power load would lead to speed collapse, while too a large reactive power load would cause voltage collapse.

When a generator acts alone on a load, or it is by far the strongest in an area of a power system, its frequency may be controlled via generator speed, to remain constant with load (isochronous control). On the contrary, when the SG is part of a large power system, and electric generation is shared by two or more SGs, the frequency (speed) cannot be controlled to remain constant because it would forbid generation sharing between various SGs. Control with speed droop is the solution, to allow for fair generation sharing.

Automatic generation control (AGC) distributes the generation task between SGs; and based on this as input, the speed control system of each SG controls its speed (frequency) with an adequate speed droop so that generation "desired" sharing is obtained.

By fair sharing, we mean either power delivery proportional to ratings of various SGs or based on some cost function optimization, such as minimum cost of energy.

Speed (frequency) control quality depends on the speed control of the SG itself, but also on the other "induced" influences, besides the load dependence on frequency.

In addition, torsional shafts oscillations—due to turbine shaft, couplings, generator shaft elasticity, and damping effects—and subsynchronous resonance (SSR) (due to transmission lines series capacitor compensation to increase transmission power capacity at long distance) influence the quality of speed (active power) control. Measures to counteract such effects are required. Some of them are presented in this chapter.

In principle, reactive power flow of an SG may be controlled through SG output voltage control, which in turn is performed through excitation (current or voltage) control. SG voltage control quality depends on the SG parameters, excitation power source dynamics with its ceiling voltage, available to "forcing" the excitation current when needed in order to obtain fast voltage recovery upon reactive power load severe variations. The knowledge of load reactive power dependence on voltage is essential to voltage control system design.

Though active and reactive power control interactions are small, in principle, they may influence the control stability of each other.

To decouple them, power system stabilizers (PSS) have been added to the automatic voltage regulators (AVRs). PSSs have inputs such as speed or active power deviations and have steered lately extraordinary interest. In addition, various limiters—such as overexcitation (OEL) and underexcitation (UEL)—are required to ensure stability and avoid overheating of the SG. Load shedding and generator tripping are also included to match power demand to offer.

In phase of the utmost complexity of SG control, with power quality as a paramount objective, SG models, speed governor models (Chapter 3), excitation systems and their control models, and PSSs have all been standardized through IEEE recommendations.

The developments of powerful DSP systems and of advanced power electronics converters with IGBTs, GTIs, and MCTs, together with new nonlinear control systems such as variable structure systems, fuzzy-logic-neural-networks, and self-learning systems, may lead in the near future to the integration of active and reactive power controls into unique digital multi-input self-learning control systems. The few attempts along this path so far are very encouraging.

In what follows, the basics of speed and voltage control are given while ample reference to recent solutions is made, with some sample results. For more on power system stability and control, References see [1–3].

In Figure 6.1, we distinguish the following components:

- AGC
- Automatic reactive power control (AQC)
- Speed/power and the voltage/reactive power droop curves
- Speed governor (Chapter 3) and the excitation system
- Prime mover/turbine (Chapter 3) and SG (Chapter 5)
- Speed, voltage, and current sensors
- Step-up transformer, transmission line (X_T), and the power system electromagnetic force (emf) Es
- PSS added to the voltage controller input

In the basic SG control system, the active and reactive power control subsystems are rather independent with only the PSS as a "weak link" between them.

The active power reference P^* is obtained through the AGC system. A speed (frequency)/power curve (straight line) leads to the speed reference ω_r^*. The speed error $\omega_r^*-\omega_r$ then enters a speed governor control system whose output drives the valves and, respectively, the gates of various types of turbine speed-governor servomotors. AGC is part of load–frequency control of the power system that the SG is part of. In the so-called supplementary control, AGC moves the ω_r/P curves for desired load sharing between generators. On the other hand, AQC may provide the reactive power reference of the respective generator $Q^* <> 0$.

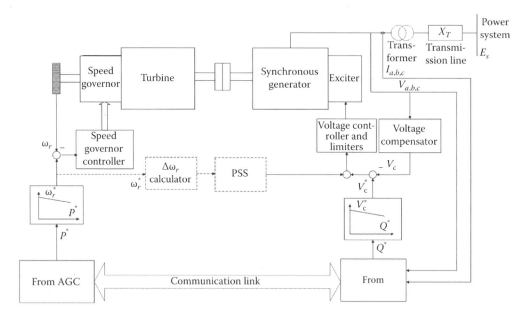

FIGURE 6.1 Generic SG control system.

A voltage/reactive power curve (straight line) will lead to voltage reference V_C^*. The measured voltage V_G is augmented by an impedance voltage drop $I_G(R_C + jX_C)$ to obtain the compensated voltage V_C. The voltage error $V_C^*-V_C$ then enters the excitation voltage control (AVR) to control the excitation voltage V_f in such a manner that the reference voltage V_C^* is dynamically maintained.

The PSS adds to the input of AVR a signal that is meant to provide a positive damping effect of AVR upon the speed (active power) low-frequency local pulsations.

The speed governor controller (SGC), the AVR, and the PSS may be implemented in various ways from PI, PID to variable structure, fuzzy logic, artificial neural networks (ANNs), μ_∞, etc. Also, there are various built-in limiters and protection measures.

In order to design SGC, AVR, PSS, proper turbine, speed governor, and SG simplified models are required. As for large SGs in power systems, the speed and excitation voltage control takes place within a bandwidth of only 3 Hz, simplified models are really feasible.

6.2 Speed Governing Basics

Speed governing is dedicated to generator response to load changes. An isolated SG with a rigid shaft system and its load are considered to illustrate the speed governing concept (Figure 6.2).

The motion equation is

$$2H\frac{d\omega_r}{dt} = T_m - T_e \tag{6.1}$$

where
T_m—turbine torque (p.u.)
T_e—SG torque (p.u.)
H (seconds)—inertia

We may use powers instead of torques in the equation of motion. For small deviations,

$$P = \omega_r T = P_0 + \Delta P$$

$$T_m = T_{m0} + \Delta T_m; \quad T_e = T_{e0} + \Delta T_e \tag{6.2}$$

$$\omega_r = \omega_{r0} + \Delta\omega_r$$

For steady state $T_{m0} = T_{en}$, and thus from Equations 6.1 and 6.2

$$\Delta P_m - \Delta P_e = \omega_0 \left(\Delta T_m - \Delta T_e \right) \tag{6.3}$$

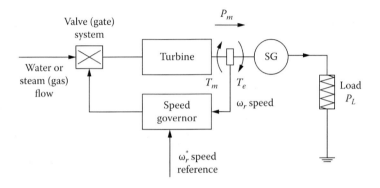

FIGURE 6.2 SG with its own load.

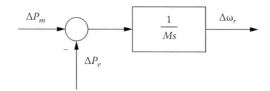

FIGURE 6.3 Power/speed transfer function (in p.u.).

For rated speed $\omega_0 = 1$ (p.u.)

$$2H\frac{d\Delta\omega_r}{dt} = \frac{(\Delta P_m - \Delta P_e)}{\omega_0}; \quad M = 2H\omega_0 \tag{6.4}$$

The transfer function in Equation 6.4 is illustrated in Figure 6.3.

The electromagnetic power P_e is delivered to composite loads. Some loads are frequency independent (lighting and heating loads). In contrast, motor loads depend notably on frequency. Consequently,

$$\Delta P_e = \Delta P_L + D\Delta\omega_r \tag{6.5}$$

where
ΔP_L—load power change that is independent of frequency
D—a load damping constant

Introducing Equation 6.5 into Equation 6.4 leads to

$$2H\omega_0 d\frac{\Delta\omega_r}{dt} + D\Delta\omega_r = \Delta P_m - \Delta P_L \tag{6.6}$$

The new speed/mechanical power transfer functions are as in Figure 6.4.

The steady-state speed deviation $\Delta\omega_r$, when the load varies, depends on the load frequency sensitivity. For a step variation in load power (ΔP_L), the final speed deviation $\Delta\omega_r = \Delta P_L/D$ (Figure 6.4). The simplest (primitive) speed governor would be an integrator of speed error that will drive the speed to its reference value in the presence of load torque variations. This is called the isochronous speed governor (Figure 6.5a).

The primitive (isochronous) speed governor cannot be used when more SGs are connected to a power system because it will not allow for load sharing. Speed droop or speed regulation is required: in principle, a steady-state feedback loop in parallel with the integrator (Figure 6.6a and b) will do.

It is basically a proportional speed controller with R providing the steady-state speed versus load power (Figure 6.6c) straight-line dependence:

$$R = \frac{-\Delta f}{\Delta P_L} \tag{6.7}$$

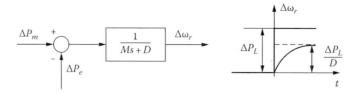

FIGURE 6.4 Power/speed transfer function with load frequency dependence.

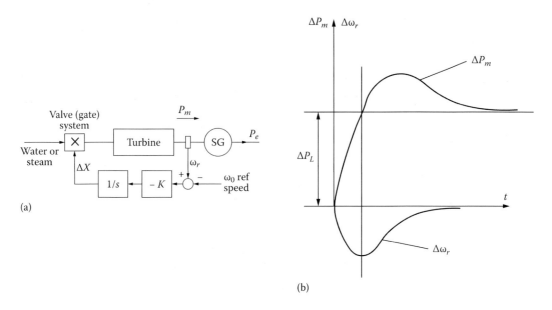

FIGURE 6.5 Isochronous (integral) speed governor: (a) the schematics and (b) response to step load increase.

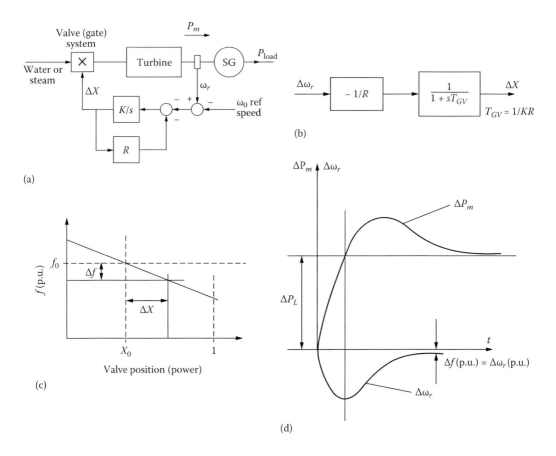

FIGURE 6.6 The primitive speed-droop governor: (a) the schematics, (b) reduced structural diagram, (c) frequency/power droop, and (d) response to step load power.

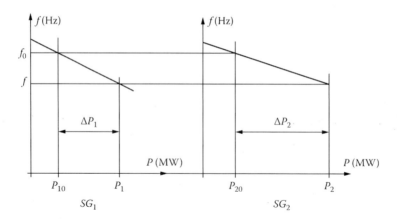

FIGURE 6.7 Load sharing between two SGs with speed-droop governor.

The time response of a primitive speed-droop governor to a step load increase is characterized now by speed steady-state deviation (Figure 6.6d).

With two (or more) generators in parallel, the frequency will be same for all of them, and thus the load sharing depends on their speed-droop characteristics (Figure 6.7).

As

$$-\Delta P_1 R_1 = \Delta f;$$
$$-\Delta P_2 R_2 = \Delta f$$

(6.8)

it follows that

$$\frac{\Delta P_2}{\Delta P_1} = \frac{R_1}{R_2}$$

(6.9)

Only if the speed droop is the same ($R_1 = R_2$), the two SGs are loaded proportionally to their rating. The speed/load characteristic may be moved up and down by the load reference setpoint (Figure 6.8).

By moving the straight line up and down, the power delivered by the SG for given frequency goes up and down (Figure 6.9).

The example in Figure 6.9 is related to a 50 Hz power system. It is similar for 60 Hz power systems. In essence, the same SG may deliver at 50 Hz, zero power (point *A*), 50% (point *B*), and 100% power (point *C*).

In strong power systems, the load reference signal changes the power output and not its speed, as the latter is determined by the strong power system.

It is also to be noted that, in reality, the frequency (speed) power characteristics depart from a straight line but still have negative slopes, for stability reasons.

This departure is due to valve (gate) nonlinear characteristics; when the latter are linearized, the straight line $f(P)$ is restored.

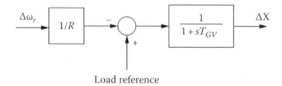

FIGURE 6.8 Speed-droop speed governor with load reference control.

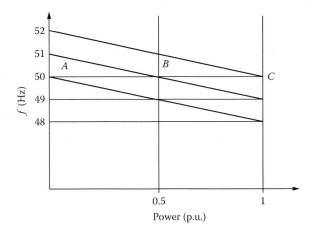

FIGURE 6.9 Moving the frequency (speed)/power characteristics up and down.

6.3 Time Response of Speed Governors

In Chapter 3, we have introduced models that are typical for steam reheat or nonreheat turbines (Figures 3.9 and 3.10) and hydraulic turbines (Figure 3.40 and Equation 3.43). Here, we add them to the speed-droop primitive governor with load reference as discussed in the previous section (Figure 6.10a and b).

T_{CH}—Inlet and steam chest delay (typically 0.3 s)
T_{RH}—Reheater delay (typically 6 s)
T_{HP}—HP flow fraction (typical $F_{HP} = 0.3$)
With nonreheater steam turbines, $T_{RH} = 0$.

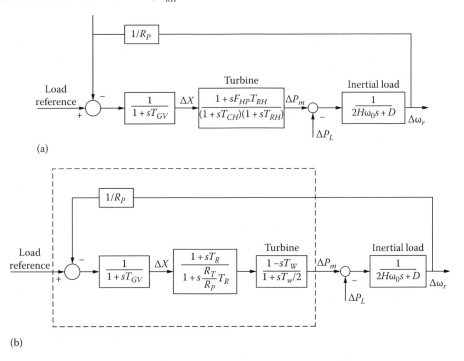

FIGURE 6.10 (a) Basic speed governor and steam turbine—generator and (b) basic speed governor and hydraulic turbine—generator.

For hydraulic turbines, the speed governor has to contain transient droop compensation. This is so because a change in the position of the gate, at the foot of the penstock, produces first a short-term turbine power change opposite to the expected one. For stable frequency response, long resetting times are required in standalone operation.

A typical such system is shown in Figure 6.10b.

T_W—Water starting constant (typically, $T_W = 1$ s)
R_p—Steady-state speed droop (typically, 0.05)
T_{GV}—Main gate servomotor time constant (typically, 0.2 s)
T_R—Reset time (typically, 5 s)
R_T—Transient speed droop (typically, 0.4)
D—Load damping coefficient (typically, $D = 2$)

Typical responses of the systems in Figure 6.10a and b to a step load (ΔP_L) increase are shown in Figure 6.11 for speed deviation $\Delta \omega_r$ (in p.u.).

As expected, the speed deviation response is rather slow for hydraulic turbines, average with reheat steam turbine generators and rather fast (but oscillatory) for nonreheat steam turbine generators.

The speed governor turbine models in Figure 6.10 are rather standard. More complete nonlinear models are closer to reality. Also nonlinear more robust SGCs are to be used to improve speed (or power angle) deviation response to various load perturbations (ΔP_L).

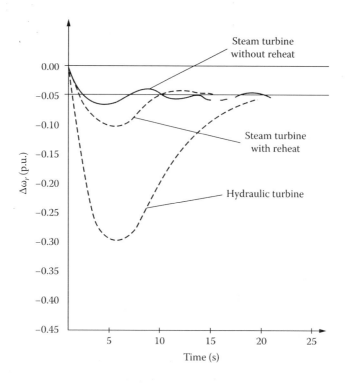

FIGURE 6.11 Speed deviation response of basic speed governor-turbine—generator systems to step load power change.

6.4 Automatic Generation Control

In a power system, when load changes, all SGs contribute to the change in power generation. The restoration of power system frequency requires additional control action that adjusts the load reference set-points. Load reference set-points modification leads to automatic change of power delivered by each generator.

AGC has three main tasks:

1. Regulating frequency to a specified value
2. Maintaining intertie power (exchange between control areas) at scheduled values
3. Distributing the required change in power generation amongst SGs such that the operating costs are minimized.

The first two tasks are also called "load-frequency control."

In an isolated power system, the function of AGC is to restore frequency as intertie power exchange is not present.

This function is performed by adding an integral control on the load reference settings of the speed governors for the SGs with AGC. This way the steady-state frequency error becomes zero. This integral action is slow and thus overrides the effects of composite frequency regulation characteristics of the isolated power system (made of all SGs in parallel). Thus the generation share of SGs that are not under the AGC is restored to scheduled values (Figure 6.12).

For an interconnected power system AGC is accomplished through the so-called tie-line control.

Each subsystem (area) has its own central regulator (Figure 6.13a).

The interconnected power system in Figure 6.13 is in equilibrium if for each area:

$$P_{Gen} = P_{load} + P_{tie} \tag{6.10}$$

The intertie power exchange reference $(P_{tie})_{ref}$ is set at a higher level of power system control, based on economical and safety reasons.

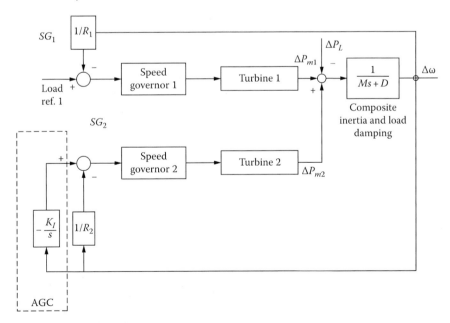

FIGURE 6.12   AGC control of one SG in a two SGs isolated power system.

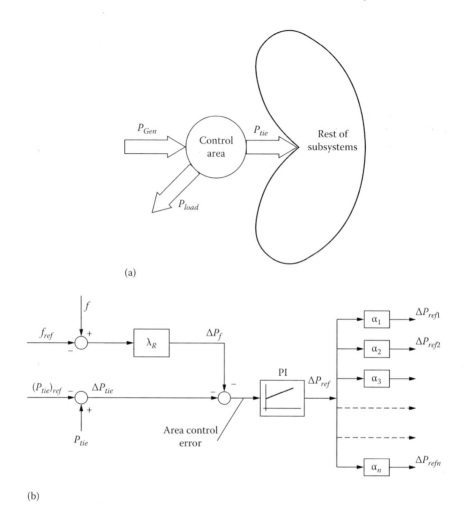

FIGURE 6.13 Central subsystem (tie-line): (a) power balance and (b) structural diagram.

The central subsystem (area) regulator has to maintain frequency at f_{ref} and the net tie-line power (tie-line control) from the subsystem area at a scheduled value $P_{tie\ ref}$.

In fact (Figure 6.13b), the tie-line control changes the power output of the turbines by varying the load reference (P_{ref}) in their speed governor systems.

The area control error (ACE) is (Figure 6.13b):

$$\text{ACE} = -\Delta P_{tie} - \lambda_R \Delta f \qquad (6.11)$$

ACE is aggregated from tie-line power error and frequency error. The frequency error component is amplified by the so-called frequency bias factor λ_R.

The frequency bias factor is not easy to adopt as the power imbalance is not entirely represented by load changes in power demand but also in the tie-line power exchange.

A PI controller is applied on ACE to secure zero steady-state error. Other nonlinear (robust) regulators may be used. The regulator output signal is ΔP_{ref}, which is distributed over participating generators with participating factors $\alpha_1 \dots \alpha_n$. Some participating factors may be zero. The control signal acts upon load reference settings (Figure 6.12).

Intertie power exchange and participation factors are allocated based on security assessment and economic dispatch via a central computer.

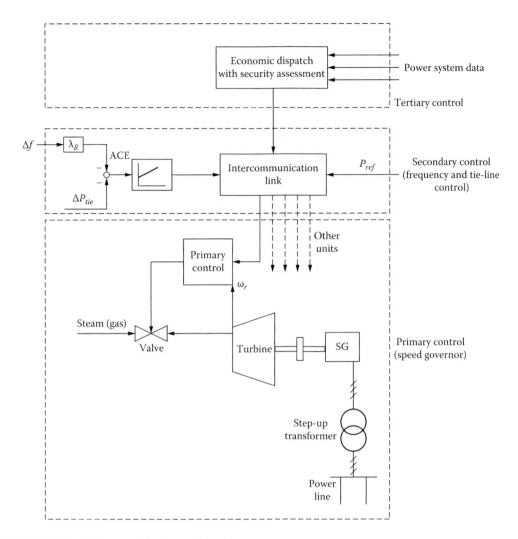

FIGURE 6.14 AGC as a multilevel control system.

AGC may be treated as a multilevel control system (Figure 6.14). The primary control level is represented by the speed governors with their load reference point. Frequency and tie-line control represent secondary control that forces the primary control to bring the frequency and tie-line power deviations to zero.

Economic dispatch with security assessment represents the tertiary control. Tertiary control is the slowest (minutes) of all control stages, as expected.

6.5 Time Response of Speed (Frequency) and Power Angle

Thus far, we did describe the AGC as containing three control levels in an interconnected power system.

Based on this, the response in frequency, power angle, and power of a power system to a power imbalance situation may be approached. If a quantitative investigation is necessary, all the components have to be introduced with their mathematical models. But if a qualitative analysis is sought, then the AVRs are supposed to maintain constant voltage, while electromagnetic transients are neglected. Basically, the power system moves from a steady state to another steady-state regime, while the equation of motion applies to provide the response in speed and power angle.

Power system disturbances are numerous but consumer load variation, disconnection, or connection of an SG from (to) the power system are representative examples.

Four time stages in the response to a power system imbalance may be distinguished:

1. Rotor swings in the SGs (the first few seconds)
2. Frequency drops (several seconds)
3. Primary control by speed governors (several seconds)
4. Secondary control by central subsystem (area) regulators (up to a minute)

During periodic rotor swings, the mechanical power of the remaining SGs may be considered constant. Therefore, if one generator, out of two, is shut off, the power system mechanical power is reduced twice.

The capacity of the remaining generators to deliver power to loads is reduced from

$$P_-\left(\delta_0'\right) = \frac{E'V_S}{\left(X_d' + X_T\right)/2 + X_S}\sin\delta_0' \tag{6.12}$$

to

$$P_+\left(\delta\right) = \frac{E'V_S\sin\delta_0'}{X_d' + X_T + X_S},\text{(p.u.)} \tag{6.13}$$

in the first moments after one generator is disconnected. Note that X_T is the transmission line reactance (there are two lines in parallel) and X_S is the power system reactance. X_d' is the transient reactance of the generator, E' is the generator transient emf, and V_S is the power system voltage. The situation is illustrated in Figure 6.15.

Let us note that the load power has not been changed. Both the remaining generator (ΔP_{RI}) and the power system have to cover for the deficit ΔP_0:

$$\Delta P_{RI} = P_+(\delta') - P_{m+}$$
$$\Delta P_{SI} = \Delta P_0 - \Delta P_{RI} \tag{6.14}$$

While the motion equation leads to the rotor swings in Figure 6.15, the power system still has to cover for the power $\Delta P_{SI}(t)$. Therefore, the transient response to the power system imbalance (by disconnecting a generator out of two) continues with stage two: frequency control.

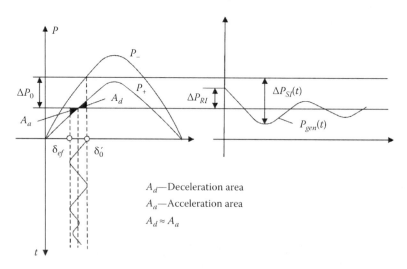

FIGURE 6.15 Rotor swings and power system contributing power change.

Due to the additional power system contribution requirement during this second stage, the generators in the power system are slow and the system frequency drops. During this stage, the share from ΔP_{SI} is determined by generators, inertia. The basic element is that the power angle of the studied generator goes further down, while the SG is still in synchronism.

When this drop in power angle and frequency occurs, we enter the third stage when primary (SGs) control takes action, based on the frequency/power characteristics.

The increase in mechanical power required from each turbine is, as known, inversely proportional to the droop in the $f(P)$ curve (straight line). When the disconnection of one of the two generators occurred, the $f(P)$ composite curve is changing from P_{T-} to P_{T+} (Figure 6.16).

The operating point moves from A to B as one generator was shut off. The load/frequency characteristic is $f(P_L)$ in Figure 6.16. Along the trajectory BC, the SG decelerates until it hits the load curve in C, then accelerates up to D, and so on until it reaches the new stable point E.

The straight line characteristics $f(P)$ will remain valid—power increases with frequency (speed) reduction—up to a certain power when frequency collapses. In general, if enough power (spinning) reserve exists in the system, the straight line characteristic holds. Spinning reserve is the difference between rated power and load power in the system.

Frequency collapse is illustrated in Figure 6.17.

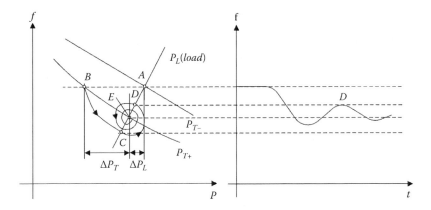

FIGURE 6.16 Frequency response for power imbalance.

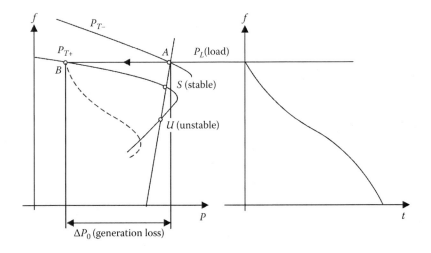

FIGURE 6.17 Extended $f(P)$ curves with frequency collapse when large power imbalance occurs.

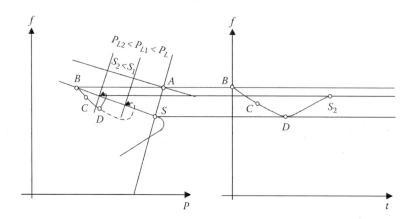

FIGURE 6.18 Frequency restoration via two-stage load shedding.

Because of the small spinning reserve, the frequency decreases initially so much that it intercepts the load curve in U, an unstable equilibrium point. Therefore, the frequency decreases steadily and finally collapses. To prevent frequency collapsing, load shedding is performed. At given frequency level, underfrequency relays in substations shut down scheduled loads in 2–3 steps in an attempt to restore frequency (Figure 6.18).

When frequency reaches point C, the first stage of load (P_{L1}) shedding is operated. The frequency still decreases but at a slower rate until it reaches level D, when the second load shedding is performed. This time (as D is at the right side of S_2), the generator accelerates and restores frequency at S_2.

In the last stage of response dynamics, frequency and the tie-line power flow control through the AGC takes action. In islanded system AGC actually moves up stepwise the $f(P)$ characteristics of generators such that to restore frequency at its initial value. Details on frequency dynamics in interconnected power systems are to be found in References [1,2].

6.6 Voltage and Reactive Power Control Basics

Dynamically maintaining constant (or controlled) voltage in a power system is a fundamental requirement of power quality. Passive (resistive—inductive, resistive—capacitive) loads and active loads (motors) require both active and reactive power flows in the power system.

While composite load power dependence on frequency is mild, the reactive load power dependency on voltage is very important.

Typical shapes of composite load (active and reactive power) dependence on voltage are shown in Figure 6.19.

As loads "require" reactive power, the power system has to provide for it.

In essence, reactive power may be provided or absorbed by

- Control of excitation voltage of SGs by automatic voltage regulation (AVR)
- Power electronics controlled capacitors and inductors by static voltage controllers: SVC, placed at various locations in a power system.

As voltage control is related to reactive power balance in a power system, to reduce losses due to increased power line currents, it is appropriate to "produce" the reactive power as close as possible to the place of its "utilization." Decentralized voltage (reactive power) control should thus be favored.

As the voltage variation changes, both the active and reactive power that can be transmitted over a power network vary, it follows that voltage control interferes with active power (speed) control. The separate treatment of voltage and speed control is based on their weak coupling but also on necessity.

One way to treat this coupling is to add to the AVR the so-called PSS whose input is speed or active power deviation.

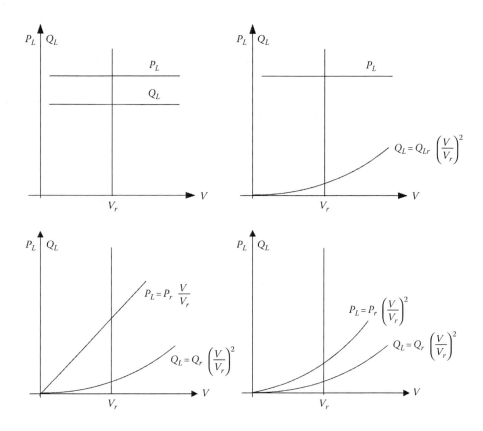

FIGURE 6.19   Typical P_L, Q_L load powers versus voltage.

6.7 Automatic Voltage Regulation Concept

AVR acts upon the DC voltage V_f that supplies the excitation winding of SGs. The variation of field current in the SG increases or decreases the emf (no-load voltage) and thus finally, for given load, the generator voltage is controlled as required.

The excitation system of an SG contains the exciter and the AVR (Figure 6.20).

The exciter is in fact the power supply that delivers controlled power to SG excitation (field) winding.

As such, the exciters may be classified into the following:

- DC exciters
- AC exciters
- Static exciters (power electronics)

The DC and AC exciters contain an electric generator placed on the main (turbine—generator) shaft and have low power electronics control of their excitation current. The static exciters take energy from a separate AC source or from a step-down transformer (Figure 6.20) and convert it into DC-controlled power transmitted to the field winding of the SG through slip rings and brushes.

The AVR collects information on generator current and voltage (V_g, I_g) and on field current, and based on the voltage error controls the V_f (the voltage of the field winding) through the control voltage V_{con}, which acts on the controlled variable in the exciter.

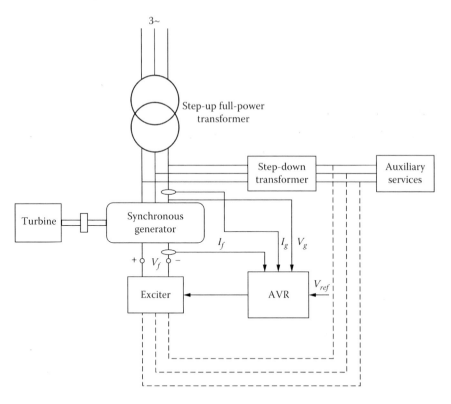

FIGURE 6.20   Exciter with AVR.

6.8 Exciters

As already mentioned, exciters are of three types, each with numerous embodiments in industry.

The DC exciter (Figure 6.21), still in existence for many SGs below 100 MVA per unit, consists of two DC comutator electric generators: the main one (main exciter or ME) and the auxiliary one (auxiliary exciter or AE). Both are placed on the SG main shaft. The ME supplies the SG field winding (V_f), while the AE supplies the ME field winding.

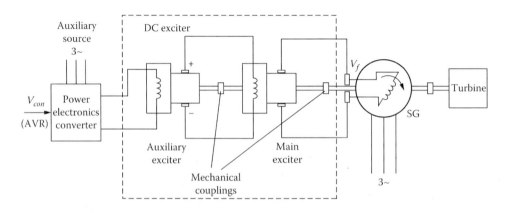

FIGURE 6.21   Typical DC exciter.

The field winding of the AE is supplied with the voltage V_{con} controlled by the AVR. The power electronics source required to supply the AE field winding is of very low power ratings as the two DC commutator generators provide a total power amplification ratio around 600/1 in general.

The advantage of low-power electronics external supply required for the scope is paid for by

- Rather slow time response due to the large field winding time constants of the two excitation circuits plus the moderate time constants of the two armature windings
- Problems with brush wearing in the ME and AE
- Still all excitation power (the peak value may be 4%–5% of rated SG power) of the SG has to transmitted through the slip ring brush mechanism
- The flexibility of the exciter shafts and mechanical couplings adds at least one additional shaft torsional frequency to the turbine-generator shaft

Though still present in industry DC exciters are gradually replaced now with AC exciters or (and static exciters).

6.8.1 AC Exciters

AC exciters are basically making use of inside-out SGs with diode rectifiers on their rotor. As both the AC exciter and the SG use the same shaft, the full excitation power diode rectifier is connected directly to the field winding of SG (Figure 6.22). The stator-based field winding of the AC exciter is controlled from the AVR.

The static power converter has now a rating that is about 1/20(30) of the SG excitation winding power ratings as only one step of power amplification is performed through the AC exciter.

The AC exciter in Figure 6.22 is characterized by

- The absence of electric brushes in the exciter and in the SG
- A single machine addition on the main SG-turbine shaft
- Moderate time response in V_f (SG field-winding voltage) as only one (transient) time constant (T'_{d0}) delays the response; the static power converter delay is small, in comparison
- Addition of one torsional shaft frequency due to the flexibility of the AC exciter machine shaft and mechanical coupling
- Small controlled power in the static power converter: (1/20 [30] of the field-winding power rating)

The brushless AC exciter (as in Figure 6.22) is used frequently in industry, even for new SGs because it does not need an additional sizeable power source to supply the exciter's field winding.

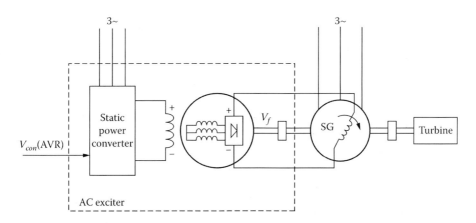

FIGURE 6.22 AC exciter.

6.8.2 Static Exciters

Modern electric power plants are provided with emergency power groups for auxiliary services that may be used to start the former from blackout. Therefore, an auxiliary power system is available in general.

This trend gave way to static exciters, mostly in the form of controlled rectifiers supplying directly the field winding of the SG through slip rings and brushes (Figure 6.23).

The excitation transformer is required to adapt the voltage from the auxiliary power source or from the SG terminals (Figure 6.23a).

It is also feasible to supply the controlled rectifier from a combined voltage (VT) and current (CT) transformer connected in parallel and, respectively, in series with the SG stator windings (Figure 6.23b). This solution provides a kind of basic AC voltage stabilization at the rectifier input terminals. This way short circuits or short voltage sags at SG terminals do not influence very much the excitation voltage ceiling produced by the controlled rectifier.

In order to cope with fast SG excitation current control, the latter has to be forced by an overvoltage available to be applied to the field winding. The voltage ceiling ratio (V_{fmax}/V_{frated}) characterizes the exciter.

Power electronics (static) exciters are characterized by fast voltage response, but still the T_d' time constant of the SG delays the field current response. Consequently, a high-voltage ceiling is required for all exciters.

To exploit with minimum losses, the static exciters, two separate controlled rectifiers may be used, one for "steady state" and one for field forcing (Figure 6.24). There is a switch that has to be kept open unless the field forcing (higher voltage) rectifier has to be put to work. When V_{fmax}/V_{frated} is notably larger than 2, such a solution may be considered.

The development of IGBT PWM converters up to 3 MVA per unit (for electric drives) at low voltages (690 VAC, line voltage) provides for new, efficient, lower volume static exciters.

The controlled thyristor rectifiers in Figure 6.24 may be replaced by diode rectifiers plus DC–DC IGBT converters (Figure 6.25).

A few such four quadrant DC–DC converters may be paralleled to fulfill the power level required for the excitation of SGs in the hundreds of MVAs per unit. The transmission of all excitation power through slip rings and brushes still remains a problem. However, with the today's doubly fed induction generators at 400 MVA/unit, 30 MVA is transmitted to the rotor through slip rings and brushes. The solution is thus here for the rather lower power ratings of exciters (less than 3%–4% of SG rating).

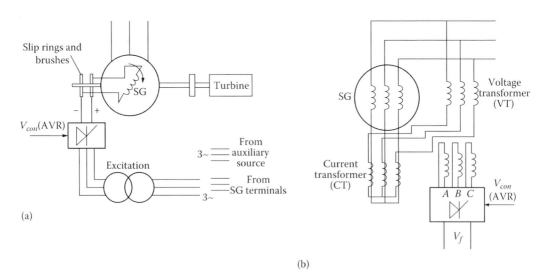

FIGURE 6.23 Static exciter: (a) voltage fed and (b) voltage and current fed.

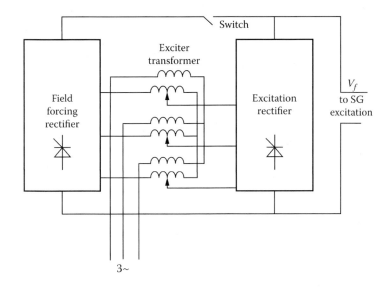

FIGURE 6.24 Dual rectifier static exciter.

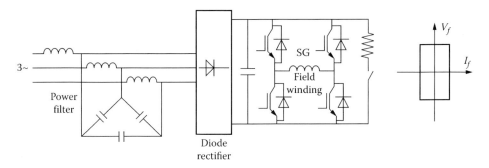

FIGURE 6.25 Diode-rectifier + four-quadrant DC–DC converter as static exciter.

The four-quadrant chopper static exciter has the following features:

- It produces fast current response with smaller ripple in the field winding current of the SG
- It can handle positive and negative field current that may occur during transients as a result of stator current transients
- The AC input currents (in front of the diode rectifier) are almost sinusoidal (with proper filtering), while the power factor is close to unity irrespective of load (field) current
- The current response is even faster than with controlled rectifiers
- Active front-end IGBT rectifiers may also be used for static exciters.

6.9 Exciter's Modeling

While it is possible to derive complete models for exciters—as they are interconnected electric generators and (or) static power converters, for power system stability studies, simplified models have to be used.

IEEE standard 421.5 from 1992 contains "IEEE Recommended Practice for Excitation System Models for Power Systems."

Moreover "Computer Models for Representation of Digital-Based Excitation Systems" have also been recommended by IEEE in 1996.

6.9.1 New PU System

The so-called reciprocal PU system used for the SG where the base voltage for the field winding voltage V_f is the SG terminal rated voltage $V_n \times$ sqrt(2) leads to PU value of V_f in the range of 0.003 or so. Such values are too small to handle in designing the AVR.

A new, nonreciprocal, PU system is by now widely used to handle this situation. Within this PU system, the base voltage for V_f is V_{fb}, the field winding voltage required to produce the air gap line (nonsaturated) no load voltage at the generator terminals.

For the SG in PU, at no load:

$$V_{d0} = +\Psi_{q0} = +l_q I_{q0} = 0$$
$$Vq_0 = -\Psi_{d0} = -l_{dm}I_f$$
$$|V_{q0}| = V_0 = l_{dm}I_f = 1.0 \tag{6.15}$$

Therefore,

$$I_f = 1/l_{dm}(\text{p.u.}) \tag{6.16}$$

The field voltage V_f corresponding to I_f is

$$V_f = r_f \times I_f = \frac{r_f}{l_{dm}}(\text{p.u.}) \tag{6.17}$$

This is the reciprocal PU system.

In the nonreciprocal PU system, the corresponding field current $I_{fb} = 1.0$, and thus

$$I_{fb} = l_{dm}I_f \tag{6.18}$$

The exciter voltage in the new PU system is thus

$$V_{fb} = \frac{l_{dm}}{r_f}V_f \tag{6.19}$$

Using Equation 6.16 in Equation 6.18, we evidently find $V_{fb} = 1.0$ as we are at no-load conditions of Equation 6.15.

In Chapter 5, the operational flux Ψ_d at no load has been defined as follows:

$$\Delta\Psi_d(s) = \frac{l_{dm}}{r_f}\frac{(1+sT_D)\Delta V_f}{(1+sT'_{d0})(1+sT''_{d0})} \tag{6.20}$$

in the reciprocal PU system.

In the new, nonreciprocal, PU system, by using Equation 6.18 in Equation 6.20, we obtain the following:

$$\Delta\Psi_{db}(s) = \frac{(1+sT_D)\Delta V_{fb}}{(1+sT'_{d0})(1+sT''_{d0})} \tag{6.21}$$

However at no load,

$$\Delta\Psi_{db} = \Delta V \tag{6.22}$$

Consequently, with the damping winding eliminated (T''_{d0}, $T_D = 0$):

$$\frac{\Delta V_0(s)}{\Delta V_{fb}(s)} = \frac{1}{1+sT'_{d0}} \tag{6.23}$$

The open-circuit transfer function of the generator has a gain equal to unity and the time constant T'_{d0}:

$$T'_{d0} = \frac{1}{\omega_{base}} \times \frac{(l_{fl}+l_{dm})}{r_f} \tag{6.24}$$

Example 6.1

Let us consider an SG with the following PU parameters: $l_{dm} = l_{qm} = 1.6$, $l_{sl} = 0.12$, $l_{fl} = 0.17$, $r_f = 0.006E_f$.

The rated voltage $V_0 = 24/\text{sqrt}(3)$ kV, $f_1 = 60$ Hz. The field current and voltage required to produce the rated generator voltage at no load on the air gap-line are $I_f = 1500$ A, $V_f = 100$ V.

Calculate the following:

(a) The base values of V_f and I_f in the reciprocal and nonreciprocal (V_{fb}, I_{fb}) PU system
(b) The open-circuit generator transfer function $\Delta V_0/\Delta V_{fb}$

Solution

(a) Evidently, $V_{fb} = 100$ V, $I_{fb} = 1500$ A, by definition, in the nonreciprocal PU system.
For the reciprocal PU system, we make use of Equations 6.17 and 6.18:

$$I_f = l_{dm}I_{fb} = 1.6 \cdot 1500 = 2400 \text{ A}$$

$$V_f = \frac{l_{dm}}{r_f}V_{fb} = \left(\frac{1.6}{0.00064}\right) \cdot 100 = 250 \text{ kV}$$

(b) In the absence damper winding, only the time constant T'_{d0} remains to be determined in Equation 6.24:

$$T'_{d0} = \frac{1}{2\pi60} \cdot \frac{(0.17+1.6)}{0.00664} = 7.34 \text{ s}$$

$$\frac{\Delta V_0(s)}{\Delta V_{fb}(s)} = \frac{1}{1+7.34 \cdot s}$$

When temperature varies r_f varies, and thus all base variables vary. And so does the time constant T'_{d0}.

6.9.2 DC Exciter Model

Let us consider the separately excited DC commutator generator exciter (Figure 6.26), with its no-load and on-load saturation curves at constant speed.

Due to magnetic saturation, the relationship between DC exciter field current I_{ef} and the output voltage V_{ex} is nonlinear (Figure 6.26).

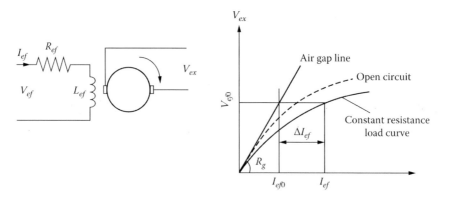

FIGURE 6.26 DC exciter and load–saturation curve.

The air gap line slope in Figure 6.26 is R_g (as in a resistance). In the IEEE standard 451.2, the magnetic saturation is defined by the saturation factor $S_e(V_{ex})$.

$$I_{ef} = \frac{V_{ex}}{R_g} + \Delta I_{ef}$$

$$\Delta I_{ef} = V_{ex} \cdot S_e(V_{ex}) \tag{6.25}$$

The saturation factor is approximated in general with an exponential function:

$$S_e(V_{ex}) = \frac{1}{R_g} e^{B_e V_{ex}} \tag{6.26}$$

Other approximations are also feasible.

The no-load DC exciter voltage V_{ex} is proportional to its excitation field Ψ_{ef}. For constant speed,

$$V_{ex} = K_e \cdot \Psi_{ef} = K_e \cdot L_{ef} \cdot I_{ef} \tag{6.27}$$

$$V_{ef} = R_{ef} I_{ef} + \frac{d\Psi_{ef}}{dt} = R_{ef} I_{ef} + L_{ef} \frac{dI_{ef}}{dt} \tag{6.28}$$

With Equations 6.25 through 6.27, Equation 6.28 becomes

$$V_{ef} = \left(\frac{R_{ef}}{R_g} + R_{ef} S_e(V_{ex}) \right) V_{ex} + \frac{1}{K_e} \frac{dV_{ex}}{dt}. \tag{6.29}$$

This is basically the voltage transfer function of the DC exciter on no load, considering magnetic saturation.

Again PU variables are used with base voltage equal to the SG base field voltage V_{fb}:

$$V_{exb} = V_{fb}$$

$$I_{efb} = \frac{V_{fb}}{R_g}; \quad R_{gb} = R_g \tag{6.30}$$

In PU, the saturation factor becomes thus:

$$s_e(V_{ex}) = R_g S_e(V_{ex}) \tag{6.31}$$

Finally, Equation 6.29 in PU is

$$V_{ef} = \frac{R_{ef}}{R_g} V_{ex} \left(1 + s_e\left(V_{ex}\right)\right) + \frac{1}{K_e}\frac{dV_{ex}}{dt} \tag{6.32}$$

It is obvious that $1/K_e$ has the dimension of a time constant:

$$\frac{1}{K_e} \approx \frac{L_{ef}}{R_g}\frac{I_{ef0}}{V_{ex0}} = \frac{L_{ef}}{R_g} = T_{ex} \tag{6.33}$$

The values I_{ef0} and V_{ex0} in PU, now correspond to a given operating point.
  Finally, Equation 6.32 becomes

$$V_{ef} = V_{ex}\left(K_E + S_E\left(V_{ex}\right)\right) + T_E\frac{dV_{ex}}{dt} \tag{6.34}$$

where

$$K_E = \frac{R_{ef}}{R_g};$$

$$S_E\left(V_{ex}\right) = s_e\left(V_{ex}\right)\cdot\frac{R_{ef}}{R_g} \tag{6.35}$$

This is the rather widely accepted DC exciter model used for AVR design and power system stability studies. It may be expressed in a structural diagram as in Figure 6.27.
  It is evident that for small-signal analysis, the structural diagram in Figure 6.27 may be simplified to

$$\Delta V_{ef} = \Delta V_f \frac{\left(1 + sT\right)}{K}$$

$$K = \frac{1}{S_E\left(V_{ef0}\right) + K_E}; \quad T = KT_E \tag{6.36}$$

The corresponding structural diagram is shown in Figure 6.28.

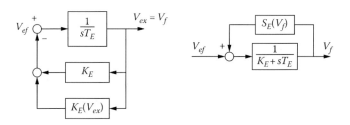

FIGURE 6.27   DC exciter structural diagram.

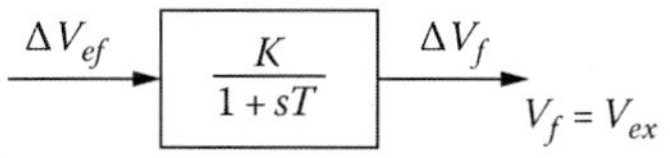

FIGURE 6.28   Small-signal deviation structural diagram of DC exciters with separate excitation.

As expected, in its most simplified form, for small-signal deviations, the DC exciter is represented by a gain and a single time constant. Both K and T, however, vary with the operating point (V_{f0}).

Note: The self-excited DC exciter's model is similar, but $K_E = R_{ef}/R_g - 1$ instead of $K_E = R_{ef}/R_g$. Also K_E now varies with the operating point.

6.9.3 AC Exciter

The AC exciter is, in general, a synchronous generator (inside-out for brushless excitation systems). Its control is again through its excitation and thus, in a way, is similar to the DC exciter. If a diode rectifier is used at the output of the AC exciter, the output DC current I_f is proportional to the armature current as almost unity power factor operation takes place with diode rectification.

Therefore, what is additional in the AC exciter is a longitudinal demagnetizing armature reaction that tends to reduce the terminal voltage of the AC exciter.

Consequently, one more feedback is added to the DC exciter model (Figure 6.29) to obtain the model of the AC exciter.

The saturation factor $S_E(V_f)$ should now be calculated from the no-load saturation curve and the air gap line of the AC exciter. The armature reaction feedback coefficient K_d is related to the d-axis coupling inductance of the AC exciter (l_{dm}, when the field winding of the AC exciter is reduced to its armature winding). It is obvious that the influence of armature resistance and damper cage (if any) are neglected and speed is considered constant.

It is V_{ex} and not V_f in Figure 6.29, because a rectifier is used between the AC exciter and the SG field winding to change AC to DC.

The uncontrolled rectifier that is part of the AC exciter is shown in Figure 6.30.

The $V_f(I_f)$ output curve of the diode rectifier is, in general, nonlinear and depends on the diode commutation overlapping. The alternator reactance (inductance) x_{ex} plays a key role in the commutation process. Three main operation modes may be identified from no-load to short circuit [4]:

Stage 1: two diodes conducting before commutation takes place (low load)

$$\frac{V_f}{V_{ex}} = 1 - \frac{1}{\sqrt{3}}\frac{I_f}{I_{sc}}, \quad \text{for} \quad \frac{I_f}{I_{sc}} < \left(1 - \frac{1}{\sqrt{3}}\right) \tag{6.37}$$

$$I_{sc} = \frac{V_{ex}\sqrt{2}}{x_{ex}} \tag{6.38}$$

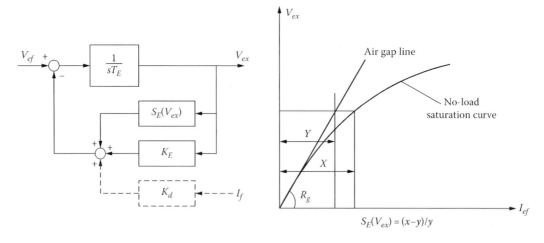

FIGURE 6.29 Structural diagram of an AC exciter alternator.

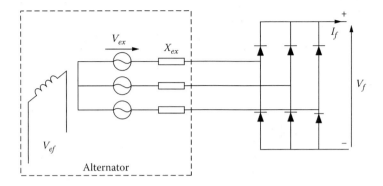

FIGURE 6.30 Diode rectifier plus alternator = AC exciter.

Stage 2: when each diode can conduct only when the counterconnected diode of the same phase has ended its conduction interval:

$$\frac{V_f}{V_{ex}} = \sqrt{\frac{3}{4} - \left(\frac{i_f}{i_{sc}}\right)^2};$$

$$1 - \frac{1}{\sqrt{3}} < \frac{I_f}{I_{sc}} < \frac{3}{4}$$

(6.39)

Stage 3: four diodes conduct at the same time:

$$\frac{V_f}{V_{ex}} = \sqrt{3}\left(1 - \frac{i_f}{i_{sc}}\right);$$

$$\frac{3}{4} < \frac{i_f}{i_{sc}} < 1$$

(6.40)

$V_f/V_{ex}(I_f/I_{sc})$ functions in Equations 6.27, 6.29, and 6.30 are illustrated in Figure 6.31.

This is a steady-state characteristic. The response of the rectifier is so fast, in comparison with the alternator or to the SG field current response, that the steady-state characteristic suffices to model the rectifier.

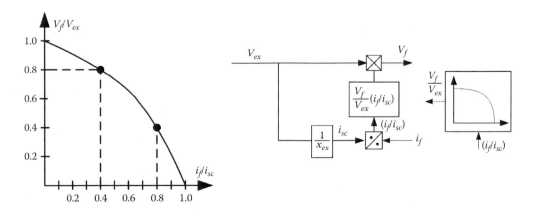

FIGURE 6.31 Diode rectifier voltage/current characteristic and structural diagram.

6.9.4 Static Exciter

Among the static exciter configurations, let us consider here the controlled three-phase rectifier (Figure 6.32).

The average value (steady-state) characteristic represents the output voltage of the V_f as a function of input voltage V_{ex} and the load (I_f) current [4,5]:

$$V_f = \frac{3\sqrt{2}}{\pi} V_{ex} \sqrt{3} \cos\alpha - \frac{3}{\pi} x_{ex} I_f$$

$$I_{sc} = \frac{V_{ex}\sqrt{2}}{x_{ex}}$$

(6.41)

In PU values,

$$\frac{V_f}{V_{ex}} = \cos\alpha - \frac{1}{\sqrt{3}} \frac{I_f}{I_{sc}} = F_{CR}\left(\alpha, \frac{I_f}{I_{sc}}\right)$$

(6.42)

The characteristic in Equation 6.42 is very similar to the first stage of the diode rectifier characteristic. The structural diagram is also similar (Figure 6.33). For $\alpha = 0$, it degenerates into stage 1 of the diode rectifier case of Equation 6.37.

This time the voltage V_f applied to the field winding may be either positive or negative, while the field current I_f is always positive. Faster response in I_f is expected, while the control is done through the firing delay angle α.

The power source (a transformer in general) has a rather constant emf V_{ex} and an internal reactance x_{ex}, therefore $\cos\alpha$ is input to the rectifier and is produced as the output of the AVR.

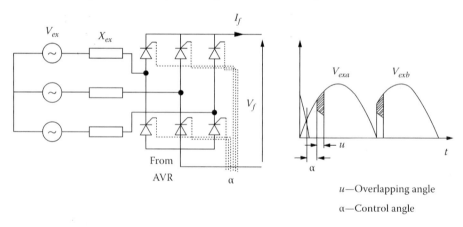

FIGURE 6.32 The controlled rectifier.

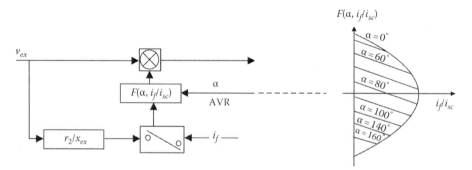

FIGURE 6.33 The structure of a controlled rectifier.

6.10 Basic AVRs

The basic AVR has to provide closed-loop control of the SG terminal voltage by acting upon the exciter input with a voltage, V_{con}. It may have 1, 2, 3 stabilization loops and additional inputs, besides reference voltage V_{ref} of SG and its measured value with load compensation V_c:

$$V_c = \left| \overline{V}_g + (r_c + jX_c)\overline{I}_g \right| \cdot \frac{1}{1 + sT_T} \tag{6.43}$$

The load compensator introduces the compensation of generator voltage variation due to load and also the delay T_T due to the voltage sensor.

Other than that, a major field winding voltage V_f loop is introduced. The voltage regulator may be of many types—a lead-lag compensator, for example—with various limiters. Figure 6.34 shows the IEEE 1992 AC1A excitation system (with AVR).

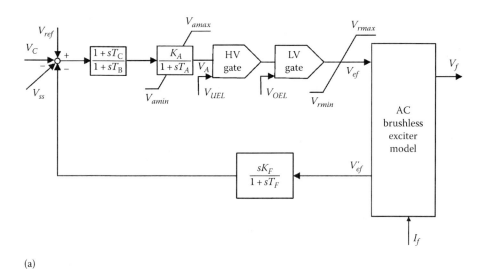

(a)

(b)

FIGURE 6.34 (a) IEEE 1992, AC1A (brushless) excitation system and (b) AC exciter with diode rectifier.

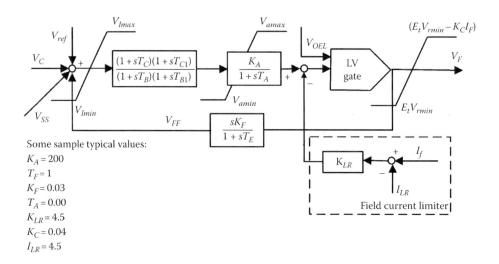

FIGURE 6.35 IEEE 1992 type ST1A excitation system (with AVR).

A few remarks are in order:

- The feedback loop uses V'_{ef} instead of V_{ex} (exciter's excitation voltage) or SG field winding V_f
- A windup limited single constant block with gain K_A is imposed to limit the output variable V_A
- V_{UEL} is underexcitation limiter input
- V_{OEL}—is the overexcitation limiter (OEL) input
- $(1 + sT_C)/(1 + sT_B)$ is in fact the voltage regulator implemented as a simple lead-lag compensator
- A nonwindup limiter V_{Rmax}, V_{Rmin} is applied to the exciter excitation supply voltage

The IEEE 1992-type ST1A excitation system model is shown in Figure 6.35. It represents a potential source controlled rectifier.

A transformer takes the power from the SG terminals and supplies the controlled rectifier. The exciter ceiling voltage is thus proportional to SG terminal voltage E_t. The rectifier-voltage regulation is represented by K_C ($K_C = x_{ex}$ in previous structural diagrams). The field current I_F is limited through gain K_{LR} at the current limit I_{LR}.

Again nonwindup and windup limiters are included. And so are underexcitation (V_{UEL}) and over-excitation (V_{OEL}) limiters.

The controlled rectifier model is considered only through the nonwindup limiter V_{Rmax}, V_{Rmin}.

The IEEE 451.2 standard from 1992 contains a myriad of models for existing excitation systems. More are added in Reference 6.

In the same time, more sophisticated and robust AVRs are presented, proposed, and tested. In what follows, a case study of a digital VAR design is presented.

Example 6.2: Digital Excitation System Design

Let us consider here a PID AVR to be used with a IEEE-1992 standard 421.5, type AC5A alternator and diode rectifier brushless exciter for turbine—SG sets of low-to-medium power (say up to 50 MW).

Provide a direct design method for the PID type AVR.

Solution

The IEEE-1992, type AC5A analogous excitation system model is modified to introduce the PID type AVR (Figure 6.36).

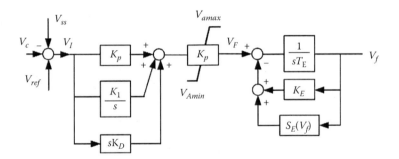

FIGURE 6.36 Simplified AC5A excitation system with PID-type AVR.

The diode rectifier voltage regulation (reduction) with load is neglected here for simplicity.

Although the AVR is to be implemented digitally, the design of the PID controller may be done as if it were continuous, because the sampling frequency is more than 20 times the damped frequency of the closed-loop system.

The transfer function of a PID controller is:

$$G_C(s) = K_p + \frac{K_I}{s} + sK_D \tag{6.44}$$

with K_p—proportional gain, K_I—integral gain, K_D—derivative gain.

Selecting K_P, K_I, and K_D is called controller tuning.

The AC exciter model in Figure 6.36 may be considered as a first-order model, at least for small deviations (Figure 6.29). The SG may also be modeled as a first-order system represented by the excitation time constant T'_{d0} at constant speed, in the absence of the damper windings. Therefore, the AC exciter plus the SG may be modeled through a second-order transfer function $G(s)$:

$$G(s) = \frac{l_{dm}/\left(r_f\left(S_E + K_E\right)\right)}{\left(1 + sT'_{d0}\right)\left(1 + sT_e\right)} \tag{6.45}$$

$$T_e = \frac{1}{\left(S_E + K_E\right)} \cdot T_E$$

$$T'_{d0} = \frac{\left(l_{dm} + l_{fl}\right)}{\left(\omega_b r_f\right)} \tag{6.46}$$

The closed-loop system characteristic is

$$G(s)G_C(s) + 1 = 0 \tag{6.47}$$

Let us consider for simplicity: $l_{dm}/\left(r_f\left(S_E + K_E\right)\right) = 1$

With Equations 6.44 and 6.45, Equation 6.47 becomes

$$K_D^2 s^2 + K_P s + K_I = -s\left(1 + sT'_{d0}\right)\left(1 + sT_e\right) \tag{6.48}$$

It is desirable that the closed-loop system be almost of second order. To do so, let us select a real negative pole $s_3 = c$ in the far left-half plane with the other two ones as complex conjugates $s_{1,2} = a \pm jb$. The peak overshoot and settling time represent the basis for the pole placement (a, b, c). In this pole placement design method, from the three equations we find three unknowns: K_P, K_I, and K_D. The controller settings of K_P, K_I, and K_D give rise to two zeroes that might be real or complex conjugates. The zeroes affect the transient response, and thus some trial and error is required to complete the design. Overdesigning the specifications, such as the choosing of a smaller-than-desired value to overshooting, leads, eventually, to an adequate design.

For $T'_{d0} = 1.5$ s, $T_e = 0.3$ s, $f_1 = 60$ Hz, settling time = 1.5 s, peak overshoot = 10%, a good analogous PID controller gain set is: $K_P = 39.33$, $K_I = 76.50$, $K_D = 5.4$ [7].

The conversion of PID analogous settings into discrete form is straightforward with the trapezoidal integration method:

$$s \Rightarrow \frac{\left(1 - z^{-1}\right)}{T};$$

$$\frac{1}{s} \Rightarrow \frac{T}{2} \frac{\left(1 + z^{-1}\right)}{1 - z^{-1}}$$

(6.49)

z^{-1} represents the unit delay.

Consequently, $G_C(t)$ is

$$G_C(z) = \left[K_{PD} + \frac{K_{ID}}{1 - z^{-1}} + K_{DD}\left(1 - z^{-1}\right) \right] \times K_{AA} = \frac{\Delta V_F(z)}{\Delta V_I(z)}$$

(6.50)

where

$$K_{PD} = K_P - \frac{K_I T}{2}$$

$$K_{ID} = K_I T$$

$$K_{DD} = \frac{K_D}{T}$$

(6.51)

A gain K_{AA} was added in Equation 6.50. For the case in point, $T = 12.5$ ms, $K_{PD} = 777$, $K_{ID} = 19$, $K_{DD} = 8640$, $S_E = 7$, for a 75 kVA, 208 V, 0.8 PF lag generator [7].

Using the known property that $Z^{-1}X(K) = X(K-1)$, the expression of $G_C(z)$ may be converted in time discrete form as follows:

$$\Delta F(K) = \Delta F(K-1) + \left(K_{PD} + K_{ID} + K_{DD}\right)\Delta V_I(K)$$
$$- \left(K_{PD} + 2K_{DD}\right)\Delta V_I(K-1) + K_{DD}\Delta V_I(K-2)$$

(6.52)

where ΔV_I is the generator voltage error (Figure 6.36).

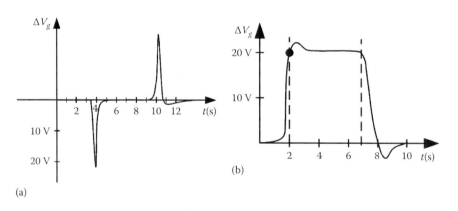

(a)

(b)

FIGURE 6.37 Generator voltage response: (a) 50 kV AR application and rejection and (b) step change in voltage set point.

A 50 kV AR reactive load application and rejection response is shown in Figure 6.37a, while a step change in voltage set point is presented in Figure 6.37b [7].

The settling time varies between 0.4 and 0.6 s, while the voltage overshoot is below ±10% (20 V) [7].

6.11 Underexcitation Voltage

The input V_{UEL} in Figure 6.34a signals the presence of the underexcitation limiter UEL.

The UEL does not interfere with the AVR under normal transient or steady-state conditions but takes over the AVR control under severe conditions. When the excitation level is too low, UEL boosts excitation, overriding or adding to the AVR.

UEL acts to prevent loss of SG synchronism due to insufficient excitation or to prevent operation loading to overheating in the stator core end region of the SG, as defined in the leading reactive power zone of the SG capability curve (Chapter 4).

There are many causes for excitation reduction such as the following:

- Increases in the power system voltage may lead to reduction of SG excitation to keep the voltage at the SG terminals at preset level by absorbing reactive power (UEL)
- Faults in the AVR
- Inadvertent reduction of AVR setting point V_{ref}

When underexcited, the SG absorbs more and more reactive power even when the active (turbine) power is maintained constant, and thus the machine stability limit in the $P(\delta_V)$ curve (Chapter 4) is reached or the SG stator core end region gets overheated.

The UEL may input the AVR either at the generator voltage setting point V_{ref} or after the lead-lag compensator (through a HV gate).

Three main types of UEL models have been recommended by IEEE in 1995 [8]:

1. Circular type (Figure 6.38a)
2. Straight-line type (Figure 6.38b)
3. Multisegment type (Figure 6.38c)

A rather simple digital UEL is presented in Reference [9]—Figure 6.39 where the limit reactive power—from the P-Q capability curves at various voltages is tabled as a function $f(P, V)$:

$$Q_{lim} = f(P, V_c) \tag{6.53}$$

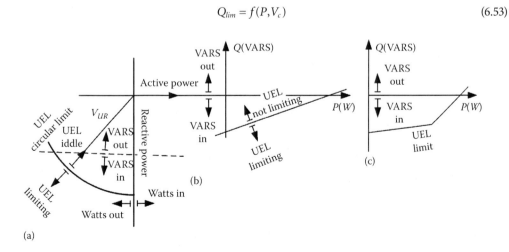

FIGURE 6.38 Underexcitation limiters: (a) circular, (b) straight line, and (c) multisegment.

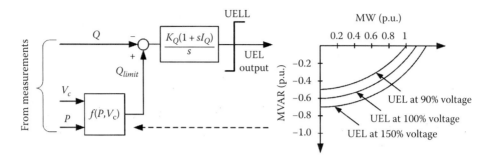

FIGURE 6.39 Simplified underexcitation limiter.

Care must be exercised to prevent UEL side effects on other limiters such as loss of excitation protection (LOE), PSSs, or the AVR, OEL, volt/hertz limiters, overvoltage limiters [10].

6.12 Power System Stabilizers

The field winding current flux Ψ_f transients may be explored rather handily by using the third-order model ($\Delta\Psi_f$, $\Delta\delta$, $\Delta\omega_r$), where the damper cage is eliminated and so are the stator transients (Chapter 4).

After linearization [1],

$$\Delta T_e = K_1\Delta\delta + K_2\Delta\Psi_f \tag{6.54}$$

$$\Delta\Psi_f = \frac{K_3}{1+sT_3}\left(\Delta V_f - K_4\Delta\delta\right) \tag{6.55}$$

where

ΔT_e—SG torque small deviation

$\Delta\delta$—power angle small deviation

$\Delta\Psi_f$—SG field winding flux linkage small deviation

The change $\Delta\Psi_f$ in field flux linkage of Equation 6.55, even for constant field winding voltage V_f ($\Delta V_f = 0$), is explained by armature reaction contribution change when the power angle changes ($\Delta\delta \neq 0$).

Combining Equations 6.54 and 6.55 yields

$$\Delta T_e = K_1\Delta\delta + \frac{K_3 K_2}{1+sT_3}\Delta V_f - \frac{K_2 K_3 K_4}{1+sT_3}\Delta\delta \tag{6.56}$$

The last term in Equation 6.56 represents the variation of Ψ_f transients caused by the electromagnetic torque, due to power angle variation. At steady state, or low oscillating frequency:

$$\Delta T_e \text{ due to } \Delta\Psi_f \text{ is } (-K_2 K_3 K_4 \Delta\delta); \quad \omega \ll 1/T_3 \tag{6.57}$$

Therefore, the field flux variation produces a negative synchronizing torque component of Equation 6.57. If

$$K_1 \leq K_2 K_3 K_4 \tag{6.58}$$

the system becomes monotonically unstable.

At higher oscillating frequencies ($\omega \gg 1/T_3$) the last term of Equation 6.56 becomes

$$\Delta T_e = \frac{-K_2 K_3 K_4}{1+j\omega T_3}\Delta\delta \approx \frac{K_2 K_3 K_4}{\omega T_3}\left(j\Delta\delta\right) \tag{6.59}$$

The air gap torque deviation caused by $\Delta\Psi_f$ is now 90° ahead of power angle deviation and thus in phase with speed deviation $\Delta\omega_r$.

Consequently, the field winding flux linkage deviation $\Delta\Psi_f$ produces a positive damping torque component. At 1 Hz $\Delta\Psi_f$ produces both, a reduction in synchronizing torque and an increase in damping torque of the SG.

Moreover, as K_2, K_3 are positive, in general, K_4 may be positive or negative [1]. With $K_4 < 0$, the synchronizing torque increasing produced by $\Delta\Psi_f$ is accompanied by a negative torque damping component.

AVRs may introduce similar effects [1].

These two phenomena have prompted the addition of PSSs as inputs to the AVRs.

Damping the SG rotor oscillations is the main role of PSS. The most obvious input of PSS should be the speed deviation $\Delta\omega_r$.

Adding motion Equation 6.1 to Equations 6.44 and 6.45, a simplified model for an AVR–PSS system is shown in Figure 6.40.

It is based on the small-signal third-order model of SG Equations 6.44 through 6.46 with the damper windings present only in the motion equation by the asynchronous (damping) torque $K_D\Delta\omega_r$.

The transfer function of the PSS would be a simple gain if the exciter and generator transfer function $\Delta T_e/\Delta V_f$ were a pure gain.

In reality, this is not so and the $G_{pss}(s)$ has to contain some kind of phase compensation, (phase lead) to produce a pure damping torque contribution. A simple such PSS transfer function is shown in Figure 6.41.

The frequency range of interest is 1–2 Hz, in general.

The washout component is a high-pass filter. Without washout contribution steady-state changes in speed would modify the voltage V_s ($T_w = 1$–20 s). Especially, fast-response static exciters need PSS contribution to increase damping.

Temporary increase of SG excitation current can significantly improve the transient stability because the synchronizing power (torque) is increased.

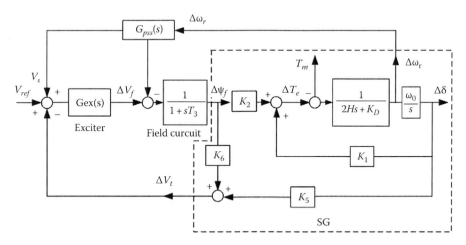

FIGURE 6.40 AVR with PSS: the small-signal model.

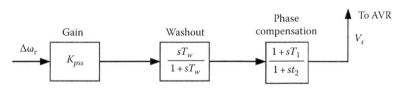

FIGURE 6.41 Basic PSS transfer function.

Ceiling excitation voltage is limited to 2.5–3.0 PU in thermal power units. But fast voltage variations lead to degraded damping as already pointed out. The PSS can improve damping if a terminal voltage limiter is added also. This is true for the first positive rotor swing. But, after the first peak of the local swing, the excitation is allowed to decrease before the highest composite peak of the swing is passed by. Keeping excitation at the ceiling value would be useful until this composite swing peak is reached.

Discontinuous excitation control may be added to the PSS to keep the excitation voltage at its maximum over the entire positive swing of the rotor (around 2 s or so). This is called transient stability excitation control (TSEC) and is applied, together or in place of other methods such as fast valving or generator tripping, in order to improve power system stability. TSEC imposes smaller duty on turbine shaft and on steam supply of the unit.

As PSS should produce electromagnetic torque variations ΔT_e in phase with speed, the measured speed is the obvious input to PSS. But what speed? It could be the turbine measured speed or the generator speed. Both of them are, however, affected by noise. Moreover, the torsional shaft dynamics causes noise that is very important and apparently difficult to filter out from the measured signal.

The search for other more adequate PSS inputs is based on the motion equation in power PU terms (Figure 6.3) redrawn here in slightly new denominations (Figure 6.42).

It should be noted that as P_m may not be measured, it could be estimated with ω_r (estimated) and P_e (measured). It is practical to estimate ω_r as the frequency of the generator voltage behind the transient reactance as it varies slowly enough. This way the speed sensor signal is not needed. Then the structural diagram in Figure 6.42 can be manipulated to estimate the mechanical power P'_m and then calculate the accelerating power (Figure 6.43).

With a single structure, both speed input and accelerating power input PSS may be investigated [11], Figure 6.44.

The speed input of PSS may be obtained with $T_p = 0$ and $M = T_f$. For $T_p \neq 0$ and $M = 2H$, the accelerating power PSS is obtained. Accelerating power PSS are claimed to perform better than speed PSSs in

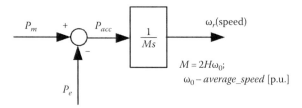

FIGURE 6.42   Accelerating power.

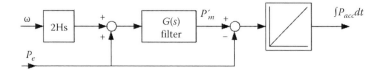

FIGURE 6.43   The integral of accelerating power as input to PSS.

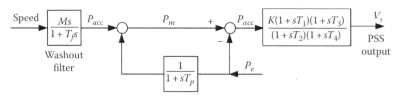

FIGURE 6.44   Accelerating power or speed input PSS.

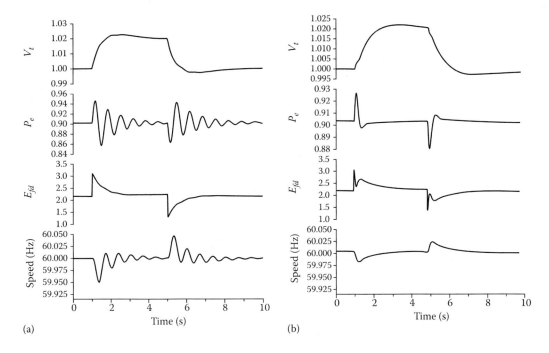

FIGURE 6.45 Simulation response to ±2% step in terminal voltage: (a) no PSS and (b) integral of accelerating power PSS.

damping local system oscillations in the interval from 0.2 to 5 Hz for 220 MVA turbogenerators [11,12]. Other inputs such as the electrical power variation itself, ΔP_e, or frequency may be used with good results [13]. A great deal of attention has attracted the optimization of PSS with solutions involving fuzzy-logic [13,14], linear optimal PSS [15], synthesis [16], variable structure control [17], and H∞ [18]. Typical effects of PSS are shown in Figure 6.45a and b [12].

The damping of electric power and speed (frequency) oscillations is evident. The excitation system design—including AVR and PSS—today in digital circuitry implementation is a complex enterprises, which is beyond our scope here. For details, see References [1,10–18].

6.13 Coordinated AVR-PSS and Speed Governor Control

Coordinated voltage and speed control of SGs requires methods of multivariable nonlinear control design. Basically, the SG of interest may be modeled by a third-order model ($\Delta\delta$, $\Delta\omega_r$ or $\Delta\sigma$, $\Delta\Psi_f$), while the power system to which it is connected might be modeled by a dynamic equivalent in order to produce a reduced-order system.

Optimization control methods are to be used to allocate proper weightings to various control variables participation.

A primitive such coordinated voltage and speed SG control system is shown in Figure 6.46 [19].

The weights K_1–K_6 are constants in Figure 6.46, but they may be adaptable.

The washout filter provides also for zero steady-state error. A digital embodiment of coordinated exciter-speed governor control intended for a low head Kaplan hydro turbine generator is introduced in References [20,21].

Given the complexity of coordinated control of SG, continually on-line trained ANN control systems seem adequate [20–22] for the scope. Coordinated control of SGs has a long way to go.

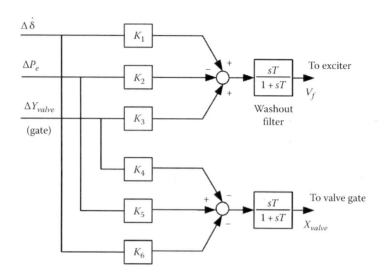

FIGURE 6.46 Primitive coordinated SG control system.

6.14 FACTS-Added Control of SG

FACTS stands for flexible AC transmission systems. FACTS contribute voltage and frequency control for enhancing power systems stability. By doing so, FACTS intervene in active and reactive power flows in power systems, that is between SGs and various loads supplied through transmission lines.

Traditionally, only voltage control was available through changing transformer taps or by switching current and adding fixed capacitors or inductors in parallel or in series.

Power electronics has changed the picture. In the early stages, PE has used back-to-back thyristors in series, which, line commutated, provided for high-voltage and power devices capable to change the voltage amplitude. The thyristor turns off only when its current goes to zero. This is a great limitation.

The gate turn-off (GTO) thyristor overcame this drawback. MOS-controlled thyristors (MCTs) or insulated-gate thyristors (IGCTs) are today the power switches of preference due to a notably larger switching frequency (kHz) for large voltage and power per unit. For MW range, IGBTs are used.

FACTS uses power electronics to increase the active and reactive power transmitted over power lines, while maintaining stability.

Integrated now in FACTS are as follows:

- Static VAR compensators (SVCs)
- Static compensators (STATCOMs)
- Thyristor-controlled resistors
- Power electronics controlled superconducting energy storage (SMES)
- Series compensators and SSR dampers
- Phase angle regulators
- Unified power flow controller
- High voltage DC (HVDC) transmission lines

SVCs deliver or drain controlled reactive power according to power system mainly local needs.

SVCs make use as basic elements of thyristor controlled inductance or capacitor energy storage elements (Figure 6.47).

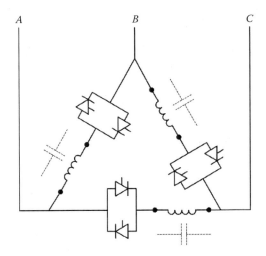

FIGURE 6.47   Basic SVC.

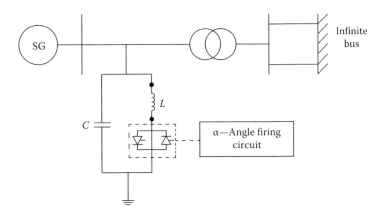

FIGURE 6.48   SG with static VAR compensation.

Adequate control of voltage amplitude on reactor SVCs and capacitor SVCs, connected in parallel to the power system (in power substations), provides for positive or negative reactive power flow and thus contributes to voltage control.

The presence of SVC at SG terminals or close to SG influences the reactive power control of the AVR-PSS system. Therefore, a coordination between AVR-PSS and SVC is required [20,21]. The SVC may enhance the *P/Q* capability at SG terminals [22]—Figure 6.48.

A simplified structure of such a coordinated exciter—SVC, based on multivariable control theory, is shown in Figure 6.49. It uses as variables: the speed variation $\dot{\delta}$, terminal voltage V_t, and active power P_e [23].

A standard solution is here considered in the form of a coordinated exciter—SG control system.

The improvement in dynamic stability boundaries is notable (Figure 6.49b) as claimed in Reference [23], with an optimal coordinated controller. STATCOM is an advanced SVC that uses a PWM converter to supply a fixed capacitor. GTOs or MCTs are used and multilevel configurations are proposed. A step-down transformer is needed (Figure 6.50).

The advanced SVC uses an advanced voltage source converter instead of a VARIAC.

It is power electronics intensive but more compact and better in power quality.

V_2 may be brought in phase with V_1, by adequate control. If V_2 is made larger than V_1 by larger PWM modulation indexing (more apparent capacitor), the advanced SVC acts as a capacitor at SG terminals. If $V_2 < V_1$ is acts as an inductor. For a transformer reactance of 0.1 PU, a $\pm 10\%$ change in V_2 may produce

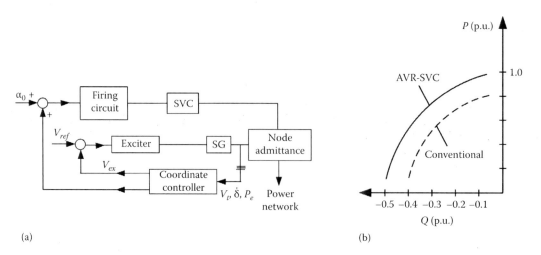

FIGURE 6.49 Coordinated exciter—SVC control: (a) structural diagram and (b) dynamic stability boundary.

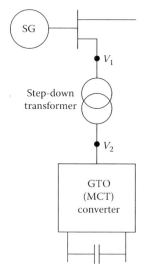

FIGURE 6.50 Advanced SVC = STATCOM.

$a \pm 1.00\%$ PU change in the reactive power flow from (to) SVC. A parallel connected thyristor-controlled resistor is used only for transient stability improvement as it can absorb power from the generator during positive swings preventing the loss of synchronism.

Note: With SVCs used as reactive power compensators and power factor controllers, the interference with the AVR–PSS controllers has to be carefully assessed as adverse effects have been reported by industry [24]. In essence the SG voltage supporting capability may be reduced.

Superconducting magnetic energy storage (SMES) may also be used for energy storage and for damping power system oscillations. The high-temperature superconductors seem a potential practical solution. They allow for current density in the superconducting coil wires of 100 A/mm². Losses and volume per MW h stored are reduced. And so is the cost.

SMES may be controlled to provide both active and reactive power control, with adequate power electronics. Four quadrant P, Q operation is feasible (Figure 6.51) as demonstrated in References [25,26].

To investigate the action of SMES, let us consider a single SG connected to the PS bus by a transmission line.

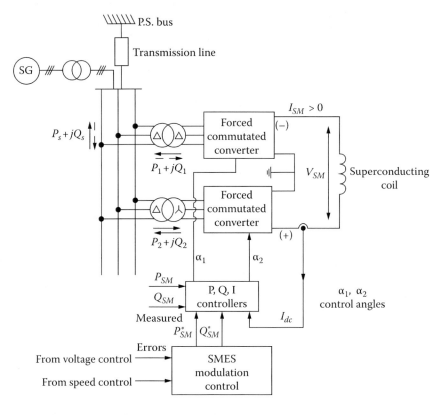

FIGURE 6.51 SMES with double forced commutated converter for four-quadrant *P, Q* operation.

The dual converter impresses $\pm V_{SM}$ volts on the superconducting coil. On the AC side of the converter, the phase angle between the voltage and current may vary from 90° when no energy is transferred from power system to SMES, to 0° when the SMES is fully charging, and to 180° for maximum discharge of SMES energy into the power system.

The SMES current I_{SM} is not reversible but the voltage V_{SM} is, and therefore is the converter output power P_{SM}:

$$P_{SM} = I_{SM} V_{SM}$$

$$V_{SM} \approx K_M V_L \cos \alpha$$

$$V_{SM} \approx L_{ds} \cdot s \cdot I_{SM} \tag{6.60}$$

$$P_{SM} = K_M V_L I_{SM} \cos \alpha$$

$$Q_{SM} = K_M V_L I_{SM} \sin \alpha$$

with α—the converter commutation angle, K_M—modulation coefficient, and V_L—line AC voltage of the converter.

For given P_{SM}^*, Q_{SM}^* demands, unique values of K_M and α are obtained [27].

The DC voltage is controlled by changing the firing angle difference in the two converters: $\alpha_{1,2}$. The transfer function of the converter angle control loop may be written as follows:

$$\Delta V_{SM} = \frac{K_0}{1 + sT_\alpha} \Delta \alpha \tag{6.61}$$

The SMES energy stored is

$$W_{SM} = \frac{1}{2}L_d I_{SM0}^2 + \int_{t_0}^{t} P_{SM}(\tau)d\tau \tag{6.62}$$

The SMES power P_{SM} intervenes in the SG motion equations:

$$Ms\Delta\omega_r = \Delta P_m - D\Delta\omega_r - \Delta P_e - \Delta P_{SM};$$
$$M = 2H \tag{6.63}$$

$$s\Delta\delta = \omega_b \times \Delta\omega_r \tag{6.64}$$

The power angle δ is now considered as the angle between the no-load voltage of the SG and the voltage V_0 of the PS bus. Essentially, the transmission line parameters r_T, x_T are added to SG stator parameters.

D is the damping provided by load frequency dependence and by the SG damper winding; ΔP_m—turbine power variation; ΔP_e—electric power variation, ΔP_{SM}—SMES power variation, all in PU at (around) base speed (frequency) ω_b. The SMES also may influence the reactive power balance by the phase shift between converter AC voltages and currents.

It was shown that optimal multivariable control, with minimum time transition of the whole system from state A to B as the cost function, and $|\Delta\alpha|<\pi/2$ as constraint, may produce speed and voltage stabilization for large PS active power load perturbations or faults (short circuit) [28]. However, using speed variation $\Delta\omega_r$ as input, instead of optimal control, to regulate the angle $\Delta\alpha$ loop, does not provide satisfactory results [28]. Notable SMES energy exchange within seconds is required for the scope [28].

Besides SG better control, the SMES has also been proposed to assist AGC that deals with interarea power exchange control [29]. This time the input of the SMES is proportional to or ACE, see Section 6.4) or is produced by a separate adaptive controller. In both cases, the superconducting inductor current deviation (derivative) is used as negative feedback in order to provide quick restoration of I_{SM} to the setpoint, following a change in the load demand [30,31].

The SMES appears to be a promising emerging technology, but new performance improvement and cost reduction are required before it will be common practice in power systems, together with other forms of energy storage, such as pump storage, batteries, fuel cells, and so forth. Meanwhile, the first 20 MW competitive inertial storage systems have been promoted for stabilization in railroad power supplying systems.

6.14.1 Series Compensators

A transmission line may be modeled as lumped capacitors in series. The line impedance may be varied through thyristor-controlled series capacitors (Figure 6.52).

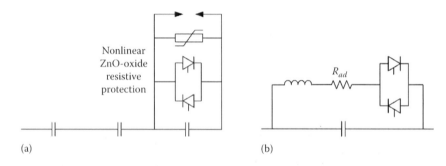

(a)   (b)

FIGURE 6.52 (a) Series compensator and (b) damper.

The capacitor is varied by PWM short-circuiting it through antiparallel thyristors. Overvoltage protection of thyristors is imperative. Nonlinear zinc-oxide resistors with paralleled air gap is the standard protection for the scope.

When an additional resistance R_{ad} is added (Figure 6.52b), an oscillation damper is built. Series compensation of long heavily loaded transmission lines might produce oscillations that need attenuation.

6.14.2 Phase-Angle Regulation and Unit Power Flow Control

In-phase and quadrature voltage regulation done traditionally with tap changers in parallel-series transformers may be accomplished by power electronics means (FACTS), Figure 6.53.

The thyristor AC voltage tap changer (Figure 6.53a) performs the variation of ΔV through the ST while the unified power controller (Figure 6.53b) may vary both the amplitude and the phase of voltage variation ΔV.

The phase angle regulator may transmit reactive power to increase voltage $V_2 > V_1$, while the unified power flow controller allows the flow of both active and reactive power, thus allowing for damping electromechanical oscillations.

HVDC transmission systems contain a high-voltage high-power rectifier at the entry, a DC transmission line and an inverter at the end of it [1]. It is power electronics intensive, but it really makes the power transmission system much more flexible. It also interferes with other SG control systems.

As the power electronics SG control systems cost goes down, the few special (cable) or long power transmission lines HVDC, now in existence, will be extended in the near future as they

- Increase transmitted power capability of the transmission lines
- Produce additional electromechanical oscillation damping in the AC system
- Improve the transient stability
- Isolate system disturbances
- Perform frequency control of small isolated systems (by the output inverter)
- Interconnect between power systems of different frequencies (50 or 60 Hz)
- Provide for active power and voltage dynamic support

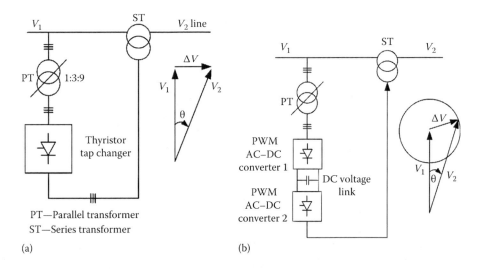

FIGURE 6.53 (a) Phase angle regulator and (b) unified power flow controller.

6.15 Subsynchronous Oscillations

In Chapter 3, dedicated to prime movers, we did mention the complexity of mechanical shaft, couplings, and of mechanical gear in relation to wind turbine generators. In essence, the turbine—generator shaft system consists of several masses (inertias): turbine sections, generator rotor, mechanical couplings, and exciter's rotor. These masses are connected by shafts of finite rigidity (flexibility). A mechanical perturbation will thus produce torsional oscillations between various sections of the shaft system that are above 5–6 Hz but, in general, below base frequency 50(60) Hz.

Hence the term: subsynchronous oscillations.

The entire turbine—generator rotor oscillates with respect to other generators at a frequency in the range of 0.2–2 Hz.

Torsional oscillations may cause:

- Torsional interaction with various power system (or SG) controls
- SSR in series—capacitor—compensated power transmission lines

Torsional characteristics of hydrogenerators units do not pose such severe problems as the generator inertia is much larger than in turbine—generator systems whose shafts have total lengths of up to 5 m.

6.15.1 Multimass Shaft Model

The various gas or steam turbine sections such as low pressure (LP), high pressure (HP), intermediate pressure (IP), generator rotor, couplings, and exciter rotor (if any), are elements of a lumped mass model.

Let us consider for start only a two mass rotor (Figure 6.54).

The motion equations of the generator and LP turbine rotors are as follows:

$$2H_1 \frac{d(\Delta\omega_1)}{dt} = K_{12}(\delta_2 - \delta_1) - T_e - D_1(\Delta\omega_1)$$

$$\frac{d\delta_1}{dt} = \Delta\omega_1 \cdot \omega_0$$

(6.65)

$$2H_2 \frac{d(\Delta\omega_2)}{dt} = T_{LP} + K_{23}(\delta_3 - \delta_2) - K_{12}(\delta_2 - \delta_1) - D_2(\Delta\omega_2)$$

$$\frac{d\delta_2}{dt} = \Delta\omega_2 \cdot \omega_0$$

(6.66)

ω_0 is the rated electrical speed in rad/s: 377 rad/s for 60 Hz and 314 rad/s for 50 Hz.

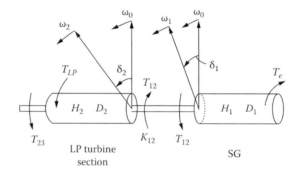

FIGURE 6.54 The two mass shaft model.

For more masses, more equations are added. For a coal-fired turbine generator with HP, IP, and LP sections (and static exciter), two more equations are added:

$$2H_3 \frac{d(\Delta\omega_3)}{dt} = T_{IP} + K_{34}(\delta_4 - \delta_3) - K_{23}(\delta_3 - \delta_2) - D_3(\Delta\omega_3)$$

$$\frac{d\delta_3}{dt} = \Delta\omega_3 \cdot \omega_0$$

(6.67)

$$2H_4 \frac{d(\Delta\omega_4)}{dt} = T_{HP} - K_{34}(\delta_4 - \delta_3) - D_4(\Delta\omega_4)$$

$$\frac{d\delta_4}{dt} = \Delta\omega_4 \cdot \omega_0$$

(6.68)

Two more additional equations are needed for a nuclear power unit with LP_1, LP_2, IP, and HP sections with static exciter.

Example 6.3

Let us consider a 4 mass coal-fired steam turbine generator with HP, IP, LP sections and a static exciter with inertias: $H_1 = 0.946$ s, $H_2 = 3.68$ s, $H_3 = 0.337$ s, $H_4 = 0.099$ s.

The generator is at nominal power and 0.9 PF ($T_e = 0.9$). The stiffness coefficients K_{12}, K_{23}, K_{34} are $K_{13} = 82.74$ PU torque/rad, $K_{23} = 81.91$, $K_{34} = 37.95$.

Calculate the steady-state torque and angle of each shaft section if the division of powers is 30% for HP, 40% IP, and 30% LP.

Solution

Note that the damping coefficients (due to blades and shafts materials, hysteresis, friction, etc.) are zero ($D_1 = D_2 = D_3 = D_4 = 0$). The angle by which the LP turbine section leads the generator rotor is Equation 6.65:

$$\delta_2 - \delta_1 = \frac{T_{12}}{K_{12}}$$

But $T_{12} = T_e = 0.9$ (from power balance) and thus

$$\delta_2 - \delta_1 = 0.9/82.74 = 0.010877 \text{ rad(electrical)}$$

The torque between LP and IP shaft sections T_{23} at steady state is

$$T_{23} = T_{12} - T_{LP} = 0.9(1 - 0.3) = 0.54$$

From Equation 6.66,

$$\delta_3 - \delta_2 = T_{23}/K_{23} = 0.54/81.91 = 6.5926 \times 10^{-3} \text{ rad(electrical)}$$

Again: $T_{34} = T_{23} - T_{IP} = 0.54 - 0.4 \cdot 0.9 = 0.18$

$$\delta_4 - \delta_3 = \frac{T_{34}}{K_{34}} = \frac{0.18}{37.95} = 4.7431 \times 10^{-3} \text{ rad(electrical)}$$

So the HP turbine rotor section leads the generator rotor section by $\delta_4 - \delta_1$

$$\delta_4 - \delta_1 = 0.010877 + 6.5926 \times 10^{-3} + 4.7431 \times 10^{-3}$$

$$= 2.22057 \times 10^{-2} \text{ rad} = 1.273°$$

at steady state.

6.15.2 Torsional Natural Frequency

The natural frequencies and modal shapes are to be obtained by finding the eigen values and vectors of the linearized free system equations from Equation 6.65 to Equation 6.68, with $T_e = K_S \Delta\delta_1$:

$$\Delta\dot{X} = A\Delta X \tag{6.69}$$

and $\Delta T_{HP} = \Delta T_{LP} = \Delta T_{IP} = 0$.

With: $\Delta X = \left[\Delta\omega_1, \Delta\delta_1, \Delta\omega_2, \Delta\delta_2, \Delta\omega_3, \Delta\delta_3, \Delta\omega_4, \Delta\delta_4\right]^T$

$$|A| = \Delta
\begin{array}{c}
\\
\Delta\omega_1 \\
\Delta\delta_1 \\
\Delta\omega_2 \\
\Delta\delta_2 \\
\Delta\omega_3 \\
\Delta\delta_3 \\
\Delta\omega_4 \\
\Delta\delta_4
\end{array}
\begin{array}{|cccccccc}
\Delta\omega_1 & \Delta\delta_1 & \Delta\omega_2 & \Delta\delta_2 & \Delta\omega_3 & \Delta\delta_3 & \Delta\omega_4 & \Delta\delta_4 \\
- & - & - & - & - & - & - & - \\
-\dfrac{D_1}{2H_1} & -\dfrac{(K_{12}+K_S)}{2H_1} & & \dfrac{K_{12}}{2H_1} & & & & \\
\omega_0 & & & & & & & \\
& \dfrac{K_{12}}{2H_2} & -\dfrac{D_2}{2H_2} & -\dfrac{K_{12}+K_{23}}{2H_2} & & \dfrac{K_{23}}{2H_2} & & \\
& & \omega_0 & & & & & \\
& & & \dfrac{K_{23}}{2H_3} & -\dfrac{D_3}{2H_3} & -\dfrac{(K_{23}+K_{34})}{2H_3} & & \dfrac{K_{34}}{2H_3} \\
& & & & \omega_0 & & & \\
& & & & & \dfrac{K_{34}}{2H_4} & \dfrac{-D_4}{2H_4} & -\dfrac{K_{34}}{2H_4} \\
& & & & & & \omega_0 &
\end{array}$$

The natural frequencies are the eigen values of $|A|$:

$$|A| - \lambda|I| = 0 \tag{6.70}$$

With zero damping coefficients ($D_i = 0$), the eigenvalues are complex numbers $j\omega_i$ ($i = 0, 1, 2, 3$). The corresponding frequencies for the case in point are 22.4, 29.6, and 52.7 Hz. The first, low (system mode) frequency is not included here as it is the oscillation of the entire rotor against the power system:

$$f_s = \frac{\omega_s}{2\pi} = \frac{\omega_0}{2\pi}\sqrt{\frac{K_s}{2(H_1 + H_2 + H_3 + H_4)}}$$

With $K_s = 1.6$ PU torque/rad,

$$f_s = 60 \times \sqrt{\frac{1.6}{2(0.946 + 3.68 + 0.33 + 0.049) \times 377}} = 1.23 \, \text{Hz}$$

The torsional free frequencies are, in general, above (6–8) Hz, so they do not interfere, in this case, with the excitation, speed-governor or intertie control (below 3–5 Hz).

The PSS may interfere with the rather high (above 8 Hz) torsional frequencies as it is designed to provide pure damping (zero phase shift at system frequency $f_s = 1.23$ Hz in our case). At 22.4 Hz, PSS may produce a large phase lag (well above 20°) and thus notable negative damping; hence instability.

To eliminate such a problem, the speed sensor should be placed between LP and IP sections to reduce torsional modes influence on speed feedback. Other possibility would be a filter with a notch at 22.4 Hz. The terminal voltage limiter may also produce torsional instability. Adequate filter is required here too.

HVDC systems may, in turn, cause torsional instabilities. But it is the series capacitor compensation of power transmission lines that most probably interacts with the torsional dynamics.

6.16 Subsynchronous Resonance

In a typical (noncompensated) transmission system, transients (or faults) produce DC attenuated, 60(50) Hz and 120(100) Hz torque components (the latter, for unbalanced loads and faults). Consequently, the torsional frequencies originate always from these two frequencies.

Series compensation is used to bring long power lines capacity closer to their thermal rating. In essence, the transmission line total reactance is partially compensated by series capacitors. Let us consider a simple radial system (Figure 6.55).

The presence of the capacitor eliminates the DC stator current transients, but it introduces offset AC currents at natural frequency ω of the LC series circuits:

$$\omega_n \approx \frac{1}{\sqrt{LC}} = \omega_0 \sqrt{\frac{X_C}{X_L}} \tag{6.71}$$

where

$$X_L \approx (X'' + X_T + X_E)\omega_0; \quad X_C = 1/\omega_0 C \tag{6.72}$$

$X'' = (X_d'' + X_q'')/2$ is the subtransient reactance of the SG.

The f_n frequency offset stator currents produces rotor currents at slip frequency: $60(50) - f_n$.

The SSR frequency f_n may depend solely on the degree of compensation X_C/X_L (Table 6.1).

Shunt compensation (SVC) tends to produce oversynchronous natural frequencies unless the degree of SVC is high and the transmission line is long.

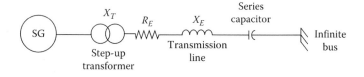

FIGURE 6.55 Radial system with series capacitor compensation.

TABLE 6.1 SSR Frequency

Compensation Ratio in % (X_C/X_L)	Natural Frequency f_n (Hz)	Slip Frequency $60-f_n$ (Hz)
10	18	42
20	26.83	33.17
30	32.86	27.14
40	37.94	22.06
50	42.46	17.54

There are two situations when series compensation may cause undamped subsyncronous oscillations:

1. Self-excitation due to induction generator effect
2. Interaction with torsional oscillations

As $f_n < f_0$, the slip S in the SG is in fact negative. Consequently, with respect to stator currents at frequency f_n, the SG behaves like an induction generator connected to the power system.

The effective synchronous resistance of the SG at $S = (f_n - f_0)/f_0 < 0$, R_{as}, is negative. As such, it may surpass the positive resistance of the transmission line. The latter becomes an LC circuit with negative resistance. Electrical oscillations will self-excite at large levels. A strong damper winding in the SG or an additional resistance in the power line would solve the problem.

The phenomenon is independent of torsional dynamics, being of purely electromechanical essence.

If the series compensation slip frequency 60(50) Hz—f_n is close to one of the torsional-free frequencies of the turbine—generator unit, torsional oscillations are initiated. This is the SSR. Torsional oscillations buildup may cause shaft fatigue or turbine—generator shaft breaking. In fact, the SSR was "discovered" after two such disastrous events took place in 1970–1971.

Some countermeasures to SSR are as follows:

- Damping circuits in parallel with the series compensation capacitors (Figure 6.52b).
- Dynamic filters: the unit power flow controller may be considered for the scope (Figure 6.53b) as it can provides an additional voltage ΔV of such a phase as to compensate the SSR voltage [32].
- Thyristor controlled shunt reactors or capacitors (SVCs).
- Selected frequency damping in the exciter control [33].
- Superconducting magnetic storage systems. The ability of the SMES to quickly inject or extract from the system active and (or) reactive power by request makes it a very strong candidate to SSR attenuation. About 1 s attenuation time is claimed in Reference 34.
- Protective relays that trip the unit when SSR is detected in speed by generator current feedback sensors.

6.17 Note on Autonomous Synchronous Generators' Control

Autonomous synchronous generators are present in on-ground and on board of vehicles local power systems.

In general, they are driven at rather constant speed irrespective of load and procure rather constant output a.c. voltages amplitude and frequency (50(60) Hz or 400 Hz in avionics).

However, to reduce fuel consumption and pollution variable speed primary movers (with power increasing with speed) are used to drive synchronous generators in vehicular an on-ground (cogeneration) applications.

In lower cost systems, variable frequency constant amplitude output voltages are acceptable, for frequency-immune consumers, and thus no inverter is added; a diode rectifier may be connected to the SG stator to yield DC voltage output fully controlled by SG excitation current control.

While inverter and diode rectifier-controlled SG will be treated in the variable speed generators volume, the constant voltage variable frequency generator control will be described here as it involves the exciter's model too.

6.17.1 Variable Frequency/Speed SG with Brushless Exciter

A typical variable frequency/speed SG with a brushless exciter involving three electric machines on a single shaft, typical for avionics or gas-turbine-driven (ship, etc.) applications is illustrated in Figure 6.56 [35].

The system comprises a permanent magnet synchronous generator (PMSG) of low power (< 0.3% p.u.) whose stator output is diode rectified and then treated via a DC–DC converter to control the main exciter's stator-placed field winding current. In turn, the latter's armature winding placed on the common rotor has the three-phase terminals connected to a diode rectifier (designed at less than 7%–8% p.u. power) which supplies the main SG excitation winding.

As there are three electric (synchronous) generators in raw, their rather nonlinear models introduce delays, instabilities, and thus a robust control system is needed. Adaptive model-based control [35] has certain virtues, but more robust (i.e., sliding mode) control might be tried in the future.

In any case, in model-based control only the *dq* (orthogonal) model in rotor coordinates suffers enough simplicity to be practical.

While the PMSG model may be neglected treating it as a gain in a first instant, a first-order delay may be introduced for better precision as

$$V_{PM} = \omega_r \Psi_{PM} - L_s \frac{di_{PM}}{dt} - R_{PM} i_{PM} \tag{6.73}$$

The speed ω_r is controlled through the prime mover's governor. On the other hand, the main SG dq model is standard (with dq stator windings, d_r, q_r rotor cages and f_d excitation winding): as in Figure 5.20, but with single d_r, q_r cage circuits).

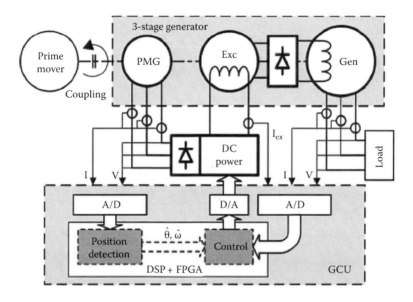

FIGURE 6.56 Brushless—exciter, variable frequency/speed SG with control unit.

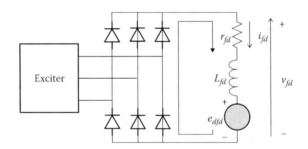

FIGURE 6.57   Main exciter with diode rectifier and load.

Finally, for the main exciter (SG) with diode rectifier (Figure 6.57), a continuous average model is preferred for control (though in reality the armature current may be discontinuous, as known):

$$i_{dc} = k_i \left[i_{ed} \sin\left(\delta_e + \phi\right) + i_{eq} \cos\left(\delta_e + \phi\right) \right]; \tag{6.74}$$

$$V_{ed} = \frac{V_{DC}}{k_v} \sin\delta_e; \quad V_{eq} = \frac{V_{DC}}{k_v} \cos\delta_e; \quad \delta_e = \tan^{-1}\left(\frac{V_{eq}}{V_{ed}}\right)$$

K_i and K_v are power equivalence (from AC to DC) coefficients; V_{ed}, V_{eq}, i_{ed}, and i_{eq} are dq armature voltages and currents of main exciter; δ_e is the power angle (taken for control as $\delta_e \cong 45°$, though it varies with load) and $\Phi \cong 0$ is the power factor angle for the diode rectifier.

The above equations are valid in rectifying mode, where the power in AC is transferred into DC and not in the freewheeling operation mode.

Freewheeling mode occurs when the armature currents (especially id) undergoes severe transients due to strong changes required in the field current of the main strong changes required in the field current of the main generator. The d-axis rotor cage is not capable of presenting freewheeling.

To avoid freewheeling, a feedforward reference field current component, $i_{exchange}$ proportional to the i_d current variation in the main SG is added. This implicitly means that rotor position and speed should be available. To avoid an encoder, the small PMSG output voltage/current phases and frequency may be used to build a robust rotor position and speed observer. The basic control system of main SG output voltage is shown in Figure 6.58.

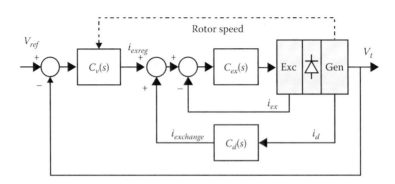

FIGURE 6.58   Output voltage regulator of SG.

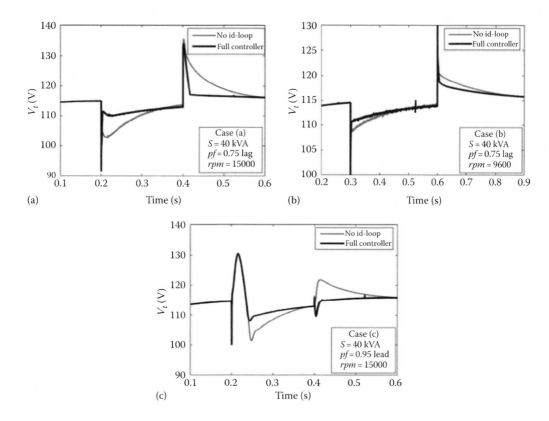

FIGURE 6.59 Controller without and with antifreewheeling (i_d) feedforward $i_{exchange}$ for inductive load at (a) low speed, (b) high speed, and (c) for capacitor load. (From Rosado, S. et al., *IEEE Trans. Energ. Convers.*, 23(1), 42, 2008.)

The control system comprises the following:

An outer AC output voltage amplitude regulator $C_v(s)$
A feedforward $i_{ex\ reg}$ corrector by feedforward action through $i_{exchange}(C_d(s))$, as explained earlier (to avoid freewheeling)
Main exciter excitation current i_{ex} regulator $(C_{ex}(s))$

Skipping the tedious zeroes and poles selection for $C_v(s)$, $C_d(s)$, and $C_{ex}(s)$, in relation to involved machines' parameters, with one speed adaptive $(C_v(s))$ and with numerous filters, we present here some final results of Reference [35] obtained on a 40 kVA, 115 V, speed 9.6 krpm—16 krpm system.

They refer to various load transients: inductive at high and low speed and capacitive at high speeds (Figure 6.59), and to inductive load connection and disconnection without and with speed adaptation of $C_d(s)$ in the voltage outer regulator (Figure 6.60).

The results in Figures 6.59 and 6.60 show very good dynamic performance for wide speed and load ranges.

It was included here, however, only as an exemplary solution. Robust (variable structure or rule-based) control, on the other hand, could offer faster control in less complex reliable systems with increased fault tolerance [36].

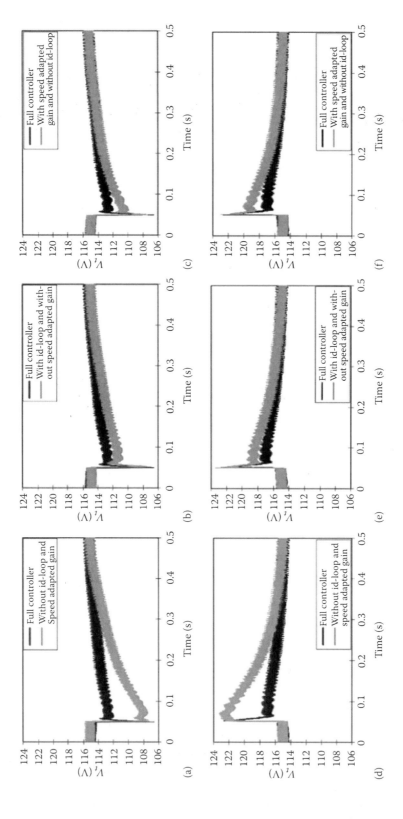

FIGURE 6.60 Controller without and with speed adaptation in the voltage outer loop regulator (Vd's): (a–c) inductive lead connection and (d–f) disconnection at $\cos \phi = 0.76$ and 12 krpm. (From Rosado, S. et al., *IEEE Trans. Energ. Convers.*, 23(1), 42, 2008.)

6.18 Summary

- Control of SGs means basically active power (or speed) and reactive power (or voltage) control.
- Active power (or speed) control of SGs is performed through turbine closed loop speed governing.
- Reactive power (or voltage) control is done through field winding voltage (current I_f) closed loop control (AVR).
- Though, in principle, weakly coupled, the two controls interact with each other. The main decoupling means used thus far is the so-called PSS. The PSS input is based on active power (or speed) deviation. Its output enters the AVR control system with the purpose of increasing the damping torque component.
- When the SG operates in connection with a power system, two more control levels are required besides primary control (speed governing and AVR–PSS). They are AGC and economic dispatch with security assessment generation allocation control.
- AGC refers to frequency-load control and intertie control.
- Frequency load control means to allocate frequency (speed)/power characteristics for each SG and move them up and down through ACE, to determine how much power contribution is asked for each SG.
- ACE is formed by frequency error multiplied by a frequency bias factor λ_R added to intertie power error ΔP_{tie}. ACE is then PI controlled to produce load frequency set points for each SG by allocating pertinent participation factors α_i (some of them may be zero).
- The amount of intertie power exchange between different areas of a power system and the participation factors of all SGs are determined in the control computer by the economic dispatch with security assessment, based on lowest operation costs per kWh or on other cost functions.
- Primary (speed and voltage) control is the fastest (seconds) while economic dispatch is the slowest (minutes).
- Active and reactive power flow in a power system may be augmented by flexible AC transmission systems (FACTS) that make use of power electronics and of various energy storage elements.
- Primary (speed and voltage) control is slow enough that third- (fourth-) order simplified SG models suffice for its investigation.
- Constant speed (frequency) closed loop is feasible only in isolated SGs.
- SGs operating in a power system have speed-droop controllers to allow for power sharing between various units.
- Speed droop is typically 4%–5%.
- Speed governors require at least second-order models, while for hydraulic turbines transient speed-droop compensation is required to compensate water starting time effect. Speed governors for hydraulic turbines are the slowest in response (up to 20 s and more for settling time), while steam turbines are faster (especially in fast valving mode), but they show an oscillatory response (settling time is less than 10 s in general).
- In an isolated power system with a few SGs, AGC means, in fact, adding an integrator to the load–frequency set point of the frequency/power control to keep the frequency constant for that generator.
- AGC in interconnected power systems means introducing the intertie power exchange error ΔP_{tie} with frequency error weighted by the frequency bias factor to form the ACE. ACE contains PI filters to produce ΔP_{ref} that provides finally the set points level of various generators in each area. Consequently, not only the frequency is controlled but also the intertie power exchange. All according to, say, minimum operation costs with security assessment.
- Time response of SG in speed and power angle for various power perturbations is qualitatively divided into four stages: rotor swings, frequency drops, primary control (speed and voltage

control), and secondary control (intertie control and economic dispatch). Frequency band of speed governor control in power systems is less than 2 Hz, in general.

- Spinning reserve is defined as rated power of all SGs in a system minus the actual power needed in certain conditions.
- If spinning reserve is not large enough, frequency does not recover; it keeps decreasing. To avoid frequency collapse, load is dumped in designated substations in 1–3 stages until frequency recovers. Underfrequency relays trigger load shedding in substations.
- Active and reactive powers of various loads depend on voltage and frequency to a larger or smaller degree.
- Equilibrium of frequency (speed) is reached for active power balance between offer (SG) and demand (loads).
- Similarly, equilibrium in voltage is reached for balance in reactive power between offer and demand. Again if not enough reactive power reserve exists, voltage collapse takes place. To avoid voltage collapse either important reactive loads are dumped, or additional reactive power injection from energy storage elements (capacitors) is performed.
- The SG contribution to reactive power (voltage) control is paramount.
- Voltage control at SG terminals AVR is done through field winding (excitation) voltage V_f (current I_f) control. Frequency band of AVRs is within 2–3 Hz, in general.
- DC excitation power for SGs is provided by exciters.
- Exciters may be of three main types: DC, AC, and static.
- DC exciters contain two DC commutator generators mounted on the SG shaft: the auxiliary exciter (AE) and the main exciter (ME). The ME armature supplies the SG excitation through brushes and slip rings at full excitation power rating. The AE armature excites the ME. The AE excitation is power electronics controlled at the command (output) of AVR. DC exciters are in existence for SGs up to 100 MW despite of their slow response—due to large time constant of AE and ME—and commutator wear, due to low control power of AE.
- AC exciters contain an inside-out synchronous generator (ME) with an output is diode-rectified and connected to the SG excitation winding. It is a brushless system, as the ME DC excitation circuit is placed on the stator and is controlled by power electronics at the command (output) of AVR. With only one machine on SG shaft, the brushless AC exciter is more rugged and almost maintenance free. The control power is 1/20 (1/30) of SG excitation power rating, but the control is faster as now only one machine (ME) time constant additional delay still exists.
- Static exciters are placed away from the SG and are connected to the excitation winding of SG through brushes and slip rings. They are power electronics (static) AC—DC converters with very fast response. Controlled rectifiers are typical but diode rectifiers with capacitor filters and 4 quadrant choppers are also feasible. Static exciters are the way of the future now that slip ring—brush energy transmission at 30 MW has been demonstrated in 400 MVA doubly fed induction generators pump storage power plants. Also converters up to this rating are already feasible.
- Also, competitive contactless power transfer to the SG rotor is feasible.
- Exciter modeling requires, for AVR design, a new, nonreciprocal, PU system where the base excitation voltage is that which produces no-load rated voltage at SG terminals at no load. The corresponding PU field current is also unity in this case:

$$I_{fb} = l_{dm}I_f = 1; \quad V_f = I_f r_f$$

$$V_{fb} = \frac{l_{dm}}{r_f} V_f = 1$$

This way working with PU voltages in the range of 10^{-3} is avoided.

- The simplest model of SG excitation is a first-order delay with the transient open-circuit time constant $T'_{d0} = \dfrac{\left(l_{dm} + l_{lf} \right)}{r_f} \times \dfrac{1}{\omega_0}$ (seconds) and a unity PU gain (Equation 6.22).
- The DC exciter model (one level; ME or AE), accounting for magnetic saturation, is a nonlinear model that contains a first-order delay and a nonlinear saturation-driven feedback. For small signal analysis a first-order model with saturation-variable gain and a time constant is obtained.
- For the AC exciter, the same model may be adopted but with one more feedback proportional to field SG current, to account for d-axis demagnetizing armature reaction in the ME (alternator).
- The diode rectifier may be represented by its steady-state voltage/current output characteristics. Three diode commutation modes are present from no load to short circuit. In essence, the diode rectifier characteristic shows a voltage regulation dependent on ME commutation (subtransient) inductance.
- The controlled rectifier may also be modeled by its steady-state voltage/current characteristic with the delay angle α as parameter and control variable.
- The basic AVR acts upon the exciter input and may have 1–3 stabilizing loops and, eventually, additional inputs. The sensed voltage is corrected by a load compensator.
- A lead-lag compensator constitutes the typical AVR stabilizing loop.
- Various exciters and AVRs are classified in IEEE standard 512.2 of 1992.
- Alternative AVR stabilizer loops such as PID (variable structure) are also practical.
- All AVR systems are provided with underexcitation (UEL) and overexcitation (OEL) limiters.
- Exciter dynamics and AVRs may introduce negative damping generator torques. To counteract such a secondary effect, PSSs have been introduced.
- PSSs have speed deviation or accelerating power deviation as input and act as an additional input to AVR.
- The basic PSS contains a gain, a washout (high pass) filter and a phase compensator in order to produce torque in phase with speed deviation (positive damping). Washout filter's role is to avoid PSS output voltage modification due to steady-state changes in speed.
- Accelerating-power-integral input PSSs have been proven better than speed or frequency or electric power deviation input PSSs.
- Besides independent speed-governing and AVR–PSS control of SGs, coordinated speed and voltage SG controls were introduced through multivariable optimal control methods.
- Advanced nonlinear digital control methods, such as fuzzy logic, ANN, μ synthesis, H∞, and sliding-mode, have been all proposed for integrated SG generator control [37,38].
- Power electronics-driven active/reactive power flow in power systems may be defined as FACTS (flexible AC transmission systems). They may use of external energy storage elements such as capacitors, resistors, inductors (normal or of superconductor material). They also assist in voltage support and regulation.
- FACTS may enhance the dynamic stability limits of SGs but interfere with their speed-governing and AVR.
- The steam (or gas) turbines have long multimass shafts of finite rigidity. Their characterization by lumped 4.5 masses is typical.
- Such flexible shaft systems are characterized by torsional natural frequencies above 6–8 Hz, in general, for large turbine generators.
- Series compensation by capacitors to increase the transport capacity of long power lines leads to the occurrence of offset AC currents at natural frequency f_n solely dependent on the degree of transmission line reactance compensation by series capacitors ($X_C/X_L < 0.5$).
- It is the difference $f_0 - f_n$, the slip (rotor) frequency of rotor currents due to this phenomenon, that may fall over a torsional-free frequency to cause SSR.
- SSR may produce shafts breaking or, at least, their premature wearing.

- The slip frequency currents of frequency $f_0 - f_n$ produced by the series compensation effect, manifest themselves as if the SG were an induction generator connected at the power grid. As slip is negative ($f_n < f_0$), an equivalent negative resistance is seen by the power grid. This negative resistance may overcompensate the transmission line resistance. With negative overall resistance, the transmission line reactance plus series compensation capacitor circuit may ignite dangerous torque pulsations. This phenomenon is called induction generator self-excitation and has to be avoided. One way of doing this is to use a strong (low resistance) damper cage in SGs.
- Various measures to counteract SSR have been proposed. Among them are damping circuit in parallel with the series capacitor, thyristor-controlled shunt reactors or capacitors, selected frequency damping in AVR, SMES, protective relays to trip the unit when SSR is detected through generator speed or by current feedback sensors oscillations.
- Coordinated digital control of both active and reactive power with various limiters, by multivariable optimal theory methods with self-learning algorithms seems the way of the future and much progress in this direction is expected in the near future.
- Emerging silicon-carbide power devices [34] may enable revolutionary changes in high-voltage static power converters for frequency and voltage control in power systems.

References

1. P. Kundur, *Power System Stability and Control*, Mc Graw-Hill, New York, 1994.
2. J. Machowski, J.W. Bialek, J.R. Bumby, *Power Systems Dynamics and Stability*, John Wiley & Sons, Chichester, U.K., 1997.
3. L.L. Grigsby (ed.), *Electric Power Engineering Handbook*, CRC Press, Boca Raton, FL, 1998.
4. N. Mohan, T. Undeland, R. Williams, *Power Electronics*, 3rd edn., John Wiley & Sons, New York, 2002
5. I. Boldea, S.A. Nasar, *Electric Drives*, CRC Press, Boca Raton, FL, 1998, Chapter 5, pp. 90–91.
6. Task Force, Computer models for representation of digital-based excitation systems, *IEEE Transactions*, EC-11(3), 607–615, 1996.
7. A. Godhwani, M.J. Basler, A digital excitation control system for use on brushless excited synchronous generators, *IEEE Transactions*, EC-11(3), 616–620, 1996.
8. Task Force, Under-excitation limiter models for power system stability studies, *IEEE Transactions*, EC-10(3), 524–531, 1995.
9. G. Roger Bérubé, L.M. Hayados, R.E. Beaulien, A utility perspective on underexcitation limiters, *IEEE Transactions*, EC-10(3), 532–537, 1995.
10. G.K. Girgis, H.D. Vu, Verification of limiter performance in modern excitation control systems, *IEEE Transactions*, EC-10(3), 538–542, 1995.
11. H. Vu, J.C. Agee, Comparison of power system stabilizers for damping local mode oscillations, *IEEE Transactions*, EC-8(3), 533–538, 1993.
12. A. Murdoch, S. Venkataraman, R.A. Lawson, W.R. Pearson, Integral of accelerating power type PSS, Part. I + II, *IEEE Transactions*, EC-14(4), 1658–1672, 1999.
13. T. Hiyama, K. Miyazaki, H. Satoh, A fuzzy logic excitation system for stability enhancement of power systems with multimode oscillations, *IEEE Transactions*, EC-11(2), 449–454, 1996.
14. P. Hoang, K. Tomsovic, Design and analysis of an adaptive fuzzy power system stabilizer, *IEEE Transactions*, EC-11(2), 455–461, 1996.
15. G.P. Chen, O.P. Malik, H.Y. Qim, G.Y. Xu, Optimization technique for the design of linear optimal power system stabilizer, *IEEE Transactions*, EC-7(3), 453–459, 1992.
16. S. Chen, O.P. Malik, Power system stabilizer design using μ synthesis, *IEEE Transactions*, EC-10(1), 175–181, 1995.
17. Y. Caoi, L. Jiang, S. Cheng, D. Chen, O.P. Malik, G.S. Hope, A nonlinear variable structure stabilizer for power system stability, *IEEE Transactions*, EC-9(3), 488–495, 1994.

18. R. Asgharian, S.A. Tavakoli, A schematic approach to performance weight selection in design of robust H ∞ PSS using genetic algorithms, *IEEE Transactions*, EC-11(1), 111–117, 1996.

19. W.J. Wilson, J.D. Applevich, Co-ordinated governor exciter stabilizer design in multimachine power systems, *IEEE Transactions*, EC-1(3), 61–67, 1986.

20. M. Djukanovic, M. Novicevic, D. Dobrojevic, B. Babic, D. Babic, Y. Pao, Neural-net based coordinated stabilizing control for exciter and governor loops of low head hydroelectric power plants, *IEEE Transactions*, EC-10(4), 760–767, 1995.

21. M.B. Djukanovic, M.S. Calovic, B.V. Vesovic, D.J. Sobajic, Neuro-fuzzy controller of low head hydropower plants using adaptive-network based fuzzy inference system, *IEEE Transactions*, EC-12(4), 375–381, 1997.

22. G.K. Venayagamoorthy, R.G. Herley, A continually online trained neurocontroller for excitation and turbine control of a turbogenerator, *IEEE Transactions*, EC-16(3), 261–269, 2001.

23. A.R. Mahran, B.W. Hegg, M.L. El-Sayed, Co-ordinated control of synchronous generator excitation and static VAR compensator, *IEEE Transactions*, EC-7(2), 615–622, 1992.

24. J.D. Hurley, L.M. Bize, C.R. Mummart, The adverse effects of excitation system VAR and power factor controllers, *IEEE Transactions*, EC-14(4), 1636–1641, 1999.

25. T. Ise, Y. Murakami, K. Tsuji, Simultaneous active and reactive power control of SMES using Gto converters, *IEEE Transactions*, PWRD-1(1), 143–150, 1986.

26. E. Handschim, T. Stephanblome, New SMES strategies as a link between network and power plant control, *International IFAC Symposium on Power Plants and Power System Control*, Munich, Germany, March 9–11, 1992.

27. Q. Jiang, M.F. Coulon, The power regulation of a PWM type superconducting magnetic energy storage unit, *IEEE Transactions*, EC-11(1), 168–174, 1996.

28. A.H.M. Rahim, A.M. Mohammad, Improvement of synchronous generator damping through superconducting magnetic energy storage systems, *IEEE Transactions*, EC-9(4), 736–742, 1996.

29. S.C. Tripathy, R. Balasubramanian, P.S. Nair, Adaptive automatic generation control with superconducting magnetic energy storage in power systems, *IEEE Transactions*, EC-7(3), 439–441, 1992.

30. S.C. Tripathy, K.P. Juengst, Sample data automatic generation control with superconducting magnetic energy storage in power systems, *IEEE Transactions*, EC-12(2), 187–192, 1997.

31. J. Chatelain, B. Kawkabani, Subsynchronous resonance (SSR) countermeasures applied to the second benchmark model, *EPCS Journal*, 21, 729–739, 1993.

32. L. Wang, Damping of torsional oscillations using excitation control of synchronous generator: The IEEE second benchmark model investigation, *IEEE Transactions*, EC-6, 47–54, 1991.

33. A.H.M.A. Rahim, A.M. Mohammad, M.R. Khan, Control of subsynchronous resonant modes in a series compensated system through superconducting magnetic energy storage units, *IEEE Transactions*, EC-11(1), 175–180, 1996.

34. A Hefner, R. Singh, J. Lai, Emerging silicon-carbide power switches enable revolutionary changes in high voltage power conversion, *IEEE Power Electronics Society Newsletter*, 16(4), 10–13, 2004.

35. S. Rosado, M. Xiangfei, G. Francis, F. Wang, D. Boroyevich, Model-based digital generator control unit for a variable frequency synchronous generator with brushless exciter, *IEEE Transactions on Energy Conversion*, 23(1), 42–52, 2008.

36. D. Ursu, V. Gradinaru, B. Fahimi, I. Boldea, BLDC multiphase reluctance machines: A revival attempt with 2D FEA, standstill and running tests, *Record of IEEE-ECCE*, Denver, CO, 2013.

37. M.A. Abido, Optimal design of power system stabilizers using particle swarm optimization, *IEEE Transactions*, EC-17(3), 406–413, 2002.

38. P. Zhao, O.P. Malik, Design of an adaptive PSS based on recurrent adaptive control theory, *IEEE Transactions*, EC-24(4), 884–892, 2009.

7

Design of Synchronous Generators

7.1 Introduction

Most of the synchronous generator (SG) power is transmitted through power systems to various loads, but there are various stand-alone applications as well.

In this chapter, the design of SGs connected to power system is dealt with in some detail.

The successful design and operation of an SG depends heavily on agreement between the SG manufacturer and user in regard to technical requirements (specifications). Published standards such as ANSI C50.13 and IEC 34-1 contain these requirements for a broad class of synchronous generators (SGs). The Institute of Electrical and Electronics Engineers (IEEE) has recently launched two new, consolidated standards for high-power SGs [1]:

- C50.12—for large salient pole generators
- C50.13—for cylindrical rotor large generators

The liberalization of electricity markets led, in the last 10 years, to the gradual separation of production, transport, and supply of electrical energy. Consequently, to provide for safe, secure, and reasonable cost supply, formal interface rules—grid codes—have been put forward recently by private utilities around the world.

Grid codes do not align in many cases with established standards such as IEEE and American National Standards Institute (ANSI). Some grid codes exceed the national and international standards "Requirements on Synchronous Generators." Such requirements may impact unnecessarily on generator costs as they may not produce notable benefits for the power system stability [2].

Harmonization of International Standards with grid codes thus becomes necessary, and it is pursued already by joint efforts of SG manufacturers and interconnectors [3] to specify the turbo- and hydrogenerator parameters. Generator specifications parameters are, in turn, related to the design principles and, ultimately, to the costs of generator and of its operation (losses, etc.).

New, stricter, grid codes have been imposed recently for wind generators.

In this chapter, a discussion of turbogenerator specifications as guided by standards and grid codes is presented, in relation to fundamental design principles. Hydrogenerators pose similar problems in power systems, but their power share is notably smaller than that of turbogenerators, except for a few countries such as Norway, Brazil, and Romania. Then the design principles and a methodology for salient pole SGs and, respectively, of cylindrical rotor generators with numerical examples are presented in considerable detail.

Special design issues related to generator-motors for pump storage plants or self-starting turbogenerators are treated in a dedicated paragraph.

7.2 Specifying Synchronous Generators for Power Systems

Turbogenerators are still at the core of electric power systems. Their prime function is to produce the active power [1–6]. However, they are also required to provide (or absorb) reactive power, in a refined controlled manner, to maintain both frequency and voltage stability in the power system (see Chapter 6).

313

As the control of SGs has became lately faster and more robust with advanced nonlinear digital control methods, the parameter specification is about to change rather markedly.

7.2.1 Short-Circuit Ratio

The short-circuit ratio (SCR) of a generator is the inverse ratio of saturated direct axis reactance in p.u.

$$SCR = \frac{1}{x_{d(sat)}} \tag{7.1}$$

SCR has a direct impact on the static stability and on the leading (absorbed) reactive power capability of the SG.

A larger SCR means a smaller $x_{d(sat)}$, and thus almost inevitably a larger air gap. In turn, this requires more ampere-turns (mmf) in the field winding to produce the same apparent power.

As the permissible temperature rise is limited by the SG insulation class (class B, in general $\Delta T = 130°$), more excitation mmf means a larger rotor volume and thus a larger SG.

Also, the SCR has an impact on SG efficiency. An increase of SCR from 0.4 to 0.5 tends to produce a 0.02%–0.04% reduction in efficiency while it increases the machine volume by 5%–10% [3].

The impact of SCR on SG static stability may be illustrated by the expression of electromagnetic torque t_e (p.u.) in a lossless SG connected to an infinite power bus:

$$t_e \approx SCR \cdot E_0 \cdot V_1 \sin \delta \tag{7.2}$$

The larger the SCR, the larger the torque for given no-load voltage (E_0), terminal voltage V_1, and power angle δ (between E_0 and V_1 per phase). If the terminal voltage decreases, a larger SCR would lead to a smaller power angle δ increase for given torque (active power) and given field current.

If the transmission line reactance—including the generator step-up transformer—is x_e and V_1 is now replaced by the infinite grid voltage V_g behind x_e, the generator torque t'_e is

$$t'_e = SCR \times E_0 \times V_g \times \frac{\sin \delta'}{\left(1 + x_e/x_d\right)} \tag{7.3}$$

The power angle δ' is the angle between E_0 of generator and V_g of infinite power grid.

The impact of improvement of a larger SCR on maximum output is diminished as x_e/x_d increases.

Increasing SCR from 0.4 to 0.5 produces the same maximum output if the transmission line reactance ratio x_e/x_d increases from 0.17 to 0.345 at a leading power factor of 0.95 and 85% rated MW output.

Historically, the trend has been toward lower SCRs, from 0.8–1.0, 70 years ago to 0.58–0.65 in the 1960s and to 0.5–0.4 today. Modern—fast-response—excitation systems compensate for the apparent loss of static stability grounds. The lower SCRs mean lower generator volume, losses, and costs.

7.2.2 SCR and x'_d Impact on Transient Stability

The critical clearing time of a three-phase fault on the high-voltage side of the SG step-up transformer is a representative performance index for the transient stability limits of the SG tied to an infinite bus-bar.

The transient d-axis reactance x'_d (in p.u.) takes the place of x_d in Equation 7.3 to approximate the generator torque transients before the fault clearing. In the case in point $x_e = x_{Tsc}$ is the short-circuit reactance (in p.u.) of the step-up transformer. A lower x'_d allows for a larger critical clearing time and so does a large inertia. Air-cooled SGs tend to have a larger inertia/MW than hydrogen-cooled SGs as their rotor size is relatively larger and so is their inertia.

7.2.3 Reactive Power Capability and Rated Power Factor

A typical family of *V* curves is shown in Figure 7.1. The reactive power capability curve (Figure 7.2) and the *V* curves are more or less equivalent in reflecting the SG capability to deliver active and reactive power, or to absorb reactive power until the various temperature limitations are met (Chapter 5).

The rated power factor determines the delivered/lagging reactive power continuous rating at rated active power of the SG.

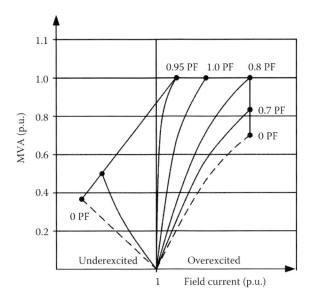

FIGURE 7.1 Typical V_{ee} curve family.

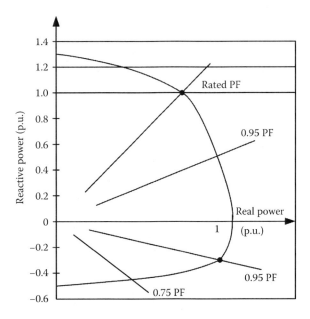

FIGURE 7.2 Reactive power capability curve.

The lower the rated (lagging) power factor, the larger the MVA per rated MW. Consequently, the excitation power is increased and the step-up transformer has to be rated higher (MVA). The rated power factor is placed, in general, in the interval 0.9–0.95 (overexcited) as a compromise between generator initial and loss capitalized costs and power system requirements. Lower values down to 0.85(0.8) may be found in air-(hydrogen)-cooled SGs. The minimum underexcited rated power factor is 0.95 at rated active power. The maximum absorbed (leading) reactive power limit is determined by the SCR and corresponds to maximum power angle and (or) to end stator core over-temperature limit.

7.2.4 Excitation Systems and Their Ceiling Voltage

Fast control of excitation current is needed to preserve SG transient stability and control its voltage. Higher ceiling excitation voltage, corroborated with low electrical time constants in the excitation system, provides for fast excitation current control.

Today's ceiling voltages are in the range of 1.6–3.0 p.u. There is a limit here dictated by the effect of magnetic saturation that makes ceiling voltages above 1.6–2.0 p.u. hardly practical.

This is more so as higher ceiling voltage means sizing the insulation system of the exciter or the rating of the static exciter voltage for maximum ceiling voltage at notably larger exciter's costs.

The debate over which is best: the AC brushless exciter or static exciter (which is specified also with a negative ceiling voltage of: –1.2 to 1.5 p.u.) is still not over. A response time of 50 ms in "producing" the maximum ceiling voltage is today fulfilled by the AC brushless exciters, but faster responses times are feasible with static exciters. However, during system faults, the AC brushless exciter is not notably disturbed as it draws its input from the kinetic energy of the turbine-generator unit.

In contrast, the static exciter is fed from the exciter transformer that is connected, in general, at SG terminals and seldom to a fully independent power source. Consequently, during faults, when the generator terminal voltage decreases, to secure fast rather undisturbed excitation current response, a higher voltage ceiling ratio is required.

Also, existing static exciters transmit all power to the SG excitation through the brush–slip-ring (mechanical) system with all the limitations and maintenance incumbent problems.

7.2.4.1 Voltage and Frequency Variation Control

As detailed in Chapter 6, the SG has to deliver active and reactive power with designed speed and voltage variations.

The size of the generator is related to the active power (frequency) and reactive power (voltage) requirements. Typical such practical requirements are shown in Figure 7.3.

In general, SGs should be thermally capable of continuous operation within the limits of *P/Q* curve (Figure 7.2) over the ranges of ±5% in voltage but not necessarily at the power level typical for rated frequency and voltage.

Voltage increase, accompanied by frequency decrease, means a higher increase in the *V/ω* ratio.

The total flux in the machine increases. A maximum of flux increase is considered practical and should be there by design. The SG has to be sized to have a reasonable magnetic saturation level (coefficient) such that the field mmf (and losses) and the core loss are not increased so much to compromise the thermal constrains in the presence of corresponding adjustments of active and reactive power delivery under these conditions.

To avoid oversizing the SG, the continuous operation is guaranteed only in the hatched are, at most 47.5–52 Hz.

In general, the +5% overvoltage is allowed only above rated frequency, to limit the flux increase in the machine to maximum 5%.

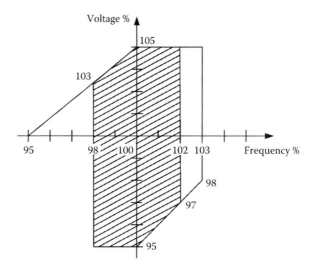

FIGURE 7.3 Voltage/frequency operation.

The rather large (±5)% voltage variation is met by SGs using tap changers on the generator step-up transformer (IEC).

7.2.4.2 Negative Phase Sequence Voltage and Currents

Grid codes tend to restrict the negative sequence voltage component at 1% (V_1/V_2 in %). Peaks up to 2% might be accepted for short duration by prior agreement between manufacturer and interconnector.

The SGs should be able to withstand such voltage imbalance that translates into negative sequence currents in the stator and rotor with negative sequence reactance $x_2 = 0.10$ (the minimum accepted by IEC) and a step-up transformer with a reactance $x_T = 0.15$ p.u. Then the 1% voltage unbalance translates into a negative sequence current i_2 (p.u. in%) of

$$i_2 = \frac{v_2}{x_2 + x_T} = \frac{0.01}{0.1 + 0.15} = 0.04 \text{ p.u.} = 4\% \tag{7.4}$$

The SG has to be designed to withstand the additional losses in the rotor damper cage, in the excitation winding and in the stator winding, produced by the negative sequence stator current. Turbogenerators above 700 MV seem to need explicit damper windings for the scope.

7.2.4.3 Harmonic Distribution

Grid codes specify the total harmonic voltage distortion (THD) at 1.5% and 2% in near 400 kV and respectively, in the near 275 kV power systems. Proposals are made to raise these values to 3(3.5)% in the voltage THD. The voltage THD may be converted into current THD and then into an equivalent current for each harmonic, considering that the inverse reactance x_2 may be applied for time harmonics as well.

For the fifth time harmonic, for example, a 3% voltage THD corresponds to a current i_5:

$$i_5 = \frac{v_5}{5 \cdot (x_2 + x_T)} = \frac{0.03}{5 \cdot (0.1 + 0.15)} = 0.024 \text{ p.u.} \tag{7.5}$$

7.2.4.4 Temperature Basis for Rating

Observable and hot-spot temperature limits appear in IEEE/ANSI standards, but only the former appears in IEC-60034 standards.

In principle, the observable temperature limits have to be set such that the hot-spot temperatures should not go above 130° for insulation class B and 155° for insulation class F.

In practice, one design could meet observable temperatures (in a few spots in the SG) but exceed the hot-spot limits of the insulation class. Or we may overrestrict the observable temperature, while the hot-spot are well below the insulation class limit.

Also, the rated cold coolant temperature has to be specified if the hot-spot temperature is maintained constant when the cold coolant temperature varies, as for temperature—following SGs where the observable temperature also varies.

Holding one of the two temperature limits as constant, with cold coolant (ambient) temperature variable, leads to different SG over (under) rating (Figure 7.4).

It seems thus reasonable to fix the observable temperature limit for a single cold coolant temperature and calculate the SG MVA capability for different cold coolant (ambient) and hot-spot temperatures.

This way, the SG is exploited optimally, especially for the "ambient following" operation mode.

7.2.4.5 Ambient: Following Machines

SGs that operate for ambient temperatures for –20° to +50° should have permissible generator output power, variable with cold—coolant temperatures. Eventually, peak (short-term) and base MVA capabilities should be set at rated power factor (Figure 7.5).

7.2.4.6 Reactances and Unusual Requirements

The already mentioned d-axis synchronous reactance x_d and d-axis transient reactance x'_d are key factors in defining static and transient stability and maximum leading reactive power rating of SG.

In general, it is practice that x_d and x'_d values are subject for agreement between vendors and purchasers of SGs, based on operating conditions (weak or strong power system area exciter performance, etc.).

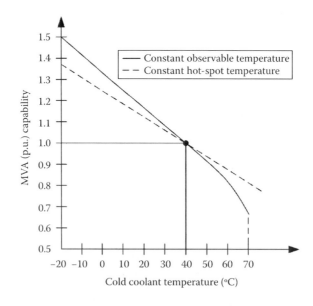

FIGURE 7.4 Synchronous generator megavoltampere rating versus cold coolant temperature.

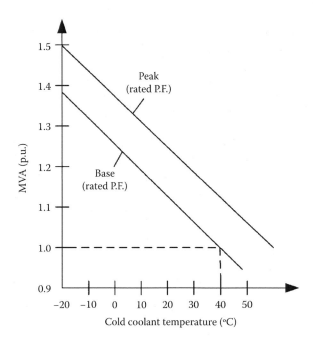

FIGURE 7.5 Ambient following SG ratings.

To limit the peak short-circuit current and circuit breaker rating it may be considered appropriate to specify (or agree upon) a minimum value of the subtransient reactances at the saturation level of rated voltage.

Also, the maximum value of unsaturated (at rated current) value of transient d-axis reactance x'_d may be limited based on unsaturated and saturated subtransient and transient reactances, see IEEE 100 (11).

There should be tolerances for these agreed-upon values of x''_d and x'_d, positive for first one ($+20\% \div 30\%$) and negative ($-20\% \div -30\%$) for the second one.

7.2.4.7 Start–Stop Cycles

The total number of starts is important to specify as the SG should by design prevent cyclic fatigue degradation.

According to IEC, IEEE, and ANSI trends, it seems that the number of starts should be as follows:

- 3,000 for base load SGs
- 10,000 for peaking load SGs or other frequently cycled units [1]

7.2.4.8 Starting and Operation as a Motor

Combustion turbines generator units may be started with the SG as a motor fed from static power converter of lower rating, in general. Power electronics rating, drivetrain losses, inertia, speed versus time, and the restart intervals have to be considered to assure that the generator temperatures are all within limits.

Pump-storage hydrogenerator units also have to be started as motors on no-load, with power electronics, or back-to-back from a dedicated generator that is accelerated simultaneously with the motor asynchronous starting.

The pumping action will force the SG to work as a synchronous motor and the hydraulic turbine–pump and generator–motor characteristics have to be optimally matched to best exploit the power unit in both operation modes.

7.2.4.9 Faulty Synchronization

SGs are also designed to survive without repairs after synchronization with ±10° initial power angle. Faulty synchronization (outside ±10°) may cause short-duration current and torque peaks larger than during sudden short circuits. As a result, internal damage of SG may result, and so inspection for damage is required.

Faulty synchronization at 120° or 180° out of phase with a low system reactance (infinite) bus might require partial rewind of the stator or (and) extensive rotor repairs.

A special attention should be paid to these aspects from design stage onward.

7.2.4.10 Forces

Forces in an SG occur due to the following reasons:

- System faults
- Thermal expansion cycles
- Double-frequency (electromagnetic) running forces

The relative number of cycles for peaking units (one start per day for 30 years) is shown in Figure 7.6 [4], together with the force level.

For system faults (short-circuit, faulty, or successful synchronization) forces have the highest level (100:1). The thermal expansion forces have an average level (1:1), while the double-frequency running forces are the smallest in intensity (1:10).

A base load unit would encounter a much smaller thermal expansion cycle count.

The mechanical design of SG should manage all these forces and secure safe operation over the entire anticipated operation life of the SG.

7.2.4.11 Armature Voltage

In principle, the armature voltage may vary in a 2:1 ratio without having to change magnetic flux or the armature reaction mmf, that is, for same machine geometry.

Choosing the voltage should be the privilege of the manufacturer, to enable him enough freedom to produce the best designs for given constraints.

The voltage level determines the insulation between the armature winding and the slot walls in an indirectly cooled SG. Not so in directly cooled stator (rotor) windings where the heat is removed through

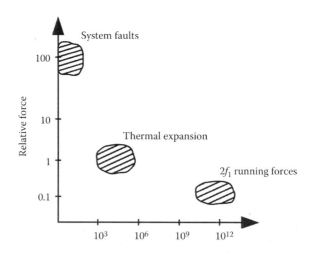

FIGURE 7.6   Forces cycles.

a cooling channel located in the slots. Consequently, directly cooled SG may be designed for higher voltages (say 28 kV instead of 22 kV) without paying a too-high price in cooling expenses.

However, for air-cooled generators, higher voltage may influence the Corona effect. Not so in hydrogen-cooled SGs, because of higher Corona-start voltage.

7.2.4.12 Runaway Speed

The runaway speed is defined as the speed the prime mover may be allowed to have if it is suddenly unloaded from full (rated) load. Steam (or gas) turbines are, in general, provided with quick action speed governors set to trip the generator at 1.1 times the rated speed.

Therefore, the runaway speed for turbogenerators may be set at 1.25 p.u. speed.

For water (hydro) turbines, the runaway speeds are much higher (at full gate opening):

- 1.8 p.u. for Pelton (impulse) turbines—SGs
- 2.0–2.2 p.u. for Francis turbines—SGs
- 2.5–2.8 p.u. for Kaplan (reaction) turbines—SGs

The SGs are designed to withstand mechanical stress at runaway speeds. The maximum peripheral speed is about 140–150 m/s for salient-pole SGs and 175–180 m/s for turbogenerators. The rotor diameter design is limited by this maximum peripheral speed.

The turbogenerator are built today in general in only two-pole configurations, either at 50 Hz or at 60 Hz.

7.2.4.13 Design Issues

SG design deals with many issues, but the most important ones are the following:

- Output coefficient and basic stator geometry
- The number of stator slots
- Design of stator winding
- Design of stator core
- Salient pole rotor design
- Cylindrical rotor design
- Open-circuit saturation curve
- Field current at full load
- Stator leakage inductance, resistance, and synchronous reactance calculation
- Losses and efficiency calculation
- Calculation of time constants and transient and subtransient reactances
- Cooling system and thermal design
- Design of brushes and slip-rings (if any)
- Design of bearings
- Brakes and jacks design
- Exciter design

Currently, design methodologies of SGs are today put in the computer codes, and they may contain optimization stages and interface with finite element software for refined calculation of electromagnetic, thermal, and mechanical stress, either for verification or for a final geometrical optimization design stage.

7.3 Output Power Coefficient and Basic Stator Geometry

The output coefficient C is defined as the SG kVA per cubic meter of rotor volume. The value of C (kVA/m^3) depends on machine power/pole, the number of pole pairs p_1, and the type of cooling, and it is much based on past experience (Figure 7.7).

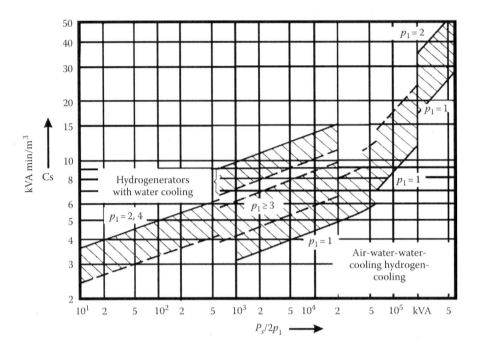

FIGURE 7.7 Output power coefficient for SGs.

The output power coefficient C may be expressed in terms of machine magnetic and electric loadings, starting from the electromagnetic power P_{elm}:

$$P_{elm} = 3 \cdot \frac{\omega_1}{\sqrt{2}} \cdot \left(W_1 \cdot K_{W1} \right) \cdot \Phi_1 \cdot I_1; \quad \omega_1 = 2\pi p_1 \cdot n_n \tag{7.6}$$

The ampereturns/m, or the electric specific loading (A_1) is

$$A_1 = \frac{6 \cdot W_1 \cdot I_1}{\pi \cdot D \cdot l_i} \ (A/m); \quad K_{W1}\text{—winding factor} \tag{7.7}$$

with l_i—the ideal stator stack length and D—the rotor (or stator bore) diameter.

The flux per pole Φ_1 is

$$\Phi_1 = \frac{2}{\pi} \cdot B_{g1} \cdot \frac{\pi \cdot D}{2 p_1} \cdot l_i \tag{7.8}$$

Using Equations 7.7 and 7.8 in Equation 7.6 yields

$$P_{elm} = \frac{\pi^2}{\sqrt{2}} \cdot K_{W1} \cdot A_1 \cdot B_{g1} \cdot l_i \cdot n_1 \cdot D^2 = C \cdot D^2 \cdot l_i \cdot n_n \tag{7.9}$$

Therefore,

$$C = \frac{\pi^2}{\sqrt{2}} \cdot K_{W1} \cdot A_1 \cdot B_{g1} \tag{7.10}$$

The air gap flux density B_{g1} relates to magnetic specific loading (saturation level), while A_1 defines the electric-specific loading.

C is not quite equal to power/rotor volume but is proportional to it. The proportionality coefficient is $\pi/4$.

Going further, we might define the average shear stress on rotor f_t (specific tangential force in N/m² or N/cm²):

$$f_t = \frac{F_t}{\pi \cdot D \cdot l_i} = \frac{T_{elm} \cdot (2/D)}{\pi \cdot D \cdot l_i} = \frac{P_{elm}}{2\pi n_1 \cdot \pi \cdot (D/2) \cdot D \cdot l_i} = \frac{1}{\pi^2} C \tag{7.11}$$

Therefore, the power output coefficient C is proportional to specific tangential force f_t exerted on the rotor exterior surface by the electromagnetic torque; with C in VA/m³, f_t comes into N/m². In general, C is given in kVA/m³.

As can be seen in Figure 7.7, C is given as a function of power per pole: $P_s/2p_1$ [5]. Direct water cooling in turbogenerators ($2p_1 = 2, 4$) allows for the highest output power coefficient.

The provisional rotor diameter D of SGs is, in general, limited by the maximum peripheral speed (140–150 m/s) with 44–55 kg/mm² yield point, typical rotor core materials.

This maximum peripheral speed U_{max} is to be reached at the runaway speed n_{max}, set by design as discussed earlier:

$$U_{max} = \pi \cdot D_{max} \cdot n_n \cdot \frac{n_{max}}{n_n} \tag{7.12}$$

For hydrogenerators, n_{max}/n_n is much larger than for turbogenerators.

It is imperative that the chosen diameter gives the desired flywheel effect required by the turbine design. As already discussed in Chapter 5, the inertia constant H in seconds is

$$H = \frac{J(\omega_1/p_1)^2}{2S_n} \tag{7.13}$$

where
J is the rotor inertia (in kg m²)
S_n is the rated apparent power in VA

H is defined in relation to the maximum speed increase allowed until the speed governor closes the fuel (water) input.

In general,

$$\frac{\Delta n_{max}}{n_n} \approx \sqrt{\frac{T_{GV}}{H} + 1} \tag{7.14}$$

T_{GV} is the speed governor (gate) time constant in seconds.

For *hydro* generators,

$$\frac{\Delta n_{max}}{n_n} < 0.3 - 0.4 \tag{7.15}$$

T_{GV} for hydrogenerators is in the orders 5–8 s.

For turbogenerators T_{GV} and $\Delta n_{max}/n_n$ are notably smaller (<0.1–0.15).

H for hydrogenerators varies, in general, in the interval from 3 to 8 s above 1 MVA per unit. *H* is often stated as GD_{ig}^2 (kg m²) where D_{ig} is twice the gyration radius of the rotor and *G* the rotor weight in kilogram:

$$H \approx \frac{1.37 \times 10^{-6} \, GD_{ig}^2 n_n^2}{S_n(\text{kVA})} \quad (\text{kWs/kVA}) \tag{7.16}$$

Approximately,

$$GD_{ig}^2 \approx \frac{\pi \gamma_{iron}}{8} \left[1 - \left(\frac{D_{ir}}{D} \right)^2 \right] \cdot D^4 \cdot l_p; \quad l_p \text{—rotor length} \tag{7.17}$$

where γ_{iron} = the iron-specific weight in kg/m³, D_{ir} = the interior rotor diameter, and D_{ir} = interior rotor diameter: zero in turbogenerators.

GD_{ig}^2 may be specified in t m². Alternatively, *H* in seconds may be specified or calculated from Equation 7.14 with T_{GV} and $\Delta n_{max}/n_n$ already specified.

With the rotor diameter provisional upper limit from Equation 7.12, the length of the stator core stack l_i may be calculated from Equation 7.9 if P_{elm} is replaced by S_n. Then with GD_{ig}^2 or *H* given, from Equations 7.16 and 7.17 and with length $l_p \approx l_i$, the internal rotor interior diameter $D_{ir} < D$ may be calculated.

The pole pitch may also be computed as follows:

$$\tau \approx \frac{\pi D}{2 p_1}; \quad p_1 = \frac{f_n}{n_n} \tag{7.18}$$

The ratio l_i/τ has to be placed in a certain interval to secure low enough stator copper losses as the stator end–connections length is proportional to the pole pitch τ. In general,

$$\lambda = l_i/\tau = 1\text{–}4 \quad \text{for } p_1 = 1$$
$$= 0.5\text{–}2.5 \quad \text{for } p_1 > 1 \tag{7.19}$$

The intervals for λ are rather large, leaving ample freedom to designer.

Though optimization design may be performed, it is good to have a good design start, and so λ has to be in the intervals suggested by Equation 7.19.

With the output power coefficient *C* given by Equation 7.10, and based on past experience, the air gap flux density fundamental B_{g1} is

$$B_{g1} = 0.75\text{–}1.05 \text{ T for cylindrical rotor SGs}$$

$$B_{g1} = 0.80\text{–}1.05 \text{ T for salient pole rotor SGs}$$

Correspondingly, with *C* from Figure 7.7 the linear current loading *A* (A/m) intervals may be calculated for various cooling methods. The orientative design current density intervals may also be specified (Table 7.1).

TABLE 7.1 Orientative Electric "Stress" Parameters

	Indirect Air Cooling	Indirect Hydrogen Cooling	Direct Cooling
A (KA/m)	30–80	90–120	160–200
Stator current density j_{cos} (A/mm²)	3–6	4–7	7–10 For water
Rotor current density j_{cor} (A/mm²)	3–5	3–5	6–13
		With stator and rotor direct cooling: (13–18) A/mm² and A = (250–300) kA/m	

7.4 Number of Stator Slots

The first requirement on the number of stator slots is to produce symmetrical (balanced) three-phase emfs.

For q equal to an integer number of slots per pole and phase, the number of stator slots N_s is

$$N_s = 2p_1 \cdot q \cdot m; \quad m = 3 \text{ phases}; \quad p_1 - \text{pole pairs} \tag{7.20}$$

A larger integer q is typical for turogenerators ($2p_1 = 2, 4$): $q > (4–6)$.

For low-speed generators, q may be as low as 3 but not less. For $q < 3$, 4 and large power hydrogenerators a fractionary q winding is adopted:

$$N_s = 2p_1 \cdot \left(b + \frac{c}{d}\right) \cdot 3; \quad q = b + \frac{c}{d} \tag{7.21}$$

To secure balanced emfs, the slot pitch number x between the start of phase A and B (C) is such that

$$\frac{2\pi}{N_s} \cdot p_1 \cdot x = K \cdot \frac{2\pi}{3}; \quad K\text{-integer} \neq 3^p \tag{7.22}$$

Replacing N_s from Equations 7.21 and 7.22 yields

$$\frac{\pi}{3} \cdot \left(\frac{d}{bd+c}\right) \cdot x = K \cdot \frac{2\pi}{3}; \quad x = 2 \cdot \frac{bd+c}{d} \cdot K \tag{7.23}$$

Now x has to be an integer and $2K$ has to be divisible by d. Also, d may not contain a 3^p factor as this is eliminated from K.

For fractionary windings not only N_s should be a multiple of 3 but also the denominator c of q should not contain 3 as a factor.

According to Equation 7.21, if N_s contains a factor of 3^p, then p_1 (pole pairs number) should contain it also so that it would not appear in c.

In large SGs, the stator core is made of segments (Chapter 4) as the lamination sheets size is limited to 1–1.1 m in width.

The number of slots per segment N_{ss}, for N_c segments is

$$N_{ss} = \frac{N_s}{N_c} \tag{7.24}$$

For details on stator core segments, revisit Chapter 3.

In general, it is advisable that N_{ss} be an even number so that N_s has to be an even number. But in such cases, apparently, only integer q values are feasible. For fractionary windings, N_{ss} may be an odd number and contain 3 as a factor.

Moreover, large stator bore diameter hydrogenerators have their stator core made of a few N_K sections that are wound at manufacturer's site and assembled together at the user's site.

Therefore, the number of slots N_s has to be divisible by both N_c and N_K.

In large SGs, the stator coil turns are made by transposed copper bars and in general there is one turn (bar) per coil. Therefore, the total number of turns for all three phases—$3 \cdot a \cdot W_a$—is equal to the number of slots N_s:

$$N_s = 3 \cdot a \cdot W_a \tag{7.25}$$

a—number of current paths in parallel; W_a—turns per path/phase.

On the other hand, the number of turns W_a per current path is related to the flux per pole and the resultant emf E_t per phase:

$$E_t = \pi\sqrt{2} \cdot f_n \cdot \left(W_a \cdot K_{W1} \right) \cdot \Phi_1 \tag{7.26}$$

$$\Phi_1 = \frac{2}{\pi} \cdot B_{g1} \cdot \tau \cdot l_{Fe}; \quad \tau \approx \pi D/2p_1 \tag{7.27}$$

τ—the pole pitch of stator winding.

At this stage, $l_{Fe} \approx l_i$, D, τ are known and B_{g1}—the air gap rated flux density is chosen in the interval given in the previous paragraph. The winding factor K_{W1} is

$$K_{W1} = K_{q_1} K_{y_1} \tag{7.28}$$

$$K_{q_1} \approx \frac{\sin \pi/6}{q \sin \pi/6q}; \quad K_{y_1} = \sin \frac{y}{\tau} \frac{\pi}{2} \tag{7.29}$$

y/τ—coil span/pole pitch.

For fractionary $q = (bd + c)/d$, q will be replaced in Equation 7.29 by $bd + c$.

From Figure 7.8, the emf (air gap emf) is

$$E_t \approx V_1 \sqrt{1 + x_{sl}(2\sin\varphi_n + x_{sl})} \approx (1.07 - 1.1) \cdot V_n \tag{7.30}$$

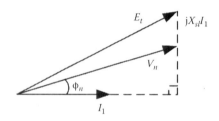

FIGURE 7.8 The total emf E_t.

The leakage reactance x_{sl} is, in general, less than 0.15 or

$$x_{sl} \approx x'_d \times (0.35 - 0.4) \tag{7.31}$$

This value of x_{sl} is only orientative and will be recalculated later in the design process.

V_n is root mean squared (RMS) of the rated phase voltage of the SG. The rated current I_n is

$$I_n = \frac{P_n}{3V_n \cos \varphi_n} \tag{7.32}$$

with φ_n—the rated power factor angle (specified).

The number of current paths in parallel depends on many factors such as the type of winding (lap or wave) and the number of stator sectors.

With tentative values for a and W_a obtained from Equations 7.26 through 7.30 and W_a rounded to a multiple of 3 (for three phases), the number of slots is calculated from Equation 7.25. Then, it is verified if N_s is divisible by the number of stator sections N_K.

The number of stator sections is, in general,

$$N_K = 2 \quad \text{for } D < 4 \text{ m}$$

$$N_K = 4 \quad \text{for } D = 4 \div 8 \text{ m} \tag{7.33}$$

$$N_K = 6 \, (8) \quad \text{for } D > 8 \text{ m}$$

To obtain a symmetrical winding for fractionary $q = b + c/d$, we need $2p_1/d =$ integer and, as pointed out earlier, $d/3 \neq$ integer.

With current paths,

$$d \leq \frac{2p_1}{a} \tag{7.34}$$

It is also appropriate to have a large value for d so that the distribution factor of higher space harmonics be small even if, by necessity, $c/d = 1/2$, $b > 3$.

For *wave* windings, the simplest configuration is obtained for

$$\frac{3c \pm 1}{d} = \text{integer} \tag{7.35}$$

Therefore, the best c/d ratios are as follows:

$$\frac{c}{d} = \frac{2}{5}, \frac{3}{5}, \frac{2}{7}, \frac{5}{7}, \frac{3}{8}, \frac{5}{8}, \frac{3}{10}, \frac{7}{10}, \frac{4}{11}, \frac{7}{11}, \frac{4}{13}, \frac{9}{13} \cdots \tag{7.36}$$

For $d = 5, 7, 11, 13, \ldots$

$$\frac{6c \pm 1}{d} = \text{integer} \tag{7.37}$$

with

$$\frac{c}{d} = \frac{1}{5}, \frac{4}{5}, \frac{1}{7}, \frac{6}{7}, \frac{2}{11}, \frac{9}{11}, \frac{2}{13}, \frac{11}{13} \cdots \tag{7.38}$$

A low level of noise with fractionary windings requires

$$3 \cdot \left(b + \frac{c}{d} \right) \pm \frac{1}{d} \neq \text{integer} \tag{7.39}$$

with $c/d = 1/2$ ($b > 3$) the subharmonics are canceled.

For more details on the choice of the number of slots for hydrogenerators, see Reference 6.

7.5 Design of Stator Winding

The main stator winding types for SGs have been introduced in Chapter 4. For turbo generators, with $q > 4$ (5), and integer q two layers windings with lap- or wave-chorded coils are typical. They are fully symmetric with 60° phase spread per pole.

Example 7.1: Integer q Turbogenerator Winding

Let us take a numerical example of a 2 pole turbogenerator with an interior stator diameter $D_{is} = 1.0$ m and with a typical slot pitch $\tau_s \approx 60\text{--}70$ mm. Let us find an appropriate number of slots for integer q and then build a two-layer winding for it.

The number of slots N_s is

$$N_s = 2p \cdot q \cdot m; \quad N_s \cdot \tau_s = \pi \cdot D_{is} \tag{7.40}$$

Therefore,

$$q = \text{integer}\left(\pi \cdot \frac{D_{is}}{\tau_s} \cdot \frac{1}{2p_1 \cdot m} \right) = \text{integer}\left(\frac{\pi \cdot 1.0}{0.07} \cdot \frac{1}{2 \cdot 1 \cdot 3} \right) \tag{7.41}$$

with $\tau_s = 0.0654$ m, $q = 8$.

With a stator stack length $l_{Fe} = 4.5$ m, $B_{g1} = 0.837$ T, $V_A = 12/\sqrt{3}$ kV, $f_n = 60$ Hz, and $E_t = 1.10$ V_n of Equation 7.30, $a = 2$ current paths, the number of turns per current path/phase, W_a, is Equation 7.26

$$W_a = \frac{E_t}{\pi\sqrt{2} \cdot f_n \cdot K_{W1} \cdot \Phi_1} \tag{7.42}$$

$$\Phi_1 = \frac{2}{\pi} \cdot B_{g1} \cdot \tau \cdot l_{Fe} = \frac{2}{\pi} \cdot 0.837 \cdot \frac{\pi}{2} \cdot 1 \cdot 4.5 = 3.7674 \text{ Wb/pole} \tag{7.43}$$

$$K_{W1} = \frac{\sin \pi/6}{8 \cdot \sin\left(\pi/(6 \cdot 8) \right)} \times \sin \frac{21}{24} \cdot \frac{\pi}{2} = 0.9556 \times 0.966 = 0.9230 \tag{7.44}$$

Therefore from Equation 7.42,

$$W_a = \frac{1.07 \times \left(12/\sqrt{3}\right)\cdot 10^3}{\pi\sqrt{2}\times 60\times 0.923\times 3.7674} = 8.012 \approx 8 = q \tag{7.45}$$

Fortunately, the number of turns per current path, which occupies just one pole of the two, is equal to the value of q. A multiple of q would be also possible. With $W_a = 8$, we do have one turn/coil, therefore the coils are made of single bars aggregated from transposed conductors.

From Equation 7.25, the number of stator slots N_s is

$$N_s = a\times W_a \times 3 = 2\times 8\times 3 = 48 \tag{7.46}$$

The condition $W_a = q$ (or kq) could in general be fulfilled with modified stator bore diameter or stack length or slot pitch.

In small machines, $W_a = kq$ with $k > 2$.

Building an integer q two-layer winding comprises the following steps:

- The electrical angle of emfs in two adjacent slots α_{es}:

$$\alpha_{es} = \frac{2\pi}{N_s}\cdot p_1 = \frac{2\pi}{48}\cdot 1 = \frac{\pi}{24} \tag{7.47}$$

- The number t of slots with in-phase emfs:

$$t = \text{largest common divisor } (N_s, p_1) = p_1 = 1 \tag{7.48}$$

- The number of distinct slot emfs:

$$N_s/t = 48/1 = 48 \tag{7.49}$$

- The angle of neighboring distinct emfs:

$$\alpha_{ef} = \frac{2\pi\cdot t}{N_s} = \frac{2\pi\cdot 1}{48} = \frac{\pi}{24} = \alpha_{es} \tag{7.50}$$

- Draw the star of slot emfs with $N_s/t = 48$ elements (Figure 7.9).
- Divide the distinct emfs in $2\times m = 6$ equal zones. Opposite zones represent the in and out slots of a phase in the first layer. The angle between the beginnings of phases A, B, and C is $2\pi/3$, clockwise.
- From each in and out slot phase coils are initiated in layer 1 and completed in layer 2, from left to right according to the coil span y: $y = \frac{20}{24}\tau = 20$ slot pitches (Figure 7.10).

Using wave–bar coils and two current paths in parallel, practically no additional connectors to complete the phase are necessary.

The single-turn bar coils with wave connections are usually used for hydrogenerators ($2p_1 > 4$) to reduce overall length of connectors—at the price of some additional labor. Here, the very large power of the SG at only 12 kV line voltage has imposed a single-turn coil winding.

Doubling the line voltage to 24 kV would lead to a 2 turn-coil winding where lap coils are preferable in general.

For the fractionary windings, so typical for hydrogenerators, after setting the most appropriate value of fractionary q in the previous paragraph, an example is worked out here.

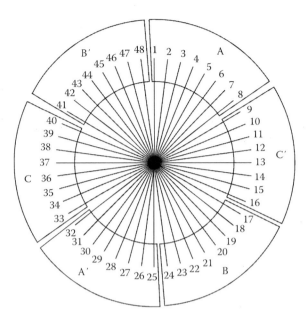

FIGURE 7.9   Electromagnetic force star for $2p_1 = 2$, $N_s = 48$.

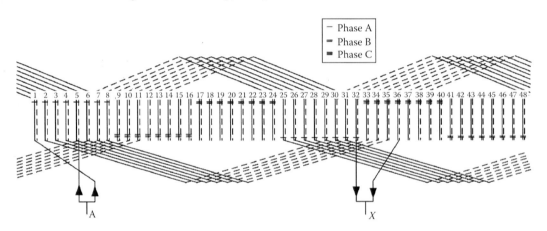

FIGURE 7.10   Two pole, wave-bar winding with $N_s = 48$ slots, $q = 8$ slot/pole/phase, $y/\tau = 20/24$.

Example 7.2: Fractionary q Winding

Consider the case of 100 MVA hydrogenerator designed at $V_{nl} = 15,500$ V (RMS line voltage, star connection), $f_n = 50$ Hz, $\cos \phi_n = 0.9$, $n_n = 150$ rpm, and $n_{max} = 250$ rpm.

Let us calculate the main stator geometry and then with $B_{g1} = 0.9$ T, the number of turns with one current path and design a single-turn (bar) coil winding with fractionary q.

Solution

For indirect air cooling, Figure 7.7, with the power per pole:

$$\frac{S_n}{2p_1} = \frac{100 \cdot 10^6}{40} = 2.5 \cdot 10^3 \text{ kVA min/pole} \tag{7.51}$$

the output power coefficient $C = 9$ kVA min/m³.

The maximum rotor diameter of Equation 7.12 for $U_{max} = 10.70$ m:

$$D = \frac{U_{max}}{\pi \cdot n_{max}} = \frac{140 \text{ m/s}}{\pi \cdot (250/60) \text{ rad/s}} = 10.70 \text{ m} \tag{7.52}$$

The ideal stack length l_i is calculated from Equation 7.9:

$$l_i \approx \frac{S_n}{C \cdot D^2 \cdot n_n} = \frac{100,000 (\text{kVA})}{9 (\text{kVA}_{min}/\text{m}^3)(10.70)^2 (\text{m}^2) 150 (\text{rpm})} = 0.647 \text{ m} \tag{7.53}$$

The flux per pole Φ_1 of Equation 7.8 is

$$\Phi_1 = \frac{2}{\pi} \cdot B_{g1} \cdot \tau \cdot l_i = \frac{2}{\pi} \cdot 0.9 \cdot \frac{\pi \cdot 10.7}{40} \cdot 0.647 = 0.3115 \text{ Wb/pole} \tag{7.54}$$

With $E_{t1}/V_{nph} = 1.07$ and an assumed winding factor $K_{W1} \approx 0.925$, the number of turns per current path W_a is Equation 7.26:

$$W_a = \frac{1.1 \cdot V_{nph}}{\pi\sqrt{2} \cdot f_n \cdot K_{W1} \cdot \Phi_1} = \frac{1.07 \times 15,500/\sqrt{3}}{\pi\sqrt{2} \cdot 50 \cdot 0.925 \cdot 0.3115} \approx 150 \text{ turns/path/phase} \tag{7.55}$$

For one current path, the total number of turns for all three phases (equal to the number of slots N_s) would be as follows:

$$N_s = 3 \cdot a \cdot W_a = 3 \cdot 1 \cdot 150 = 450 \text{ slots} \tag{7.56}$$

A tentative value of q_{ave} would be as follows:

$$q_{ave} = \frac{N_s}{2p_1 \cdot m} = \frac{450}{40 \cdot 3} = \frac{450}{120} = 3\frac{3}{4} \tag{7.57}$$

It is obvious that this value of q is not amongst those suggested in Equations 7.37 and 7.38, but still Equation 7.35 is half-fulfilled as $c = 3$, $d = 4$, and

$$\frac{3c-1}{d} = \frac{3 \cdot 3 - 1}{4} = 2 = \text{integer} \tag{7.58}$$

with 15 slots per segment ($N_{ss} = 15$), the total number of segments N_c per section is

$$N_c = \frac{N_s}{N_{ss} \cdot N_K} = \frac{450}{15 \times 6} = 5 \tag{7.59}$$

Therefore, the total number of segments of stator core is $N_c \cdot N_K = 5 \cdot 6 = 30$. For a 10.7 m rotor diameter, this is a reasonable value (lamination sheet width is less than 1.1–1.2 m).

Though $N_{ss} = 15$ slots/segment is an odd (instead of even) number, it is acceptable.

Finally, we adopt $q_{ave} = 3\frac{3}{4}$ for 40 pole single-turn bar winding with one current path.

To build the winding, we adopt a similar path as for integer q:

- Calculate the slot emf angle: $\alpha_{ec} = \frac{2\pi}{N_s} \cdot p_1 = \frac{2\pi}{450} \cdot 20 = \frac{4\pi}{45}$
- Calculate the highest common divisor t of N_s and p_1: $t = 10 = p_1/2$
- The number of distinct slot emfs: $N_s/t = 450/10 = 45$

- The angle between neighboring distinct emfs:

$$\alpha_{et} = \frac{2\pi \cdot t}{N_s} = \frac{2\pi \cdot 10}{450} = \frac{2\pi}{45} = \frac{\alpha_{ec}}{2}$$

- Draw the emf star observing that only 45 of them are distinct and every 10 of them are overlapping each other (Figure 7.11).

As there are only 45 (N_s/t) distinct emfs, it is enough to consider them, only, as the situation repeats itself identically 10 times. After 4 poles ($d = 4$ in $q = b + c/d$), the situation repeats itself.

- Calculate: $\dfrac{N_s/t}{3} = \dfrac{450/10}{3} = 15$ and start by allocating phase A 8 in—emfs (slots) and 7 out—emfs such tat the 8 ones and 7 ones are in phase opposition as much as possible. In our case in-slots for phase A are (1, 2, 3, 4, 24, 25, 26, 27) and out slots are (13, 14, 15, 35, 36, 37, 38).
- Proceed the same way for phases B and C by allowing groups of 8 and 7 neighboring slots to alternate. Also, the sequence (clockwise) to complete the circle is A, C′, B, A′, C, B′.
- The division of slots between the two layers is valid in layer 1; for layer 2, the allocation comes naturally by observing the coils span y:

$$y \leq \text{integer}\left(\frac{N_s}{2p_1}\right) = \text{integer}\left(\frac{450}{40}\right) = 11 \text{ slot pitches} \tag{7.60}$$

It is possible to choose $y = 9, 10, 11$ but $y = 10$ seems a good compromise in reducing the fifth and seventh space harmonics while not reducing the emf fundamental too much.

Note: For fractionary q, some of connections between successive bars of a bar—wave winding have to made of separate (nonwave) connectors.

For a minimum number of such additional connectors, y_1 (Figure 7.11) should be as close as possible to two times the pole pitch and $6q = $ integer.

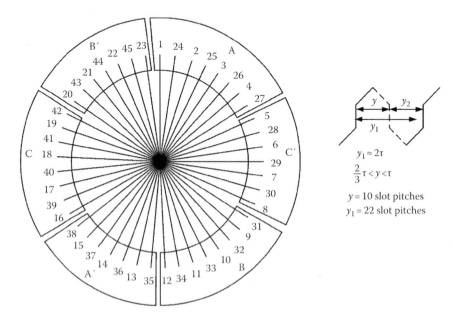

FIGURE 7.11   The emf star for $N_s = 450$ slots, $2p = 40$ poles for the first distinct 45 slots.

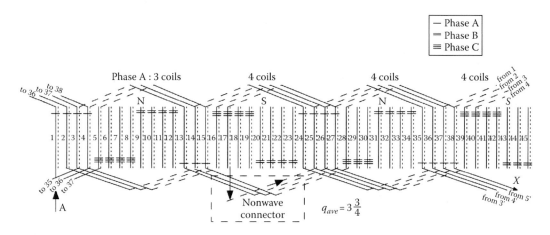

FIGURE 7.12 450 slot, 40 poles, $q = 3\frac{3}{4}$ wave bar winding/phase A for the first $N_s/t = 45$ slots.

In our case, $6q = 6 \cdot \dfrac{15}{4} = \dfrac{45}{2} \neq$ integer, thus the situation is not ideal. But the symmetry of the winding is notable as there are 10 identical zones of the winding, each spanning 4 poles.

The first 45 slots with all phases allocation completed, but only with phase. A coils shown, is given in Figure 7.12.

As Figure 7.12 shows, there is only one nonwave connector per phase for N_s/t section of machine.

A simple rule for allocation of slots per phases is apparent from Figure 7.12.

Based on the sequence A, C′, B, A′, C, B′ ..., we allocate for each phase group d slots for b groups and then c slots for one group and repeat this sequence for all the slots of the machine. Again, d should not be divisible by 3 for symmetry ($q = b + c/d$).

The allocation of slots to phase may be also done through tables [6], but the principle is the same as above.

The emf star has the added advantage of allowing simple verifications for phase balance by finding the position of resultant emf of each phase after adding the up (forward) and the opposite of down (backward) emfs.

It is also evident that the distribution factor formula (Equation 7.29) may be adopted for the purpose by noting that the number of vectors included should not be q but the denominator of q, that is, $bd + c$, in our case $3 \cdot 3 + 3 = 15 = 8 + 7$ vectors:

$$K_{q1} = \frac{\sin(\pi/6)}{(bd+c) \cdot \sin(\pi/6(bd+c))} \tag{7.61}$$

The chording factor K_{y1} formula (Equation 7.29) still holds.

7.6 Design of Stator Core

By now in our design process, the rotor diameter D, the stator core ideal length l_i, the pole pitch τ, and the number of slots N_s are all already calculated as shown in previous paragraphs.

To design the stator core, the stator bore diameter D_{is} is first required. But to accomplish this, the air gap g has to be calculated first, because

$$D_{is} = D + 2g \tag{7.62}$$

Calculating or choosing the air gap should account for the required SCR (or $1/x_d$):

- The reduction of air gap flux density harmonics due to slot openings such that to limit the emf time harmonics within standards requirements.
- Increased excitation winding losses with larger air gap.
- Reduced stator space harmonics losses in the rotor with larger air gap, for given stator slotting.
- Mechanical limitation on air gap that might vary during operation by at most 10% of its rated value.

The trend today is to impose smaller SCR (0.4–0.6), that is, smaller air gap, to reduce the excitation winding losses. Transient stability is to be preserved through fast exciter voltage forcing by adequate control.

With smaller air gap, care must be exercised in estimating the emf time harmonics and the additional rotor surface (or cage) losses.

Therefore, it seems reasonable to adopt the air gap based on a preliminary calculated value of the x_d:

$$x_d = x_{sl} + x_{ad} \tag{7.63}$$

At this point, x_{sl} may be assigned a value $x_{sl} = 0.1$–0.15 or $x_{sl} \approx (0.35 - 0.4)x'_d$ when $\left(x'_d\right)_{\max}$ is imposed as a specification.

The magnetization phase reactance X_{ad} (in Ω) is

$$X_{ad} = K_{ad} \frac{6\mu_0 \omega_1 \left(W_a \cdot K_{W1}\right)^2 \cdot \tau \cdot l_i}{\pi^2 \cdot g \cdot K_C \cdot (1 + K_{sd}) p_1 a} = x_{ad} \left(\frac{V_n}{I_n}\right)_{\text{phase}} \tag{7.64}$$

$$\left(I_n\right)_{\text{phase}} = \frac{S_n}{3\left(V_n\right)_{\text{phase}} \cos\varphi_n} \tag{7.65}$$

K_{ad} is a reduction coefficient of d axis magnetizing reactance when a salient pole rotor is used (Chapter 4):

$$K_{ad} \approx \frac{\tau_p}{\tau} + \frac{1}{\pi} \sin\left(\frac{\tau_p}{\tau} \cdot \pi\right); \quad \tau_p \text{—rotor pole shoe span} \tag{7.66}$$

Equation 7.66 is valid for constant air gap salient poles. For hydrogenerators, with reduced air gap, the air gap under the salient poles varies to yield a more sinusoidal air gap flux density distribution.

With $\tau_p/\tau \approx 0.62$–0.75 in general $K_{ad} > 0.9$ for uniform air gap, but it is lower for increased nonuniform air gap.

The Carter coefficient K_C, which includes the influence of slot openings, and the effect of radial channels in the stator core stack are also unknown at this stage of the design, but in general $K_C < 1.15$.

Finally, the magnetic saturation level is not known yet, but is known to be less than 0.25 ($K_{sd} < 0.25$). Therefore, basically, Equations 7.64 and 7.65 with assigned values of K_{ad}, K_c, and K_{sd} and known winding data W_a, K_{W1} (from the previous paragraph) provide a preliminary value for the air gap to secure the required value of x_d.

A rather traditional expression for air gap is

$$g = 4.0 \cdot 10^{-7} \frac{A \cdot \tau}{\left(x_d - 0.1\right) B_{g1}} \tag{7.67}$$

where
A = the linear current loading (A/m)
B_{g1} = the design air gap flux density fundamental (specified)
τ = pole pitch $\tau \approx \pi D/2p_1$
$x_{sl} = 0.1$ is the assigned value of stator leakage reactance in p.u.

Knowing the rated current I_n, the number of current path a, the number of turns/path, A, is

$$A \approx \frac{6 \cdot W_a \cdot a \cdot I_{na}}{\pi \cdot D} \tag{7.68}$$

For $SCR = 0.5$, $x_d = 2$ and $D = 10.7$ m, $a = 1$, $W_a = 150$, $I_{na} = 4000$ A (Example 7.2), $2p_1 = 40$ poles, $B_{g1} = 0.9$ T:

$$g = 4.0 \times 10^{-7} \cdot \frac{6 \cdot W_a \cdot a \cdot I_{na}}{(x_d - 0.1) \cdot 2p_1 \cdot B_{g1}} = \frac{4 \cdot 10^{-7} \times 6 \times 150 \times 4000}{40 \cdot 0.9 \cdot (2 - 0.1)} = 0.02105 \text{ m} \tag{7.69}$$

Now the pole pitch τ is

$$\tau = \frac{\pi(D + 2g)}{2p_1} = \frac{\pi(10.7 + 2 \cdot 0.02105)}{40} = 0.8435 \text{ m} \tag{7.70}$$

Note: The rather small specified SCR led to a high $x_d(x_d = 2.0)$ and thus to a rather small air gap for this 10.7 m rotor diameter with a 0.8453 m pole pitch.

Turbogenerators are characterized by larger air gap for same A, B_{g1}, and SCR as τ is notably larger.

Moreover, the smaller periphery length (smaller diameter) in turbogenerators imposes larger values of A than in hydrogenerators; one more reason for a larger air gap. Air gaps of 60–70 mm in 2 pole, 1.2 m rotor diameter turbogenerators are not uncommon.

This preliminary air gap value is to be modified if the desired x_d is not obtained, or some of the mechanical, emf harmonics or additional losses constraints are not met.

The stator terminal line voltage is chosen based on the following:

- Insulation costs
- Insulation maintenance costs
- Step-up transformer, power switches, and protection costs

In general, the higher the power, the higher the voltage.

Also, the voltage is higher for directly cooled windings because the transmission through conductor and slot insulation to the slot walls is not a main constraint anymore.

As starting point,

$$V_{nl} \approx 6\text{–}7 \text{ kV} \quad \text{for } S_n < 20 \text{ MVA}$$

$$10\text{–}11 \text{ kV} \quad \text{for } S_n \approx (20\text{–}60) \text{ MVA}$$

$$13\text{–}14 \text{ kV} \quad \text{for } S_n \approx (60\text{–}75) \text{ MVA} \tag{7.71}$$

$$15\text{–}16 \text{ kV} \quad \text{for } S_n \approx (175\text{–}300) \text{ MVA}$$

$$16\text{–}28 \text{ kV} \quad \text{for } S_n > 300 \text{ MVA}$$

Recently, 56 and 100 kV cable-winding SGs have been proposed.

7.6.1 Stator Stack Geometry

As radial or radial–axial cooling is used (Figure 7.13), there are n_c radial channels and each cooling channel is b_c wide.

The total iron length l_1 is

$$l_1 = l - n_c b_c \tag{7.72}$$

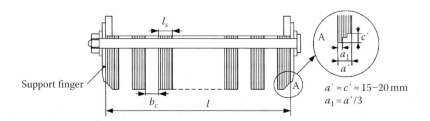

FIGURE 7.13   Stator core with radial channels.

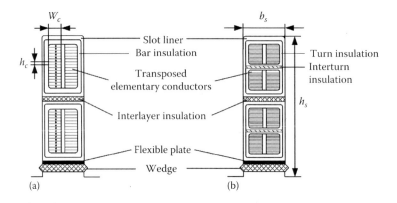

FIGURE 7.14   Stator conductors in slot (indirect cooling): (a) single-turn bar winding and (b) two-turn coil winding.

The ideal length l_i is aproximatively

$$l_i = \left[l - n_c \cdot b'_c - 2a' \cdot \left(1 - \frac{g}{g + c'/2} \right) \right] \cdot k_{Fe} \qquad (7.73)$$

b'_c is an equivalent cooling channel width, that is, smaller than b_c and dependent on air gap g.

The larger the air gap, the smaller the b'_c/b_c.

In general, $b_s = 8\text{–}12$ mm and the elementary stack width $l_s = 45\text{–}60$ mm. When the air gap g is larger than b_c, $b'_c/b_c < 0.2 - 0.3$ due to the large fringing flux.

K_{Fe} is the iron filling factor that accounts for the existing insulation layer between laminations. For 0.5 mm thick laminations, $K_{Fe} \approx 0.93\text{–}0.95$.

The open stator slots may house uni-turn (bar) or multiturn (two in general) coils placed in two layers (Figure 7.14).

The single- and two-turn coils are made of multiple rectangular cross-section conductors in parallel that have to be fully transposed (Figure 7.15) in large power SGs (Roebel bars). The elementary conductor height h_c is smaller than 2.5 mm in general.

The elementary conductors are transposed to cancel eddy currents induced by each of them in the others and thus reduce drastically the total skin effect AC resistance factor (more details in the forthcoming paragraph on stator resistance). The transposition provides for positioning each elementary conductor in all the positions of the other conductors, along the stack length. The transposition step along stack length is above 30 mm, and there should be as many transpositions as many elementary conductors are used to make a turn.

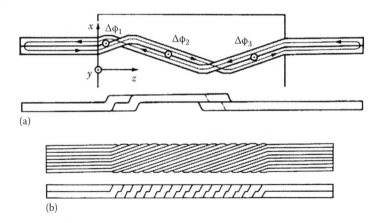

(a)

(b)

FIGURE 7.15 Roebel bar: (a) two conductors and (b) complete Roebel bar.

The thickness of various insulation layers depends on the terminal voltage and on the number of slots; but in general,

$$\frac{2W_c}{b_s} = \frac{\text{copper width}}{\text{slot width}} \approx 0.6\text{–}0.7$$

$$\frac{b_s}{\tau_s} = \frac{\text{slot width}}{\text{slot pitch}} \approx 0.35\text{–}0.55$$

$$\tau_s = \pi \cdot D_{is}/6p_1q_1 \tag{7.74}$$

$$\frac{h_s}{b_s} = \frac{\text{slot height}}{\text{slot width}} = 4\text{–}10$$

For direct cooling, the copper area per slot area is smaller than for indirect cooling because each elementary conductor has an interior channel for the coolant.

A single layer single turn per coil winding, as shown in Figure 7.16, exhibits sequences of two solid elementary conductors followed by a tubular one.

It is also possible to use only tubular elementary conductors.

The slot area A_{slotu} may be calculated by knowing the total current per slot, the design current density j_{cos}, and the total copper filling factor K_{fill}:

$$A_{slotu} = \frac{6 \cdot W_a \cdot I_n}{N_s \cdot K_{fill} \cdot j_{cos}}; \quad A_{slot} = b_s \cdot h_s \tag{7.75}$$

In general, the output power coefficient secures values of ampere-turns per slot that lead to fulfilling constraints Equation 7.74.

The design current density depends on the adopted cooling system; and for start values, Table 7.1 may be used. Also, it should be noted that the terminal voltage impresses lower limits on slot width with orientative values from 15 mm below 6 kV (line voltage) up to 35–40 mm at 24 kV. The slot total filling factor goes down from values of up to 0.55 below 6 kV to less than 0.3–0.35 at 20 kV and higher, for indirect cooling. Smaller values of K_{fill} are practical for direct cooling windings.

With stator bore diameter D_{is}, number of stator slots N_s, rated path current I_n, number of turns per current paths in parallel W_a, already assigned K_{fill}, j_{cos}, from Equations 7.75 with 7.74, the rectangular

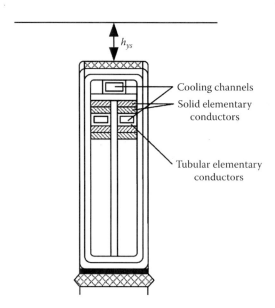

FIGURE 7.16   Single-layer winding with direct cooling.

stator slot may be sized by calculating h_s and b_s. Finally, all insulation layers are accounted for and a more exact filling factor is obtained.

The stator yoke height h_{ys} is simply

$$h_{ys} \approx \frac{B_{g1}}{B_{y1}} \cdot \frac{\tau}{\pi}; \quad B_{y1} = 1.3 - 1.7 \text{ T} \tag{7.76}$$

where τ is the stator pole pitch and B_{y1} is the design stator yoke flux density.

As the slots are rectangular, the teeth are rather trapezoidal, therefore the tooth flux density B_{t1} varies along the radial direction. The maximum value B_{tmax} occurs approximately at the slot top:

$$B_{tmax} \approx B_{g1} \cdot \frac{\tau_s}{\tau_s - b_s} \approx 1.6 - 2.0 \text{ T} \tag{7.77}$$

In this equation, the reduction of tooth flux density due to the fringing flux lines through the slots is neglected.

Example 7.3: Stator Slot and Yoke Sizing

For same hydrogenerator as in Example 7.2 with S_n = 100 MVA, I_n = 4000 A, U_{nl} = 15 kV, $2p_1$ = 40, D = 10.7 m, l_i = 0.647 m, N_s = 450, air gap g = 2.1 × 10^{-2} m, B_{g1} = 0.9 T, W_a = 150 turns/current path, a = 1 current paths, indirect air cooling, the size of the stator slot, yoke, find the stator core outer diameter D_{os}.

Solution

For indirect air cooling, a total slot filling factor is adopted K_{fill} = 0.4.

The current density (Table 7.1) is j_{cos} = 6.0 A/mm².

From Equation 7.75, the slot useful area A_{slotu} is

$$A_{slotu} = \frac{6 \cdot W_a \cdot I_n}{N_s \cdot K_{fill} \cdot j_{cos}} = \frac{6 \cdot 150 \cdot 4000}{450 \cdot 0.4 \cdot 6 \cdot 10^6} = 3333 \cdot 10^{-6} \text{ m}^2$$

The slot pitch τ_s is

$$\tau_s = \frac{\pi(D+2g)}{N_s} = \frac{\pi(10.7 + 2 \cdot 0.021)}{450} = 74.95 \cdot 10^{-3} \text{ m}$$

The slot width is selected according to Equation 7.74

$$b_s = \tau_s \cdot 0.4 = 74.95 \cdot 10^{-3} \approx 30 \cdot 10^{-3} \text{ m}$$

The maximum tooth flux density is

$$B_{tmax} = B_{g1} \cdot \frac{\tau_s}{\tau_s - b_s} = 0.9 \cdot \frac{74.95}{74.95 - 30} = 1.5 \text{ T}$$

The slot height h_s may now determined from Equation 7.75:

$$h_s \approx \frac{A_{slotu}}{b_s} = \frac{3333 \cdot 10^{-6}}{30 \times 10^{-3}} = 111 \cdot 10^{-3} \text{ m}$$

The ratio $h_s/b_s = 111/30 = 3.7033 < 4$ as suggested in Equation 7.74.

The rather low h_s/b_s ratio tends to produce a low stator slot leakage inductance, that is, a reduction in x'_d also. As the maximum value of x'_d is limited for transient stability reasons, it may be adequate to retain this slot geometry.

The moderate B_{tmax} does not account for the further reduction of the tooth width in the wedge area. With stator yoke flux density $B_{ys} = 1.4$ T, the stator yoke height h_{ys} of Equation 7.76 is

$$h_{ys} = \frac{B_{g1}}{B_{ys}} \cdot \frac{\tau}{\pi} = \frac{0.9}{1.4} \cdot 74.95 \times 10^{-3} \cdot \frac{450}{40} \cdot \frac{1}{\pi} = 172.74 \times 10^{-3} \text{ m}$$

The external stator diameter:

$$D_{os} = D + 2g + 2h_s + 2h_{ys} = 10.7 + 2 \cdot 0.0201 + (2 \times 111 + 2 \times 172) \cdot 10^{-3} = 11.309 \text{ m}$$

In general, the stator yoke height h_{ys} should be larger than the slot height h_s to avoid large noise and vibration at $2f_n$ frequency.

7.7 Salient: Pole Rotor Design

Hydrogenerators and most industrial generators make use of salient pole rotors. They are also to be found in some wind-generators above 2 MW/unit.

The air gap under the rotor pole shoe gets larger toward the pole shoe ends (Figure 7.17).

In general, $g_{max}/g = 1.5$–2.5 to make the air gap flux density, produced by the field current, rather sinusoidal.

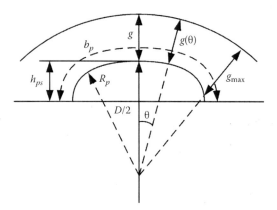

FIGURE 7.17 Variable air gap salient pole.

Ideally,

$$g(\theta) = \frac{g}{\cos(\theta p_1)} \tag{7.78}$$

In reality, the pole shoe may be cut from 1 to 1.8 mm laminations along a circle with radius $R_p < D/2$ where D is the rotor diameter at minimum air gap (g).

The pole shoe b_p per pole pitch τ is, in general, a compromise between leaving enough room for field coils and limiting the interpole flux linkage:

$$\alpha_i = \frac{b_p}{\tau} \approx 0.66 - 0.75 \tag{7.79}$$

In general, α_i increases with the pole pitch τ reaching 0.75 at $\tau = 0.8$ m.

Also, the ratio between the pole shoe span b_p and the stator slot pitch τ_s should be $b_p/\tau_s > 5.5$ to avoid notable pulsations in the emf due to stator slotting. With $q \geq 3$, this condition is met automatically for all values of α_i in Equation 7.79.

Given the central and maximum air gaps (g and g_{max}), rotor diameter D and the pole shoe span b_p, the radius R_p of the pole shoe shape is approximately

$$R_p = \frac{D}{2} \cdot \frac{1}{\left(1 + \left(4 \cdot D \cdot (g_{max} - g)\right)/b_p^2\right)} < \frac{D}{2} \tag{7.80}$$

The cross-section through a salient rotor pole is shown in Figure 7.18.

The length of pole body made of 1–2 mm thick diecasting laminations l_p is made smaller than stator core total length l by around 50–80 mm, while the end plates (Figure 7.18), l_{ep}, made of solid iron, are $l_{ep} = 50$–120 mm.

Therefore, the total ideal length of rotor pole l_{pi} is

$$l_{pi} = l - (0.05 - 0.08) + l_{ep} > l \tag{7.81}$$

The effective iron length of rotor l_{pFe} is

$$l_{pFe} \approx l_{pi} \cdot K_{Fe} \tag{7.82}$$

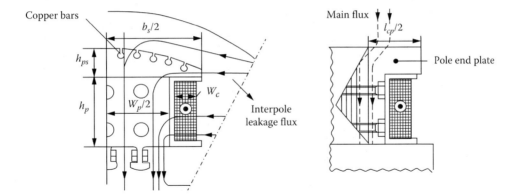

FIGURE 7.18 Salient rotor pole construction.

The lamination filling factor (due to insulation layers) $K_{Fe} \approx 0.95$–0.97, for lamination thickness going from 1 mm to 1.8 mm.

The total length of rotor l_{pi} is still larger than the stator stack length l, to reduce further the flux density in the rotor pole body whose width W_p (Figure 7.18) is, in general,

$$W_p \approx (0.45 - 0.55) \cdot \tau \tag{7.83}$$

The wound rotor pole height h_p per pole pitch τ ratio K_h decreases with the pole pitch τ and with increased average air gap flux density:

$$K_h = \frac{h_p}{\tau} \tag{7.84}$$

In general, for $B_{gav} = 0.7$ ($B_{gl} = 0.9$ T), K_h starts from 0.3 at $\tau = 0.4$ m and ends at 0.1 for $\tau = 1$ m. Higher values of K_h may be used for smaller air gap flux densities.

To design the field winding, the rated, V_{fn}, and peak, V_{fmax}, voltages have to be known, together with field pole mmf $W_f I_f$. By I_{fn}, we mean the excitation current required to produce full voltage at full-load and rated power factor. At this stage of the design method, I_{fn} is not known and it may not be calculated rigorously, because the rotor pole and yoke design is not finished.

But a preliminary design of rotor pole and yoke is feasible here.

Example 7.4: Salient Pole Rotor Preliminary Design

For the data in Example 7.3, let us design the rotor salient pole. The ratio $g_{max}/g = 2.5$.

Solution

Knowing the pole pitch τ and choosing a conservative $\alpha_i = 0.7$, from Equation 7.79 the pole width b_p is (Example 7.3):

$$b_p \approx \alpha_i \cdot \tau = 0.7 \cdot \frac{\pi(10.7 + 2 \cdot 0.021)}{40} = 0.7 \cdot 0.843 = 0.59 \text{ m}$$

The radius of rotor pole shoe R_p of Equation 7.80 is

$$R_p = \frac{D}{2} \cdot \frac{1}{\left(1 + \left(4 \cdot D \cdot \left(g_{max} - g\right)\right)\right)/b_p^2} = \frac{10.7}{2} \cdot \frac{1}{1 + \left(4 \cdot 10.7\right)/0.59^2 \cdot \left(2.5 - 1\right) \cdot 2.1 \cdot 10^{-2}} = 1.0978 \, \text{m}$$

The rotor pole shoe height at center h_{ps} (Figure 7.17) should be large enough to accommodate the damper winding and is proportional to the pole pitch:

$$\frac{h_{ps}}{\tau} \approx 0.1 \quad \text{for } \tau = 0.3 \, \text{m}$$

$$\approx 0.2 \quad \text{for } \tau = 1 \, \text{m}$$

The pole body width W_p is chosen from Equation 7.83:

$$W_p = 0.5 \cdot \tau = 0.5 \cdot 0.843 = 0.4215 \, \text{m}$$

Consequently, the space left for coil width W_c is

$$W_c = \frac{b_p - W_p}{2} = \frac{0.59 - 0.4215}{2} = 0.08425 \, \text{m}$$

The pole body (and coil) height $h_p = K_h$, $\tau = 0.18$, $0.843 = 0.1517$ m.

Therefore, with a total coil filling $K_{fill} = 0.62$ design current density $j_{cor} = 10$ A/mm^2, the ampere-turns of field coil per pole are as follows:

$$W_F I_{Fn} = j_{cor} \cdot W_c \cdot h_p \cdot K_{fill} = 10 \cdot 151.7 \cdot 84.25 \cdot 0.62 = 79240 \, \text{At}$$

On the other hand, the stator rated mmf per pole F_{1n} is

$$F_{1n} = \frac{3\sqrt{2} \cdot W_a \cdot K_{W1} \cdot I_n}{\pi \cdot p_1} = \frac{3\sqrt{2} \cdot 150 \cdot 0.925 \cdot 4000}{\pi \cdot 20} = 37383, \, \text{At/pole}$$

As

$$\left(F_{1n}/W_F I_{Fn}\right)^{-1} = \frac{79240}{37383} = 2.12$$

there are chances that the calculated rated field pole m.m.f. $W_F I_{Fn}$ will suffice for rated power, rated voltage, and rated power factor.

However, let us also note that the rated current density was raised to 10 A/mm^2 in the rotor, in contrast to 6 A/mm^2 in the stator. The much shorter end connections justify this choice. Later in the design, the rather exact $W_F I_{Fn}$ value will be calculated.

The rotor yoke design is basically similar to the stator yoke design, but there is an additional, leakage (interpole) magnetic flux to consider. Later on, it will be calculated in detail but for now a 10%–15% increase in polar flux is enough to allow for the rotor yoke radial height h_{yr} preliminary calculation:

$$h_{yr} \approx \frac{B_{g1}}{B_{yr}} \cdot \frac{\tau}{\pi} \left(1 + K_{leak}\right) = \frac{0.9}{1.5} \cdot \frac{0.843}{\pi} \cdot \left(1 + 0.12\right) = 0.1804 \, \text{m}$$

This is a conservative value.

Though the design methodology can produce a detailed analytical calculation of no-load and on-load magnetization curves, only FEM can provide rather exact distributions of flux density in the various parts of the machine for given operation conditions.

7.8 Damper Cage Design

Stator space mmf harmonics of order 5, 7, 11, 13, 17, 19, … as well as air gap permeance harmonics due to slot openings, induce voltages and thus produce currents in the rotor damper winding. These stator mmf-aggregated space harmonics are reduced drastically by fractionary windings ($q = b + c/d$) whose first slot harmonics is $6(bd + c) \pm 1$. When $bd + c > 9$, these harmonics are negligible; and thus, it is feasible to use the same slot pitch in the stator τ_s and in the rotor τ_d: $\tau s = \tau_d$. However, for integer q or $bd + c < 9$ or $q = b + 1/2$:

$$\tau_s \neq \tau_d \tag{7.85}$$

Otherwise, the induced currents in the damper windings by the stator slotting harmonics are augmented when $\tau_s = \tau_r$.

For these cases, it is recommended [7] that

$$\tau_r < \tau_s$$

$$\frac{2\tau}{\tau_r} \geq 6q \pm 1 + c; \quad c = 0 \text{ for } q = \text{integer} \tag{7.86}$$

$$b_p - 2\tau_r = \frac{2k\tau}{6q + 1}$$

In Reference [6], the condition $N_s/p_1 = 2K_1 \pm 1/2$ is demonstrated to lead to the reduction of bar-to-bar currents due to stator first slot opening harmonic. But the second slot opening harmonic ($\nu = 2$) may violate this condition.

The number of damper bars per pole N_2 is

$$N_2 \approx \frac{b_p}{\tau_r} - 1 \tag{7.87}$$

In some cases, the damper cage may be left out, but then the pole shoe (at least) should be made of solid mild steel.

The cross-section of the damper cage bars per pole represents a fraction of stator slot area per pole:

$$A_{\text{bar}} = \frac{1}{N_2} \cdot (0.15 - 0.3) \cdot \frac{6 \cdot W_a \cdot I_n}{2p_1 \cdot j_{\cos}} \tag{7.88}$$

The cage bars are round and made of copper or brass, therefore their diameters d_{bar} are standardized:

$$d_{\text{bar}} = \sqrt{\frac{4}{\pi} \cdot A_{\text{bar}}} \tag{7.89}$$

The cage bars are connected through partial or integral end rings.

The cross-section of end ring A_{ring} is about half the cross-section area of all bars under a pole:

$$A_{\text{ring}} \approx (0.5 - 0.6) \cdot N_2 \cdot A_{\text{bar}} \tag{7.90}$$

The complete end rings, though useful in providing $x_d'' \approx x_q''$, $V_0 \approx \varepsilon_1 \left(\frac{2x_q''}{x_d''} - 1 \right)$ (thus limiting the open phase voltage in line to line short circuits) and q axis current damping during transients, hamper the free axial circulation of cooling agent between rotor poles.

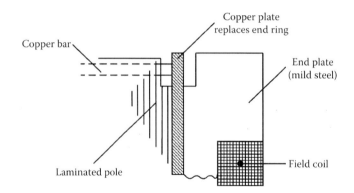

FIGURE 7.19   Copper end plate replaces end ring.

It is thus practical to use copper end plates that follow the shape of the poles and extend below the first row of pole bolts. They are located right between the laminated rotor pole core and the end plate made of steel (Figure 7.19).

For good contact with the copper bars, the copper end-plate should have a thickness of about 10 mm, in general [6].

The copper plate plays the role of the complete end ring but without obstructing the cooler axial flow between the rotor poles. Also, it is mechanically more rugged than the latter.

7.9 Design of Cylindrical Rotors

The cylindrical rotor is made, in general, from solid iron with milled slots over about 2/3 of periphery such that to produce $2p_1$ poles with distributed field coils in slots (Chapter 4)—Figure 7.20. Slots are radial and open in Figure 7.20.

According to Chapter 4, Equation 4.23, the air gap flux density produced by the distributed field winding is

$$B_{g\nu}(x) = K_{f\nu} \cdot B_{gav} \cdot \cos \nu \frac{\pi}{\tau} x \qquad (7.91)$$

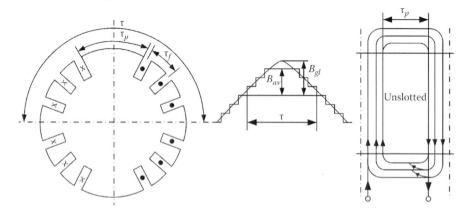

FIGURE 7.20   Two-pole cylindrical rotor with field coils.

in rotor coordinates, with K_{fv}

$$K_{fv} \cong \frac{8}{v^2 \pi^2} \cdot \frac{\cos v\left(\tau_p/\tau\right) \cdot \left(\pi/2\right)}{1 - v\tau_p/\tau} \tag{7.92}$$

Only odd harmonics are present if all poles are balanced. For the fundamental, the form factor K_{fv}

$$\left(K_{f1}\right)_{\left(\tau_p/\tau\right)=1/3} = \frac{8}{\pi^2} \cdot \frac{\cos\left(\tau_p/\tau\right) \cdot \left(\pi/2\right)}{1 - \tau_p/\tau} = \frac{8}{\pi^2} \cdot \frac{\sqrt{3}/2}{1 - 1/3} = 1.0528 \tag{7.93}$$

The third harmonic is already reduced:

$$\left(K_{f3}\right)_{\left(\tau_p/\tau\right)=1/3} = \frac{8}{9 \cdot \pi^2} \cdot \frac{\cos 3\left(\tau_p/\tau\right) \cdot \left(\pi/2\right)}{1 - 3\tau_p/\tau} = -0.1415 \tag{7.94}$$

The stator and rotor slot opening air gap permeance influence on the excitation air gap flux density harmonics is not considered in Equation 7.92.

The rotor excitation slot pitch τ_f should be chosen in relation to stator slot pitch τ_s such that the stator emf harmonics and solid rotor eddy current losses are minimized.

Further on,

$$B_{av} = \frac{\mu_0 \cdot W_{fc} \cdot I_{fa} \times \left(N_{fp}/2\right)}{g \cdot K_c \cdot \left(1 + K_s\right)} \tag{7.95}$$

N_{fp} is the number of field winding slots per rotor pole; K_c is the Carter coefficient accounting for the apparent increase of air gap due to stator and rotor slotting and for the presence of radial cooling channels (if any). Magnetic saturation is accounted for by K_s, which in general is $K_s < 0.2$–0.25.

Example 7.5: Field Winding Design

Let us consider a 2 pole 30 MVA, 50 Hz, $\cos \varphi_n = 0.9$ lagging turbo generator. With a stator bore diameter $D_{is} = 0.85$ m, $N_{fp(\text{in the rotor})} = 12$ slots/pole, $B_{g1} = 0.825$ T, $A = 56,000$ A/m, SCR = 0.55, $V_{fn} = 500$ V, $V_{f\max} = 2V_{fn}$.

Design the pertinent field winding after calculating the necessary air gap g.

Solution

The ampere-turns/m A may be turned into mmf per pole F_{1n}:

$$\tau = \pi \cdot \frac{D_{is}}{2} = \pi \cdot \frac{0.83}{2} = 1.303 \text{ m}$$

$$F_{1n} \approx \frac{A \cdot \tau}{2} = \frac{56000 \cdot 1.303}{2} = 36349 \text{ At/pole}$$

The no-load equivalent field winding mmf fundamental per pole F_{f10} is

$$F_{f10} \approx SCR \cdot F_{1n} = 0.5 \times 36349 = 18242 \text{ At/pole}$$

With $K_s = 0.25$, $K_c = 1.1$, $N_{fp} = 12$, the air gap flux density produced at no-load by the field winding is

$$B_{g10} = \frac{\mu_0 \cdot F_{f10}}{g \cdot K_c \cdot (1 + K_s)} \qquad (7.96)$$

Therefore, the air gap g becomes

$$g = \frac{1.256 \cdot 10^{-6} \cdot 18242}{1.1 \cdot 1.25 \cdot 0.825} = 20.2 \times 10^{-3} \text{ m}$$

Consequently, from Equation 7.89 with Equations 7.85 and 7.87, the rotor pole slot mmf $W_{Fc}I_{fo}$ at no-load is

$$I_{fo}W_{fc} = \frac{2}{N_{fp}} \cdot \frac{F_{f10}}{K_{f1}} = \frac{2}{12} \cdot \frac{18242}{1.0528} = 2887.8 \text{ At/slot} \qquad (7.97)$$

At full load and rated power factor, the excitation mmf requirement is about two times larger than for no-load:

$$W_{fc} \cdot I_{fn} = 2W_{fc} \cdot I_{f0} = 5775.7 \text{ At/slot} \qquad (7.98)$$

The slot pitch of rotor slots τ_{fr} is

$$\tau_{fr} = \frac{\pi (D_{is} - 2g) \cdot (1 - \tau_p/\tau)}{2p_1 \cdot N_{fp}} = \frac{\pi (0.83 - 2 \cdot 0.02) \cdot (1 - 0.3)}{2 \cdot 12} = 0.07235 \text{ m}$$

The value of $\tau_p/\tau = 0.3$ is taken to avoid $\tau_{fr} = \tau_s$ (slot pitch in the stator with $q = 6$).

With the slot fill factor $K_{fill} = 0.5$ (profiled conductors), and a design current density $j_{cor} = 6$ A/mm², the rotor slot useful area A_{slotr} is

$$A_{slotr} = \frac{W_{fc}I_{fn}}{j_{cor} \cdot k_{fill}} = \frac{5775.7}{6 \cdot 0.5} = 1925.23 \text{ mm}^2$$

The slot width W_{sr} is

$$W_{sr} = 0.4 \cdot \tau_{fr} = 0.4 \times 0.07235 = 0.02894 \approx 0.029 \text{ m}$$

The slot useful height h_{sr} is

$$h_{sr} = \frac{A_{slotr}}{W_{sr}} = \frac{1925.23 \cdot 10^{-6}}{0.029} = 66.38 \cdot 10^{-3} \text{ m}$$

The aspect ratio of the slot is rather small ($h_{sr}/W_{sr} = 2.289$), and thus either the current density might be reduced or, if needed, higher then in Equation 7.98 field mmfs are feasible.

To finish the design, let us calculate the number of turns per coil and the conductor cross-section.

The field circuit rated voltage V_{fn} should be considered for designing the field winding, with the voltage ceiling left for field current forcing during transients to enhance transient stability limits with a rather small SCR = 0.5.

First the field winding resistance per pole R_{fp} has to be calculated:

$$R_{fp} = \rho_{co} \frac{N_{fp} \cdot l_{ave}}{I_{fn}} \cdot W_{fc} \cdot a_p \cdot j_{cor} \tag{7.99}$$

l_{ave}—the average length of turn. Approximately,

$$l_{ave} \approx 2 \cdot l_{pi} + \pi \cdot K_{av} \cdot \frac{\tau}{2}; \quad K_{av} \approx 0.5 - 0.7 \tag{7.100}$$

Considering a_p current paths in the rotor, the field voltage equation under steady state is

$$V_{fn} = R_f \cdot I_{fn} \tag{7.101}$$

where

$$R_f = \frac{R_{fp} \times 2p_1}{a_p^2} \tag{7.102}$$

Using Equations 7.99 and 7.100 in Equation 7.101 yields

$$V_{fn} = \rho_{co} \cdot \frac{N_{fp} \cdot l_{ave} \cdot j_{cor}}{W_{fp} \cdot \left(I_{fn}/a_p\right)} \cdot W_{fc}^2 \times \frac{2p_1}{a_p^2} \cdot I_{fn} = \rho_{co} \cdot N_{fp} \cdot l_{ave} \cdot j_{cor} \frac{2p_1}{a_p} \cdot W_{fc} \tag{7.103}$$

Equation 7.103 provides for the direct computation of the number of turns per field coil. The copper resistivity should be considered at rated temperature.

Example 7.6: Field Coils Sizing

Let us calculate, for the rotor in Example 7.5, the number of turns per field coil and the wire cross-section if the stator core total length l is 2.5 m. Also, $I_{fn}W_{fc} = 5775$ At/coil

Solution

The turn average length of Equation 7.100 is

$$l_{avef} = 2 \cdot \left(2.6 + \pi \cdot 0.6 \cdot \frac{1.303}{2} \right) \cong 7.65 \text{ m}$$

with $\rho_{co} = 2.15 \times 10^{-8}$ Ω m, $j_{cor} = 6$ A/mm^2, $a_p = 2$ current paths in parallel, $2p_1 = 2$, $N_{fp} = 12$ slots/rotor pole, $V_{fn} = 500$ V, from Equation 7.103 the number of turns per field coil (same for all) W_{fc} is

$$W_{fc} = \frac{V_{fn} \cdot a_p}{\rho_{CO} \cdot N_{fp} \cdot l_{ave} \cdot j_{cor} \cdot 2p_1} = \frac{500 \cdot 2}{2.15 \times 10^{-8} \cdot 12 \cdot 7.65 \cdot 6 \times 10^6 \cdot 2 \cdot 1} = 42.22 \text{ turn/coil}$$

Let us adopt $W_{fc} = 42$ turns/coil

The total field current I_{fn} comes from the known $I_{fn}W_c$:

$$I_{fn} = \frac{W_{fc} \cdot I_{fn}}{W_{fc}} = \frac{5755}{42} = 137.50 \text{ A}$$

The current per path (in the coils) I_{fna} is

$$I_{fna} = \frac{I_{fn}}{a_p} = \frac{137.5}{2} = 68.75 \text{ A}$$

The copper conductor cross-section A_{co} is

$$A_{co} = \frac{I_{fna}}{j_{cor}} = \frac{68.75}{6 \times 10^6} = 11.458 \cdot 10^{-6} \text{ m}^2$$

A single rectangular cross-section wire may be used.

The total rated power in the excitation winding P_{exn} is

$$P_{exn} = V_{fn} \cdot I_{fn} = 500 \times 137.5 = 68750 \text{ W} \tag{7.104}$$

For a 30 MW SG, this means only 0.229%.

The rather small air gap ($g = 20 \times 10^{-3}$ m), the moderate rated current density ($j_{cor} = 6 \times 10^6$ A/m^2) and the 2/1 ratio between full-load and no-load field mmf may justify the rather small power (0.229%) in the field winding.

7.10 Open-Circuit Saturation Curve

The open-circuit saturation curve represents basically the no-load generator phase voltage E_{10} as function of excitation current (or m.m.f.) I_f, at rated frequency:

$$E_{10} = \sqrt{2}\pi \cdot f_n \cdot W_a \cdot K_{W1} \cdot \frac{2}{\pi} \cdot B_{g1} \cdot \tau \cdot l_i = K_E \cdot \Phi_{1g} \tag{7.105}$$

Also at no-load, from Equations 7.27, 7.91, and 7.97, we obtain

$$\Phi_{1g} = \frac{2}{\pi} \cdot l_i \cdot \tau \cdot B_{g1} \; ; \; B_{g1} = K_{f1} \cdot B_{av}; \quad B_{av} = \frac{\mu_0 \left(W_f I_f \right)_{pole}}{g_a K_c \left(1 + K_s \right)} \tag{7.106}$$

The saturation factor K_s depends on I_f, that is, on B_{av} and the machine stator and rotor core geometry and the $B(H)$ curves of stator and rotor core materials. The form factor K_{f1} is (Chapter 4).

$$K_{f1} \approx \frac{4}{\pi} \cdot \sin\frac{\tau_p}{\tau} \cdot \frac{\pi}{2} \quad \text{for salient rotor poles}$$

$$K_{f1} \approx \frac{8}{\pi^2} \cdot \frac{\cos(\tau_p/\tau) \cdot (\pi/2)}{1 - (\tau_p/\tau) \cdot \pi} \quad \text{for cylindrical rotor poles} \tag{7.107}$$

The equivalent stator stack iron length l_i is Equation 7.73:

$$l_i \approx \left(l - n_c \cdot b_c' - 2a' \cdot \left(1 - \frac{g}{g + c'/2} \right) \right) \tag{7.108}$$

The Carter coefficient K_C is, in general, the product of at least two of three terms:

- K_{C1}—accounting for air gap increase due to stator slot openings
- K_{C2}—accounting for air gap increase due to rotor slotting (caused for damper cage slots or by the field winding slots)
- K_{C3}—accounting for the air gap increase due to radial channels opening b_c

When the air gap varies under the salient rotor pole shoe from g to g_{max}, in calculating K_{C1}, K_{C2}, and K_{C3}, an average air gap g_a is used:

$$g_a \approx \frac{2}{3}g + \frac{1}{3}g_{max} \tag{7.109}$$

The literature on induction machines abounds in analytical formulae for Carter coefficients. A simplified practical version is given here:

$$K_{C_1} = \frac{\tau_s + 10g_a}{(\tau_s - W_{ss}) + 10g_a}; \quad W_{ss}\text{—slot opening, } \tau_s\text{—stator slot pitch}$$

$$K_{C_2} = \frac{\tau_r + 10g_a}{(\tau_r - W_{rr}) + 10g_a}; \quad \tau_r\text{—rotor slotting pitch, } W_{rr}\text{—rotor slot opening} \tag{7.110}$$

$$K_{C_3} = \frac{\tau_c + 10g_a}{(\tau_c - \tau_b) + 10g_a}; \quad \tau_c\text{—radial channel average pitch, } b_c\text{—radial channel width}$$

$$K_C = K_{C_1} \cdot K_{C_2} \cdot (K_{C_3})^i \tag{7.111}$$

The value of i is $i = 1$ if the radial channels are present only in the stator, but it is $i = 2$ when they are present both on the stator and the rotor.

When the air gap is constant (cylindrical rotors) $g_a = g$, as expected.

We will proceed to an analytical calculation of open-circuit magnetization curve because we have all the components, for given air gap flux density B_{g1} (or B_{av}) defined in previous paragraphs. All but one: the interpole rotor leakage flux Φ_{rl}, which is dependent on the mmf drop along the air gap + stator teeth + yoke: $F_{rl} = F_{AA'}$ (Figure 7.21).

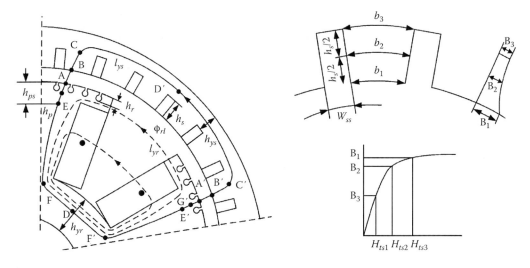

FIGURE 7.21 Average no-load flux path in the SG.

According to Ampere's law, for the average flux line:

$$\left(F_{10}\right)_{pole} = F_g + F_{ts} + F_{ys} + F_{tr} + F_{ps} + F_p + F_{yr}$$

$$AB \quad BC \quad CD \quad AG \quad GE \quad EF \quad FD'$$

(7.112)

also

$$F_{rl} = 2 \cdot \left(F_g + F_{ts} + F_{ys}\right) = F_{AA'}$$

$$F_{10} = F_{rl} + F_{rr}; \quad F_{rr} = 2\left(F_{tr} + F_{ps} + F_p + F_{yr}\right)$$

(7.113)

with

$$F_g = \frac{B_{g1}}{\mu_0} \cdot g_a \cdot K_C$$

$$F_{ts} = H_{ts}^{av} \cdot h_s$$

$$F_{ys} = H_{ys}^{av} \cdot l_{ys}^{av}$$

$$F_{tr} = H_{tr}^{av} \cdot h_r$$

$$F_{ps} = H_{ps}^{av} \cdot h_{ps}$$

$$F_p = H_p^{av} \cdot h_p$$

$$F_{yr} = H_{yr}^{av} \cdot l_{yr}^{av}$$

(7.114)

For given B_{g1} and machine geometry, the flux densities in various stator regions may be calculated.
The average value of field H_{ts}^{av} is calculated as follows for the trapezoidal stator teeth:

$$H_{ts}^{av} = \frac{H_{ts1} + 4H_{ts2} + H_{ts3}}{6}$$

(7.115)

H_{ts1}, H_{ts2}, and H_{ts3} correspond to the tooth flux densities B_1, B_2, B_3 in the three locations indicated in Figure 7.21.

$$B_1 \approx B_{g1} \cdot \frac{\tau_s}{b_1}; \quad b_1 = \tau_s - W_{ss}$$

$$B_2 \approx B_1 \cdot \frac{b_1}{b_2}$$

(7.116)

$$B_3 \approx B_1 \cdot \frac{b_1}{b_3}$$

The widths of the stator tooth b_1, b_2, b_3 at tooth top, middle, and bottom are straightforward. From the known lamination magnetization curves, H_{ts1}, H_{ts2}, and H_{ts3}, corresponding to B_1, B_2, and B_3, are obtained. A similar procedure may be used to calculate the mmf drop F_{tr} in the rotor tooth.
For the stator yoke the maximum value of the flux density is used

$$B_{ys}^{max} \approx \frac{B_{g1}}{h_{ys}} \cdot \frac{\tau}{\pi}$$

(7.117)

to obtain H_{ys} from same magnetization curve.

However, to account for the fact that lower flux density levels exist in the yoke and the lengths of various flux lines in this zone are different from each other, flux line length l_{ys}^{av} must be defined:

$$l_{ys}^{av} \approx K_{ys} \frac{\pi \left(D + 2g + 2h_s + h_{ys} \right)}{4 p_1}; \quad K_{ys} \approx \left(0.66 - 0.8 \right) \tag{7.118}$$

Also

$$l_{ys}^{av} \approx \frac{\pi \left(D - 2h_{ps} - 2h_p \right)}{4 p_1} \tag{7.119}$$

Realistic values of K_{ys} may be obtained through FEM or multiple magnetic circuit field distribution calculation methods [8].

The total pole flux in the rotor includes also the interpole leakage flux Φ_{rl} besides the air gap flux

Φ_{1g}:

$$\Phi_{1g} = \frac{2}{\pi} \cdot B_{g1} \cdot \tau \cdot l_i \tag{7.120}$$

$$\Phi_{1r} = \Phi_{1g} + 2\Phi_{rl}$$

It is to be recognized that not all the sections of the rotor pole encounter the entire rotor leakage flux Φ_{rl}, but the rotor yoke does. Calculating the dependence of leakage rotor flux Φ_{rl} on the mmf $\frac{1}{2} F_{AB} = F_{rl}$ Equation 7.113 needs either analytical or numerical flux distribution investigation.

However, as the tangential distance between neighboring rotor poles in air is notable, to a first approximation, we have

$$\Phi_{rl} = P_{rl} \cdot F_{rl} \tag{7.121}$$

There are quite a few analytical approximations for P_r—the permeance of the leakage interpolar flux [6,7].

Here, we use the similitude of the interpolar space with a semiclosed slot plus the air gap flux permeance known as zigzag (air gap) leakage:

$$P_{rl} = 2\mu_0 \cdot \left(\lambda_p + \lambda_f \right) \cdot l_{pi} \tag{7.122}$$

$$\lambda_p \approx \frac{h_{ps1}}{\left(b_{r1} + b_{r2} \right)} + \frac{1}{3} \frac{h_{ps2}}{\left(b_{r2} + b_{r3} \right)} + \frac{1}{3} \frac{h_{p1}}{\left(b_{r3} + b_{r4} \right)} \tag{7.123}$$

$$\lambda_f = \frac{5 g_a K_c / 2 b_{r1}}{5 + \left(4 g_a K_c / 2 b_{r1} \right)} \tag{7.124}$$

Once the geometry of the rotor is known, all variables in Equations 7.123 and 7.124 are given, and with F_{rl}—the corresponding mmf of Equation 7.113—also calculated, the interpolar leakage flux is obtained.

Now for the rotor pole shoe, pole body, rotor yoke average flux densities B_{ps}^{av}, B_p^{av}, B_{yr} calculation, the leakage flux has to be added to the air gap flux per pole:

$$B_{ps}^{av} \approx \frac{1/\pi \cdot B_{g1} \cdot \tau \cdot l_i + C_{ps} \cdot \Phi_{rl}}{l_{pi} \cdot (b_p/2)}; \quad C_{ps} = 0.3\text{--}0.5$$

$$B_p^{av} \approx \frac{1/\pi \cdot B_{g1} \cdot \tau \cdot l_i + C_p \cdot \Phi_{rl}}{l_{pi} \cdot (W_p/2)}; \quad C_p \approx 0.75\text{--}0.85 \tag{7.125}$$

$$B_{yr} \approx \frac{1/\pi \cdot B_{g1} \cdot \tau \cdot l_i + \Phi_{rl}}{l_{pi} \cdot h_{yr}}$$

l_{pi} is the total rotor iron length of Equation 7.81; all other dimensions are visible in Figure 7.22.

The coefficients C_{ps} and C_p account for the fact that only a part of the leakage flux adds in the pole shoe and in the pole regions.

With the rotor flux densities known, the corresponding rotor mmf F_{ps}, F_p, F_{yr} per pole may be calculated. Therefore, all the terms in Equation 7.112 may be calculated for given air gap flux density fundamental. Let us note that the field pole mmf fundamental F_{10} is related to the field pole mmf $(W_f I_f)_{pole}$ by

$$F_{10} = K_{f1} \cdot (W_f I_f)_{pole} \tag{7.126}$$

The translation of this mmf per pole into an equivalent stator mmf per pole F_{1d} is

$$F_{10} = F_{1d} \cdot K_{ad} = \frac{3\sqrt{2} \cdot W_1 \cdot K_{W1} \cdot I_d}{\pi \cdot p_1} \cdot K_{ad} \tag{7.127}$$

$$K_{ad} \approx \frac{\tau_p}{\tau} + \frac{1}{\pi} \cdot \sin\left(\pi \frac{\tau_p}{\tau}\right); \quad 0.8 < K_{ad} < 1.0 \tag{7.128}$$

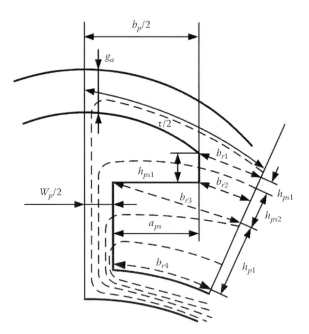

FIGURE 7.22 Rotor leakage flux permenace P_{rl} calculation.

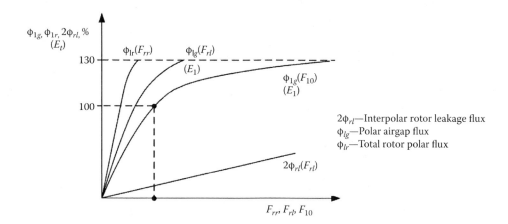

FIGURE 7.23 Partial no-load magnetization curves.

The d axis magnetization reactance reduction coefficient K_{ad} (Chapter 4) accounts for the rotor saliency, and it is equal to unity for cylindrical rotor.

The whole open-circuit saturation curve may be calculated without any iteration by just repeating the above computation sequence for ever higher values of B_{g1} until the no-load voltage E_1 reaches about 130% of the generator rated terminal phase voltage. The acquired data allow also for the representation of the so-called partial no-load magnetization curves (Figure 7.23).

The partial magnetization curves are in general used to calculate the rated field mmf at rated power, voltage and rated lagging power factor. Also, the no-load magnetization curve is essential to design and control of autonomous SGs.

The horizontal variable is either *total field mmf* F_{10} or the partial *stator $\pm$ air gap mmf*, F_{rl}, and, respectively, *the rotor mmf* F_{rr}. Also, Φ_{1g} is the air gap flux per pole, $2\Phi_{rl}$ is the total interpolar rotor leakage flux, and Φ_{1r} is the total flux per pole in the rotor ($\Phi_{1r} = 2\Phi_{rl} + \Phi_{1g}$).

The air gap flux Φ_{1g} is proportional to emf E_1 of Equation 7.105. Therefore, in p.u. values Φ_{1g} and E_1 are superimposed in Figure 7.23.

7.11 On-Load Excitation mmf F_{1n}

Calculating the on-load excitation F_{1n} per pole is essential in designing the SG in relation to field winding losses and overtemperatures.

Traditionally, there are two methods to calculate F_{1n}:

1. The Potier diagram method
2. The partial magnetization curve method

The Potier diagram is meant for cylindrical rotors, while the partial magnetization curve method is necessary for the salient-pole rotor.

What is needed in both methods is the rated armature mmf F_{a1}, the rated voltage, and the leakage stator reactance, x_{sl}. At this stage in the design, x_{sl} may be calculated.

For now, we consider it known (in general $x_{sl} = 0.09-0.15$ for all SGs and increases with power). The Potier reactance is

$$x_p \approx x_{sl} + 0.02 \text{ p.u.} \tag{7.129}$$

7.11.1 Potier Diagram Method

The diagram in Figure 7.24 is drawn in p.u. with rated terminal voltage V_n as base voltage. Also, the base field mmf corresponds to field mmf at rated voltage under no-load.

With the rated voltage along the vertical axis and the rated power factor angle, the phase of rated current is visualized.

Then, with x_p in p.u.—the segment AB—90° ahead of I_{1n}—the total (air gap) emf at rated load E_t is found:

$$E_t = \sqrt{v_1^2 + x_p^2 i_1^2 + 2 \cdot x_p \cdot i_1 \cdot v_1 \cdot \sin \varphi_n} \ \text{(p.u.)} \tag{7.130}$$

Then, the segment CD in the no-load saturation curve represents exactly E_t and OD—the corresponding field mmf.

Now we only have to add vectorially to $\overline{OD}$ the $\overline{ED}$ in phase with I_{1n} but with its value (in p.u.).

F_{an} is defined as the ratio between the stator pole rated mmf divided by the field mmf F_{10n} corresponding to rated voltage $v_n = 1$ (p.u.) at no-load.

Rotating $\overline{OE}$ until it reaches the abscise: $\overline{OF} = \overline{OE} = \overline{F_{1n}}$—the rated field mmf.

It is well understood that the whole method could be put into algebraic form and integrated into a rather simple computer program that calculates first the no-load saturation curve by advancing in 0.05 (or so) steps from zero to 130% of rated air gap flux/pole (or no-load voltage). This way, the graphical errors are removed to a great extent.

The vectorial addition of field (OD) and armature (DE) mmfs per pole to reach the resultant mmf is implicitly valid only if the SG rotor has no magnetic saliency. Even the cylindrical rotor, with slots in axis q (only), to house the field coils, has an up to 5% saliency, $x_{dm}/x_{qm} = 1.05$ under full-load and rated power factor.

For the salient pole rotor, the saliency $x_{dm}/x_{qm} = 1.3$–1.5 and such an approximation is not practical anymore.

This is how the partial magnetization curve method becomes necessary.

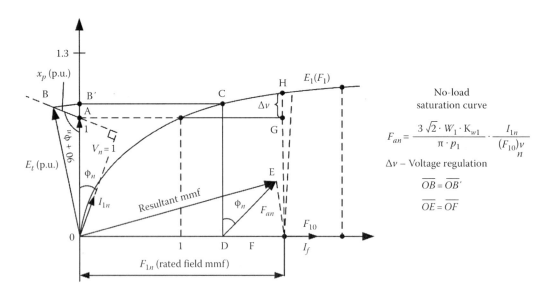

FIGURE 7.24 Rated field mmf calculation.

7.11.2 Partial Magnetization Curve Method

Within the frame of this method, the partial no-load magnetization curves (Figure 7.25) are first determined point by point up to 1.3 p.u. voltage or (flux).

It is well known that magnetic saturation and saliency produce an angle shifting between the resultant air gap flux and armature and resultant mmf. The cross-coupling magnetic saturation is responsible for this phenomenon (Chapter 4).

In Reference 11, a rather lucrative procedure to account for this phenomenon is introduced.

Figure 7.25 shows the phasor diagram similar to Figure 7.24, with E_t the rated air gap emf $E_t = \overline{OB'} = \overline{OB}$. To E_t, the unsaturated (straight line) Φ_{1g} (F_{rl}) retains the unsaturated $\overline{B'C'}$ and saturated $\overline{B'C''}$ stator + air gap mmfs F_{rlu}, F_{rls}.

$$\frac{F_{rlu}}{F_{rls}} = \frac{\overline{B'C'}}{\overline{B'C''}} \tag{7.131}$$

At this ratio, from Figure 7.26 [7], we extract the saturation coefficients K_{sd}, K_{sq}, K_1 for constant air gap ($g_{max} = g$) and K'_{sd}, K'_{sq}, K'_1 for variable air gap ($g_{max}/g = 1.5$–2.5).

Then, for rated armature mmf F_{an} (p.u.), Figure 7.24, a q-axis equivalent armature reaction occurring for saliency and saturation is calculated:

$$F_{an}^{qs} = F_{an} \text{ (p.u.)} \cdot K_{sq} \cdot K_{aq} \tag{7.132}$$

For F_{an}^{qs}, we read on the Φ_{1g} (F_{rl}) curve the fictitious emf $E_{qo} = \overline{HH'}$. E_{qo} is projected as BD along the leakage reactance voltage drop direction $(\overline{AB})$.

The direction OD corresponds to the q axis.

The perpendicular from B to $\overline{OD}$ corresponds to the d axis.

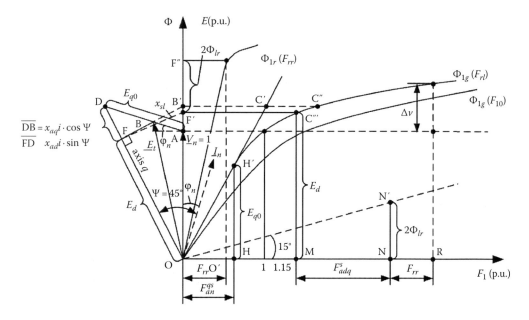

FIGURE 7.25 The partial magnetization curve method.

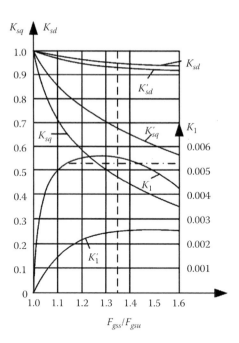

FIGURE 7.26 Saturation coefficients to account for cross-coupling magnetic saturation. (From Dombrowski, V.V. et al., *Design of Hydro-Generators*, Vol. I + II, Energy Publishers, Moscow, Russia, 1965 (in Russian).)

The perpendicular from B to $\overline{OD}$ touches the latter in F and $\overline{OF} = E_d$ (resultant emf along axis d). To E_d, on the Φ_{1g} (F_{rl}) curve, the mmf $\overline{F'C'''} = OM$ corresponds.

The magnetic saturation corresponding effect is considered by an equivalent F_{adq}^s component [9]:

$$F_{adq}^s = K_{sd} \cdot K_{ad} \cdot F_a \text{ (p.u.)} \cdot \sin \Psi + K_1 \cdot \frac{\tau_p}{g} \cdot F_a \text{ (p.u.)} \cdot \cos \Psi$$

for constant air gap, and

$$F_{adq}^s = K'_{sd} \cdot K_{ad} \cdot F_a \text{ (p.u.)} \cdot \sin \Psi + K'_1 \cdot \frac{\tau_p}{g} \cdot F_a \text{ (p.u.)} \cdot \cos \Psi \tag{7.133}$$

for variable air gap;

with K_{sd}, K'_{sd}, K_1, K'_1—from Figure 7.26.

The angle Ψ is the phase angle between the armature current vector and axis q in the dq model. Equations 7.133 contain the influence of both axes along the axis d.

With F_a (p.u.) known (for rated point, Figure 7.24) and Ψ determined from Figure 7.25, F_{adq}^s is calculated and added to OM along the horizontal axis ($F_{adq}^s = \overline{MM}$).

Adding the rotor interpole leakage flux per pole ($2\Phi_{lr} = \overline{NN'}$), from the Φ_{1r} (F_{rr}) partial magnetization curve, the rotor mmf contribution $F_{rr} = \overline{OO'}$, the total, on-load, excitation mmf is obtained as $\overline{OR} = F_{1n}$.

Corresponding to F_{1n}, the voltage regulation Δv (p.u.) is also obtained.

The graphical procedure seems at first a bit complicated, but it may be acquired after 1–2 examples.

Also, the procedure may be mechanized into a computer program including the calculation of partial magnetization curves.

It may be argued that the whole problem may be solved directly through FEM To do so, one needs to calculate first E_t (from Equation 7.122) must be first calculated, and then with given (rated) stator current and assigned values of Ψ to calculate the air gap flux Φ_{1g} must be calculated and then put Φ_{1g} into p.u. values and redo the FEM process again with ever new values of Ψ until Φ_{1g} (p.u.) = E_t.

The advantage of FEM is the possibility to calculate also the slot leakage inductance. Therefore, only the end connection leakage inductance is needed in order to obtain a rather exact value of x_{ls}.

Still, FEM seems practical only in the design refinement stages rather than in the general optimization design process itself, due to high computation times and lack in generality of results.

Example 7.7: Rated On-Load Excitation Current

Let us consider the no-load saturation curves in Figure 7.25 as pertaining to a real SG with a leakage reactance $x_{sl} = 0.11$ and a rated power factor angle $\varphi_n = 20°$.

Also the no-load excitation mmf F_{10n} that produces the rated voltage at zero load is (from Example 7.5): $F_{10n} = 18242$ At/pole, SCR = 0.7.

Calculate, for a salient pole rotor $\tau_p/\tau = 0.7$, $g/\tau = 0.04$, with constant air gap, the rated load excitation mmf in p.u. and in At/pole.

Solution

First let us apply the Potier diagram method despite of the fact that SG has salient poles. From Equation 7.129,

$$x_p \approx x_{sl} + 0.02 = 0.11 + 0.02 = 0.13$$

The air gap emf at full load E_t, for $V_1 = 1$ (p.u.), $i_1 = 1$ (p.u.), is from Equation 7.130:

$$E_t = \sqrt{1 + 0.13^2 \cdot 1^2 + 2 \cdot 0.13 \cdot 1 \cdot 1 \cdot \sin 20°} = 1.0515 \text{ (p.u.)}$$

From Figure 7.24, at scale for $E_t = 1.0515$, from the no-load saturation curve, the resultant mmf OD = 1.3 (due to magnetic saturation).

Now the rated armature F_{an} is $F_{an} = \dfrac{F_{10n}}{SCR} = \dfrac{18,242}{0.7} = 26,060$ At/pole $= 1.35 \cdot F_{10n}$
Therefore, in p.u. $F_{an} = 1.35$ (p.u.)
Finally from Figure 7.24, solving for triangle CDF the total load field mmf $F_{1n} = \overline{OE} = \overline{OF}$ is

$$F_{1n} = \sqrt{\overline{OD}^2 + \overline{DE}^2 + 2 \cdot \overline{OD} \cdot \overline{DE} \cdot \cos\varphi_n} = \sqrt{1.3^2 + 1.428^2 + 2 \cdot 1.3 \cdot 1.428 \cdot \cos 20°} = 2.68 \text{ (p.u.)}$$

The value of field rated mmf is thus

$$F_{1n} = F_{1n}(\text{p.u.}) \cdot F_{10n} = 2.68 \cdot 18,242 = 48,876 \text{ At/pole}$$

As expected, no reference has been made to the rotor saliency as the Potier diagram method has been used.

Let us now turn to Figure 7.25 and consider that for E_t again the mmf saturation ratio of Equation 7.131 is

$$\frac{F_{qss}}{F_{qsu}} = \frac{\overline{B'C''}}{\overline{B'C'}} = 1.35$$

This saturation ratio corresponds in Figure 7.26, for constant air gap under rotor pole, to:

$$K_{sd} = 0.95, \quad K_{sq} = 0.46, \quad K_1 = 0.55 \times 10^{-2}$$

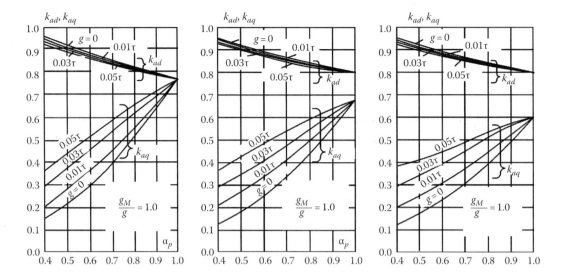

FIGURE 7.27 K_{ad} and K_{aq} (saliency) reactance factors for various τ_p/τ, g_{max}/g, and g/τ values.

Now, for the rated mmf F_{an} (1.35 in p.u.), the equivalent armature reaction allowing for saturation and saliency F_{an}^{qs} is, Equation 7.132:

$$F_{an}^{qs} = F_{an}\,(p.u.)\cdot K_{sq}\cdot K_{aq} = 1.35\times0.46\cdot0.57 = 0.354\ (p.u.)$$

The values of K_{ad} and K_{aq} are given in Figure 7.27 for constant and variable air gap under rotor pole, for given τ_p/τ.

For $g/\tau = 0.04$, $\tau_p/\tau = 0.7$, $K_{ad} = 0.84$, $K_{aq} = 0.57$ (Figure 7.27).

For $F_{an}^{qs} = 0.354$, we read from the $\Phi_{1g}\,(F_{g0})$ the value of emf $E_{q0} = \overline{HH'} \approx 0.40 = \overline{BD}$. From triangle OBD, we determine point F (Figure 7.25) and thus $E_d = \overline{OF} = \overline{OF'}$. For E_d, again, from Φ_{1g} (F_{rl}) we determine point M, which corresponds to about 1.15 p.u. (Figure 7.25).

With current angle Ψ to q axis known, we are now able to calculate from Equation 7.133 the cross-coupling global magnetic saturation effect mmf F_{adq}^s:

$$F_{adq}^s = K_{sd}\cdot K_{ad}\cdot F_a(p.u.)\cdot\sin\Psi + K_1\cdot\frac{\tau_p}{g}\cdot F_a\,(p.u.)\cdot\cos\Psi$$

$$= 0.95\cdot0.84\cdot1.35\cdot\sin45 + 0.55\cdot10^{-2}\cdot\frac{2}{3}\cdot\frac{1}{0.04}\cdot1.3\cdot\cos45 = 0.7616 + 0.0834 = 0.845$$

$F_{adq}^s = \overline{MN}$ along the horizontal axis in Figure 7.25, therefore the rotor leakage flux $2\Phi_{lr} = \overline{NN'}$ that corresponds to about 0.35. This value is equal to $\overline{F'F''}$ along the vertical axis ($\overline{F'F''} = 0.35$) which, from the $\Phi_{1r}(F_{rr})$ curves, leads to a rotor mmf $F_{rr} \approx 0.30$.

Now F_{rr} is added along the horizontal axis as $\overline{NR}$.

The load field mmf is thus $F_{1n} = \overline{OR} = \overline{OM} + \overline{MN} + \overline{NR} = 1.15 + 0.845 + 0.30 = 2.295$.

The value (2.295) obtained is different from (smaller than) that obtained with Potier diagram method (2.68). The three no-load magnetization curves have not been calculated point by point, so there is no guarantee of which is better.

However, in terms of precision, it is no doubt that the partial magnetization curves method is better, especially as it accounts for cross-coupling saturation and magnetic saliency of the rotor.

7.12 Inductances and Resistances

The inductances and resistances of an SG refer to the following:

- Synchronous magnetizing inductances (reactances): $L_{ad}(X_{ad})$, $L_{aq}(X_{aq})$
- Stator phase leakage inductance (reactance): $L_{sl}(X_{sl})$
- Homopolar stator inductance: $L_o(H_o)$
- Stator phase resistance R_s $[\Omega]$
- Rotor cage leakage inductances (reactances): $L_{Dl}(X_{Dl})$, $L_{Ql}(X_{Ql})$
- Rotor cage resistances R_D, R_Q
- Excitation leakage inductance (reactance): $L_{fl}(X_{fl})$
- Excitation resistance R_f

7.12.1 Magnetization Inductances L_{ad}, L_{aq}

Simplified formulae for L_{ad}, L_{aq} now have been derived in Chapter 4:

$$L_{ad} = L_m \cdot \frac{\left(K_{ad} \cdot K_f \right)}{1 + K_{sd}^e} \qquad X_{ad} = \omega_1 \cdot L_{ad}; \quad x_{ad} = X_{dm} \cdot \frac{I_n}{V_n}$$

$$L_{aq} = L_m \cdot \frac{\left(K_{aq} \cdot K_f \right)}{1 + K_{sq}^e} \qquad X_{aq} = \omega_1 \cdot L_{aq}; \quad x_{aq} = X_{qm} \cdot \frac{I_n}{V_n} \tag{7.134}$$

$$L_m = \frac{6\mu_0}{\pi^2} \cdot \frac{\tau \cdot l_i \cdot \left(W_1 \cdot K_{W1} \right)^2}{p_1 g \cdot K_c}; \quad K_f\text{—from Equation 7.107}$$

I_n, V_n are the base (rated) RMS values of SG phase current and voltage. L_m represents the air gap inductance for the uniform air gap machine. The air gap g in L_m expression is the minimum air gap for variable air gap under rotor poles. The saliency coefficients K_{ad}, K_{aq} are given in Figure 7.27.

The total magnetic saturation coefficients K_{sd}^e and K_{sq}^e are distinct from those in Figure 7.26 as the latter ones are in relation to the stator plus air gap part of mmf.

K_{sd}^e and K_{sq}^e are both dependent on stator current I_1 and the current angle Ψ with q axis; that is, on both $I_d = I_1 \cdot \sin\Psi$ and $I_q = I_1 \cdot \cos\Psi$. Apparently, only FEM is precise enough to calculate K_{sd}^e, K_{sq}^e family curves as pointed out also in Chapter 5 on transients. When $I_q = 0$ ($\Psi = 90°$), $K_{sd}^e\left(I_f\right)$ may be determined directly from the no-load saturation curve. But this is only a particular case.

From the no-load curve, the direct and quadrature ($d - q$) axes magnetization reactances x_{ad}, x_{aq} in p.u. may be obtained directly from Figure 7.25 as follows:

$$x_{aq} \cdot i = \overline{DB}$$

$$x_{ad} \cdot i \cdot \sin \Psi = \overline{FD} \tag{7.135}$$

As Figure 7.25 may be drawn (or translated in algebraic form), from any value of current i (p.u.) and power factor angle φ_n, the values of x_{ad} and x_{aq} for pairs of i_d, i_q ($i_d = i \cdot \sin\Psi$, $i_q = i \cdot \cos\Psi$) may be obtained for given voltage. From same rationale, the required excitation mmf may be calculated for each situation. And so is the power angle $\delta_v = \Psi - \varphi$.

7.12.2 Stator Leakage Inductance L_{sl}

The stator leakage inductance of SG has four components as it is the case for induction machines [10]:

$$L_{sl} = L_{ssl} + L_{szl} + L_{sel} + L_{sdl}; \quad X_{sl} = \omega_1 \cdot L_{sl}; \quad x_{sl} = L_{sl} \cdot \omega_1 \cdot \frac{I_n}{V_n} \tag{7.136}$$

where

    L_{ssl}—slot leakage
    L_{szl}—(air gap) leakage
    L_{sel}—end connection leakage
    L_{sdl}—differential (harmonics) leakage.

Let us consider the two-layer winding with rectangular stator slots, so typical for SGs (Figure 7.28).

In general, shorted coils are used, and thus the currents in the upper and lower layer coils are dephased by γ_K (in general, $\gamma_K = 60°$ or zero). The angle γ_K is zero in slots where the same phase occupies both layers.

As both $\gamma_K = 0°$ and $\gamma_K = 60°$ exist, an average of the slot geometrical permeance is to be used. L_{sl} may be written as [10] follows:

$$L_{sl} \approx 2\mu_0 \cdot \frac{W_1^2 \cdot l}{p_1 \cdot q} \left(\lambda_{ss} + \lambda_{sz} + \lambda_{se} + \lambda_{sd} \right) \tag{7.137}$$

with the slot leakage permeance ratio:

$$\lambda_{ss} = \frac{1}{4} \left[\frac{h_{sl} + h_{su} \cos^2 \gamma_K}{3W_{ss}} + \frac{h_{su}}{W_{ss}} + \frac{h_{su} \cos \gamma_K}{W_{ss}} + \frac{h_i}{W_{ss}} + \left(1 + \cos \gamma_K \right)^2 \left(\frac{h_{os}}{W_{ss}} + \frac{h_W}{W_{ss}} + \frac{h_o}{W_{ss}} \right) \right] \tag{7.138}$$

When the number of turns per coil is different in the two layers ($n_{cl} \neq n_{cu}$), Equation 7.138 may be corrected by

- Replacing $\cos \gamma_K$ by $K \cos \gamma_K$ with $K = n_{cu}/n_{cl}$ and
- Replacing number 4 by $(1 + K)^2$

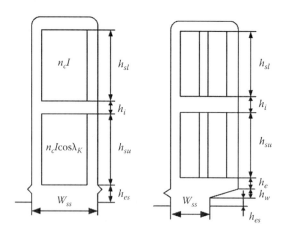

FIGURE 7.28 Typical high- and medium-power SG stator slots.

For the zigzag (z), air gap permeance ratio λ_{sz}, the so-called Richter formula is applied:

$$\lambda_{sz} \approx \frac{5g \cdot K_c/W_{ss}}{5 + 4g \cdot K_c/W_{ss}} \frac{(3\beta_y + 1)}{4}; \quad \beta_y = \frac{y}{\tau} = \text{coil span ratio} \tag{7.139}$$

It is sometimes inferred that the differential (harmonics) permeance ratio λ_{sd} is included in Equation 7.139.

The end-connection permeance ratio λ_{se} (for double layer winding) is

$$\lambda_{se} \approx 0.34 \cdot \frac{g}{l_i} \cdot (l_{ec} - 0.64 \cdot y); \quad y\text{—coil span} \tag{7.140}$$

l_{ec}—end connection length per machine side. A few remarks are in order:

- The total geometrical (rather than iron) length of stator l was used—though it includes the radial channels length—to account for the "slot leakage" of coil ports corresponding to cooling channels. It is an overestimation and better approximations are welcome.
- Alternative formulae for z permeance ratio have been proposed [11]:

$$\lambda_{sz} = \frac{1}{\pi} \cdot l_{er} \cdot \frac{\sqrt{1 + (2g \cdot K_C/W_{ss})^2}}{2} + \frac{g \cdot K_C}{W_{ss}} \left(1 - \frac{2}{\pi} \cdot \tan^{-1} \frac{2 \cdot g \cdot K_C}{W_{ss}}\right) \tag{7.141}$$

Equation 7.141 of λ_{sz} produces large errors for small values of $g \cdot K_C/W_{ss}$, which is not the case, in general, for SG.

- There are reasons to consider separately the differential harmonics permeance ratio λ_{sd}. For standard chorded coil winding [10],

$$\sigma_{dso} \approx \frac{2\pi^2}{9 \cdot K_{W1}^2} \cdot \frac{5g^2 + 1 - 3/4(1 - y/\tau)\left[9q^2\left(1 - y^2/\tau^2\right) + 1\right]}{12q^2} \tag{7.142}$$

$$\lambda_{sd} = \sigma_{dso} \cdot \frac{3}{\pi^2} \cdot \frac{\tau \cdot q}{K_C \cdot g} \cdot \frac{l_i}{l} \tag{7.143}$$

- It should be noted that, as for IMs, the solid rotor or the rotor cage may attenuate the differential leakage coefficient λ_{sd} and σ_{dso} "in exchange" for additional losses in the rotor under SG load.
- However, as the slot pitch is about the same in the rotor cage and in the stator, and the air gap tends to be large, this attenuation is limited. Values of attenuation factor of 0.8–0.5 might be expected in special cases.
- The end-connection leakage permeance λ_{se} depends very much on the exact shape of end connections, their vicinity to iron (metallic) parts of frame and to the stator stack end plate, and even on the value of field mmf (the power factor).
- Saturation of stator end core laminations, that produces additional losses in the under excited machine, also causes a reduction in end-connection leakage permeance ratio λ_{se}, due to induced currents effects.
- And so do the currents induced in the neighboring metallic parts.

- Three-dimensional FEM has been used to compute the end-connection leakage flux and losses in the end core and in the metallic parts nearby. The results seem satisfactory for the case in the point but lack generality.
- The case of variable air gap above the rotor pole shoe complicates further the computation of both zigzag and differential leakage inductance. As usual, FEM is the solution for refined calculations, but in the design preliminary stages the above formulae give results with ±10% error, in general, and are fairly reliable.

7.13 Excitation Winding Inductances

As already pointed out in Chapter 4, when reduction to stator is performed, the mutual inductance (reactance) between the field winding and stator phases, $L_{fa}(x_{fa})$, equals the d axis magnetization inductance, $L_{ad}(x_{ad})$. However, in the design and in the testing process, it is useful to calculate first the main and leakage excitation inductance, before the reduction to stator.

The main field inductance L_{fg} and the leakage field inductance L_{fl} make up the excitation total inductance as seen (or measured) from the rotor:

$$L_f = L_{fg} + L_{fl} \tag{7.144}$$

L_{fg} is simply (as for L_{ad})

$$L_{fg} = \frac{2p_1 \cdot \mu_0 \cdot W_f^2 \cdot \tau \cdot l_i \cdot \left(\tau_p / \tau\right)}{g \cdot K_C \cdot \left(1 + K_{sd}\right)} \tag{7.145}$$

The stator reduction coefficient K_{fa} is

$$K_{fa} = \frac{3}{2} \cdot \left(\frac{W_1 \cdot K_{W1}}{2p_1 \cdot W_f}\right)^2 \left(\frac{4}{\pi} \cdot K_{ad}\right)^2 \tag{7.146}$$

On the other hand, the actual mutual inductance between the excitation and armature windings, M_{af}, is

$$M_{af} \approx 6 \cdot W_f \cdot W_1 \cdot K_{W1} \cdot K_f \cdot \frac{2\mu_0 \cdot \tau \cdot l_i}{\pi \cdot g \cdot K_C \cdot \left(1 + K_{sd}\right)} \tag{7.147}$$

The leakage excitation inductance L_{fl} contains also a few terms:

$$L_{fl} = 2p_1 \cdot \mu_0 \cdot W_f^2 \cdot l_p \cdot \left(\lambda_p + \lambda_{zf} + \lambda_{ef}\right) \tag{7.148}$$

L_{fl} components have similar formulae to those derived for the stator.

The slot leakage permeance ratio λ_p has been calculated in Equation 7.138; λ_{zf} in Equation 7.139. The end-connection permence ratio λ_{ef} is [11]

$$\lambda_{ef} \approx 2.55 \cdot \frac{a_{ps}}{l_i} \cdot \ln\left(1 + \frac{\pi}{4} \cdot \frac{\tau_p}{2b_{r1}}\right) \tag{7.149}$$

b_{r1}, a_{ps}—from Figure 7.22.

The reduction of L_{fl} to the stator makes use of K_{fa} of Equation 7.146, while the p.u. translation is done by dividing to $L_B = (V_n/I_n \cdot \omega_1)$.

$$l_{fl} = L_{fl} \cdot K_{fa} / L_B = \frac{L_{ad} \cdot \pi \cdot g \cdot q \cdot K_C}{3 \cdot \mu_0 \cdot K_{W1}^2 \cdot K_{ad} \cdot l_i \cdot \tau} \cdot \left(\lambda_p + \lambda_{zf} + \lambda_{ef} \right) \tag{7.150}$$

Now it is easier to add the differential component missing in Equation 7.148. The differential leakage is very similar to the case of stator winding:

$$l_{fl} \rightarrow l_{fl} + l_{ad} \cdot \left(2 \frac{\tau_p}{\tau} \cdot \frac{K_{ad}}{K_{f1}^2} - 1 \right) \tag{7.151}$$

K_{fl} has been calculated in Equation 7.107 for constant air gap under the pole shoes. It is rather common practice to express the base inductance (reactance) L_B (X_B) based on the approximations:

$$V_n \approx \pi\sqrt{2} \cdot f_1 \cdot W_1 \cdot K_{W1} \cdot \Phi_{10} \tag{7.152}$$

with

$$\Phi_{10} = \frac{2}{\pi} \cdot B_{g10} \cdot \tau \cdot l_i$$

B_{g10} corresponds to no-load conditions at rated voltage.

$$F_{an1} = \frac{3\sqrt{2} \cdot W_1 \cdot K_{W1} \cdot I_n}{\pi \cdot p_1} \tag{7.153}$$

as

$$X_b = \frac{V_n}{I_n} = \omega_1 \cdot l_b = \frac{3}{\pi} \cdot \frac{W_1}{p_1} \cdot \frac{\left(W_1 K_{W1}\right)^2}{p_1} \cdot \frac{\Phi_{10}}{F_{an1}} \tag{7.154}$$

By denoting F_{gn}, the air gap mmf requirements

$$F_{gn1} = \frac{B_{g1}}{K_{f1}} \cdot \frac{g \cdot K_C}{\mu_0} \tag{7.155}$$

The magnetization reactance x_{ad} and x_{aq} in (p.u.) become

$$x_{ad} = \frac{K_{ad}}{K_{f1}} \cdot \frac{F_{an1}}{F_{gn1}} \tag{7.156}$$

$$x_{aq} = \frac{K_{aq}}{K_{f1}} \cdot \frac{F_{an1}}{F_{gn1}} \tag{7.157}$$

with x_{ad} given in Equation 7.156, together with the air gap flux density B_{g1} and linear current loading $A_n = 3W \cdot I_n/(p_1 \cdot \tau)$, the air gap g may be found:

$$g = \frac{2 \cdot \mu_0 \cdot K_{ad} \cdot K_{W1}}{x_{ad} \cdot K_{f1} \cdot K_C} \cdot \frac{A_n \cdot \tau}{B_{gn1}} \tag{7.158}$$

This expression has been used to size the air gap earlier in this chapter Equation 7.67.

Note that for the cylindrical rotor, the leakage inductance formula is similar to Equation 7.148; but instead of $2p_1 \cdot W_f^2$, it will be $2p_1 \cdot N_{fp} \cdot W_{fc}^2$ where W_{fc} is the number of turns per rotor slot and N_{fp}—the number of field winding slots per pole in the rotor.

While the slot and zigzag permeances are straightforward, the end-connection permeance ratio λ_{ef} is

$$\lambda_{ef} \approx \frac{0.2 \times \tau_r}{l_i} \tag{7.159}$$

τ_r—the rotor excitation slot pitch

7.14 Damper Winding Parameters

The d and q axis damper winding leakage reactances $x_{D\sigma}$ and $x_{Q\sigma}$ in p.u. have expressions similar to those of excitation (l_{fl}) [12]:

$$x_{Dl} \approx x_{ad} \cdot \left[\frac{K_{ad}}{\pi \cdot K_D^2} \frac{\alpha_b \cdot N_2 \cdot (1-K_b)}{4\sin^2(\alpha_b/2)} \cdot \left(2 + \frac{\cos N_2\alpha_b - K_b\cos\alpha_b}{1-K_b} \right) \cdot \right.$$
$$\left. \cdot \left(1 + \frac{2 \cdot \lambda_{De} \cdot g \cdot K_C}{l_i} \right) - 1 + \frac{N_2 \cdot (1-K_b) \cdot K_{ad} \cdot l_{pi} \cdot g \cdot K_C}{l_i \cdot \tau \cdot K_D^2} \cdot (\lambda_{sD} + \lambda_{zD}) \right] \tag{7.160}$$

$$\alpha_b = \frac{\pi \cdot \tau_b}{\tau}; \quad \tau_b\text{—damper bar pitch}$$

$$K_b = \frac{\sin(N_2\alpha_b)}{N_2\sin\alpha_b} \tag{7.161}$$

$$K_D \approx \frac{N_2}{\pi} \cdot (1-K_b)$$

λ_{De}—end-ring permeance ratio. Complete end-ring presence is supposed.

$$\lambda_{De} \approx \frac{3.32}{\pi} \cdot \log\frac{2.35 \cdot D_{ring}}{(a+2b)} \tag{7.162}$$

D_{ring}—the average ring diameter a, b cross-section ring dimensions: a—radial and b—axial.

The zigzag and damper slot permence ratios λ_{ZD} and λ_{sD} are straightforward.

For axis q,

$$x_{Ql} = x_{ql} \cdot \left\{ \frac{K_{aq}}{\pi K_Q^2} \frac{\alpha_b N_2}{4\sin^2(\alpha_b/2)} \cdot \left[\cos N_2\alpha_b - K_b\cos\alpha_b + \frac{\alpha_i\pi + \pi(1-\alpha_i)(gK_c/2_{g\max})}{N_2 \cdot \alpha_b} \right. \right.$$
$$\left. \left. \times \left(1 + 2 \cdot \lambda_{De} \cdot \frac{g \cdot K_c}{l_i} \right) \right] - 1 + \frac{N_2 \cdot (1+K_b) \cdot K_{aq} \cdot l_{pi} \cdot g \cdot K_c}{l_i \cdot \tau \cdot K_Q^2}(\lambda_{SQ} + \lambda_{ZQ}) \right\} \tag{7.163}$$

with

$$K_Q \approx \frac{N_2}{\pi} \cdot (1 + K_b) - \frac{4}{\pi} \cdot \left(1 - \frac{4}{\pi} \frac{\sigma \cdot g \cdot K_c}{g_{max}}\right) \cdot \cos \frac{\alpha_i \cdot \pi}{2} \cdot \frac{\sin\left((N_2/2) \cdot \alpha_b\right)}{2 \sin(\alpha_b/2)} \qquad (7.164)$$

$\alpha_i = \tau_p/\tau$—rotor pole span per pole pitch.

The damper cage resistances in p.u., r_D and r_Q, are

$$r_D \approx \frac{6(W_1 \cdot K_{W1} \cdot K_{ad})^2 \cdot \rho_D \cdot N_2 \cdot (1 - K_b)}{\pi^2 \cdot p_1 \cdot K_D^2 \cdot X_b} \cdot \left[\frac{l_{pi}}{A_{bar}} + \frac{\alpha_b \cdot \tau}{2\pi \cdot A_{ring} \cdot \sin^2(\alpha_b/2)} \right.$$

$$\left. \times \left(2 + \frac{\cos N_2 \alpha_b - K_b \cos \alpha_b}{1 - K_b}\right) \right] \qquad (7.165)$$

$$r_Q \approx \frac{6(W_1 \cdot K_{W1} \cdot K_{aq})^2 \cdot \rho_D \cdot N_2}{\pi^2 \cdot p_1 \cdot K_Q^2 \cdot X_b} \cdot \left\{ \frac{l_{pi}}{A_{bar}}(1 + K_b) + \frac{\alpha_b \cdot \tau}{2\pi \cdot A_{ring} \cdot \sin^2 \frac{\alpha_b}{2}} \right.$$

$$\left. \times \left[\cos N_2 \alpha_b - K_b \cos \alpha_b + \frac{\pi}{N_2 \cdot \alpha_b} \cdot (1 - \cos N_2 \alpha_b)\right] \right\} \qquad (7.166)$$

Note: For incomplete end rings, in general

$$x_{Ql} = (2 - 3)x_{Dl}; \quad R_{Ql} = (4 - 6)R_{Dl} \qquad (7.167)$$

- ρ_D—damper cage resistivity (Ω m)
- A_{bar}—damper bar cross-section (m²)
- A_{ring}—end-ring cross-section (m²)

The field winding and stator resistances in p.u. are as follows:

$$r_f = \frac{R_f}{X_b} \cdot \frac{6(W_1 \cdot K_{W1} \cdot K_{ad})^2}{(\pi \cdot p_1 \cdot W_f \cdot K_{f1})^2}; \quad R_f = \rho_{cor} \cdot l_f \cdot \frac{W_f}{A_{cof}} \cdot 2p_1 \qquad (7.168)$$

R_f—actual field winding resistance (Ω); $W_f = N_{fp} \cdot W_{fc}$ for the cylindrical rotor; l_f—average field turn length; and A_{cof}—field turn cross-section

$$r_s = \frac{R_s}{X_b}; \quad R_s = \rho_{cos} \cdot W_a \cdot \frac{l_{coil} \cdot W_a}{A_{cos} \cdot a} \cdot K_R \qquad (7.169)$$

l_{coil}—turn length, A_{cos}—stator turn cross-section, K_R—skin effect coefficient (to be determined in the paragraph on losses), a—stator current paths number, and W_a—turns/path.

7.15 Solid Rotor Parameters

Cylindrical rotors are built of solid iron and for some salient pole SG also solid rotor poles may be used.

The solid iron of rotor acts as a damper cage with variable resistance and inductance (reactance).

The solid iron parameters depend on the frequency of induced rotor currents—Sf_1 for the fundamental and $2f_1$ for the inverse components.

The presence of field winding slots in the solid iron poles makes the d and q axes equivalent parameters of solid iron dampers different from each other and difficult to calculate.

Comprehensive analytical formulae with widespread acceptance for r_D, r_Q, l_{Dl}, l_{Ql} for solid iron poles are not yet available, but

$$x_{Dl} \approx 0.6 \cdot r_{Dl}; \quad x_{Ql} \approx 0.6 \cdot r_{Ql} \tag{7.170}$$

The trajectory of eddy currents induced by the mmf space harmonics and stator slot opening air gap conductance harmonics or during transients depend on the pole pitch of the respective harmonics, the frequency as "sensed" by the rotor currents, level of flux density in the rotor solid iron body.

For the mmf fundamental only, the frequency of rotor induced currents is Sf_1 ($S = (\omega_1 - \omega_r)/\omega_1$).

The depth of field penetration for a traveling wave with pole pitch τ_1 is [10]

$$\delta_1 = \text{Real} \sqrt{\frac{1}{\left(\pi/\tau\right)^2 + j\mu_{Fe}\sigma_{Fe} \cdot 2\pi\left(Sf_1/K_t\right)}} \tag{7.171}$$

K_t takes into account the conductivity reduction due to the eddy currents tangential closure [10]—the so-called transverse cage effect:

$$K_t \approx \frac{1}{1 - \left[\tan\left(\left(\pi/\tau\right)\cdot\left(l_{pi}/(2\cdot n)\right)\right)\right]\big/\left(\left(\pi/\tau\right)\cdot\left(l_{pi}/(2\cdot n)\right)\right)} \tag{7.172}$$

n is number of solid iron rings put together axially to form the rotor (Figure 7.29).

The iron permeability μ_{Fe} is rather "dictated" by the "steady state" conditions. Therefore, the procedures used to calculate the magnetic saturation curves and rated excitation mmf yield also at least an average value for μ_{Fe}. To a first approximation, on the rotor surface, the flux density is, in general, around 1–1.2 T in the unslotted region and 1.5–1.8 T in the slotted region of cylindrical rotors. Magnetic saturation leads to a decrease in μ_{Fe} and thus to a larger field penetration depth. That is, to a smaller equivalent rotor-iron resistance.

For an unslotted rotor the solid iron resistance reduced to the stator is [10]

$$R_{Ds} = \frac{6\left(W_1 \cdot K_{W1}\right)^2 \cdot l_{pi} \cdot K_t}{\sigma_{Fe} \cdot p_1 \cdot \tau \cdot \delta_1}; \quad r_D = \frac{R_{Ds}}{X_B} \tag{7.173}$$

We may, to a first approximation consider $R_{Ds} = R_{Qs}$ and allow also for the air gap leakage:

$$x_{Ds} \approx x_{Qs} \approx 0.6 \cdot r_{Ds} + 0.035 \tag{7.174}$$

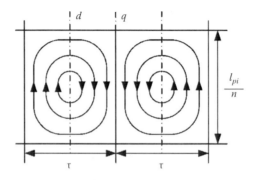

FIGURE 7.29 Solid iron rotor eddy currents trajectory in axis d.

7.16 SG Transient Parameters and Time Constants

The d—axis transient reactance x'_d is

$$x'_d = x_{sl} + \frac{1}{\left(1/x_{ad}\right)+\left(1/x_{fl}\right)} \text{ (p.u.)} \tag{7.175}$$

The transient reactance of d-axis damper cage, x'_D, is

$$x'_D = x_{Dl} + \frac{1}{\left(1/x_{ad}\right)+\left(1/x_{fl}\right)} \text{ (p.u.)} \tag{7.176}$$

The subtransient d–q axis reactance x''_d:

$$x''_d = x_{sl} + \frac{1}{\left(1/x_{ad}\right)+\left(1/x_{Dl}\right)+\left(1/x_{fl}\right)} \text{ (p.u.)} \tag{7.177}$$

For the q-axis x''_q is

$$x''_q = x_{sl} + \frac{1}{\left(1/x_{aq}\right)+\left(1/x_{Ql}\right)} \text{ (p.u.)} \tag{7.178}$$

The total excitation time constant T_f is

$$T_f \approx \frac{x_{ad} + x_{fl}}{r_f \cdot \omega_n} \text{ (s)} \tag{7.179}$$

The transient short circuit d-axis time constant T'_d is

$$T'_d = T_f \cdot \frac{x'_d}{x_d} \text{ (s)}; \quad x_d = x_{sl} + x_{ad} \tag{7.180}$$

The d-axis damping time constant T_D—with open stator and short-circuited superconducting ($r_f = 0$) field circuit:

$$T_D = \frac{x'_D}{r_D \cdot \omega_n} \text{ (s)} \tag{7.181}$$

The subtransient d-axis time constant T''_d is

$$T''_d = T_D \cdot \frac{x''_d}{x'_d} = \frac{x'_D}{r_D \cdot \omega_n} \cdot \frac{x''_d}{x'_d} \text{ (s)} \tag{7.182}$$

If the q-axis damper winding is considered in isolation, its time constant T_Q is

$$T_Q \approx \frac{x_{Ql} + x_{aq}}{r_Q \cdot \omega_n} \text{ (s)} \tag{7.183}$$

The subtransient q axis time constant (the stator and the field windings are short-circuited and superconducting) T_q'' is

$$T_q'' = T_Q \cdot \frac{x_q''}{x_q} = \frac{\left(x_{Ql} + x_{aq}\right)}{\left(x_{sl} + x_{aq}\right)} \cdot \frac{x_q''}{r_Q \cdot \omega_n} \; (\text{s}) \tag{7.184}$$

The negative sequence reactance x_2 is

$$x_2 = \frac{x_d'' + x_q''}{2} \tag{7.185}$$

or

$$x_2 = \sqrt{x_d'' \cdot x_q''} \tag{7.186}$$

depending on the fact if a negative sequence voltages or currents are present in the SG stator.

The time constant T_a of the stator currents when all other windings are superconducting and short-circuited is

$$T_a = \frac{2 x_d'' \cdot x_q''}{r_s \left(x_d'' + x_q''\right) \cdot \omega_n} \tag{7.187}$$

All the above parameters appear in the sudden short circuit time response. As the sudden three short circuit is used to measure (estimate) the above parameters, their expressions, as derived in the previous paragraph, serve for checking the design accuracy and predict the SG transient behavior.

7.16.1 Homopolar Reactance and Resistance

The homopolar sequence (Chapter 4) does not produce interference with the rotor in terms of the fundamental. However, it produces a fixed (AC) magnetic field in the air gap with a pole pitch $\tau_3 = \tau/3$, similar to a third space harmonic. Therefore, AC currents are induced in the rotor cage through this AC third harmonic field. In general, this effect is neglected, and the homopolar reactance x_o is assimilated to a stator leakage inductance calculated with slot total currents either zero or twice the homopolar mmf: $2n_c \cdot I_o$. This is so, as homopolar currents in all phases are all the same.

To a first approximation,

$$x_o \approx x_{sl} \qquad \text{for diametrical coils}$$
$$x_o \approx 3 \cdot \left(1 - \frac{y}{\tau}\right) \cdot x_{sl} \quad \text{for chorded coils} \tag{7.188}$$

More complicated expressions are to be found in the literature [6].

It is evident that Equation 7.188 ignores the damping effect of rotor damper induced currents produced by the homopolar stator mmf. In such conditions, the homopolar resistance $r_o \approx r_s$.

Example 7.8: Transient Reactances and Time Constants

A designed salient pole SG has the following calculated parameters (in p.u.):

$$r_s = 0.003, \; r_f = 0.006, \; x_{ad} = 1.5, \; x_{aq} = 0.9, \; x_{sl} = 0.12, \; x_{fl} = 0.15, \; x_{Dl} = 0.04, \; x_{Ql} = 0.05, \; r_D = 0.02,$$

$$r_Q = 0.022.$$

Calculate the transient x_d' and subtransient reactances x_d'', x_q'', and the time constants T_d', T_d'', T_q'', T_a, and T_D in seconds for $\omega_r = 2\pi \cdot 60$ rad/s.

Solution

From Equation 7.175, the subtransient d-axis reactance x_d' is

$$x_d' = x_{sl} + \frac{1}{\left(1/x_{ad}\right) + \left(1/x_{fl}\right)} = 0.12 + \frac{1}{\left(1/1.5\right) + \left(1/0.15\right)} = 0.25636 \text{ p.u.}$$

From Equation 7.177, x_d'' is

$$x_d'' = x_{sl} + \frac{1}{\left(1/x_{ad}\right) + \left(1/x_{Dl}\right) + \left(1/x_{fl}\right)} = 0.12 + \frac{1}{\left(1/1.5\right) + \left(1/0.04\right) + \left(1/0.15\right)} = 0.1509 \text{ p.u.}$$

From Equation 7.178,

$$x_q'' = x_{sl} + \frac{1}{\left(1/x_{aq}\right) + \left(1/x_{Ql}\right)} = 0.12 + \frac{1}{\left(1/0.9\right) + \left(1/0.05\right)} = 0.16736 \text{ p.u.}$$

$$x_2 = \frac{x_d'' + x_q''}{2} = \frac{0.1509 + 0.16736}{2} = 0.15913 \text{ p.u.}$$

or

$$x_2 = \sqrt{x_d'' \cdot x_q''} = \sqrt{0.1509 \cdot 0.16736} = 0.158917 \text{ p.u.}$$

The time constants are calculated from Equations 7.180 through 7.184 and 7.187:

$$T_d' = T_f \cdot \frac{x_d'}{x_d} = \frac{1.5 \cdot 0.15}{0.006 \cdot 2\pi \cdot 60} \cdot \frac{0.25636}{\left(1.5 + 0.12\right)} = 0.115 \text{ (s)}$$

$$T_D = \frac{\left(x_{Dl} + \left(1/\left(1/x_{ad} + 1/x_{fl}\right)\right)\right)}{r_D \cdot \omega_n} = \frac{\left(0.05 + \left(1/\left(1/1.5 + 1/0.15\right)\right)\right)}{0.02 \cdot 2\pi \cdot 60} = 0.024716 \text{ (s)}$$

$$T_d'' = T_D \cdot \frac{x_d''}{x_d'} = 0.024716 \cdot \frac{0.1509}{0.25636} = 0.01458 \text{ (s)}$$

$$T_q'' = \frac{\left(x_{Ql} + x_{aq}\right)}{r_Q \cdot \omega_n} \cdot \frac{x_q''}{x_q} = \frac{\left(0.05 + 0.9\right)}{0.022 \cdot 2\pi \cdot 60} \cdot \frac{0.1676}{0.9} = 0.02133 \text{ (s)}$$

$$T_a = \frac{2x_d'' \cdot x_q''}{r_s \left(x_d'' + x_q''\right) \cdot \omega_n} = \frac{0.1509 \cdot 0.16736}{0.003 \times 0.15913 \cdot 2\pi \cdot 60} = 0.140 \text{ (s)}$$

The total no-load excitation time constant T_f of Equation 7.179 is

$$T_f = \frac{x_{ad} + x_{fl}}{r_f \cdot \omega_n} = \frac{1.5 + 0.15}{0.006 \cdot 2\pi \cdot 60} = 0.7254 \text{ (s)}$$

7.17 Electromagnetic Field Time Harmonics

In order to perform well, when connected to the power system, the open-circuit voltage waveform has to be very close to a sine wave.

Standards limit the time harmonics traditionally through the so-called telephonic harmonic factor (THF): 5% up to 1 MW, 3% up to 5 MW and 1.5% above 5 MW per unit.

Today "grid codes" specify total harmonic distortion (THD) to 1.5% for SGs in near 400 kV power systems and 2% in near 275 kV power systems.

There are proposals to raise these limits to 3(3.5)%.

To analyze possibilities to reduce emf THD let us start with its expression:

$$E = \omega_1 \cdot \sum_{v=1}^{\infty} K_{W_v} \cdot W_a \cdot \Phi_v$$

$$\Phi_v = \frac{2}{\pi} \cdot \frac{\tau}{v} \cdot B_{gv} \cdot l_i \tag{7.189}$$

$$K_{W_v} = K_{dv} K_{yv} = \frac{\sin(\upsilon\pi/6)}{q \cdot \sin(\upsilon\pi/6q)} \cdot \sin\left(\frac{v \cdot y}{\tau} \cdot \frac{\pi}{2}\right)$$

For fractionary q windings,

$$q = b + c/d = (bd + c)/d$$

$$K_{dv} = \frac{\sin(\upsilon\pi/6)}{(bd + c) \cdot \sin(\upsilon\pi/6(bd + c))} \tag{7.190}$$

Therefore, the harmonics occur in the air gap flux density (B_{g1}) first.

They may be reduced by

- Adjusting the ratio of rotor pole shoe span τ_p per pole pitch τ: $\alpha_i = \tau_p/\tau$ in salient pole rotors
- Varying the air gap from g to g_{max} from center to margins in the salient pole rotor shoe
- Adjusting the large (central) tooth τ_p per pole pitch τ ratio in nonsalient pole rotors with uniform air gap

The form (harmonics) coefficients in the excitation air gap flux density for constant air gap are as follows:

$$K_{fv} \approx \frac{4}{\pi \cdot v} \sin\left(v \cdot \frac{\tau_p}{\tau} \cdot \frac{\pi}{2}\right) \quad \text{salient poles} \tag{7.191}$$

$$K_{fv} \approx \frac{4}{\pi^2 \cdot v^2} \frac{\cos\left(v \cdot \frac{\tau_p}{\tau} \cdot \frac{\pi}{2}\right)}{1 - v \cdot \frac{\tau_p}{\tau} \cdot \frac{2}{\pi}} \quad \text{nonsalient poles} \tag{7.192}$$

It seems that the minimum of harmonics contents,

$$\frac{\sum_{v>1} E_v^2}{E_1^2} = \frac{\sum K_{fv}^2}{K_{f1}^2} \tag{7.193}$$

caused solely by the air gap flux density harmonics, is obtained in the first case (salient poles) for $\alpha_p = \tau_p/\tau = 0.77$–$0.8$. However, these large values produce too large interpole excitation flux leakage and $\alpha_i \approx 0.7$ is practiced; occasionally larger.

For the nonsalient poles $\tau_p/\tau \approx 1/3$ seems adequate, but the precise ratio τ_p/τ may be used to destroy a certain harmonic ν:

$$\nu \cdot \frac{\tau_p}{\tau} \cdot \frac{\pi}{2} = (2K+1) \cdot \frac{\pi}{2} \tag{7.194}$$

As even harmonics do not normally occur (all poles geometry, air gap, and excitation coils are identical), the harmonics of interest are $\nu = 3, 5, 7, 9, 11, 13, 15, 17, 19, \ldots$.

The first slot-opening-caused harmonic pair $\nu_c = 6q \pm 1$ may also be the target, especially, with integer $q < 5$.

For salient pole machines, the air gap may be varied ideally as

$$\delta(\theta) = \frac{\delta}{\cos p_1 \theta} \tag{7.195}$$

For such a case,

$$\left(K_{f\nu} \right)_{g\sin} \approx \frac{2}{\pi} \cdot \left[\frac{\sin(\nu-1)\alpha_p \pi/2}{\nu-1} + \frac{\sin(\nu+1)\alpha_p \pi/2}{\nu+1} \right] \tag{7.196}$$

A notable reduction in the fifth and seventh harmonics is obtained. The augmentation of the third harmonic is not a problem in three-phase machine (Table 7.2).

In reality, air gap variation under pole shoes is done through lower radius rotor pole machining. FEM should be used to quantify the "exact" effect on air gap flux harmonics.

Next step in reduction emf harmonics is the proper design of the stator winding. One solution is the reducing of $K_{d\nu}$ and(or) $K_{y\nu}$ for key harmonics.

An increase in integer q above 5 does not produce a notable decrease in $\left(K_{d\nu} \right)_{\nu \geq 5}$, not to mention K_{d3} which is almost independent of q and large (around 0.67). But $q = 2, 3$ produces significantly higher harmonics. The worst situation, $K_{d\nu_c} = 1$, takes place for the slot harmonics $\nu_c = 6K_q \pm 1$.

Raising the order of the first slot harmonics implies larger integer q ($q \geq 5$) or fractionary $q = b + c/d = (bd + c)/d$. In this case, the first slot harmonic is $\nu_c = 6(bd + c) \pm 1$.

One more way to reduce the emf harmonics is to cancel (reduce) the fifth and seventh harmonics by two-layer windings with chorded coils:

$$\sin\left(\nu \cdot \frac{y}{\tau} \cdot \frac{\pi}{2} \right) = 0 \tag{7.197}$$

$$\nu \cdot \frac{y}{\tau} \cdot \frac{\pi}{2} = K_n \cdot \pi \tag{7.198}$$

TABLE 7.2 Air Gap Flux Density Harmonics ($\alpha_p = 2/3$)

	K_{f1}	K_{f3}	K_{f5}	K_{f7}
Constant gap	1.105	0	−0.221	+0.158
Inverse sine gap	0.941	0.137	−0.137	+0.069

For $y/\tau = 0.833 = 5/6$, the best reduction of 5th and 7th harmonics takes place:

$$K_{y5} = \sin\left(5 \cdot 0.833 \cdot \frac{\pi}{2}\right) = 0.256$$

$$K_{y7} = \sin\left(7 \cdot 0.833 \cdot \frac{\pi}{2}\right) = 0.2624$$

A coil chording of about 5/6 may be obtained easily with a large q.

The third harmonic has to be kept below 5% of the fundamental to avoid false tripping of relays for differential protection when an SG is connected or disconnected to/from a bus bar, with both systems earthed to allow for triple frequency currents to flow to ground through neutral points.

It should be mentioned that the stator mmf harmonics due to the location of stator coils in slots produces mmf harmonics of orders $\nu = 6K \pm 1$, that is, −5, +7, −11, +13. These space harmonics do not produce time harmonics in the emf but produce additional losses in the rotor. This is why they have to be reduced through increased or fractionary q.

7.18 Slot Ripple Time Harmonics

In certain conditions, the air gap permeance variation due to slot openings may produce higher harmonics, known as slot ripple, in the open-circuit emf wave.

It is the stator slot opening air gap magnetic permeance in interaction to the field mmf fundamental that produces the slot ripple emf effect (Figure 7.30):

The excitation (rotor) mmf fundamental F_{f1} with respect to stator writes:

$$F_{f1}(\theta_s) = F_{f1m} \cdot \sin(p_1 \cdot \theta_s - \omega_r \cdot t) \tag{7.199}$$

The air gap magnetic conductance variation is

$$P_{g f1} = \frac{\mu_0 \cdot K_{fm}}{g} \cdot \sin \nu_1 N_s \theta_s \tag{7.200}$$

The air gap flux density B_g is

$$B_g = P_{g f1} \cdot F_{f1}$$

$$= \frac{\mu_0 \cdot K_{fm}}{g} \cdot F_{f1m} \cdot \left\{ \cos\left[(\nu_1 N_s - p_1)\theta_s + p_1 \cdot \omega \cdot t\right] - \cos\left[(\nu_1 N_s + p_1)\theta_s - p_1 \cdot \omega \cdot t\right] \right\} \tag{7.201}$$

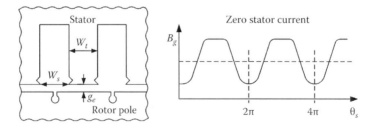

FIGURE 7.30 Slot-opening permeance variation caused flux density pulsations.

For $\nu = 1$, the first air gap permeance harmonic is considered; while for $\nu_1 = 2$, the second one comes into play. Both are important.

If the rotor damper bar pitch τ_d corresponds to a division of circle by $N_s - p_1$ or $N_s + p_1$, then the currents induced by the flux density pulsation due to $\nu_1 = 1$ in the rotor bars are zero.

This is why $\tau_s = \tau_r$ (stator slot pitch $\approx$ damper bar pitch) produces good results.

In addition, with a complete end-ring, rotor damper bar induced currents may flow between adjacent poles if

$$\frac{N_s}{p_1} \pm 1 = 2K_o \pm 1 \tag{7.202}$$

That is, when N_s/p_1 is even integer number, there will be induced currents in the rotor bars—due to stator slotting—between poles. These rotor bar-induced interpolar currents produce their own mmf:

$$F_{1d} = K_{1d} \cdot \left[\cos\left(p_1\theta_s - \nu_1 N_s \omega t \right) - \cos\left(p_1\theta_s + \nu_1 N_s \omega t \right) \right] \tag{7.203}$$

The mmf produces its own reaction flux density in the air gap:

$$B_{d1} = K_{1d} \cdot \left[\cos\left(p_1\theta_s - \left(\nu_1 N_s + p_1\right)\omega t \right) - \cos\left(p_1\theta_s - \left(\nu_1 N_s - p_1\right)\omega t \right) \right] \tag{7.204}$$

It is now clear that this density produces emf time harmonics of the frequency f':

$$f' = \frac{\omega}{2\pi} \cdot \left(\frac{\nu_1 N_s}{p_1} \pm 1 \right) \tag{7.205}$$

To avoid that such time emf harmonics would, in turn, induce pole to pole currents in the rotor cage, we need to fulfill the condition:

$$\frac{\nu_1 N_s}{p_1} \pm 1 = 2K_o \pm \frac{1}{2} \tag{7.206}$$

Such a condition may be met by displacing the groups of bars on each pole from the center line such that the displacement of cage bars between adjacent poles should be 1/2 stator slot pitch.

Now if N_s/p_1 ratio is an odd number, condition of Equation 7.206 is fulfilled automatically.

Not so far, the second slot harmonic ($\nu_1 = 2$) when $2N_s/p_1$ = even number.

With $W_s/g_e > 2$ and $W_t/W_s > 1.5$, the second permeance harmonic may be large. To avoid the slot ripple mmf for $\nu_1 = 2$, this simply means choosing the rotor bars pitch, τ_d to be half the stator slot pitch τ_s. When there would seem to be too many rotor bars, skewing the stator slots by one-half slot pitch might be the solution.

Fractionary q tends to eliminate slot ripple influence on the mmf, with $\tau_s \approx \tau_r$.

Note: While the above rationale produces intuitive suggestions to reduce emf time harmonics, their verification is now possible using FEM.

7.19 Losses and Efficiency

Four main categories of losses occur in SGs:

- Winding (copper) losses
- Core (iron) losses
- Mechanical losses
- Brush-ring electrical losses

The winding (copper) losses include stator winding losses and excitation (plus exciter losses if the exciter is mounted on the SG shaft) losses.

Core (iron) losses include the stator core fundamental (50(60) Hz) hysteresis, eddy current losses, and additional core losses.

Additional core losses are produced both by the field winding mmf (under no-load, also):

- On the rotor surface
- On the stator end laminations stacks, in the tightening plate and, by the stator mmf space harmonics, again on the rotor core surface

Another way of classifying the electromagnetic losses is to consider them as *no-load losses* and *short-circuit losses* (at given current). Such a division of losses corresponds to practical tests on SGs where such losses may be measured.

Therefore, despite the fact that there are notable differences between short circuit and full load in terms of magnetic saturation, the combination of no-load and short-circuit losses proves to be rather adequate in industry.

There are approximate analytical expressions to calculate the various components of no-load and short-circuit losses that have stood the test of time in industry [13].

In parallel, FEM has been recently applied to calculate various components of losses in SGs.

At least for rotor surface losses, it has been recently proved that analytical methods [14] correlate very well with FEM results [15].

7.19.1 No-Load Core Losses of Excited SGs

The no-load core losses of SG consist of the following:

- Fundamental stator core hysteresis losses, p_h
- Eddy current losses, p_{edo}
- Additional rotor pole shoe losses, p_{pso}
- Additional stator core losses in the lamination stacks p_{eFe10}

There are also additional "core" losses in the stator tightening end plates or pressure fingers and in the axial stator bolts that tighten the stator stack, but these are relatively smaller in a proper design.

The fundamental stator core losses P_{Fe10} are

$$p_{Fe10} = p_h + p_{edy} = \left[K_h \cdot p_{h0} \cdot \left(\frac{f}{50} \right) + K_{ed} \cdot p_{edo} \cdot \left(\frac{f}{50} \right)^2 \right] \cdot \left(\frac{B}{1.0} \right)^2 \cdot G \qquad (7.207)$$

where p_{ho} and p_{edo} are the specific hysteresis and, respectively, eddy current losses in 1 kg of lamination at 1.0 T and 50 Hz, and G is the weight of laminations material.

As known, the flux density B is far from uniform in the stator teeth or yoke; still an average has to be considered in Equation 7.207.

Even Equation 7.207 is a bit impractical as it needs separation of hysteresis and eddy current losses, which is not directly available in single-frequency tests.

Therefore, in general

$$p_{Fe10} \approx p_{10} \cdot \left(\frac{f}{50} \right)^{a_f} \cdot \left(\frac{B}{1.0} \right)^{a_b} \cdot G \qquad (7.208)$$

where the coefficients a_f and a_b may found from curve fitting the losses for a wide range of f (frequency) and B values.

Considering a uniform flux density distribution in stator teeth and yoke, the total stator core losses is

$$(p_{Fe10})_t = p_{10} \cdot \left(\frac{f}{50}\right)^{1.3} \cdot \left[K_{y1}\left(\frac{B_{y1}}{1.0}\right)^2 G_{y1} + K_{t1}\left(\frac{B_{t1/30}}{1.0}\right)^2 G_{t1}\right] \tag{7.209}$$

$B_{t1}/30$—tooth flux density at 30% slot height.

K_{y1} and K_{t1} are empirical factors allowing for loss augmentation due to machining the laminations ($K_{y1} \approx 1.3$–1.6, $K_{t1} \approx 1.8$–2.4). G_{y1}, G_{t1} are the stator yoke and teeth weights. Finally, p_{10} is the specific losses (in W/kg) at 1.0 T and 50 Hz.

When FEM flux distribution is calculated, it is possible to consider the instantaneous values of flux density in each volume element and to add up the losses in these elements.

Still, as the machining factors are so important FEM results in core losses have not yet produced spectacular results in terms of precision over analytical expressions that have been corrected overtime, based on industrial experience.

Though it is known that for same frequency and flux density the specific iron losses (in W/kg) differ in AC and traveling fields, p_{10} is still obtained from AC field measurements. At least in the stator yoke, the field is mainly of traveling character.

Therefore, one more open question remains here: Is it worth and how to consider the traveling field and AC field loss regions when determining the stator core losses?

The rotor surface no-load losses p_{pso} are produced by the field current caused air gap flux density variation on the rotor pole surface due to stator slot openings.

The frequency of currents induced on the rotor surface is $f_{ps} = N_s \cdot n$, where n is the rotor speed.

To reduce these losses, the pole shoes are made, whenever possible, from 1 to 1.8 mm thick laminations.

However, in turbogenerators and in high-speed hydrogenerators the rotors are made of solid iron.

In this case, large air gap g per stator slot opening W_{ss} ratios ($g/W_{ss} \geq 1$) are adopted to secure low flux density pulsations due to slot opening, and thus moderate pole surface losses result.

Though the situation differs to some extent in salient pole, in contrast to cylindrical pole, rotors, general analytical expressions have been proposed:

$$p_{pso} = 0.232 \cdot 10^6 \Delta \left[(K_{C_1} - 1) \cdot B_{go} \cdot \tau_s \right]^2 \cdot 2p_1 \cdot A_{pole} \cdot (N_s \cdot n)^{1.5} \tag{7.210}$$

where
    Δ—lamination thickness in the pole shoe [m]
    K_{c1}—stator slotting Carter coefficient
    B_{go}—no-load air gap flux density [T]
    τ_s—stator slot pitch [m]
    A_{pole}—rotor pole shoe area [m^2]
    N_s—number of stator slots
    n—rotor speed in rps

This expression is hardly valid for solid rotor poles when the depth of field penetration is dependent not only on the frequency ($N_s \cdot n$) but also on flux density level (due to magnetic saturation).

As the stator mmf space harmonics also produce eddy currents on the surface of rotor solid poles, we will treat this case in the paragraph on short-circuit (load) losses.

7.19.2 No-Load Losses in the Stator Core End Stacks

As can be seen in Figure 7.31, a part of air gap excitation flux paths close through the rotor pole tightening plates. If the latter are long enough, the flux lines meet the end-stack laminations in the stator at 90°. Consequently, the lamination effect does not work anymore and considerable iron losses are produced in the end-stacks.

Stepping the air gap in the end-stack reduces these losses to some extent.

Considering that both the stator and the rotor surfaces are magnetically equipotential, the flux densities B_{g1}, B_{g2}, B_{g3}, corresponding to the air gaps g_1, g_2, g_3 are

$$B_{g1} = B_{go} \frac{g \cdot K_c}{g_1}$$

$$B_{g2} = B_{go} \cdot \frac{g \cdot K_c}{g_2} \tag{7.211}$$

$$B_{g3} = B_{go} \cdot \frac{g \cdot K_c}{g_3}$$

The air gaps g_1, g_2, g_3 are considered half-circles, and thus their lengths are straightforward if the width of end-stack step length a/n is known:

An average axial flux density B_m is

$$B_m = \sqrt{\frac{B_{g1}^2 + 4B_{g2}^2 + B_{g3}^2}{6}} \tag{7.212}$$

The approximate (classical) formula for the end-stack losses in the teeth zone is

$$p_{ezo} \approx \frac{142 \times 10^{-3}}{\rho_{lam}} \cdot \left(B_m \cdot W_{ta} \right)^2 \cdot G_{t1} \tag{7.213}$$

W_{ta} is the width of stator teeth—an average of it.

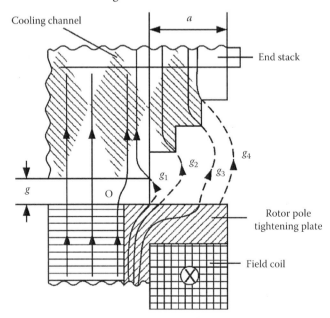

FIGURE 7.31 Excitation flux lines in the end-stack of stator.

The teeth may be splitted in the middle, radially, and then $W_{ta} = W_{tl}/2$ and thus the end-stack core losses are notably reduced. The lamination resistivity, $\rho_{lam} \approx 0.5 \times 10^{-6}$ Ω m and increases with temperature.

Therefore, the total no-load core losses p_{Feo} are as follows:

$$p_{Feo} = p_{Fe10} + p_{pso} + p_{ezo} \tag{7.214}$$

7.19.3 Short-Circuit Losses

The short-circuit losses contain quite a few components. The first one is the stator winding loss.

The stator winding (upper) AC resistance losses is

$$p_{co_{sc}} = 3(R_s)_{DC} \cdot K_R \cdot I_1^2 \tag{7.215}$$

The coefficient $K_R > 1$ accounts for the frequency (skin) effect on stator resistance.

By design in general $K_R < 1.33$ even in large power SGs. In large power SGs, to keep K_R less than 1.33, Roebel bar single-turn windings are used.

In Roebel bars, the single turn, divided into many elementary conductors in parallel, uses their transposition along the stack length such that all of them occupy all positions within the turn location and thus the circulation currents between elementary conductors is zero.

With the circulating currents made zero through transposition, the skin effect coefficient K_R is the same as for the elementary conductors in series (same current in all of them):

$$K_{Ra} = \varphi(\xi) + \frac{\left(n_l^2 - 1\right)}{3} \cdot \Psi(\xi) \tag{7.216}$$

$$\varphi(\xi) = \xi \cdot \frac{\sin h 2\xi + \sin 2\xi}{\cos h 2\xi - \cos 2\xi}; \quad \Psi(\xi) = 2\xi \cdot \frac{\sin h 2 - \sin 2\xi}{\cos h 2\xi - \cos 2\xi} \tag{7.217}$$

$$\xi = \alpha \cdot h_1; \quad \alpha = \sqrt{\pi \cdot f \cdot \mu_0 \cdot \sigma_{co} \cdot \frac{b_1}{W_s}}$$

where
b_1 is the elementary conductors width per layer in slot
n_l is the number of layers in a turn
h_1 is the elementary conductor height
σ_{co} is the copper electrical conductivity

For two-layer winding with chorded coils,

$$K_{Ra} = \varphi(\xi) + \frac{\left(n_l^2 (5 + 3\sin\beta) - 8\right)}{24} \cdot \Psi(\xi) \tag{7.218}$$

$\beta = \dfrac{y}{\tau} \cdot \dfrac{\pi}{2}$ is the coil span angle.

With some slots hosting same phase and some different phases, a kind of average value for K_{Ra} has to be calculated. Note that for the slots hosting, the same phase in both winding layers, the number of elementary conductor layers is $2n_l$ in Equation 7.216.

The skin effect tends to be much smaller in the end-connection, and even without transposition in this zone, the skin effect coefficient is smaller than in the stack (slot) zone.

To secure $K_{Ra} \leq 1.33$, the radial size h_1 of the elementary conductor in the Roebel bar has to be limited.

With the usual number of elementary conductor layers n_l going up to 12, even more, ξ has to be smaller than unity $\xi < 1$.

For this case,

$$K_{Ra} \approx 1 + \frac{n_l^2}{9} \cdot \xi^4 \quad \text{for single-layer winding} \tag{7.219}$$

and, respectively

$$K_{Ra} = 1 + n_l^2 \cdot \left(\frac{5}{72} + \frac{1}{24} \sin\beta \right) \cdot \xi^4 \quad \text{for double-layer winding} \tag{7.220}$$

For $\sigma_{co} = 4.8 \cdot 10^7 \ (\Omega \ \text{m})^{-1}, f = 60 \ \text{Hz}, b_1/W_s = 0.64$

$$\alpha = \sqrt{\pi \cdot f \cdot \mu_0 \cdot \sigma_{co} \cdot \frac{b_1}{W_s}} = \sqrt{\pi \cdot 60 \cdot 4.8 \cdot 10^7 \cdot 1.256 \cdot 10^{-6} \cdot 0.64} = 85.26 \ \text{m}^{-1} \tag{7.221}$$

To limit K_{Ra} to 1.33, it may be demonstrated that the maximum elementary conductor height h_1 is [10]

$$h_1 \approx \frac{15 \ \text{mm}}{\sqrt{n_l}} \tag{7.222}$$

The total Roebel bar copper height is $n_l \cdot h_1$:

$$n_l \cdot h_1 = h_{bar} \approx 15 \ \text{mm} \times \sqrt{n_l} \tag{7.223}$$

For example, for the number of elementary conductors $n_l = 16$ per bar, $h_{bar} = 15 \ \text{mm} \times \sqrt{16} = 60 \ \text{mm}$, $h_1 = 15/4 = 3.5 \ \text{mm}$.

With two layers in the windings, the total copper height is $2h_{bar} = 120 \ \text{mm}$. Considering all insulation layers, the total slot height could reach up to 200 mm. As slot width $W_s \leq (30–40)$ mm, it means that for the usual $h_{slot}/W_s < 6.5$.

Most of practical cases can be handled by Roebel bars up to highest power per unit (2100 MVA at 28 kV line voltage).

In reality, the number of elementary conductor layers may be increased to 48 when their radial height goes down to almost 2 mm, so even deeper slots may be allowed once conditions of Equations 7.222 and 7.223 are met.

Note that there are many other formulae corresponding to cases different from Roebel bar (characterized by the zero circulating currents).

For example, in Reference [6] for the skin effect losses per coil:

$$K'_{Ra} = \frac{n_l^2 \cdot \xi^2}{9} \cdot n_c^2 \quad \text{for bottom coil side} \tag{7.224}$$

$$K'_{Ra} = \frac{7 \cdot n_l^2 \cdot \xi^2}{9} \cdot n_c^2 \quad \text{for top coil side} \tag{7.225}$$

where n_c is the turns/coil ($n_c = 1$ for single-turn/bar coils).

Besides this, there is an extra loss coefficient due to circulating current K'_{rc}:

$$K'_{Rc} = \frac{C_e^2 \cdot \xi^4 \cdot n_l^2 \cdot b_e^2}{180} \cdot \left(1 + 15 n_c^2\right) \cdot K_{trans} \tag{7.226}$$

with C_e is the insulated strand height/bare conductor strand height. Also

$$b_e = \frac{l - 1/2\, n_c \cdot l_c}{l_{ec}} \tag{7.227}$$

where

l—total stator stack length
$1/2 \cdot n_c \cdot l_c$—half of cooling channels length
l_{ec}—end-connection coil length per machine side

There is no transposition in multiturn coils ($n_c > 1$), and then $K_{trans} = 1$.
For two turn coils (semi Roebel-transposition) $K_{trans} \approx 0.0784$.
As expected for full Roebel bar transposition ($n_c = 1$) $K_{trans} = 0$, as discussed earlier in this paragraph.
The total skin effect coefficient K_{Ra} is

$$K_{Ra} = 1 + K'_{Ra} + K'_{Rc} \tag{7.228}$$

7.19.4 Third Flux Harmonic Stator Teeth Losses

The third harmonic in air gap flux density, caused jointly by the stator mmf, magnetic saturation and the field winding mmf, produces extra iron losses in the stator teeth.
An approximate expression of these losses p_{Fe3} is

$$p_{Fe3} \approx 10.7 \times p_{10} \left(\frac{f}{50}\right)^2 \cdot \left(\frac{B_{3\sim}}{1}\right)^{1.25} \cdot G_{t1} \tag{7.229}$$

p_{10}—core losses (W/kg) at 1.0 T and 50 Hz, G_{t1}—stator teeth weight, $B_{3\sim}$ the third harmonic of air gap flux.
To a first approximation, only

$$B_{3\sim} \approx \frac{\left(B_{t1}\right)_{av}}{A_{1s}} \cdot \left(A_{3s} x_d + 1.3 A_{3r} \cdot x_{ad}\right) \tag{7.230}$$

where

$(B_{t1})_{av}$—average flux density in the stator teeth
A_{3s}—the third harmonic of stator mmf per pole (A/m)
A_{3r}—the third harmonic of field mmf per pole (A/m)
A_{1s}—the fundamental of stator mmf per pole (A/m)

Still, the magnetic saturation produced third harmonic, in phase with the fundamental, is not included in Equation 7.230. Also, the third flux density harmonic may be damped by the currents induced in the rotor poles or in damper windings.
This attention of $B_{3\sim}$ is not considered either.

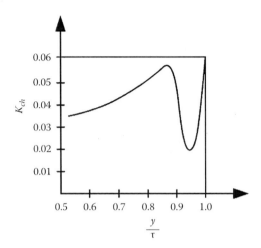

FIGURE 7.32 Coefficient K_{ch} in Equation 7.232.

The slot opening air gap permeance harmonics in the air gap flux density, now produced by the stator mmf, cause rotor surface (and cage) losses p_{pss} (at rated current):

$$p_{pss} \approx K' \cdot \left(\frac{1}{(K_{C1}-1)} \cdot \frac{2p_1}{N_s} \cdot x_{ad} \right)^2 \cdot p_{pso} \tag{7.231}$$

where p_{pso} are same losses but produced by the excitation mmf (at no-load).

$K' \approx 0.31$, for $g_{max}/g = 1$; 0.2 for $g_{max}/g = 1.5$; 0.15 for $g_{max}/g = 2.0$; 0.12 for $g_{max}/g = 2.5$. The 5th, 7th, 11th, 13th, …, stator mmf space harmonics produce the losses:

$$p_{pss}^a \approx \frac{2.1}{\sqrt[3]{q}} \cdot \left(\frac{x_{ad}}{(K_{C1}-1)} \cdot K_{ch} \right)^2 \cdot p_{pso} \tag{7.232}$$

q—slots per pole per phase, K_{C1}—Carter coefficient due to stator slotting, and K_{ch} (depending on coil chording) are shown in Figure 7.32.

In Reference 14, a more exact analytical approach, now verified by FEM and by some experiments, is given for the no-load and on-load solid rotor surface losses: (p_{pso}, p_{pss}, p_{pss}^a).

We will deal with this approach in some detail here as it seems to be notable progress in the art.

7.19.5 No-Load and On-Load Solid Rotor Surface Losses

A two-dimensional multilayer field theory was used to develop an analytical formula for solid rotor pole losses in SGs [14]: The losses per unit area of rotor solid pole are

$$\frac{p_{ps}}{A_{ps}} = \frac{B_{g1p}^2 \cdot \tau_s^2 \cdot f_{cl}^2 \cdot \sigma_{Fe}}{4 \cdot \mathrm{Re}(\underline{\gamma}_p)} \cdot K_L \tag{7.233}$$

where B_{g1p} is the first harmonic oscillation flux density in the rotor solid pole. It is a function of air gap no-load flux density pulsation, B_{g10p}:

$$B_{g1p} = B_{g10p} \cdot \frac{\left[\left(1 - \mu_{r3}\right) e^{-(2\pi \cdot g/\tau_s)} \left(\mu_{r1} - 1\right) + \left(1 + \mu_{r3}\right) e^{(2\pi \cdot g/\tau_s)} \left(\mu_{r1} + 1\right) \right]}{\left(1 - \mu_{r3}\right) e^{-(2\pi \cdot g/\tau_s)} \left(\mu_{r1} - \tau_s \left|\underline{\gamma}_p\right| / 2\pi\right) + \left(1 + \mu_{r3}\right) e^{(2\pi \cdot g/\tau_s)} \left(\mu_{r1} + \tau_s \left|\underline{\gamma}_p\right| / 2\pi\right)} \quad (7.234)$$

$$B_{g10p} = \alpha_1 \cdot B_m$$

where

$$\underline{\gamma}_p^2 = \left(\frac{2\pi}{\tau_s}\right)^2 - j \cdot \sigma_{Fe} \cdot 2\pi \cdot f_{el} \cdot \mu_0 \cdot \mu_{r,ab} \quad (7.235)$$

The relative magnetic permeabilities μ_{r1}, μ_{r3} refer to rotor pole, and, respectively, to stator core for B_{g10p}. In fact, B_{g1p} replaces B_{g10p} in Equation 7.234 to account for rotor pole induced current damping effect.

The ratio K_L is the so-called harmonic factor:

$$K_L = \left(\frac{B_m}{B_{g10p}}\right)^2 \sum_{\nu=1}^{\infty} \frac{1}{\sqrt{\nu}} \cdot \left(\frac{B_{g\nu op}}{B_m}\right)^2 \quad (7.236)$$

where
   α_1—flux density oscillation factor (Figure 7.33)
   B_m—mean flux density in the air gap over one stator slot pitch
   f_{el}—electrical frequency of induced currents
   $B_{g\nu op}$—peak value of the νth harmonic flux density at the rotor pole surface

Typical values for K_L are shown in Figure 7.34 [15].

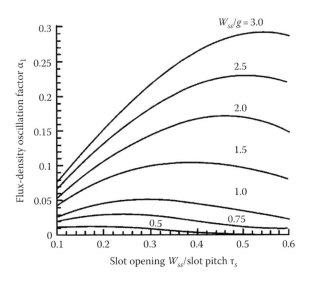

FIGURE 7.33 $\alpha_1 = B_{g10p}/B_m$ ratio versus slot opening/slot pitch.

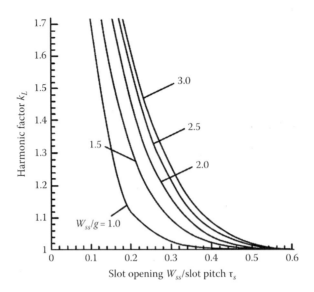

FIGURE 7.34 Harmonic factor K_L versus slot opening/air gap (W_{ss}/g) and slot pitch W_s/τ_s.

Still, in the inverse of depth of penetration formula (γ_p in Equation 7.235), there is the unique relative permeability of iron $\mu_{r,ab}$ to be determined iteratively from

$$H_o \cdot \sqrt{\mu_0 \cdot \mu_{r,ab}} = \frac{\tau_s}{2\pi} \cdot \left(0.85 \cdot B_{g1p}\right) \cdot \left|\underline{\gamma}_p\right| \cdot \frac{1}{\sqrt{\mu_0 \cdot \mu_{r,ab}}} \tag{7.237}$$

Therefore, the value of $\mu_{r,ab}(K)$ is calculated for $0.85\,B_{g1p}$, for a given value of B_{g10p}. With this value, from the magnetic material curve B_n/H_n, introduced on the right-hand side of Equation 7.237, a new value $\mu_{r,ab}(K+1)$ is obtained. After a few iterations, with some under relaxation

$$\mu_{r,ab}(K+1) = \mu_{r,ab}(K) + K_{un}\left[\mu_{r,ab}(K+1) - \mu_{r,ab}(K)\right]$$
$$K_{un} = 0.2 - 0.3 \tag{7.238}$$

sufficient convergence is obtained.

It has been also shown that the ripple loss per unit area of solid rotor is proportional to the following:

$$\frac{p_{ps}}{A_{ps}} \approx B_{g1p}^2 \cdot \tau_s^2 \cdot f_{el}^{1.5} \cdot K_L \cdot \sqrt{\sigma_{Fe}} \tag{7.239}$$

As the tangential field in the rotor pole dominates, the saturation effect is represented by

$$0.85 \cdot B_{g1p} \cdot \frac{\tau_s}{2\pi} \approx \mu_0 \cdot \mu_{r,ab} \cdot H_o \cdot \frac{1}{\sqrt{\pi \cdot f_{el} \cdot \sigma_{Fe} \cdot \mu_0 \cdot \mu_{r,ab}}} \tag{7.240}$$

$$H_o \cdot \sqrt{\mu_0 \cdot \mu_{r,ab}} \approx \sqrt{f_{el} \cdot \sigma_{Fe}} \cdot \tau_s \cdot B_{g1p} = S_D$$

In general, the ripple loss per unit area of solid rotor increases linearly with S_D [15].

It is recommended here to check, whenever possible, the flux density values in the air gap and in the solid rotor via FEM eddy current software, in order to secure, by fudge factors, good estimation level of rotor surface losses.

Negative sequence currents (limited by standards too) may also produce in solid rotor poles with smaller air gap surface rotor additional losses due to slot opening. Such losses may be reduced by damper cages but at the cost of higher excitation current (and losses), to produce the ideal emf value [16].

For special (rated) operation modes FEM may be used directly.

Additional losses occur in the stator tightening plates made of solid metal and in the teeth pressure fingers. There are again traditional analytical expressions for these losses [6]; but after FEM investigations, it seems that their use should be made with extreme care. The total short-circuit losses p_{sc} are

$$p_{sc} \approx p_{cos} + p_{Fe3} + p_{pss} \qquad (7.241)$$

7.19.6 Excitation Losses

Basically, the excitation losses contain the following:

- Field winding DC losses:

$$p_{ex} = R_f \cdot I_f^2 \qquad (7.242)$$

- Brush-slip ring losses (if any):

$$p_b = 2 \cdot \Delta V \cdot I_f; \quad \Delta V \approx 1 \text{ V by design} \qquad (7.243)$$

If a machine exciter is used, the total excitation power as seen from the SG, p_{exsh}, is

$$p_{exsh} = \frac{R_f \cdot I_f^2 + 2 \cdot \Delta V \cdot I_f}{\eta_{ex}} \qquad (7.244)$$

η_{ex} is the total efficiency of the exciter system.

In case of static exciter, η_{ex} is the static exciter efficiency. For a brushless exciter, the brush and slip-ring losses are replaced by the rectifier losses which are, in general, of the same orders of magnitude.

7.19.7 Mechanical Losses

Mechanical losses may be approximated to losses at no-load, with the SG unexcited ($I_1 = I_f = 0$):

- Bearing losses (axial and guidance bearings in vertical axial SGs)
- Ventilation losses (in the ventilator, its circuit, and shaft air friction)

When pumps are used, in direct hydrogen or water cooling, their input power has to be considered also.

$$p_{mec} = p_{bear} + p_{vent} + p_{air} \qquad (7.245)$$

Simplified formulae for losses have also been introduced over the years based on experience.

For air-cooled hydrogenerators with vertical shaft, here are a few such expressions:

- The bearing losses p_{bear} (includes both the axial and guidance bearing)

$$p_{\text{bear}} = p_{\text{bear}}^{\text{axial}} + p_{\text{bear}}^{\text{guide}} \approx \left[9.81 \cdot G_V \cdot U_{\text{bear}} + F_m + F_d \cdot U_{\text{bear}}^a \right] \cdot \mu_f \tag{7.246}$$

where

G_V—the total vertical mass of turbine and generator rotors plus the axial water-push mass in kg
U_{bear}—radial bearing peripheral speed
U_{bear}^a—the axial bearing peripheral speed
F_m—the magnetic pull due to estimated rotor eccentricity
F_d—the mechanical radial force due to eccentricity and mechanical rotational imbalance

$$F_m \approx 0.02 \cdot G_{MT+SG} \cdot 9.81$$

$$F_d \approx 1.256 \times 10^6 \cdot B_{gn}^2 \cdot \alpha_i \cdot D \cdot l_i \cdot \frac{e}{g} \tag{7.247}$$

where

B_{MT+SG}—the turbine plus SG rotor mass
α_i—the ideal rotor pole span ratio
D—the stator bore diameter
l—stator stack iron length
e—the eccentricity
g—the air gap
B_{gn}—the rated air gap flux density; and in general $e/g \approx 0.1$–0.2

The friction coefficient μ_f depends essentially on hydrogenerator speed (in rpm) and on coil temperature at 50°C, $\mu_f = 0.0035$ at $n = 50$ rpm and $\mu_f = 0.0108$ at $n = 500$ rpm.
- The ventilator + circuit losses, p_{vent}, may be written as follows:

$$p_{\text{vent}} \approx 1.1 \cdot Q_{air} \cdot U_{\text{vent}}^2 \tag{7.248}$$

Q_{air} is the air flow rate in m³/s required to evacuate the losses from the SG and U_{vent} is the peripheral average speed of the ventilator.
The rotor windage loss p_{air} is

$$p_{\text{air}} \approx C_a \cdot \left(2\pi n \right)^3 \cdot D_r^5 \cdot \left(1 + \frac{5 \cdot l_p}{D_r} \right) \tag{7.249}$$

- $C_a \approx (1.5$–$3) \cdot 10^{-3}$—a machine constant depending on rotor smoothness
- n—rotor speed in rps
- D_r—rotor external diameter
- l_p—total length of rotor poles

The total mechanical loss p_{mec} is

$$p_{\text{mec}} = p_{\text{bear}} + p_{\text{vent}} + p_{\text{air}} \tag{7.250}$$

The losses in the machine contain the no-load losses at rated voltage and short-circuit losses at rated current plus the mechanical losses and the excitation losses:

$$\sum \text{losses} = p_{Feo} + p_{sc} + p_{mec} + p_{excsh} \tag{7.251}$$

The load losses may be calculated based on no-load and short-circuit losses by adopting an appropriate value for field current I_f, stator current I_1 and the power factor angle φ.

On the other hand, the air flow rate Q_{air} may be calculated from:

$$Q_{air} = \frac{\sum \text{losses}}{C_a \cdot \Delta T_{air}} \tag{7.252}$$

where Σlosses is the total SG losses that have to be evacuated by the cooling air. $C_a \approx 1.1 \times 10^3$ W s/m² °C is the heat capacity factor of air and ΔT_{air} is the air warming temperature differential.

7.19.8 SG Efficiency

The SG efficiency is defined as the output (electrical)/input (mechanical) power:

$$\eta_{SG} = \frac{\left(P_2\right)_{electric}}{\left(P_1\right)_{mechanical}} = \frac{\left(P_2\right)_{electric}}{\left(P_2\right)_{electric} + \sum \text{losses}} \tag{7.253}$$

In industry, the efficiency is calculated for five load ratios: 25%, 50%, 75%, 100%, and 125%.

The voltage and speed are constant, so the no-load losses (or iron losses, mainly) p_{Feo} and mechanical losses p_{mec} remain constant. The "on-load" or the "short-circuit" losses p_{sc}—stator current related—and the excitation losses p_{exch} vary with their current squared:

$$p_{sc} = p_{scn}\left(\frac{I_1}{I_n}\right)^2; \quad K_{load} = \frac{I_1}{I_n} \tag{7.254}$$

$$p_{excsh} = p_{exchn}\left(\frac{I_f}{I_{fn}}\right)^2 + p_{brush} \cdot \left(\frac{I_f}{I_{fn}}\right); \quad P_2 = S_n \cdot \frac{I_1}{I_{1n}} \cdot \cos\varphi$$

Therefore, the efficiency at part load is

$$\eta_{SG} = \frac{K_{load} \cdot S_n \cdot \cos\varphi}{K_{load} \cdot S_n \cdot \cos\varphi + p_{Feo} + p_{mec} + p_{scn} \cdot K_{load}^2 + p_{exch} \cdot \left(I_f / I_{fn}\right)^2 + p_{brush} \cdot \left(I_f / I_{fn}\right) + p_{stray}} \tag{7.255}$$

Stray load losses are added.

Knowing all steady-state machine parameters, for each value of output power and cos φ, stator and field currents may be calculated after the power angle δ_V is determined along the way (Chapter 4). With losses calculated as in previous paragraphs, the total losses and efficiency at different load levels may be predetermined to assess the design goodness from this crucial viewpoint.

Note: In direct cooling SGs, the winding losses tend to dominate the losses in the machine, while for indirect cooling the nonwinding losses tend to be predominant.

It follows that for indirect cooling, the efficiency tends to be maximum above rated load and below rated load for direct cooling.

7.20 Exciter Design Issues

The excitation system supplies the SG field current to control either the terminal voltage or the reactive power to set point.

As already detailed in Chapter 6, the excitation system is provided with protective limiters.

For operation tied to weak power systems, power system stabilizers (PSS) are used to damp the oscillations in the range from 0.5 Hz to $2 \div 5$ Hz.

The specification of excitation systems is guided by IEEE standards 421 [17]. There are two key factors that define an excitation system: the transient gain and the ceiling forcing ratio (maximum/rated voltage at SG excitation winding terminal).

The transient gain has a direct impact on small signal and dynamic stability.

Too a small transient gain may fail to give the desired performance, while too a high transient gain produces faster response but may result in dynamic instability during faults. In this latter case, the PSS may have to be considered.

The transient gain refers to frequency of generator oscillations in the range of a few tenth of Hz to a few Hz.

Traditionally, excitation system controllers consist of a high steady-state gain to secure low steady-state control error and a lag-lead compensator to provide transient gain reduction.

In today's digital control systems PI, PID, or more advanced controllers such as Fuzzy Logic, H_∞, and sliding-mode are used.

Steady-state gains of 200 p.u. provide for $\pm 0.5\%$ steady-state regulation.

Transient gains vary in general from 20 to 100 with lower values for weak power systems, to avoid local mode power system oscillations (in 1–2 Hz range).

The excitation system ceiling voltage has also an impact on transient stability.

For static excitation systems 160%–200% of rated field voltage is set for design, to preserve stability after the clearing of a three-phase fault on the higher voltage side of the SG step-up transformer.

As lowering the SCR seems the trend in SG design today, to increase SG efficiency, increasing the excitation voltage ceiling becomes the pertinent option, to preserve transient stability desired performance.

The three-phase fault critical time (CCT), for constant ceiling forcing, decreases when the SCR decreases, both for lagging and for leading power factor. For leading power factor, however, the CCT is even smaller.

The required field voltage ceiling ratio thus has to be specified after critical clearing time limits are checked for lagging and leading power factor for the actual power system area where the SG will work. Such a local system is characterized by the % reactance looking out from the SG: 10% is a strong system and 40% is already a weak system.

Practical limits of ceiling (maximum) voltage is about 500 V_{DC} in general. But this maximum voltage may correspond to a 3 p.u. ceiling voltage ratio.

AC brushless exciters might allow smaller ceiling voltage ratio as they are less affected by power system faults.

In respect to exciter design we will treat here only the AC brushless exciter as it comprises two synchronous generators of smaller power rating.

An alternative brushless exciter, operational from zero speed, is made of wound rotor induction machine whose rotor winding output is diode rectified and supplies the SG excitation, while its stator winding is supplied at variable (controlled) voltage through a thyristor voltage amplitude controller (Figure 7.35b).

The frequency f_2 of the voltage induced in the AC exciter rotor is $f_2 = f_1 + n \cdot p_{1ex}$ and increases with speed, above $f_1 = 50(60)$ Hz.

As the induced rotor voltage amplitude V_2 also tends to increase with f_2, when speed increases, the voltage V_1 in the stator has to be reduced through adequate control.

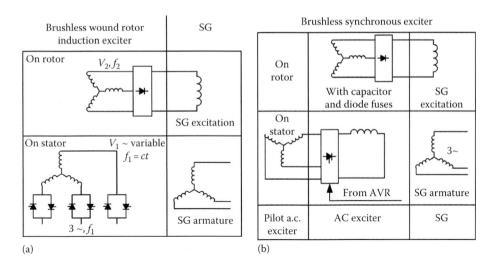

FIGURE 7.35 Brushless AC exciter: (a) synchronous exciter and (b) induction exciter.

At zero speed, all the power transmitted to the SG excitation comes from the exciter's stator; it is electrical, in other words.

As the speed increases, more and more of the excitation power comes from the shaft, especially with increasing the number of exciter's pole pair number $p_{1exc} > 3$.

The induction exciter voltage response is expected to be rather fast. However, the exciter stator winding has to deliver the whole power of SG excitation, should it be needed at zero speed as in the case of pump storage plant.

The induction exciter is used in industry also.

A typical AC brushless exciter contains a PM synchronous pilot exciter whose output is electronically controlled after rectification to supply the DC excitation of the inside-out synchronous exciter whose diode rectified output (with the rectification on the rotor) is connected directly to the excitation circuit of the SG (Figure 7.35a).

This way, neither brushes nor an additional power source is involved. However, if faster response is needed, the excitation circuit of the pilot exciter may be supplied directly from a separate power source through power electronics whose control variable is the output of the SG excitation control.

The pilot PM alternator is designed for frequency in the range of (180–450) Hz to cut the rectified output voltage pulsations. Also, the AC exciter is designed with a rather large number of poles $2p_1 = 6–8$, again to secure a rated armature frequency higher than 100 Hz, and thus low pulsations in the field winding current and small volume of the AC exciter magnetic circuit (armature yoke, especially).

The relationship between AC exciter phase voltage per phase V_{1s} (RMS value) and the rectified field winding voltage V_f is

$$V_{1s} \approx \frac{V_f}{2.34} \tag{7.256}$$

$$I_{1s} \approx 0.780 \cdot I_f \tag{7.257}$$

The voltage regulation of the diode rectifier in SG excitation circuit is neglected.

If better precision is needed, rectifier voltage regulation has to be considered, as in Figure 6.33.

Equations 7.256 through 7.257 imply an ideal (lossless) rectifier.

The AC exciter should be available to produce the ceiling voltage V_{fmax} at the SG excitation terminals, but also, during steady state, to allow for the I_{fn} (the rated load excitation current that corresponds to rated SG output power and power factor).

Besides rated and maximum field voltage and current ratings V_{fn}, I_{fn} for steady state and V_{fmax}, I_{fmax} for transients, the AC exciter has to provide a certain response time in providing the ceiling voltage V_{fmax} from the rated value V_{fn}.

Equally important, the excitation system has to have high efficiency at moderate weight and costs and be reliable. Here, a general design method for a brushless AC exciter is detailed. Optimization design issues will be treated later in this chapter for the SG in general.

7.20.1 Excitation Rating

As already alluded to, the AC exciter rating is given by V_{fn}, I_n, for steady state, and V_{fmax}, I_{fmax} for transients.

The exciter response time is also specified. The SG may be required to operate at rated MVA for 105% terminal voltage. As in such a case, the magnetic saturation level in the SG increases notably, the continuous power rating of the AC exciter will increase more than +5%.

Sometimes, the exciters have redundant circuitry.

The voltage response time (VRT) is the time in seconds for the excitation voltage to travel 95% the difference between V_{fmax} and V_{fn} (Figure 7.36).

Today's AC exciter systems are required to have VRT < 100 ms under full field current load.

7.20.2 Sizing the Exciter

Sizing the exciter is very similar to that of the main SG as already developed in this chapter. There are some peculiarities, as in Reference [18].

The reduction in VRT leads to severe limitation in the AC exciter subtransient reactance x'_d below 0.3. It may also be needed to provide for a rather strong cage in its stator to reduce the commutation reactance $x_c = \left(x''_d + x''_q\right)/2$ in the diode rectifier for lower voltage regulation and allowance for higher rated frequency (low yoke height).

As a good part of x'_d is the armature leakage inductance, reducing the latter is paramount. Wide and not-so-deep armature slots and short end-connections (chorded coils) are practical ways of achieving this goal. The leakage inductance of the AC exciter circuit has to be reduced, and star connection of AC exciter armature phases is preferred, in order to eliminate the third harmonic current and thus avoid limitations on coil span. It also requires a smaller number of slots (with two turns/slot) for same output DC voltage V_{fn}.

A smaller number of slots means lower manufacturing costs.

Two or more current paths in parallel may be necessary to produce the required V_{fmax} (V_{fn}).

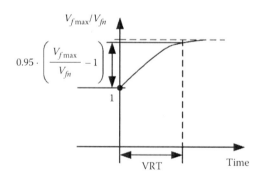

FIGURE 7.36 Voltage response time (VRT).

7.20.3 Note on Thermal and Mechanical Design

Thus far, the various no-load and load losses have been calculated. To remove the heat produced by losses, it is necessary to adopt and design appropriate cooling systems.

As already mentioned, SG cooling may be indirect—with air or hydrogen and direct (with water or hydrogen). Evaporative cooling is getting ground, also. There are many specific solutions as the power and speed (in rpm) goes up.

Analytical models are used in general for thermal design but FEM is more and more used for the scope.

The same rationale is valid for the mechanical design.

As the thermal and mechanical limitations observation influences the machine reliability and safety, their summarily treatment is not to be accepted. A detailed such analysis, on the other hand, needs rich space. This is why we stop here, drawing attention to specialized design books [5,6] while noticing that all manufacturers have their proprietary thermal design methods for SGs.

Forces, noise, and vibration are another important issues in the SG design. See for starters [6,19].

7.21 Optimization Design Issues

Thus far, we did present analytical expressions to do the dimensioning of an SG that fulfills given specifications. Also, the computation of resistances, inductances, and losses has been investigated.

Finally, the cost of active materials may be computed.

This process may be described as the general design and computer programs may be produced to mechanize it.

The general design produces a realistic (feasible) SG. It is a good starting point for optimization design. Optimization design requires the following:

- An initial practical design
- A machine model to calculate the parameters and performance (and objective functions) for given geometry and various design parameters
- A set of design variables such as the following:
  1. Stator bore diameter: D_{is}
  2. Slot width: W_{ss}
  3. Slot/tooth width ratio: W_{ss}/W_t
  4. A (A/m)—stator current loading
  5. W/m²—stator surface loading
  6. Flux density in stator teeth: B_{ts}
  7. Flux density in stator yoke: B_{ys}
  8. Flux density in the rotor pole body: B_p
  9. Thickness of field coil: W_c
  10. W/m² rotor surface loading

For each variable, the minimum and maximum values are set:

- A single- or multiple-cost function, such that generator costs plus the loss capitalized costs, is a way of accomodating various constraints (i.e., $X_d' < X_{d\max}'$, various temperature limitations, and SCR < SCR max) through penalty terms added to the cost function (Figure 7.37).
- A mathematical method (or more) to search for the optimal design [20] that produces the lowest cost function, observing the constraints and being able to deal with some design variables that are integer numbers (number of slots N_s, number of turns in the coil, etc.).
- There is a plethora of optimization methods of linear and nonlinear programming, indirect and direct, with constraints, deterministic, stochastic, and evolutionary.

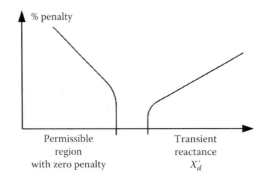

FIGURE 7.37 Penalty function for X'_d.

Such a complete optimization design software, with long use, based on Monte-Carlo optimization method is described in Reference 21.

More recently, evolution strategies and genetic algorithms have been applied to optimization subdesign of SGs. By subdesign, we mean here a partial problem in design such as the modification of rotor pole or of stator slot geometry to minimize losses and production costs [22].

When the cost (objective) function may not be stated in terms of a scalar type function Pareto optimal sets are used.

Pareto sets contain a number of solutions (designs) that are equivalent to each other in terms of trade-offs, that are so common in SG design.

Finding Pareto optimal sets of solutions to the problem, with stator slot and rotor pole dimensions as design variables, and losses and production costs to be simultaneously minimized, constitutes such an optimization subdesign problem in SGs.

For such a problem, the design variables could be as follows:

- Geometry and number of copper strands
- Dimensions of rotor pole
- Dimensions of field coil
- Pitch and diameter of damper cage bars
- The air gap

Typical constraints to be inforced through penalty functions are

$$X'_d < X'_{dmax}, \quad X''_d > X''_{dmin}$$

- Temperature rise
- Clearance between adjacent field coils
- Mechanical stresses in rotor poles and their dovetails
- Admissible flux density and current density limits

Evolutionary methods algorithms of optimization become necessary in nonsmooth and model-based objective (cost) functions. This is the case for SGs.

Figure 7.38 [22] shows the Pareto front (squares and line), with the initial solution (solid circles), while the other points are dominated solutions.

Two simultaneous goals—minimum cost and minimum losses (5000 $/KW of losses)—have been followed in this case.

The G.A. have been proved to produce better results [20] than other evolutionary strategies.

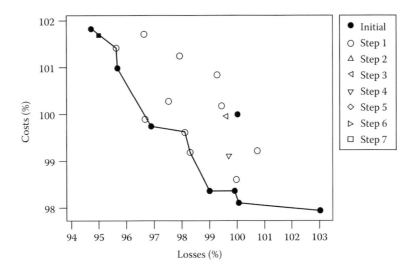

FIGURE 7.38 Pareto-front (squares and line) and the initial solution (solid circle).

7.21.1 Optimal Design of a Large Wind Generator by Hooke–Jeeves Method

Directly driven DC-excited synchronous wind-generators have been considered recently for 7.6 MW at 11.7 rpm, 400 V line voltage.

The machine model for design is nonlinear analytical one as developed in previous chapters.

There are 11 variables as in Table 7.1.

The objective function is complex as it contains terms such as initial (active materials) cost, loss energy capitalized cost, and overtemperature penalty cost (which has to be zero for the optimal solution): a simplified thermal model of the machine with an equivalent thermal transmissivity coefficient in W/(m² °C) is also introduced.

The modified Hooke–Jeeves optimization method is deterministic [20], and thus the probability to reach a global optimum is enhanced by starting the process from 15 to 20 different random initial variable vector values. When quite a few optimal results are very close, it means that most probably a global optimum was reached.

As the voltage is rather small, the number of current paths $a = 2p$ (number of poles), in order to use thin wire in making the coils, though circulatory currents between current paths (due to machine inherent/fabrication small asymmetries) have to be considered, though.

Being an integer, even number $2p$ is not included in the optimization variable vector; therefore, the optimization routine is run for a few chosen values of $2p = 140, 160, 180, 200, 220$. A large number of poles lead in general to better efficiency and lower weight, but it also leads to a larger outer diameter of rotor (Figure 7.39).

A few optimal design results are shown in Table 7.3.

In general, 40 iterations are enough to obtain a final solution, while the computation time on a contemporary desktop is a few tens of seconds.

From Table 7.3, as a compromise, we may choose solution 4, because of lower outer stator diameter (12 m, as in [23]).

The variation of efficiency versus load current in p.u. for the optimal solution is shown in Figure 7.40, for various load characters (power factors).

As the analytical/nonlinear model, though complex (including cross-coupling load-saturation) may not guarantee that the optimal solutions with calculated (i_d, i_q, i_F, power angle δ_V) can actually produce the rated average torque, FEM validation is mandatory. Besides, no-load air gap flux density average value and α_d and L_q inductance for rated power conditions should also be FEM calculated, for validation.

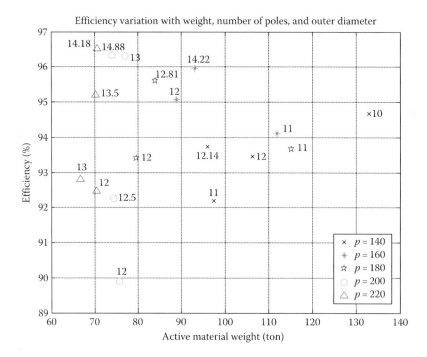

FIGURE 7.39 Rated generator efficiency dependence on the active material weight, number of poles, and generator outer diameter, which is indicated nearest to each point in meters.

TABLE 7.3 Representative Generators

No.	Poles	Outer Diameter (m)	Efficiency (%)	Weight (ton)	Comments
1	220	14.18	96.53	70.67	Better efficiency
2	220	13	92.81	66.81	Lowest weight
3	140	10	94.68	132.9	Lowest diameter
4	200	13	96.32	77.02	Optimum 1
5	180	12	93.4	80.0	Optimum 2

If errors larger than 3%–5% are perceived, fudge coefficients should be added in no-load air gap flux density formula, in electromagnetic torque and in L_d and L_q expressions (in an underrelaxed manner), and thus the FEM may be integrated (embedded) in the optimization loop. For a reasonable (101 s) computation time increase, a satisfactory solution is obtained without complicating too much the nonlinear analytical model.

As an example of FEM validation, the radial component of air gap flux density at no-load/rated voltage and for full load are shown in Figure 7.41.

A notable armature reaction is visible, as "dictated" by the initial cost reduction strong aim in the optimization function. A compromise between initial cost and loss capitalized cost (rather than efficiency) was observed in the optimal design case study of this directly driven large synchronous wind generator.

7.21.2 Magnetic Equivalent Circuit Population–Based Optimal Design of Synchronous Generator

Magnetic equivalent circuit (MEC) methods (model and mesh based) have been introduced to consider magnetic saturation more locally but also slotting, in order to calculate flux density distribution in electric machines with acceptable precision, but at a computation time at least 10 times smaller than with coupled-circuit–field (in full) finite element methods (FEMs) [8,24,25].

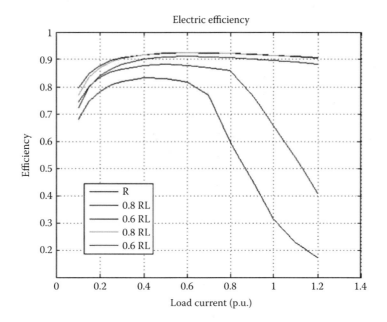

FIGURE 7.40 Efficiency versus p.u. current for various power factor loads.

A mesh-based MEC optimal design methodology of SG that uses a population-based (evolutionary) optimization algorithm is introduced in Reference 25.

Typical cross-section geometrical data and the MEC used—in the mesh variant, as it saves notable computation time—are shown in Figure 7.42a and b. Not less than 17 "genes", or variables, are used (Table 7.4).

The optimization is run for two distinct fitting functions:

$$\text{fit1} = -\text{mass}; \quad \text{fit2} = -p_{loss} \tag{7.258}$$

The sign (–) in Equation 7.258 is required if (already well tested) the genetic algorithm optimal design toolbox GOSET—Matlab is applied, as the latter looks for "maximae" and not "minimae" constraints are added also and, for a 2 kW Diesel engine small SG at 3600 rpm, the total computation time for one optimal solution is claimed to be in the 5 s range.

Constraints on SG design and loss (W)—mass (kg) Pareto—fronts are shown in Table 7.5 and in Figure 7.43.

A distinct merit of MEC analytical model, besides better local flux density calculation, is the possibility to calculate not only the average torque but also the torque pulsations as well.

Still the cross-coupling saturation and more involved machine geometry and windings require very skilled designers, capable of producing MEC for such cases.

And, anyway, FEM verifications are required.

More so with thermal and mechanical design constraints.

Therefore, the question arises: should we put more and more talent/time in developing ever more refined (MEC or other) nonlinear analytical model for optimal design, to reduce computation time, or we better invest more time in efficiently embedding FEM either at the end of optimization cycle to reduce errors in key performance indexes (no-load voltage, torque, L_d, L_q under critical load conditions) which lead to correction factors in the less-refined analytical model?

For PMSMs, already, the analytical model was replaced by limited (smart) FEM usage, for sinusoidal time/space distributions, in evolutionary multiobjective optimal design recent codes.

This path may be good also for SG optimal design.

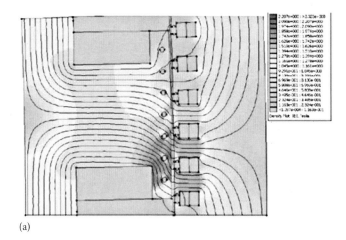

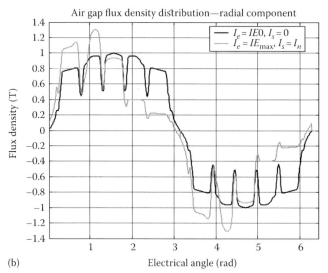

(a)

(b)

FIGURE 7.41 FEM flux lines for rated conditions (a) and radial air gap flux density (b).

7.22 Generator/Motor Issues

Pump-storage is used in some hydropower plants to replenish the water in the forebay each day and then deliver electric power during the peak demand hours.

During the pumping mode, the SG acts as a motor. To start such a large power motor, even on no-load, a pony motor has been traditionally used. It is also practical to use the back-to-back generator and motor starting when a unit works as a generator and, during its acceleration to speed, it supplies a motor unit and the two machines advance synchronously to the rated synchronous speed.

A single generator may supply one after the other all pump-storage units. Induction—motor—mode starting with self-synchronization is also feasible, at reduced voltage for large power units.

As the friction torque may be reduced, by refined high-pressure oil bearings, to less than 1% of rated torque, the power required to start the motor, with the water—less turbine—pump, has been notably reduced.

It is thus feasible to use a lower relative power rating static power converter to start the motor and disconnect the converter after the motor has been self-synchronized.

An SG designed to be used also as a motor for water pump storage undergoes about one start a day as a motor and one synchronization a day as a generator.

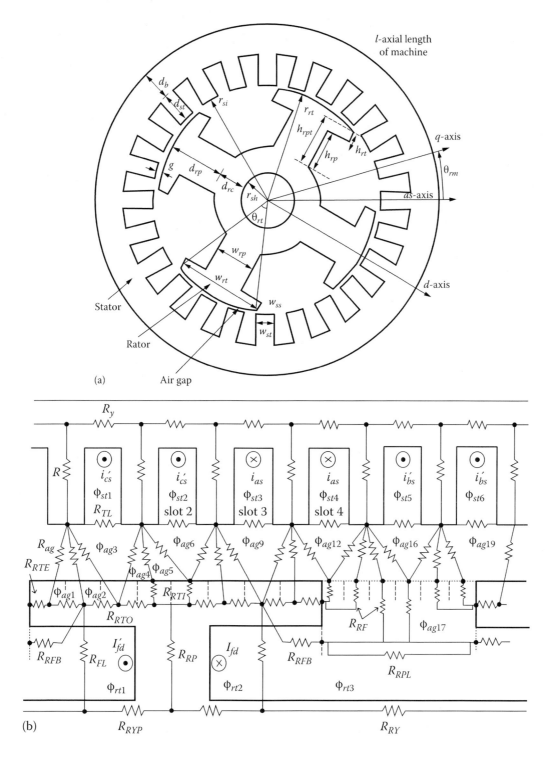

FIGURE 7.42 Cross-section geometry (a) and its MEC (*R*—reluctances, Φ—fluxes) and (b) for an SG with salient rotor poles.

TABLE 7.4   List of the Genes Used in the WRSM Design Program

#	Gene	Gene Description
1	r_{sh}	Shaft radius (m)
2	d_{rc}	Rotor core depth (m)
3	d_{rp}	Rotor pole depth (m)
4	g	Air gap length (m)
5	d_{st}	Stator tooth depth (m)
6	d_b	Stator yoke depth (m)
7	l	Stack length (m)
8	fw_{ss}	Fraction to find w_{ss}
9	fh_{rt}	Fraction to find h_{rt}
10	fw_{rt}	Fraction to find w_{rt}
11	fw_{rp}	Fraction to find w_{rp}
12	N_s	Turns per slot
13	N_{fd}	Number of field turns
14	I_s	Stator current, rms (A)
15	β	Stator phase angle (rad)
16	I_{fd}	Field current (A)
17	P_p	Number of pole pairs

TABLE 7.5   Constraints Used in Machine Design

Constraint On	Limit
Power output	$P_{elec} > 2.2$ kW
Stator/rotor current density	$J_s < 7.6$ A/mm^2, $J_r < 7.6$ A/mm^2
Phase voltage	$V_{as,pk} < 115$ V
Outer radius to length ratio	$r_o/l < 0.91$
Field conductor area	Afd < 1.3 mm^2 (16 gauge wire)
Relative permeability	$\mu_r > 200$

Self-synchronization for motoring and automatic synchronization for generation imply notable transients in currents and torque, and the machine has to be designed to withstand such demanding conditions.

Typically, the direction of motion is changed for motoring as required by the turbine-pump. High wicket gate water leakage may cause the "waterless" unit to start rotating still in the generator direction. To avoid the transients that the static power converter would incur in initiating the starting as a motor, the machine has to be stalled before beginning to accelerate it on no-load, in the motoring direction.

Asynchronous self-starting is used for low- and medium-power units, albeit using a step-down autotransformer or a variable reactor to reduce starting currents. For units above 20 MW pony motor, generator–motor back-to-back or static power converter starting should be used.

The development of multiple-level voltage source converters with IGBTs (up to 4.5 MW/unit) [22] and with GTOs or MCTs (up to 10 MW/unit and above) [27] might change the picture.

For the low power range (up to 3 MW), it is possible to use a bidirectional voltage—source IGBT converter, designed at generator–motor rating, that will not only allow startup for the motoring but also provide for generating and motoring at variable speed (to exploit optimally the turbine-pump). It is known that the optimal speed ratio between pumping and turbining is about 1.24. For high power, the multilevel voltage source GTO and MCT converters may replace the controlled rectifier current source converters that are used today [24] to accelerate pump-storage units (above 20 MW or so) and then self-synchronize them to the power grid.

The multilevel PWM back-to-back IGBT, GTO, MCT converters provide quality starting while inflicting lower harmonics both on the power line and on the motor side (Figure 7.44a and b). They are smaller inside and provide controllable (unity) power factor on the line side.

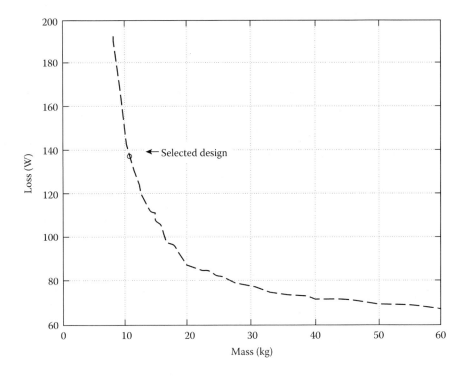

FIGURE 7.43 Pareto-optimal loss (W)–mass (kg), front for the 2 kW, 3600 rpm SG design. (From Bash, M.L. and Pekarek, S.D., *IEEE Trans. Emerg. Convers.*, EC-27(3), 603, 2012.)

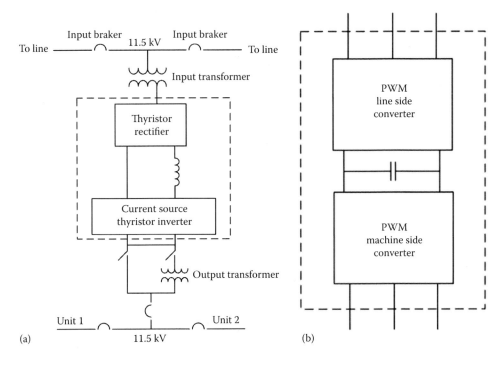

FIGURE 7.44 Static converter starting in pump-storage plants: (a) with rectifier—current (one phase shown) source inverter and (b) with voltage source converter.

As there is a step-up transformer between the static power converter and the machine, the starting cannot be initiated from zero frequency but from 1 to 2 Hz. This will cause transients that explain the necessary oversizing of the static power converter for the scope.

It goes without saying that to start the motor as a synchronous motor, a static exciter is required because full excitation is needed from 1 to 2 Hz speed. The static converters might be controlled for asynchronous starting (up to a certain speed), and thus their kVA ratings have to be increased.

Note that the "ideal" solution for pump storage is the doubly fed induction–generator motor that will be studied in *Variable Speed Generators*.

7.23 Summary

- By SG design, we mean here dimensioning to match given specifications of performance with constraints.
- Specifying SGs to be connected to power system is object of standardization. The proposed consolidated C 50.12, C 50.13 ANSI standards are typical for the scope.
- Besides manufacturers' standard, grid codes have been issued to tailor SG performance for power system typical and extreme operation modes. There may be conflicts between those two kinds of specifications and mitigation efforts are required to harmonize them.
- The first specification item is the SCR—the ratio between the short circuit and ratio current or the inverse of *d*-axis synchronous reactance in p.u. ($1/x_{dsat}$).

A larger SCR requires a larger air gap and, finally, a larger excitation mmf and losses, while it increases the peak power and thus the static stability.

As the transmission line reactive x_c at generator terminals increases, the beneficial effect of a larger SCR on static stability diminishes. A large x_c characterizes a weak power system.

Today's trend is to reduce SCR to 0.4–0.6 while the static stability is preserved by the ever-faster response of contemporary exciter systems.

- Stand-alone SGs are to be designed with higher SCR to secure reasonable voltage regulation with 0.8–0.9 lagging power factor loads.
- The second specification issue is the transient reactance x_d' that influences the transient stability. The smaller the x_d', the better, but there are limits to it due to machine geometry required to produce the design power.
- Rated power factor is essential in SG design: the lower the rated power factor, the larger the excitation power. In general, cos φ_n: 0.9–0.95 (overexcited).
- The maximum absorbed (leading) reactive power limit is determined by the SCR and corresponds to maximum power angle and (or) to stator end-core temperature limit.
- Fast control of excitation current is required to preserve SG stability and control the terminal voltage. Today's ceiling voltage ratio for excitation systems is 1.6–3.0. The limit is, in fact, determined by the heavy saturation of SG magnetic circuit. A response time of 50 ms "producing" the maximum ceiling voltage is required to qualify the response of excitation system as fast. Even with brushless rotary-machine configurations such a response is feasible. Static exciters are faster.
- Voltage and frequency variations are limited to contain the saturation level in the SG. A $\pm$ 2%–3% in both should be provided at rated power. Higher values are to be accepted for a limited amount of time to avoid SG overheating. These requirements are standardized or agreed upon between manufacturers and users.
- Negative phase-sequence voltages and currents are limited by standards to limit rotor damper cage and excitation additional losses (and overtemperatures).

Negative sequence voltage is limited to about 1% ($V_2/V_1\%$), while the current negative sequence i_2/i_1 may reach 4%–5% or more depending on the SG negative sequence reactance x_2 plus the step-up transformer short-circuit reactance x_T.

- Total voltage harmonic distortion (THD) limits are prescribed by grid codes to 1.5%–2%, with proposals to raise it to 3%–3.5% in the future.
- Temperature basis for rating power is paramount in SG design. It is between observable and hot-spot temperature. The observable temperature limit is set such that the hot-spot temperature is below 130°C for Class B and 155° for Class C insulation systems. The rated coolant temperature should also be specified. As the cold coolant temperature varies for ambient—following SGs, it seems reasonable to fix the observable temperature limit for a single cold-coolant temperature and calculate the SG MVA capability for different ambient temperatures.

The hot-spot temperature too will vary.

As the ambient temperature goes up, the SG MVA capability goes down.

- Start–stop cycles have to be specified to prevent cyclic fatigue degradation. For base load SG, 3,000 starts are typical, while 10,000 starts are characteristic to peaking-load or other frequently cycled units.
- Starting and operation as a motor (in pump storage units) faulty synchronizations, forces, armature voltage, runaway speed ratio all have to be specified in a pertinent design.
- The design process consists in sizing the SG in relation to the aforementioned specifications. The design refers to electromagnetic, thermal, and mechanical aspects. Only electromagnetic design has been followed in detail in this chapter.

  The electromagnetic design essentially comprises the following:

  - The output coefficient stator geometry
  - The selection of number of stator slots
  - The design of stator core and winding
  - The sizing of salient pole rotors
  - The sizing of cylindrical pole rotors
  - Open circuit saturation curve computation
  - Stator leakage reactance and resistance computation
  - Field mmf (current) at full load
  - Computation of synchronous reactances x_d, x_q
  - Sizing of damper cage
  - Computation of time constants and transient, subtrasient reactances
  - Exciter design
  - Generator/motor design issues
  - Optimization design issues

- The output coefficient C (min kVA/m³) is defined as the SG kVA per m³. It is a result of accumulated experience (art) and varies with the power/pole, number of poles pairs and type of cooling system.

The rotor diameter D is computed basically based on runaway speed U_{max}, with given SG synchronous speed and number of pole pairs p_1, for given frequency.

The rotor inertia H (s) has to be above a limit value such that to limit the maximum speed of SG upon loss of electric load, until the speed governor closes the fuel (water) input (unless special fast braking is provided as is the case of turbo generators).

- SGs may have integer or fractionary two-layer windings. We refer to q slots/pole/phase as integer or fractionary. Turbogenerators ($2p_1 = 2$, 4) use integer q windings in general ($q > 5$–6), while hydrogenerators today have fractionary windings.

- The choice of q determines the number of slots N_s of stator as $N_s = 6 \cdot p_1 q$. The selection of q also depends on the number of lamination segments N_c that constitute the stator core and on the number of detachable sections N_K in which the core may be divided and wound separately due to the large diameters considered:

$$N_K = 2 \quad \text{for } D < 4 \text{ m}$$

$$N_K = 4 \quad \text{for } D = 4\text{–}8 \text{ m}$$

$$N_K = 6(8) \quad \text{for } D > 8 \text{ m}$$

- It is imperative that all N_c segments should spread over an integer number of poles N_{ss}.
- For even N_{ss}, N_s is even and integer q is feasible. For fractionary q, N_{ss} may be an odd number and contain 3 as a factor.
- For fractionary $q = b + c/d$ symmetrical windings are obtained in general for $d/3 \neq$ integer. Windings are made with wave coils and one turn per coil and with lap coils when two or more turn per coil are used.

  For wave windings,

$$\frac{3c \pm 1}{d} = \text{integer}$$

The noise level is reduced if

$$3\left(b + \frac{c}{d}\right) \pm \frac{1}{d} \neq \text{integer}$$

with $c/d = 1/2$ $(b > 3)$ the subharmonics are eliminated.
- Apparently, once the value of q is settled, the number of turns per current path is obtained from the stator geometry and the total rated emf $e_t \approx 1.03\text{–}1.10$ p.u.

  For one turn (bar) per coil, the number of slots is directly related to the number of turns/path W_a and number of current paths a:

$$N_s = a \cdot W_a \cdot 3$$

It is now clear how crucial the choice of N_s is as W_a—integer—comes directly from the emf e_t value.
- To size the stator core, the air gap g has to be computed first, as the rotor diameter D_r is already known.

  Sizing g is directly related to stator rated linear current loading A, pole pitch τ, air gap flux density fundamental B_{g1} and the desired x_d (SCR).
- The rated voltage increases with SG power/unit and is larger for direct cooling. It is, otherwise, decided by the manufacturer.

  In general, $V_{nline} < 28$ kV, but cable winding (at 40–100 kV) SGs are in the advance development stage. They lead to the elimination of the step-up transformer that accompanies practically all SG tied to power systems.

- The salient pole rotors are designed for variable air gap under the pole, to render the air gap flux density more sinusoidal. The maximum to minimum air gap g_{max}/g ratio is up to 2.5.

  In practice, the radius of the pole shoe shape R_p is notably smaller than rotor radius $D/2$.

- Damper cage design is based on the main assertion that the cage bars total cross area represents (15%–30%) of stator slots area.

  The damper cage slot pitch τ_d is chosen to reduce losses in the damper cage due to stator mmf space harmonics. For fractionary stator windings ($q = (bd + c)/d$) when $bd + c > q$, it is feasible to choose $\tau_s = \tau_d$ (τ_s, stator slot pitch). For $bd + c < q$ or $q = b +1/2$, $\tau_s \neq \tau_d$.

  The damper bars of copper or brass are welded to end rings whose cross-section is about half the bars cross-section per rotor pole.

- The cylindrical rotor, made of single or multiple (axially) solid iron rings, is slotted along about 2/3 of its periphery to host the field coils. Lower air gap flux harmonics are obtained by placing the field coils in slots, while also this is the only mechanically feasible solution to withstand the centrifugal force stresses.

  To design the field winding, the rated field mmf is required. As at this stage in design this is not known a value of 2–2.5 times the rated stator mmf per pole is adopted.

- The open-circuit saturation curve may be obtained through an analytical model accounting for magnetic saturation or through FEM.

- The rated field mmf is obtained traditionally through the Potier diagram (Figure 7.24) or through the partial magnetization curves (Figure 7.25).

  The second method allows for cross-coupling saturation, while the first method does not and is more adequate for cylindrical rotor SGs.

  Based on the partial magnetization curves method, the synchronous reactances x_d may be determined as functions of i_d and i_q as families of curves. Alternatively, $\Psi_d(i_d, i_q)$ and $\Psi_q(i_d, i_q)$ may be obtained.

  Such information is essential in calculating with good precision the rated power angle and in general steady state performance of SGs.

- The computation of rotor resistances and various leakage inductances is rather straightforward if approximate formulae are accepted.

  They proved adequate in industry though today they may be computed by FEM also.

- The solid rotor behaves as an equivalent damper cage with variable parameters. A reliable analytical model to calculate these parameters is presented. This model has been recently validated by FEM.

- The various time constants: T'_{do}, T_{do}, T'_d, T''_d, T_a, T_D, and the transient and subtransient reactances (in p.u.) are straightforward once the synchronous reactance x_d, x_q, all leakage reactances and resistances are defined by analytical expressions.

  Note that $x'_d < x'_{d\,max}$.

  The negative sequence reactance x_2 is also calculated easily as follows:

$$x_2 = \left(x''_d + x''_q\right)/2 \quad \text{or} \quad x_2 = \sqrt{x''_d \cdot x''_q}$$

- Emf time harmonics are limited by standards. Higher q, coil chording, and fractionary q are all used to reduce emf time harmonics. The variable air gap obtained by shaping the rotor salient poles reduces the fifth and seventh harmonics while it introduces the third harmonic at the 10%–14% level.

- The interaction of field mmf fundamental with stator slot openings may produce the so-called slot-ripple emf effect: that is, additional damper cage losses at no-load. With N_s/p_1 as an odd number ideally zero damper cage additional losses at no-load occur.

- There are various losses in an SG, but they may be grouped into no-load and short-circuit electromagnetic losses and mechanical losses.
- Analytical formulae are given for most components of losses and, finally, the efficiency expression is obtained.

    The designer may thus calculate the total efficiency for various load conditions.

    Efficiency and losses are paramount performance indexes in an SG.
- Exciter design specifications and issues are treated in some detail for the brushless exciter: note the VRT for the excitation voltage to reach 95% of the ceiling (maximum) value (starting from the rated one).
- Once the general design is completed, a sound initial solution exists and a reliable model of the machine is in place. Optimization design may start once a set of variables is defined, a cost function is derived, and a mathematical optimization algorithm is chosen. A case study results with Hooke–Jeeves and genetic algorithm methods is presented.
- For generator/motor units, a rather detailed discussion of motor starting and synchronization methods is given. The unit has to withstand the rather daily start–stop encountered in generator/motor cycles for pump-storage operation modes.
- Numerical examples are given throughout this chapter to illustrate the various design issues.
- Inverter-connected SGs at variable speed [26,27] will be dealt with in *Variable Speed Generators*.

References

1. B.E.B. Gott, W.R. Mc Cown, J.R. Michalec, Progress in revision of IEEE/ANSI C 50 series of standards for steam and combustion turbine generators and harmonization with the IEC 60034 series, *Record of IEEE—IEMDC—2001 Conference*, MIT, 2001.
2. R.J. Nelson, Conflicting requirements for turbogenerators from grid codes and relevant generator standard, *Record of IEEE—IEMDC—2001, Meeting*, MIT, 2001, pp. 63–68.
3. C.E. Stephan, Z. Baba, Specifying a turbognerator's electrical parameters by standards and grid codes, *Record of IEEE—IEMDC—2001, Meeting*, MIT, 2001, pp. 57–62.
4. J.M. Fogarty, Connection between generator specifications and fundamental design principles, *Record of IEEE—IEMDC—2001, Meeting*, MIT, 2001, pp. 51–56.
5. K. Vogt, *Electrical Machines*, 4th edn., VEB Verlag Technik, Berlin, Germany, 1988, pp. 416 (in German).
6. J.H. Walker, *Large Synchronous Machines*, Clarendon Press, Oxford, U.K., 1981.
7. V.V. Dombrowski, A.G. Eremeev, N.P. Ivanov, P.M. Ipatov, M.I. Kaplan, G.B. Pinskii, *Design of Hydro-generators*, Vol. I + II, Energy Publishers, Moskow, Russia, 1965 (in Russian).
8. V. Ostovic, *Dynamics of Saturated Electric Machines*, Springer Verlag, New York, 1989.
9. R. Richter, *Electric Machines*, Vol. 2, Veralg Birkhäuser, Basel, Germany, 1963 (in German).
10. I. Boldea, S.A. Nassar, *The Induction Machine Handbook*, Chapter 6, CRC Press, Boc Raton, FL, 2001.
11. A.B. Danilevici, V.V. Dombrovski, E.I.A. Kazovski, *A. C. Machinery Parameters*, Science Publishers, Moskow, Russia, 1965 (in Russian).
12. I.S. Gheorghiu, A. Fransua, *Treatise of Electric Machines*, Vol. 4, Synchronous Machine, Technical Publications, Bucharest, Romania, 1972 (in Romanian).
13. E. Levi, *Polyphase Motors: A Direct Approach to Their Design*, John Wiley & Sons, New York, 1985.
14. K. Oberretl, Eddy current losses in solid pole shoes of synchronous machines at no-load and on load, *IEEE Transactions*, PAS-91, 152–160, 1972.
15. O. Drubel, R.L. Stall, Comparison between analytical and numerical methods for calculating tooth ripple losses in salient pole synchronous machines, *IEEE Transactions*, EC-16(1), 61–67, 2001.

16. D. Hiramatsu, T. Tokumasu, M. Fujita, M. Kakiuchi, T. Otaka, O. Sato, K. Nagasaka, A study of rotor surface losses in small-to-medium cylindrical synchronous machine, *IEEE Transactions on Energy Conversion*, EC-27(4), 813–820, 2012.

17. A. Murdoch, M.J. D'Antonio, Generator excitation systems—Performance specification to meet interconnection requirements, *Record of IEEE—IEMDC—2001*, MIT, June 2001.

18. M. Tartibi, A. Domijan, Optimizing A.C.—Exciter design, *IEEE Transactions*, EC-11(1), 16–24, 1996.

19. A. M. Knight, N. Karmaker, K. Weeber, Use of a permeance model to predict force harmonic components and damper winding effect in salient—Pole synchronous machines, *IEEE Transactions*, EC-17(4), 478–484, 2002.

20. S.S. Rao, *Optimization Methods in Engineering*, John Wiley & Sons Inc., New York, 1996.

21. O.W. Andersen, Optimized design of salient pole synchronous generators, *Record of ICEM-200*, Espoo, Finland, August 28–30, 2000, pp. 987–989.

22. E. Schlemmer, W. Harb, G. Lichtenecker, F. Muller, R. Kleinhaentz, Optimization of large salient pole generators using evolution strategies and genetic algorithms, *Record of ICEM—2000*, Espoo, Finland, 2000, pp. 1030–1034.

23. H. Polinder, D. Bang, R.P.J.O.M van Rooij, A.S. McDonald, M.A. Mueller, 10 MW wind turbine direct drive generator design with pitch and active stall control, *IEEE—IEMDC*, 2, 1390–1395, 2007.

24. M.L. Bash, S.D. Pekarek, Modeling of salient pole wound rotor synchronous machines for population-based design, *IEEE Transactions on Energy Conversion*, EC-26(2), 381–392, 2011.

25. M.L. Bash, S.D. Pekarek, Analysis and validation of a population-based design of a wound-rotor synchronous machine, *IEEE Transactions on Energy Conversion*, EC-27(3), 603–614, 2012.

26. J. Bäcker, J. Janning, H. Jebenstreit, High dynamic control of a three-level voltage-source-converter drive for a main strip mill, *IEEE Transactions*, EC-12(1), 66–72, 1997.

27. T.A. Meynard, H. Foch, P. Thomas, J. Courault, R. Iakob, M. Mahrstaedt, Multilevel converters: Basic concepts and industry applications, *IEEE Transactions*, IE-49(5), 955–964, 2002.

8

Testing of Synchronous Generators

Testing of synchronous generators (SGs) is performed to obtain the steady-state performance characteristics and the circuit parameters for dynamic (transients) analysis. The testing methods may be divided into standard and research types. Tests of a more general nature are included in standards that are renewed from time to time to include recent well-documented progress in the art. IEEE standards 115-1995 present a rather comprehensive plethora of tests for synchronous machines (SMs).

New procedures start as research tests. Some of them end up later on as standard tests. Standstill frequency-response (SSFR) testing of SGs for parameter estimation is such a happy case. In what follows, a review of standard testing methods and the incumbent theory to calculate the steady-state performance and, respectively, the parameter estimation for dynamics analysis is presented. In addition, a few new (research) testing methods with strong potential to become standard in the future are also treated in some detail.

Note that the term *research testing* may be used also with the meaning as "tests to research for new performance features of synchronous generators." Determination of flux density distribution in the air gap via search coil or that Hall probes is such an example. We will not dwell on such "research testing methods" in this chapter.

The standard testing methods are divided into the following aspects:

- Acceptance tests
- (Steady-state) performance tests
- Parameter estimation tests (for dynamic analysis)

From the nonstandard research tests, we will treat mainly "standstill step voltage response" and the on-load parameter estimation methods.

8.1 Acceptance Testing

According to IEEE Std. 115-1995 SG, acceptance tests are classified as follows:

A1. Insulation resistance testing
A2. Dielectric and partial discharge tests
A3. Resistance measurements
A4. Tests for short-circuited field turns
A5. Polarity test for field insulation
A6. Shaft current and bearing insulation
A7. Phase sequence
A8. Telephone-influence factor
A9. Balanced telephone-influence factor
A10. Residual component telephone-influence factor
A11. Line-to-neutral telephone-influence factor
A12. Stator terminal voltage waveform deviation and distortion factors

A13. Overspeed tests
A14. Line charging capacity
A15. Acoustic noise

8.1.1 A1. Insulation Resistance Testing

Testing for insulation resistance including polarization index, influences of temperature, moisture, and voltage duration are all covered in IEEE Std. 43-1974. If the moisture is too high in the windings, the insulation resistance is very low and the machine has to be dried out before further testing is performed on it.

8.1.2 A2. Dielectric and Partial Discharge Tests

The magnitude, wave shape, and duration of the test voltage are given in ANSI-NEMA MGI-1978. As the applied voltage is high, procedures to avoid injury to personnel are prescribed in IEEE Std 4-1978. The test voltage is applied to each electrical circuit with all the other circuits and metal parts grounded. During the testing of the field winding, the brushes are lifted. In brushless excitation SGs, the DC excitation leads should be disconnected unless the exciter is to be tested simultaneously. The eventual diodes (thyristors) to be tested should be short-circuited but not grounded. The applied voltage may be as follows:

- Alternating voltage at rated frequency
- Direct voltage (1.7 times the rated SGs voltage)—with the winding thoroughly grounded to dissipate the charge
- Very low-frequency voltage 0.1 Hz, 1.63 times the rated SG voltage

8.1.3 A3. Resistance Measurement

DC stator and field winding resistance measurement procedures are given in IEEE Std.-118-1978.

The measured resistance R_{test} at temperature t_{test} may be corrected to a specified temperature t_s:

$$R_s = R_{\text{test}} \frac{t_s + k}{t_{\text{test}} + k} \tag{8.1}$$

k = 234.5 for pure copper, in °C.

The reference field winding resistance may be DC measured either at standstill, with the rotor at ambient temperature, and the current applied through clamping rings or from a running test at normal speed. The brush voltage drop has to be eliminated from voltage measurement.

If the same DC measurement is made at standstill, right after SG running at rated field current, the result may be used to determine the field winding temperature at rated conditions, provided the brush voltage drop is eliminated from measurements.

8.1.4 A4–5. Tests for Short-Circuited Field Turns and Polarity Test for Field Insulation

The purpose of these tests is to check for field coils short-circuited turns, number of turns/coil, and short-circuit conductor size. Besides tests at standstill, a test at rated speed is required as short-circuited turns may occur at speed. There are DC and AC voltage tests for the scope. The DC or AC voltage drop across each field coil is measured. A more than +2% difference between the coil voltage drops indicates possible short circuits in the respective coils. The method is adequate for salient-pole rotors.

For cylindrical rotors, the DC field winding resistance is measured and compared with values from previous tests. A smaller resistance indicates that short-circuited turns may be present.

Also, a short-circuited coil with U-shaped core may be placed to bridge one coil slot. The U-shaped core coil is placed successively on all rotor slots. The field winding voltage or the impedance of the winding voltage or the impedance of the exciting coil decreases in case if there are some short-circuited turns in the respective field coil. Alternatively, a Hall flux probe may be moved in the air gap from pole to pole and the flux density value and polarity at standstill may be measured, with the field coil DC fed at (5%–10%) of rated current value.

If the flux density amplitude is higher or smaller than for neighboring poles, some field coil turns are short-circuited (or the air gap is larger/smaller) for the corresponding rotor pole. If the flux density does not switch polarity regularly (after each pole), the field coil connections are not correct.

8.1.5 A6. Shaft Current and Bearing Insulation

Irregularities in the SG magnetic circuit lead to a small axial flux that links the shaft. A parasitic current occurs in the shaft bearings and in the machine frame unless the bearings are insulated from stator core or from rotor shaft. The presence of PWM static converters in the stator (or rotor) of SG augments this phenomenon. The pertinent testing is performed with the machine at no-load and rated voltage. The voltage between shaft ends is measured with a high impedance voltmeter. The same current flows through the bearing radially to the stator frame.

The presence of voltage across bearing oil film (in uninsulated bearings) is also an indication of the shaft voltage.

If insulated bearings are used, their effectiveness is checked by shorting the insulation and observing an increased shaft voltage. Shaft voltage above a few volts, with insulated bearings, is considered unacceptable due to bearing in-time damaging and, in general, grounded brushes in shaft ends are necessary to prevent it.

8.1.6 A7. Phase Sequence

Phase sequencing is required for securing given rotation direction or for correct phasing of a generator prepared for power bus connection. As known, phase sequencing can be reversed by interchanging any two armature (stator) terminals.

There are a few procedures to check phase sequence:

- Using a phase sequence indicator (or induction machine)
- Using a neon-lamp phase-sequence indicator (Figure 8.1)
- Using the lamp method (Figure 8.1b)

When the SG no-load voltage sequence is 1–2–3 (clockwise), the neon lamp 1 will glow; while for the 1–3–2 sequence, the neon lamp 2 will glow. The test switch is open during these checks. The apparatus works correctly if, when the test switch is closed, both lamps glow with same intensity (Figure 8.1a).

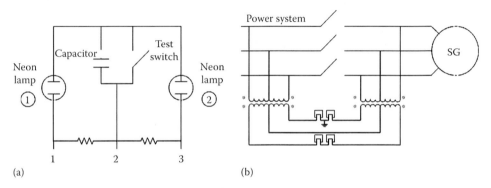

(a)   (b)

FIGURE 8.1 Phase sequence indicators (a) independent (1-2-3 or 1-3-2) and (b) relative to power grid.

With four voltage transformers and four lamps (Figure 8.1b), the relative sequence of SG phases to power grid is checked. For direct voltage sequence, all four lamps brighten and dim simultaneously; while for opposite sequence the two groups of lamps brighten and dim one after the other.

8.1.7 A8. Telephone-Influence Factor

Telephone-influence factor (TIF) is measured for the SG alone, with the excitation supply replaced by a ripple-free supply. The step-up transformers connected to SG terminals are disconnected. TIF is the ratio between the weighted rms value of SG no-load voltage fundamental plus harmonics E_{TIF} and the "rms" of the fundamental E_{rms}:

$$\text{TIF} = \frac{E_{TIF}}{E_{rms}}; \quad E_{TIF} = \sqrt{\sum_{n=1}^{\infty}\left(T_n E_n\right)} \tag{8.2}$$

T_n is the TIF weighting factor for the nth harmonic. If potential (voltage) transformers are used to reduce the terminal voltage for measurements, care must be exercised to eliminate influences on the harmonics content of the SG no-load voltage.

8.1.8 A9. Balanced Telephone-Influence Factor

For definition, see IEEE Std. 100-1992.

In essence, for a three-phase wye-connected stator, the TIF for two line voltages is measured at rated speed and voltage on no-load. The same factor may be computed (for wye connection) for the line-to-neutral voltages, excluding the harmonics 3, 6, 9, 12,….

8.1.9 A10. Residual-Component Telephone-Influence Factor

For machines connected in delta, a corner of delta may be open, at no load, rated speed, rated voltage. The TIF is calculated across the open delta corner.

$$\text{Residual TIF} = \frac{E_{TIF(opendelta)}}{3E_{rms(onephase)}} \tag{8.3}$$

Protection from accidental measured overvoltage is necessary and the use of protection gap and fuses to ground the instruments is recommended.

For machines that cannot be connected in delta, three identical potential transformers connected in wye in primary are open-delta-connected in their secondary. The neutral of the potential transformer is connected to the SG neutral point.

All measurements are now made as discussed, but in the open-delta secondary of the potential transformers.

8.1.10 A11. Line-to-Neutral Telephone-Influence Factor

The line-to-neutral TIF is measured in the secondary of a potential transformer whose primary is connected between an SG phase terminals and its neutral points. A check of values balanced, residual, and line-to-neutral TIFs is obtained from the following:

$$\text{Line-of-neutral TIF} = \sqrt{(\text{balanced TIF})^2 + (\text{residual TIF})^2} \tag{8.4}$$

8.1.11 A12. Stator Terminal Voltage Waveform Deviation and Distortion Factors

Definitions of deviation factor and distortion factor are given in IEEE Std. 100-1992. In principle, the no-load SG terminals voltage is acquired (recorded) with a digital scope (or digital data acquisition system) at high speed, and only a half-period is retained (Figure 8.2).

The half-period time is divided in J (at least 18) equal parts. The interval j is characterized by E_j. Consequently, the zero-to-peak amplitude of the equivalent sine wave E_{OM} is

$$E_{OM} = \sqrt{\frac{2}{J} \sum_{j=1}^{J} E_j^2} \tag{8.5}$$

A complete cycle is needed when even harmonics are present (fractionary windings). Waveform analysis may be carried out by software codes to implement the above method. The maximum deviation is ΔE (Figure 8.2). Then the deviation factor $F_{\Delta EV}$ is

$$F_{\Delta EV} = \frac{\Delta E}{E_{OM}} \tag{8.6}$$

Any DC component E_o in the terminal voltage waveform has to be eliminated before completing waveform analysis:

$$E_o = \frac{\sum_{i=1}^{N} E_i}{N} \tag{8.7}$$

With N samples per period.

By subtracting the DC component E_o from the waveform E_i, E_j can be obtained:

$$E_j = E_i - E_o; \quad j = 1,\ldots, N \tag{8.8}$$

The rms value E_{rms} is as follows:

$$E_{rms} = \sqrt{\frac{1}{N} \sum_{j=1}^{n} E_j^2}; \quad E_{OM} = \sqrt{2} E_{rms} \tag{8.9}$$

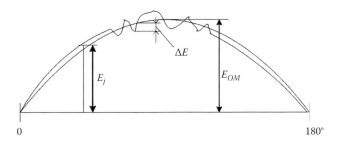

FIGURE 8.2 No-load voltage waveform for deviation factor.

The maximum deviation is searched for after the zero-crossing points of the actual waveform and of its fundamental are overlapped. A Fourier analysis of the voltage waveform is performed:

$$a_n = \frac{2}{N} \sum_{j=1}^{n} E_j \cos \frac{2\pi nj}{N}$$

$$b_n = \frac{2}{N} \sum_{j=1}^{n} E_j \sin \frac{2\pi nj}{N}$$

$$E_n = \sqrt{a_n^2 + b_n^2} \tag{8.10}$$

$$\phi_n = \tan^{-1}\left(b_n/a_n\right) \quad \text{for } a_n > 0$$

$$\phi_n = \tan^{-1}\left(b_n/a_n\right) + \pi \quad \text{for } a_n < 0$$

The distortion factor $F_{\Delta i}$ represents the ratio between the rms harmonic content and the rms fundamental:

$$F_{\Delta i} = \frac{\sqrt{\sum_{n=2}^{\infty} E_n^2}}{E_{\text{rms}}} \tag{8.11}$$

There are harmonic analyzers that output directly the distortion factor $F_{\Delta i}$. It should be mentioned that $F_{\Delta i}$ is limited by standards to rather small values detailed in Chapter 7 on SG design.

8.1.12 A13. Overspeed Tests

Overspeed tests are not mandatory but are performed upon request, especially for hydro or thermal turbine-driven generators that experience transient overspeed upon loss of load. The SG has to be carefully checked for mechanical integrity before overspeeding it by a motor (it could be the turbine (prime-mover) itself).

If overspeeding above 115% is required, it is necessary to pause briefly at various speed steps to make sure the machine is still OK. If the machine has to be excited, the level of excitation has to be reduced to limit the terminal voltage at about 105%. Detailed inspection checks of the machine are recommended after overspeeding and before starting it again.

8.1.13 A14. Line Charging Capacity

Line charging represents the SG reactive power capacity when at synchronism, at zero-power-factor, rated voltage and zero field current. In other words, the SG behaves as a reluctance generator at no load.
    Approximately,

$$Q_{\text{charge}} \approx \frac{3V_{\text{ph}}^2}{X_d} \tag{8.12}$$

X_d—d-axis synchronous reactance
V_{ph}—phase voltage (RMS)

The SG is driven at rated speed, while connected either to a no-load running overexcited SM or to an infinite power source.

8.1.14 A15. Acoustic Noise

Airborne sound tests are given in IEEE Std. 85-1973 and in ANSI Std. C50.12-1982. Noise is the undesired sound. The durations in hours of human exposure per day to various noise levels is regulated by Health Administration Agencies.

An omnidirectional microphone with amplifier weighting filters, processing electronics and an indicating dial makes a sound-level measuring device. ANSI "A," "B," "C" frequency domain is required for noise control and its suppression according to pertinent standards.

8.2 Testing for Performance (Saturation Curves, Segregated Losses, and Efficiency)

In large SGs, the efficiency is, in general, calculated based on segregated losses, measured in special tests that avoid direct loading.

Individual losses are as follows:

1. Windage and friction loss
2. Core losses (on open circuit)
3. Stray-load losses (on short circuit)
4. Stator (armature) winding loss: $3I_s^2 R_a$ with R_a calculated at specified temperature.
5. Field winding loss $I_{fd}^2 R_{fd}$ with R_{fd} calculated at a specified temperature

Among the widely accepted loss measurement methods, four of them are mentioned here:

1. The separate drive method
2. The electric input method
3. The deceleration (retardation) method
4. The heat transfer method

For methods (1), (2), and (3), two tests are run: one with open circuit and the other with short circuit at SG terminals. In open-circuit tests, the windage–friction plus core losses plus field winding losses occur. In short-circuit tests, the stator winding losses, windage–friction losses, and stray load losses, besides field winding losses are present.

During all these tests, the bearings temperature should be held rather constant. Also, the coolant temperature, humidity, gas density should be known and their appropriate influences on losses be considered. If a brushless exciter is used, its input power has to be known and subtracted from SG losses. When the SG is driven by a prime-mover that may not be uncoupled from SG, the prime-mover input and losses have to be known. In vertical shaft SGs with hydraulic turbine runners, only the thrust-bearing loss corresponding to SG weight should be attributed to the SG.

Dewatering with runner seal cooling water shut-off of the hydraulic turbine generator is required. Francis and propeller turbines may be dewatered at standstill and, in general, with manufacturer's approval. On the way of segregating open-circuit and short-circuit loss components, the no-load and short-circuit saturation curves have to be obtained from measurements, also.

8.2.1 Separate Driving for Saturation Curves and Losses

If the speed can be controlled accurately, the SG prime-mover itself can be used to drive the SG for open-circuit and short-circuit tests, but only to determine the saturation open-circuit and short-circuit curves, not for loss measurements.

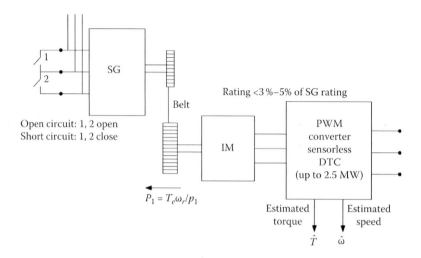

FIGURE 8.3 Driving the SG for open-circuit and short-circuit tests.

Therefore, in general, a "separate" directly or through belt gear coupled to the SG motor has to be used. If the exciter is capable of doing it, the best case is met. In general, the driving motor 3%–5% rating corresponds to open-circuit test. For small- and medium-power SGs, a dynamometer driver is adequate as the torque and speed of latter are measured, and thus the input power to the tested SG is known.

But today, when the torque and speed is estimated, in commercial direct torque-controlled (DTC) IM drives with PWM converters, the input to the SG for testing is also known, thus eliminating the dynamometer and providing for precise speed control (Figure 8.3).

- The open-circuit saturation curve

  The open-circuit saturation curve is obtained when driving the SG at rated speed, on open circuit and acquiring the SG terminal voltage, frequency, and field current.

  At least six readings below 60%, ten from 60% to 110%, two from 110% to 120%, and one at about 120% of rated speed voltage are required. Monotonous increase in field current should be observed. The step-up power transformer at SG terminals should be, in general, disconnected to avoid unintended high-voltage operation (and excessive core losses) in the latter.

  When the tests are performed at lower-than-rated speed (i.e., in hydraulic units), corrections for frequency (speed) have to be made. A typical open-circuit saturation curve is shown in Figure 8.4. The air gap line corresponds to the maximum slope from origin that is tangent to the saturation curve.

- The core, friction, and windage losses

  The aggregated core, friction, and windage losses may be measured as the input power P_{10} (Figure 8.3) for each open-circuit voltage-level reading. As the speed is kept constant, the windage and friction losses are constant (P_{fw} = const). Only the core losses P_{core} increase approximately with voltage squared (Figure 8.5).

- The short-circuit saturation curve

  The SG is driven at rated speed with short-circuited armature, while acquiring the stator and field currents; I_{sc} and I_f values should be read at rated 25%, 50%, 75%, and 100%. Data at 125% rated current should be given by the manufacturer, to avoid overheating the stator. The high current points should be taken first so that the temperature during testing stays almost constant. The short-circuit saturation curve (Figure 8.4) is a rather straight line as expected, because the machine is unsaturated during steady-state short circuit.

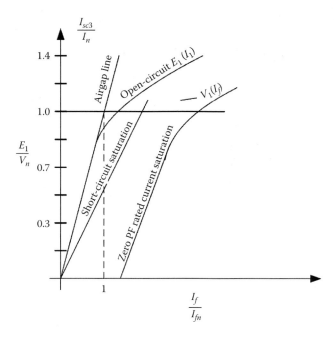

FIGURE 8.4 Saturation curves.

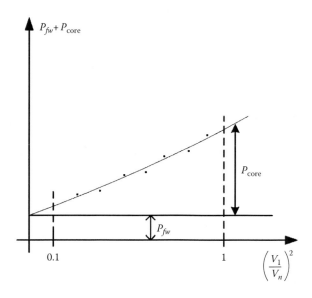

FIGURE 8.5 Core (P_{core}) and friction windage (P_{fw}) losses versus armature voltage squared at constant speed.

- The short-circuit and stray-load losses
  At each value of short-circuit stator current, I_{sc}, the input power to the tested SG (or the output power of the drive motor) P_{1sc} is measured. Their power contains the friction, windage losses, the stator-winding DC losses ($3I_{sc3}^2 R_{adc}$) and the stray-load loss P_{stray} load (Figure 8.6).

$$P_{isc} = P_{fw} + 3I_{sc3}^2 R_{sdc} + P_{strayload}$$ (8.13)

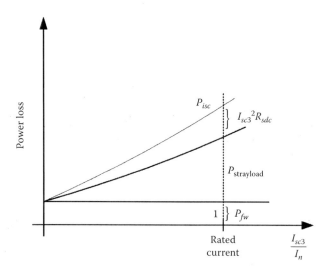

FIGURE 8.6 Short-circuit test losses breakdown.

During the tests, it may happen that the friction windage loss modifies because temperature rises. For a specified time interval, an open-circuit test with zero field current is performed when the whole loss is the friction windage loss ($P_{10} = P_{fw}$). If P_{fw} varies with more than 10%, corrections have to be made for successive tests.

Advantage may be taken by the presence of driving motor (rated at less than 5% SG ratings) to run zero-power-factor load tests at rated current and measure the field current I_f, terminal voltage V_1; from rated voltage downward.

A variable reactance is required to load the SG at zero power factor. A running underexcited SMSM may constitute such a reactance made variable through its field current. Adjusting the field current of the SG and SM leads to voltage increasing points on the zero-power-factor saturation curve (Figure 8.4).

8.2.2 Electric Input (Idle-Motoring) Method for Saturation Curves and Losses

According to this method, the SG performs as an unloaded synchronous motor supplied from a variable voltage constant frequency power rating supply. Though standards indicate doing these tests at rated speed only, there are generators that work as motors also. Gas-turbine generators with bidirectional static converters that use variable speed for generation and turbine starting as a motor are a typical example. The availability of PWM static converters with close to sinusoidal current waveforms recommends them for the no-load motoring of SG. Alternatively, a lower rating SG (below 3% of SG rating) may provide for the variable voltage supply.

The testing scheme for the electric input method is described in Figure 8.7.

When supplied from the PWM static converter, the SG acting as an idling motor is accelerated to the desired speed by a sensorless control system. The tested machine is vector controlled, and thus it is "in synchronism" at all speeds.

In contrast, when the power supply is a nearby SG, the tested SG is started either as an asynchronous motor or by accelerating the power supply generator simultaneously with the tested machine. Let us suppose that SG has been brought to rated speed and acts as no-load motor. To segregate the no-load loss components, the idling motor is supplied with descending stator voltage and descending field current such that to keep unity power factor conditions (minimum stator current). The loss components of (input electric power) P_{on} are:

$$P_{om} = P_{cu1o} + P_{coreo} + P_{fw} \tag{8.14}$$

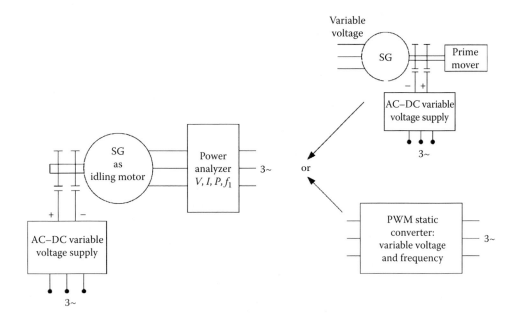

FIGURE 8.7 Idle-motoring test for loss segregation and open-circuit saturation curve.

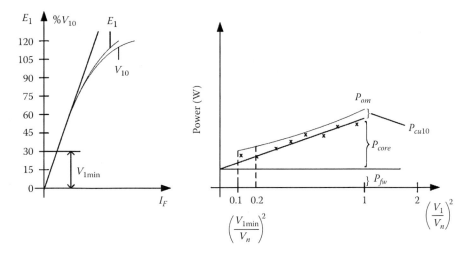

FIGURE 8.8 Loss segregation for the idle-running motor testing.

The stator winding loss P_{cul0} is

$$P_{cul0} = 3R_{adc}I_o^2 \tag{8.15}$$

and may be subtracted from the electric input P_{om} (Figure 8.8).

There is a minimum stator voltage V_{1min}, at unity power factor, for which the idling synchronous motor remains at synchronism. The difference $P_{om} - P_{cul0}$ is represented in Figure 8.8 as a function of voltage squared to underscore the core loss almost proportionally to voltage squared at given frequency (or to V/f in general). A straight line is obtained through curve fitting. This straight line is prolonged to the vertical axis, and thus the mechanical loss P_{fw} is obtained. Therefore, the P_{core} and P_{fw} have been segregated. The open-circuit saturation curve may be obtained as a bonus (down to 30% rated voltage) by neglecting the voltage drop over the synchronous reactance (current is small) in general and over the stator resistance, which is even smaller. Moreover, if the synchronous reactance X_s (an "average" of X_d and X_q) is known from design data, at unity power factor, the no-load voltage (the emf E_1) is

$$E_1 \approx \sqrt{\left(V_1 - R_a I_o^2\right) + X_s^2 I_o^2} \tag{8.16}$$

The precision in E_1 is thus improved and the obtained open-circuit saturation curve ($E_1(I_f)$) is more reliable. The initial 30% part of the open-circuit saturation curve is drawn as the air gap line (the tangent through origin to the measured open-circuit magnetic curve section). To determine the short-circuit and stray load losses, the idling motor is left to run at about 30% voltage (even at a lower value but for stable operation, Figure 8.9). By controlling the field current at this low, but constant, voltage, about six current step measurements are made from 125% to 25% of rated stator current. At least two points with very low stator current are also required. Again total losses for this idling test are

$$\left(P_{om}\right)_{\text{lowvoltage}} = P_{fw} + P_{\text{core}} + P_{cul} + P_{\text{strayload}} \tag{8.17}$$

This time, the test is performed at constant voltage, but the field current is decreased to increase the stator current up to 125%. Therefore, the stray load losses become important. As the field current is reduced, the power factor decreases, so care must be exercised to measure the input electric power with good precision. As the P_{fw} loss is already known from the previous testing, speed is constant, P_{core} is known from same source at same low voltage at unity power factor conditions, only $P_{\text{core}} + P_{\text{strayload}}$ have to be determined as a function of stator current.

Additionally, the dependence of I_a on I_f may be plotted from this low-voltage test (Figure 8.9). The intersection of this curve side with the abscissa delivers the field current that corresponds to the testing voltage $V_{1\,\text{min}}$ on the open-circuit magnetization curve. The short-circuit saturation curve is just a parallel to V curve side $I_a(I_f)$ (see Figure 8.10).

We may conclude that both separate driving and electric power input test allow for the segregation of all loss components in the machine and thus provide for the SG conventional efficiency computation:

$$\eta_c = \frac{P_1 - \sum P}{P_1} \tag{8.18}$$

$$\sum P = P_{fw} + P_{\text{core}} + P_{cul}\left(\frac{I_a}{I_n}\right)^2 + P_{\text{strayload}}\left(\frac{I_a}{I_n}\right)^2 + R_{fd} I_F^2$$

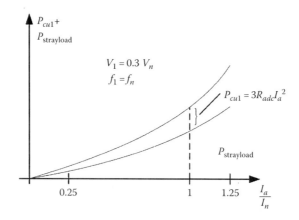

FIGURE 8.9 $P_{cul} + P_{\text{strayload}}$.

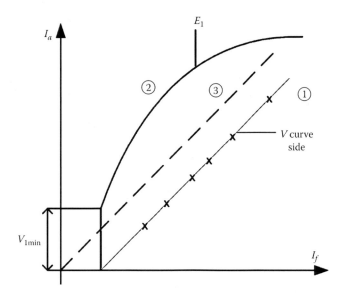

FIGURE 8.10 V curve at low voltage $V_{1\,\text{min}}$ (1), open-circuit saturation curve (2), and short-circuit saturation curve (3).

The rated stator winding loss P_{culn} and the rated stray load loss $P_{\text{strayload}}$ are determined in short-circuit tests at rated current, while P_{core} is determined from the open-circuit test at rated voltage. It is disputable if the core losses calculated in the no-load test and stray load losses from the short-circuit test are the same when the SG operates on loads of various active and reactive power level.

8.2.3 Retardation (Free Deceleration Tests)

In essence, after the SG operates as uncoupled motor at steady state to reach normal temperatures, its speed is raised to 110%. Evidently, a separate SG supply capable of producing 110% rated frequency is required. Alternatively, a lower rated PWM converter may be used to supply the SG to slowly accelerate the SG as a motor. Then, the source is disconnected. The prime mover of the SG has been decoupled or "dewatered."

The deceleration tests are performed with I_f, $I_a = 0$, then with $I_f \neq 0$, $I_a = 0$ (open circuit), and, respectively, for $I_f = $ const and $V_1 = 0$ (short circuit). In the three cases, the motion equation leads to

$$\frac{J}{p_1}\left(\frac{\omega_r}{p_1}\right)\frac{d\omega_r}{dt} = \frac{d}{dt}\left[\frac{J}{2}\left(\frac{\omega_r}{p_1}\right)^2\right] = -P_{fw}(\omega_r)$$

$$= -P_{fw}(\omega_r) - P_{core}(\omega_r)$$

$$= -P_{sc}(I_{sc3}) \tag{8.19}$$

The speed versus time during deceleration is measured, but its derivation with time has to be estimated through an adequate digital filter to secure a smooth signal.

Provided the inertia J is known a priori, at about rated speed, the speed ω_{rn} and its derivative $d\omega_r/dt$ are acquired and then used to calculate the losses for that rated speed as shown on the right-hand side of Equation 8.19 (Figure 8.11).

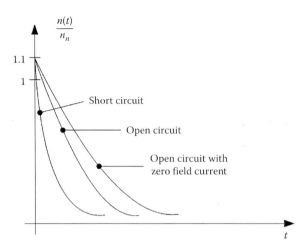

FIGURE 8.11 Retardation tests.

With the retardation tests performed at various field current levels, respectively, at different values of short-circuit current, at rated speed, the dependence of $E_1(I_F)$, $P_{core}(I_F)$, and $P_{sc}(I_{sc3})$ may be obtained. Also

$$P_{sc}\left(I_{sc3}\right) = 3R_{adc}I_{sc3}^2 + P_{strayload}\left(I_{sc3}\right) \tag{8.20}$$

This way, the open-circuit saturation curve $E_1(I_F)$ is obtained, provided the terminal voltage is also acquired.

Note: If the SG is excited from its exciter (brushless, in general), care must be exercised to keep the excitation current constant, and also the exciter input power should be deducted from losses.

If overspeeding is not permitted, the data are collected at lower than rated speed with the losses corrected to rated speed (frequency).

A tachometer, a speed recorder, or a frequency digital electronic detector may be used.

As already pointed out, the inertia J has to be known a priori for retardation tests. Inertia may be computed by quite a few methods such as through computation by manufacturer or from Equation 8.19, provided the friction and windage loss at rated speed $P_{fw}(\omega_{rn})$ are already known. With the same test set, the SG is run as an idling motor at rated speed and voltage for unity power factor (minimum current). Subtracting from input powers, the stator winding loss, $P_{fw} + P_{core}$, corresponding to no load at same field current, I_f is obtained. Then Equation 8.19 is used again to obtain J. Finally, the physical pendulum method may be applied to calculate J (see IEEE Std.115-1995, paragraph 4.4.15).

For SGs with closed-loop water coolers the calorimetric method may be used to measure directly the losses. Finally, the efficiency may be calculated from the measured output to measured input to SG. This direct approach is suitable for low- and medium-power SGs that can be fully loaded by the manufacturer to measure directly the input and output with good precision (better than 0.1%–0.2%).

8.3 Excitation Current under Load and Voltage Regulation

The excitation (field) current required to operate the SG at rated steady-state active power, power factor, and voltage is a paramount factor in the thermal design of a machine.

Two essentially graphical methods, the Potier reactance and the partial saturation curves, have been introduced in Chapter 7 on design. Here we will treat, basically, in more detail, variants of Potier reactance method.

To determine the excitation current under specified load conditions, the Potier (or leakage) reactance X_p, the unsaturated d and q reactance X_{du}, X_{qu}, armature resistance R_a and the open-circuit saturation curve are all needed. Methods for determining the Potier and leakage reactance are given first.

8.3.1 The Armature Leakage Reactance

We can safely say that there is not yet a widely accepted (standardized) direct method to measure the stator leakage (reactance) of SGs. To the valuable heritage of analytical expressions for the various components of X_l (see Chapter 7) finite elements method (FEM) calculation procedures have been added lately [1,2].

The stator leakage inductance may be calculated by subtracting two measured inductances:

$$L_l = L_{du} - L_{adu} \tag{8.21}$$

$$L_{adu} = L_{afdu} \cdot \frac{2}{3} \cdot \frac{1}{N_{af}}; \quad L_{afdu} = \frac{V_n \sqrt{2}}{\sqrt{3}\omega_n I_{fdbase}} \tag{8.22}$$

where

L_{du} is the unsaturated axis synchronous inductance

L_{adu} is the stator to field circuit mutual inductance reduced to the stator

L_{afdu} is the same mutual inductance but before reduction to stator

I_{fd} (base value) is the field current that produces, on the air gap straight line, the rated stator voltage on open stator circuit

Finally, N_{af} is the field to armature equivalent turn ratio, which may be extracted from design data or measured as shown later in this chapter.

The N_{af} ratio may be directly calculated from design data as follows:

$$N_{af} = \frac{3}{2} \frac{I_{abase}(A)}{i_{fd(base)}(A)}; \quad i_{fdbase} = I_{fdbase} \cdot l_{adu} \tag{8.23}$$

where i_{fdbase}, I_{fdbase}, and I_{abase} are all in amperes, but l_{adu} is in p.u.

A method to measure directly the leakage inductance (reactance) is given in Reference [3]. The reduction of Potier reactance when the terminal voltage increases is documented in Reference [3]. A simpler approach to estimate X_l would be the average of homopolar reactance X_o and reactance of the machine without the rotor in place, X_{lair}.

$$X_l \approx \frac{(X_o + X_{\text{lair}})}{2} \tag{8.24}$$

In general $X_o < X_l$ and $X_{\text{lair}} > X_l$, so an average of them seems realistic.

Alternatively

$$X_l \approx X_{\text{lair}} - X_{\text{air}} \tag{8.25}$$

X_{air} represents the reactance of the magnetic field that is closed through the stator bore when the rotor is not in place. From two-dimensional field and analysis, it was found that X_{air} corresponds to an equivalent air gap of τ/π (axial flux lines are neglected):

$$X_{\text{air}} = \frac{6\mu_o\omega_1 (W_1 kw_1)^2 \tau l_i}{\pi^2 p_1 (\tau/\pi)} \cong \frac{\omega_1 L_{adu}}{K_{ad}} \cdot \left(\frac{g\pi K_c}{\tau} \right) \tag{8.26}$$

where

τ is the pole pitch

g is the air gap

l_i is the stator stack length

$K_{ad} = L_{adu}/L_{mu} > 0.9$ (see Chapter 7)

The measurement of X_o will be presented later in this chapter, while X_{lair} may be measured through a three phase AC test at low voltage level, with the rotor out of place. As expected, magnetic saturation is not present in measuring X_o and X_{lair}. In reality, for large values of stator currents and (or) for very high levels of magnetic saturation of stator teeth or rotor pole, the leakage flux paths get saturated and X_l slightly decreases. FEM inquires [1,2] suggest that such a phenomenon is notable.

When identifying the machine model under various conditions, a rather realistic, even though not exact, value of leakage reactance is a priori given. The aforementioned methods may serve this purpose well, as saturation will be accounted through other components of machine model.

8.3.2 Potier Reactance

Difficulties in measuring the leakage reactance have led shortly after the year 1900 to the introduction by Potier as an alternative reactance (Potier reactance), which can be measured from the zero-power-factor load tests, at given stator voltage. At rated voltage tests, the Potier reactance X_p may be, in general, larger than the actual leakage reactance by as much as 20%–30%.

The open-circuit saturation and zero-power-factor rated current saturation curves are required to determine the value of X_p (Figure 8.12).

At rated voltage level, the segment $a'd' = ab$ is marked. A parallel to the air gap line through a' intersects the open-circuit saturation curve at point b'. The segment $b'c'$ is as follows:

$$\overline{b'c'} = X_p I_a \tag{8.27}$$

It is argued that the value of X_p obtained at rated voltage level may be notably larger than the leakage reactance X_l, at least for salient-pole rotor SGs. A simple way of correcting this situation is to apply the

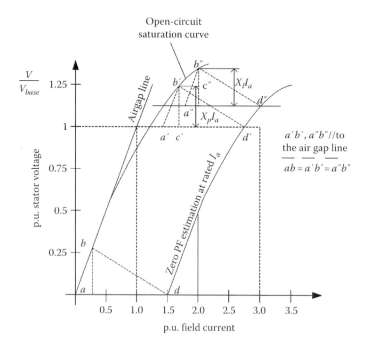

FIGURE 8.12 Potier reactance from zero-power-factor saturation curve.

same method but at higher level of voltage where the level of saturation is higher and thus the segment $\overline{b''c''} < \overline{b'c'}$ and thus $X'_p < X_p$ and approximately

$$X_l \approx X'_p \approx \frac{\overline{b''c''}}{I_a} \tag{8.28}$$

It is not yet clear what overvoltage level can be considered, but less than 110% is rather feasible if the SG may be run at such overvoltage without excessive overheating, even if only for obtaining the zero-power-factor saturation curve up to 110% rated voltage.

When the SM is operated as an SG on full load, other methods to calculate X_p from measurements are applicable [4].

8.3.3 Excitation Current for Specified Load

The excitation field current for specified electric load conditions (voltage, current, power factor) may be calculated by using the phasor diagram (Figure 8.13).

E_a—terminal phase voltage
I_a—terminal phase current
E_{as}—voltage back of X_{qu}
E_{Gu}—voltage back of X_{du}
δ—power angle
φ—power factor angle
R_a—stator phase resistance

For given stator current I_a, terminal voltage E_a, and power factor angle φ, the power angle δ may be calculated from the phasor diagram as follows:

$$\delta = \tan^{-1}\left[\frac{I_a R_a \sin\varphi + I_a X_{qu}\cos\varphi}{E_a + I_a R_a \cos\varphi + I_a X_{qu}\sin\varphi}\right] \tag{8.29}$$

$$I_d = I_a \sin(\delta+\varphi); \quad I_q = I_a \cos(\delta+\varphi) \tag{8.30}$$

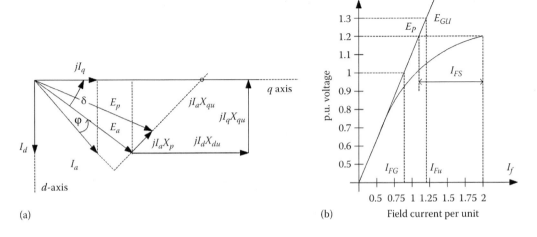

FIGURE 8.13 Phasor diagram with unsaturated reactances X_{du} and X_{qu}, (a) and the open-circuit saturation curve, (b).

Once the power angle is calculated, for given unsaturated reactances X_{du}, X_{qu} and stator resistance, the computation of voltages E_{QD} and E_{Gu}, with the machine considered as unsaturated, is feasible:

$$E_{Gu} = E_a + R_s I_a + j I_q X_{qu} + j I_d X_{du} \tag{8.31}$$

$$E_{Gu} = (E_a + R_a I_a)\cos\delta + X_{du}I_a \sin(\delta + \varphi) \tag{8.32}$$

Corresponding to E_{Gu}, from the open-circuit saturation curve (Figure 8.13b), the excitation current I_{FU} is found. The voltage back of Potier reactance E_p is simply (Figure 8.12)

$$E_p = E_a + R_s I_a + j X_p I_a$$
$$E_p = \sqrt{(E_a \sin\varphi + I_a X_p)^2 + (E_a \cos\varphi + I_a R_a)^2} \tag{8.33}$$

The excitation current under saturated conditions that produces E_p along the open-circuit saturation curve is (Figure 8.13b)

$$I_F = I_{Fu} + I_{FS} \tag{8.34}$$

The "saturation" field current supplement is I_{FS}. The field current I_F corresponds to the saturated machine and is the excitation current under specified load. This information is crucial for the thermal design of SG. The procedure is similar for the cylindrical rotor machine where the difference between X_{du} and X_{qu} is rather small (less than 10%).

For variants of this method, see Reference 4.

All methods in Reference 4 have in common a critical simplification: the magnetic saturation influence is the same in axes d and q, while the power angle δ calculated with unsaturated reactance X_{qu} is considered to hold also for all load conditions.

The reality of saturation is much more complicated, but these simplifications are still widely accepted, as they apparently allowed for acceptable results thus far.

The consideration of different magnetization curves along axes d and q, even for cylindrical rotors, and the presence of cross-coupling saturation, have been discussed in Chapter 7 on design, via the partial magnetization curve method. This is not the only approach to the computation of excitation current under load in a saturated SG, and new simplified methods to account for saturation under steady state are being produced [5].

8.3.4 Excitation Current for Stability Studies

When investigating stability, the torque during transients is mandatory. Its formula is still:

$$T_e = \Psi_d I_q - \Psi_q I_d \tag{8.35}$$

When damping windings effects are neglected, the transient model and phasor diagram may be used, with X'_d replacing X_d while X_q holds steady (Figure 8.14).

As seen from Figure 8.14, the total open-circuit voltage E_{total}, which defines the required field current I_{ftotal} is

$$E_{\text{total}} = E'_q + (X_{du} - X'_d)I_d + X_{adu}I_{fd} \tag{8.36}$$

$$E'_q = E_a \cos\delta + I_a X'_d \sin(\delta + \varphi) \tag{8.37}$$

This time, at the level of E'_q (rather than E_p), the saturation increment in excitation (in p.u.), $\Delta X_{adu}{}^*$ I_{fd} is determined from the open-circuit saturation curve (Figure 8.14). The nonreciprocal system of

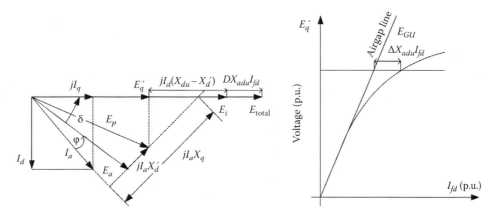

FIGURE 8.14 The transient model phasor diagram.

Equation 8.23 is used in p.u. It is again obvious that the difference in saturation levels in the d and q axis is neglected. The voltage regulation is the relative difference between the no-load voltage E_{total} (Figure 8.14) corresponding to the excitation current under load, and the SG rated terminal voltage E_{an}:

$$\text{Voltage regulation} = \frac{E_{total}}{E_{an}} - 1 \qquad (8.38)$$

8.3.5 Temperature Tests

By determining the temperature rise of various points in an SG, it is crucial to check its capability to deliver load power according to specifications. The temperature rise is calculated with respect to a reference temperature. Coolant temperature in now a widely accepted reference temperature.

Temperature rise at one (rated) or more specified load levels is required from temperature tests.

When possible, direct loading should be applied to do temperature testing, either at manufacturer's or at user's site.

Four common temperature testing methods are described here:

1. Conventional (direct) loading
2. Synchronous feedback (back-to-back M + G) loading
3. Zero-power-factor test
4. Open-circuit and short-circuit loading

8.3.5.1 Conventional Loading

The SG is loaded for specified conditions of voltage, frequency, active power, armature current, and field current (the voltage regulator is disengaged). The machine terminal voltage should be maintained within $\pm 2\%$ of rated value. If so, the temperature rise of different parts of the machine may be plotted versus p.u. squared apparent power (MVA)^2. As the voltage dependent and current dependent losses are unequal, in general, the stator winding temperature rise may be plotted versus armature current squared (A^2), while the field winding temperature, versus field winding dissipated power: $P_{exe} = R_F i_F^2 \, (\text{kW})$. Linear dependencies are expected. If temperature testing is to be done before commissioning the SG, then methods 2, 3, and 4 are to be used.

8.3.5.2 Synchronous Feedback (Back-to-Back) Loading Testing

Two identical SGs are coupled together with their rotor axes shifted with respect to each other by twice the rated power angle ($2\delta_n$). They are driven to rated speed before connecting electrically their stators together (C_1-open) (Figure 8.15).

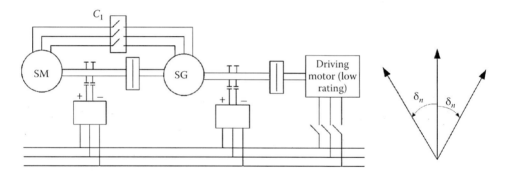

FIGURE 8.15 Back-to-back loading.

Then the excitation of both machines is raised until both SMs show the same rated voltage. With the synchronization conditions met, the power switch C_1 is closed. Further on, the excitation of one of the two identical machines is reduced. That machine becomes a motor and the other a generator. Then, simultaneously, SM excitation current is reduced and that of the SG is increased to keep the terminal voltage at rated value. The current between the two machines increases until the excitation current of the SG reaches its rated value, by now known for rated power, voltage, cos ϕ. The speed is maintained constant through all these arrangements. The net output power of the driving motor covers the losses of the two identical SMs, $2\Sigma p$, but the power exchanged between the two machines is the rated power P_n and can be measured. Therefore, even the rated efficiency can be calculated, besides offering adequate loading for temperature tests by taking measurements every half an hour until temperatures stabilize.

Two identical machines are required for this arrangement, along with the lower (6%) rating driving motor and its coupling. It is possible to use only the SM and SG, with SM driving the set, but then the local power connectors have to be sized to the full rating of the tested machines.

It is also possible, in principle, to virtually load two uncoupled SG fed from four quadrant AC–DC full power converters, one under generator and one under motor control, with a certain switching from M to G operation. Again, the power grid provides for losses in the two machines and converters system.

8.3.5.3 Zero-Power-Factor-Load Test

The SG works as a synchronous motor uncoupled at shaft; that is, a synchronous condenser (S.CON). As the active power drawn from the power grid is equal to SM losses, the method is energy efficient.

There are, however, two problems:

1. Starting and synchronizing the SM to the power source
2. Making sure that the losses in the S.CON equal to the losses in the SG at specified load conditions

Starting may be done through an existing SG supply that is accelerated in the same time with the SM, up to the rated speed. A synchronous motor starting may be used instead.

To adjust the stator winding, core losses and field windings losses, for given speed, and provide for rated mechanical losses, the supply voltage$(E_a)_{S.CON}$ and the field current may be adjusted.

In essence the voltage $(E_p)_{S.CON}$ has to provide the same voltage behind Potier reactance with the S.CON as with the voltage E_a of SG at specified load (Figure 8.16).

$$\left(E_p\right)_{S.CON} = \left(E_p\right)_{SG} \tag{8.39}$$

There are two more problems with this otherwise good test method for heating. One problem is the necessity of the variable voltage source at the level of the rated current of the SG. The second is related to the danger of too high a temperature in the field winding in SGs designed for larger than 0.9 rated power factor. The high level of E_p in the SG tests claims to a large field current (larger than for rated load in the SG design).

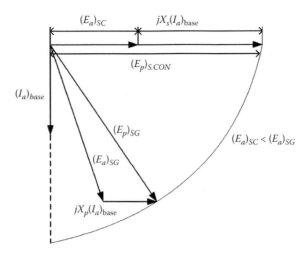

FIGURE 8.16 Equalizing the voltage back of Potier reactance for synchronous capacitor and SG operation modes.

Other adjustments have to be made for refined loss equivalence, such that the temperature rise to be very close to that in the actual SG at specified (rated load) conditions.

Open-circuit and short-circuit "loading"

As elaborated in Chapter 7 on design, the total losses are the SG under load is obtained by adding the open-circuit losses at rated voltage and the short-circuit losses at rated current and while correcting for duplication of heating due to windage losses.

In other words, the open-circuit and short-circuit tests are performed sequentially and the overtemperatures $\Delta t_t = (\Delta t)_{\text{open circuit}}$ and Δt_{sc} are added, while subtracting the additional temperature rise due to duplication of mechanical losses Δt_w:

$$\Delta t_t = \left(\Delta t\right)_{\text{open circuit}} + \left(\Delta t\right)_{\text{short circuit}} - \left(\Delta t\right)_w \tag{8.40}$$

The temperature rise $(\Delta t)_w$ due to windage losses may be determined by a zero excitation open-circuit run. For more details of practical temperature tests, see Reference [4].

8.4 Need for Determining Electrical Parameters

Prior to 1945–1965, SG transient and subtransient parameters have been developed and used to determine balanced and unbalanced fault currents. For stability response, constant voltage back-transient reactance model has been applied in same period.

The development of power electronics-controlled exciters has led, after 1965, to high initial excitation response. Considerably, more sophisticated SG and excitation control systems models have thus become necessary. Time-domain digital simulation tools have been developed and small-signal linear eigenvalue analysis has thus become the norm in SG stability and control studies. Besides second-order (two-rotor circuits in parallel along each orthogonal axis) SG models, third and higher rotor-order models have been developed to accommodate the wider frequency spectrum encountered by some power electronics excitation systems. These practical requirements led to the IEEE Std. 115A-1987 on standstill frequency testing to deal with third rotor-order SG models identification.

Tests to determine the SG parameters for steady states and for transients have been developed and standardized since 1965 at rather high pace. Steady-state parameters: X_d—unsaturated (X_{du}) and saturated (X_{ds})—and X_q—unsaturated (X_{qu}) and saturated (X_{qs})—are required first to compute the active and reactive power delivered by the SG at given power angle, voltage, armature current, and field current.

The field current required for given active, reactive powers, power factor, and voltage—as described in previous paragraphs—is necessary for calculating the maximum reactive power that the SG can deliver within given (rated) temperature constraints. The line charging-maximum absorbed reactive power of the SG at zero power factor (zero active power) is also calculated based on steady-state parameters.

Load-flow studies are based on steady-state parameters as influenced by magnetic saturation and temperature (resistances R_a and R_f). The subtransient and transient parameters (x_d'', x_d', T_d'', T_d', x_q'', x_q', T_{do}', T_{do}', T_{qo}'), determined by processing the three-phase short-circuit tests, are in general used to study the power system protection and circuit-breaker fault interruption requirements. The magnetic saturation influence on these parameters is also needed for better precision when they are applied at rated voltage and higher current conditions. Empirical corrections for saturation are still the norm.

Standstill frequency tests (SSFR) are used to determine mainly third-order rotor model sub-subtransient, subtransient and transient reactances, and time constants at low values of stator current (0.5% of rated current). They may be identified through various regression methods and some of them have been shown to fit well the SSFR from 0.001 to 200 Hz. Such a broad frequency spectrum occurs in very few transients as of such. Also, the transients occur at rather high and variable local saturation levels in the SG.

In just how many real-life SG transients such advanced SSFR methods are a must is not yet very clear. However, when lower frequency band response is required, SSFR results may be used, to produce the best "fit" transient parameters for that limited frequency band, through same regression methods.

The validation of these advanced third- (or higher) rotor-order models in most important real-time transients has led to the use of similar regression methods to identify the SG transient parameters from online admissible (provoked) transients. Such a transient is a 30% variation of excitation voltage. Limited frequency range oscillations of the exciter's voltage may also be performed to identify SG models valid for on-load transients, a posteriori.

The limits of short-circuit tests or SSFR taken separately appear clearly in such situations and their combination to identify SG models is one more way of improvising the SG modeling for on-load transients. As all parameter estimation methods use p.u values, we will revisit them here in the standardized form.

8.5 Per Unit Values

Voltages, currents, powers, torque, reactances, inductances, and resistances are required, in general, in p.u. values with the inertia and time constants left in seconds. Per unitization has to be consistent. In general, three base quantities are selected, while all the others are derived from the latter. The three commonly used quantities are three-phase base power, $S_{N\Delta}$, line-to-line base terminal voltage $E_{N\Delta}$ and the base frequency, f_N.

To express a measurable physical quantity in p.u., its physical value is divided by the pertinent base value expressed in same units. Conversion of a p.u. quantity to a new base is done by multiplying the old p.u. value to the ratio of the old to the new base quantity. The three-phase power $S_{N\Delta}$ of an SG is taken as its rated kVA (or MVA) output (apparent power).

Single-phase base powers S_N is: $S_N = S_{N\Delta}/3$.

Base voltage is the rated line to neutral voltage E_N:

$$E_N = \frac{E_{N\Delta}}{\sqrt{3}} \left(V, kV \right); \quad E_{N\Delta} = E_{LL} \tag{8.41}$$

RMS quantities are used.

When sinusoidal balanced operation is considered, the p.u. value of the line-to-line and of the phase neutral voltages is the same. Base line current I_N is that value of stator currents that corresponds to rated (base) power at rated (base) voltage:

$$I_N = \frac{S_{N\Delta}}{\sqrt{3}V_{N\Delta}} = \frac{S_N}{E_N} \left(A \right); \quad S_N = \frac{S_{N\Delta}}{3} \tag{8.42}$$

For delta-connected SGs, the phase base current $I_{N\Delta}$ is

$$I_{N\Delta} = \frac{S_{N\Delta}}{3E_{N\Delta}} \quad \text{as} \quad E_{N\Delta} = E_N \tag{8.43}$$

The base impedance Z_N is

$$Z_N = \frac{E_N}{I_N} = \frac{E_N^2}{S_N} = \frac{E_{N\Delta}^2}{S_{N\Delta}} \tag{8.44}$$

The base impedance corresponds to the balanced load phase impedance at SG terminals that requires the rated current I_N at rated (base) line-to-neutral (base) voltage E_N.

Note: In some cases, the field circuit-based impedance Z_{fdbase} is defined in a different way (Z_N is abandoned for the field circuit p.u. quantities.):

$$Z_{fbase} = \frac{S_{N\Delta}}{\left(I_{fdbase}\right)} = \frac{3S_N}{\left(I_{fdbase}\right)} \tag{8.45}$$

I_{fdbase} is the field current in Amperes required to induce at stator terminals, on open-circuit straight line, the p.u. voltage E_a:

$$E_a\left(\text{p.u.}\right) = X_{adu}\left(\text{p.u.}\right)I_a\left(\text{p.u.}\right) \tag{8.46}$$

I_a is p.u. value of stator current I_N
X_{adu} is the mutual p.u. reactance between the armature winding and field winding on the base Z_N

In general,

$$X_{du} = X_{adu} + X_l \tag{8.47}$$

where
X_{du} is the unsaturated d-axis reactance
X_l is the leakage reactance

The direct addition of terms in Equation 8.47 indicates that X_{adu} is already reduced to the stator. Rankin [6] designated i_{fdbase} as the reciprocal system.

In the conventional (nonreciprocal) system the base current of the field winding I_{fdbase} corresponds to the 1.0 p.u. volts E_a on open-circuit straight line.

$$i_{fdbase} = I_{fdbase}X_{dau} \tag{8.48}$$

The Rankin's system is characterized by equal stator/field and field/stator mutual reactances in p.u. values.

The correspondence between i_{fdbase} and I_{fdbase} is shown graphically in Figure 8.17.

All rotor quantities such as field winding voltage, reactance and resistance are expressed in p.u. values according to either the conventional (I_{fdbase}) or to reciprocal (i_{fdbase}) field current base quantity.

The base frequency is the rated frequency f_N. Sometimes the time has also a base value $t_N = 1/f_N$. The theoretical foundations and the definitions behind expressions of SG parameters for steady-state and transient conditions have been described in Chapters 5 and 6.

Here, they will be recalled at the moment of utilization.

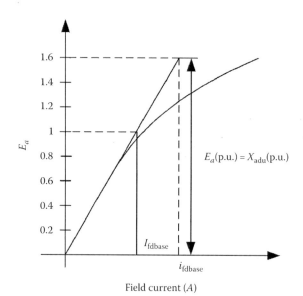

$$E_a(\text{p.u.}) = X_{\text{adu}}(\text{p.u.})$$

Field current (A)

FIGURE 8.17 I_{fdbase} and i_{fdbase} base field current definitions.

8.6 Tests for Parameters under Steady State

Steady-state operation of a three-phase SG usually takes place under balanced load conditions. That is, phase currents are equal in amplitude, but de-phased by 120° with each other. There are however situations when the SG has to operate under unbalanced steady-state conditions. As already detailed in Chapter 5 (on steady-state performance), unbalanced operation may be described through the method of symmetrical components. The steady-state reactances X_d, X_q, or X_1 correspond to positive symmetrical component, X_2 for the negative and X_o for the zero components. Together with direct sequence parameters for transients, X_2 and X_o enter the expressions of generator current under unbalanced transients.

In essence, the tests that follow are designed for three-phase SGs but, with some adaptations, they may be used for single-phase generators also. However, this latter case will be treated separately in Chapter 20, on small power single-phase linear motion generators.

The parameters to be measured for steady-state modeling of SG are as follows:

X_{du}—unsaturated direct axis reactance
X_{ds}—saturated direct axis reactance dependent on SG voltage, power (in MVA), and power factor
X_{adu}—unsaturated direct axis mutual (stator-to-excitation) reactance already reduced to the stator
$\quad$ ($X_{du} = X_{adu} + X_l$)
X_l—stator leakage reactance
X_{ads}—the saturated (main flux) direct axis magnetization reactance ($X_{ds} = X_{ads} + X_l$)
X_{qu}—unsaturated quadrature axis reactance
X_{qs}—saturated quadrature axis reactance
X_{aqs}—saturated quadrature axis magnetization reactance
X_2—negative sequence resistance
X_o—zero-sequence reactance
R_o—zero-sequence resistance
SCR—short-circuit ratio ($1/X_{du}$)
δ—internal power angle in radians or in electrical degrees

All resistances and reactances above are in p.u.

8.6.1 X_{du}, X_{ds} Measurements

The unsaturated direct axis synchronous reactance X_{du} (p.u.) can be calculated, as a ratio between two field currents:

$$X_{du} = \frac{I_{FSI}}{I_{FG}} \tag{8.49}$$

I_{FSI} is the field current on the short-circuit saturation curve that corresponds to base stator current.

I_{FG} is the field current on the open-loop saturation curve that holds for base voltage on the air gap line (Figure 8.18).

Also,

$$X_{du} = X_{adu} + X_l \tag{8.50}$$

When saturation occurs in the main flux path, X_{adu} is replaced by its saturated value X_{ads}:

$$X_{ds} = X_{ads} + X_l \tag{8.51}$$

As for steady state, the stator current in p.u. is not larger than 1.2–1.3 (for short time intervals), the leakage reactance X_l, may still be considered constant, though its differential component, may vary with load conditions, as suggested by recent FEM calculations [1,2].

8.6.2 Quadrature-Axis Magnetic Saturation X_q from Slip Tests

It is known that magnetic saturation influences also X_q and in general:

$$X_{qs} = X_{aqs} + X_l; \quad X_{qu} = X_{aqu} + X_l \tag{8.52}$$

The slip tests and the maximum lagging current test are considered here for X_q measurements.

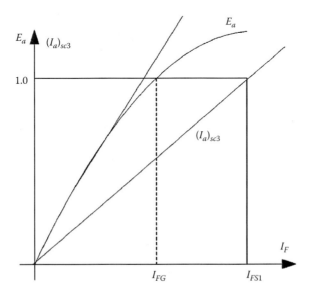

FIGURE 8.18 X_{du} calculation.

8.6.2.1 Slip Test

The SG is driven at very low slip (very close to synchronism) with open field winding. The stator is fed from three-phase power source at rated frequency. The slip is the ratio of the frequency of emf of the field winding to the rated frequency. A low-voltage spark gap across the field winding terminal will protect it from too high accidental emfs. The slip has to be very small, to avoid speed pulsations due to damper induced currents. The SG will pass from positions in direct to positions in quadrature axis, with variations in power source voltage E_a and current I_a from (E_{max}, I_{amin}) to (E_{min}, I_{amax}):

$$X_q = \frac{E_{a\,min}}{I_{a\,max}}; \quad X_d = \frac{E_{a\,max}}{I_{a\,min}} \tag{8.53}$$

The degree of saturation depends on the level of current in the machine. To determine the unsaturated values of X_d and X_q, the voltage of the power source is reduced, in general below 60% of base value V_N.

In principle, at rated voltage, notable saturation occurs, which at least for axis q may be calculated as function of I_q with $I_q = I_{amax}$. In axis d, the absence of field current makes $X_d (I_d = I_{amin})$ less representative, though still useful, for saturation consideration.

8.6.2.2 Quadrature Axis (Reactance) X_q from Maximum Lagging Current Test

The SG is run as a synchronous motor with no mechanical load at open-circuit rated voltage field current I_{FG} level, with applied voltages E_a less than 75% of base value E_N. Subsequently, the field current is reduced to zero and then reversed in polarity and increased again in small increments with the opposite polarity. During this time period, the armature current increases slowly until instability occurs at I_{as}. When the field current polarity is changed, the electromagnetic torque (in phase quantities) becomes thus:

$$T_e = 3p_1 \left(\Psi_d i_q - \Psi_q i_d \right) = 3p_1 \left[-X_{ad}I_F + \left(X_d - X_q \right) i_d \right] i_q \tag{8.54}$$

The ideal maximum negative field current I_F that produces stability is obtained for zero torque:

$$I_F = \left(X_d - X_q \right) \cdot i_d / X_{ad} \tag{8.55}$$

Also the flux linkages are now

$$\Psi_d \approx \frac{X_{ad}I_F}{\omega_1} + \frac{X_d}{\omega_1} I_{as} \tag{8.56}$$

With Equation 8.55, Equation 8.56 becomes

$$\Psi_d = \frac{X_q}{\omega_1} I_{as}; \quad \Psi_q = \frac{X_q}{\omega_1} I_q \approx 0 \tag{8.57}$$

The phase voltage E_a is

$$E_a = \omega_1 \Psi_s = \omega_1 \Psi_d = X_q I_{as} \tag{8.58}$$

See also Figure 8.19.

Finally,

$$X_q \approx \frac{E_a}{I_{as}} \tag{8.59}$$

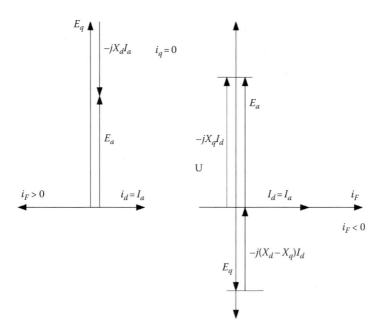

FIGURE 8.19 Phasor diagram for the maximum lagging current tests.

Though the expression of X_q is straightforward, running the machine as a motor on no-load is not always feasible. Synchronous capacitors, however, are a typical case when such a situation occurs naturally. There is some degree of saturation in the machine but mainly due to d-axis magnetizing mmfs (produced by excitation plus the armature reaction).

Catching the situation when stability is lost requires very small and slow increments in I_F, which needs special efforts in equipment.

8.6.3 Negative Sequence Impedance Z_2

Stator current harmonics may change the fundamental negative sequence voltage but without changes in the fundamental negative sequence current. This phenomenon is more pronounced in salient rotor pole machines with an incomplete damper ring or without damper winding because there is a difference between subtransient reactances X_d'' and X_q''.

Consequently, during tests, sinusoidal negative sequence currents have to be injected to the stator and the fundamental frequency component of the negative sequence voltage has to be measured for a correct estimation of negative sequence impedance Z_2. In general, corrections of the measured Z_2 are operated based on a known value of the subtransient reactance X_d''.

The negative sequence impedance is defined for a negative sequence current equal to rated current. A few steady-state methods to measure X_2 are given here.

- Applying negative sequence currents

    With the field winding short-circuited, the SG is driven at synchronous (rated) speed while supplied with negative sequence currents in the stator at frequency f_N. Values of currents around rated current are used to run a few tests and then claim an average Z_2, by measuring input power current and voltage.

    To secure sinusoidal currents, with a voltage source, linear reactors are connected in series. The waveform of one stator current should be analyzed. If current harmonics content is above 5%, the test is prone to appreciable errors. The parameters extracted from measuring power, P, voltage (E_a) and current I_a, per phase are

$$Z_2 = \frac{E_a}{I_a}, \quad \text{negative sequence impedance} \tag{8.60}$$

$$R_2 = \frac{P}{I_a^2}, \quad \text{negative sequence resistance} \tag{8.61}$$

$$X_2 = \sqrt{Z_2^2 - R_2^2}, \quad \text{negative sequence reactance} \tag{8.62}$$

- Applying negative sequence voltages

   This is a variant of the aforementioned method suitable for salient-pole rotor SG that lack damper windings. This time, the power supply has a low impedance to provide for sinusoidal voltage. Eventual harmonics in current or voltage have to be checked and left aside. Corrections to aforementioned value are [4] as follows:

$$X_{2c} = \frac{\left(X_d''\right)^2}{\left(2X_d'' - X_2\right)} \tag{8.63}$$

- Steady-state line-to-line short circuit tests

   As discussed in Chapter 4 during steady-state line-to-line short circuit at rated speed, the voltage of the open phase (E_a) is as follows:

$$\left(E_a\right)_{\text{openphase}} = \frac{2}{\sqrt{3}} Z_2 I_{sc2} \tag{8.64}$$

Harmonics are eliminated, from both, I_{sc2} and $(E_a)_{\text{openphase}}$. Their phase shift φ_{os} is measured:

$$Z_2 = \frac{E_a \sqrt{3}}{2I_{sc2}}; \quad R_2 = Z_2 \cos\varphi_{os} \tag{8.65}$$

When the stator null point of the windings is not available, the voltage E_{ab} is measured (bc line short circuit). Also the phase shift φ_{osc} between E_{ab} and I_{sc2} is measured. Consequently,

$$Z_2 = \frac{E_{ab}}{\sqrt{3}I_{sc2}} \text{ in } (\Omega); \quad R_2 = Z_2 \cos\varphi_{osc} \tag{8.66}$$

The presence of third harmonic in the short-circuit current needs to be addressed too [4]. Corrected X_{2c} value of X_2 is

$$X_{2sc} = \frac{X_2^2 + \left(X_d''\right)^2}{2X_d''} \tag{8.67}$$

Both X_2 and X_d' have to be determined at rated current level. With both E_{ab} and I_{sc2} waveforms acquired during sustained short-circuit tests, only the fundamentals are retained by post-processing and thus Z_2 and R_2 are determined with good precision.

8.6.4 Zero Sequence Impedance Z_o

With the SG at standstill, and three phases in parallel (Figure 8.20a) or in series (Figure 8.20b), the stator is AC supplied from a single-phase power source. The same tests may be performed also at rated speed.

Alternatively to phase angle φ_o measurements, the input power P_a may be measured:

$$Z_o = \frac{3E_a}{I_a}, \quad \text{for parallel connection} \tag{8.68}$$

$$R_o = \frac{3P_o}{I_a^2}, \quad \text{for parallel connection} \tag{8.69}$$

$$Z_o = \frac{E_a}{3I_a}, \quad \text{for series connection} \tag{8.70}$$

$$R_o = \frac{P_o}{3I_a^2}, \quad \text{for series connection} \tag{8.71}$$

Alternatively,

$$R_o = Z_o \cos\varphi_o \quad X_o = \sqrt{Z_o^2 - R_o^2} \tag{8.72}$$

Among other methods to determine Z_o, we mention here the two line-to-neutral sustained short circuit (Figure 8.21) with the machine at rated speed.

The zero sequence impedance Z_o is simply

$$X_o = \sqrt{Z_o^2 - R_o^2} \tag{8.73}$$

$$Z_o = \frac{E_{ao}}{I_{sc1}}; \quad R_o = Z_o \cos\varphi_{sc1} \tag{8.74}$$

In this test, the zero sequence current is, in fact, one-third of the neutral current I_{scl}. As Z_o is a small quantity care must be exercised to reduce the influence of power, voltage or current devices and of the leads in the measurements, for medium and large power SGs.

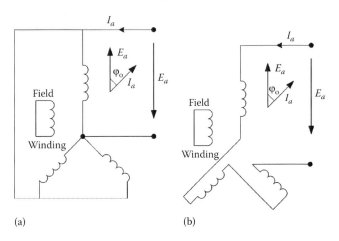

(a)   (b)

FIGURE 8.20 Single-phase supply tests to determine Z_o: (a) series parallel connection and (b) series connection.

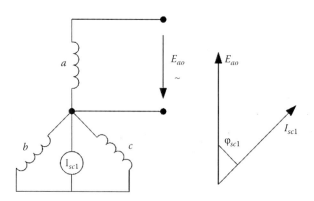

FIGURE 8.21 Sustained two line-to-neutral short circuit.

8.6.5 Short-Circuit Ratio

The short-circuit ratio SCR is obtained from the open-circuit and short-circuit saturation curves and is defined as in IEEE Std. 100-1992:

$$SCR = \frac{I_{FNL}}{I_{FSI}} \approx \frac{1}{X_d} \text{ (p.u.)} \tag{8.75}$$

I_{FNL} is the field current from the open-circuit saturation curve corresponding to rated voltage at rated frequency.

I_{FSI} is the filled current for rated armature current from the three-phase short-circuit saturation curve at rated frequency (Figure 8.22).

Though there is some degree of saturation considered in SCR, it is by no means the same as for rated load conditions; this way, SCR is only a qualitative performance index, required for preliminary design (see Chapter 7).

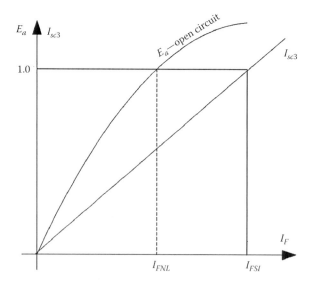

FIGURE 8.22 Extracting the SCR.

8.6.6 Angle δ, X_{ds}, X_{qs} Determination from Load Tests

The load angle δ is defined as the angular displacement of the center line of a rotor pole field axis from the axis of stator mmf wave (space vector), for a specified load. In principle, δ may be measured during transients also. When iron loss is neglected, δ is the angle between the field current produced emf and the phase voltage E_a as apparent from the phasor diagram (Figure 8.23).

With the machine loaded, if the load angle δ is measured directly by a separate sensor, the steady-state load measurements may be used to determine the steady-state parameters X_{ds}, X_{qs}, with known leakage reactance X_l ($X_{ds} = X_{ads} + X_l$) (Figure 8.21). The load angle δ may be calculated as follows:

$$\delta = \tan^{-1}\left(\frac{I_a X_{qs} \cos\varphi}{E_a + I_a X_{qs} \sin\varphi}\right) \tag{8.76}$$

As δ, I_a, E_a, φ are measured directly, the saturated reactance X_{qs} may be calculated directly for the actual saturation conditions. The I_d, I_q current components are available:

$$I_d = I_a \sin(\varphi + \delta) \tag{8.77}$$

$$I_q = I_a \cos(\varphi + \delta) \tag{8.78}$$

Also from Figure 8.21,

$$X_{ads}I_{fd} - \left(X_{ads} + X_l\right)I_d = E_a \cos\delta \tag{8.79}$$

As the leakage reactance X_l is considered already known, I_{fd}, E_a, δ measured directly (after reduction to stator), I_d from Equation 8.77, Equation 8.79 yields the saturated value of the direct-axis reactance $X_{ds} = X_l + X_{ads}$. The load angle may be measured by encoders and resolvers or by electronic angle shifting measuring devices, etc. [4].

The reduction factor of the directly measured excitation current to the stator from I_f to I_{fd} may be taken from design data or calculated from SSFR tests as shown later in this chapter. If load tests are performed for different currents and power factor angles at given voltage (P and Q) and δ, I_a, cos φ, I_{fd} are measured, families of curves $X_{ads}(I_d + I_{fd}, I_q)$ and $X_{qs}(I_q, I_d + I_{fd})$ are obtained. Alternatively, the $\Psi_d(I_q, I_d + I_{fd})$ and $\Psi_q(I_q, I_d + I_{fd})$ curve family is obtained:

$$\Psi_d = X_l i_d + X_{ads}\left(i_d + i_{fd}\right) \tag{8.80}$$

$$\Psi_q = X_{qs} i_q \tag{8.81}$$

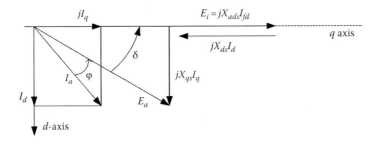

FIGURE 8.23 Phasor diagram for zero losses.

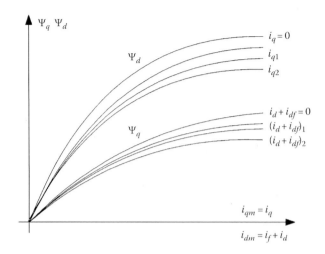

FIGURE 8.24 *d* and *q* axes flux versus current curves family.

Based on such data, curve fitting functions may be built. Same data may be obtained from FEM calculations. Such a confrontation between FEM and direct tests steady-state parameter saturation curve families (Figure 8.24) for online large SGs is still awaited for, while simplified interesting methods to account for steady-state saturation, including cross-coupling effects, are still produced [7,8].

The saturation curve family, $\Psi_d(I_q, I_d + I_{fd})$ and $\Psi_q(I_q, I_d + I_{fd})$ may be determined from standstill flux decay tests.

8.6.7 Saturated Steady-State Parameters from Standstill Flux Decay Tests

To account for magnetic saturation, high (rated) levels of current in the stator and in the field winding are required. Under standstill conditions, the cooling system of the generator might not permit long-time operation. The testing time should be reduced. Further on, with currents in both axes (I_q, $I_d + I_{fd}$) there will be torque. Therefore, the machine has to be stalled. Low-speed high-power SGs have large torque so that stalling is not easy.

If the tests are done either in axis *d* or in axis *q*, there will be no torque and the cross-coupling saturation (between orthogonal axes) is not present. Flux decay tests consist of supplying first the machine with a small DC voltage with stator phases connected such that to place the stator mmf either along axis *d* or along axis *q*. All this time, the field winding may be short-circuited, open, or supplied with a DC current. At a certain moment after the I_{ao}, V_o, I_{Fo}, are measured, the DC circuit is short-circuited by a fast switch with a very low voltage across it (when closed). A freewheeling diode may be used instead, but its voltage drop should be recorded and its integral during flux (current) decay time in the stator should be subtracted from the initial flux linkage. The *d* and *q* reactances are obtained then for the initial (DC) situation, corresponding to the initial flux linkage in the machine. The test is repeated with increasing values of initial currents along axis $d(i_d + i_{fd})$ and $q(i_q)$ and the flux current curve family as in Figure 8.22 is obtained. The same tests may be run on the field winding with open- or short-circuited stator.

A typical stator connection of phases for *d* axis flux decay tests is shown in Figure 8.25a.

To arrange the SG rotor in axis *d*, the stator windings, connected as in Figure 8.25a, are supplied a reasonably large DC current with the field winding short-circuited across the freewheeling diode. If the rotor is heavy and it will not move by itself easily, the stator is fed from an AC source with a small current and the rotor is rotated until the AC field-induced current with the freewheeling diode short-circuited will be maximum. Alternatively, if phase b and phase c in series are AC supplied then the rotor is moved until the AC field-induced current becomes zero. For fractionary stator winding, the position of axis *d* is not so clear and measurements in a few closely adjacent positions is required to average the results [4].

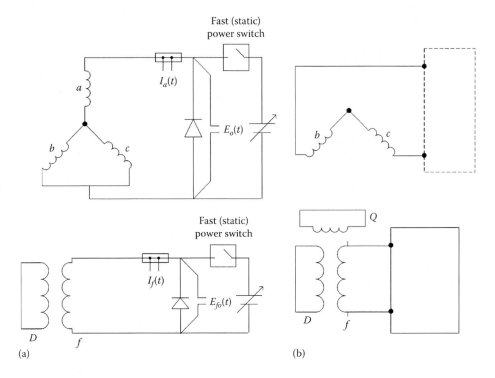

FIGURE 8.25 Flux decay tests: (a) axis *d* and (b) axis *q*.

The SG equations in axes *d* and *q* at zero speed are simply

$$I_d R_a - V_d = -\frac{d\Psi_d}{dt} \tag{8.82}$$

$$I_q R_a - V_q = -\frac{d\Psi_q}{dt} \tag{8.83}$$

In axis *d* (Figure 8.28a),

$$i_d = \frac{2}{3}\left(i_a + i_b \cos\left(\frac{2\pi}{3}\right) + i_c \cos\left(-\frac{2\pi}{3}\right)\right) = i_a \tag{8.84}$$

$$i_q = \frac{2}{3}\left(0 + i_b \sin\left(\frac{2\pi}{3}\right) + i_c \sin\left(-\frac{2\pi}{3}\right)\right) = 0; \quad i_b = i_c \tag{8.85}$$

The flux in axis *q* is zero. After short-circuiting the stator ($V_d = 0$) and integrating Equation 8.82:

$$R_a \int_0^\infty I_d dt + K_d \int_0^\infty V_{\text{diode}} dt = \left(\Psi_d\right)_{\text{initial}} - \left(\Psi_d\right)_{\text{final}} \tag{8.86}$$

$$K_d = 2/3 \text{ as } V_d = V_a \quad \text{and} \quad V_a - V_b = (2/3)\ V_a \tag{8.87}$$

Equation 8.86 provides the key to determining the initial flux linkage if the final flux linkage is known. The final flux is produced by the excitation current alone and may be obtained from a flux decay test on the excitation, from same initial field current, with the stator open but, this time, recording the stator voltage across the diode $V_o(t) = V_{abc}(t)$ is necessary. As $V_{abc}(t) = (3/2) V_d(t)$

$$\Psi_{\text{dinitial}}\left(i_{fo}\right) = \int_0^\infty \frac{2}{3} V_o(t) dt; \quad i_a = 0 \tag{8.88}$$

The initial d axis flux Ψ_{dinitial} is

$$\Psi_{\text{dinitial}}\left(i_{fdo} + i_{do}\right) = L_l i_{do} + L_{ad}\left(i_{fdo} + i_{do}\right) \tag{8.89}$$

As i_f and not i_{fd} (reduced to the stator) is measured, the value for the ratio a has to be determined first. It is possible to run a few flux decay tests on stator, with zero field current and then in the rotor, with zero stator current and use the aforementioned procedure to calculate the initial flux (final flux is zero). When the initial flux in the stator from both tests is the same (Figure 8.26), then the ratio of the corresponding currents constitute the reduction (turn ratio) value a:

$$a = \frac{i_{do}}{i_{fo}} \quad i_{fd} = a i_f \tag{8.90}$$

The variation of a with the level of saturation should be small.

When the tests are done in axis q (Figure 8.25b),

$$i_d = 0 \quad i_b = -i_c \quad i_a = 0 \tag{8.91}$$

$$i_q = \frac{2}{3}\left(i_b \sin\left(\frac{2\pi}{3}\right) + i_c \sin\left(-\frac{2\pi}{3}\right)\right) = \frac{2}{\sqrt{3}} i_b \tag{8.92}$$

$$V_q = \frac{2}{3}\left(V_a \sin\theta + \left(V_b - V_c\right)\sin\frac{2\pi}{3}\right) = \frac{2}{3}\left(V_b - V_c\right)\frac{\sqrt{3}}{2} = \frac{V_b - V_c}{\sqrt{3}} \tag{8.93}$$

Under flux decay, from Equation 8.53 with Equations 8.92 and 8.93

$$\Psi_{\text{qinitial}}\left(i_{qo}, i_{fdo}\right) = \frac{2}{\sqrt{3}} \int i_b R_a \, dt + \frac{1}{\sqrt{3}} \int_0^\infty V_{\text{diode}} \, dt \tag{8.94}$$

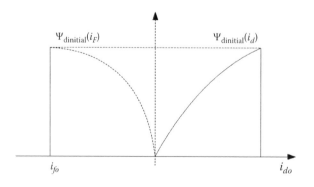

FIGURE 8.26 Turns ratio $a = i_{do}/i_{fo}$.

The final flux in axis q is zero, despite of non zero field current since the two axes are orthogonal. Doing the tests for a few values of $i_{fdo}(i_{fo})$ and for specified initial values of $i_{qo} = \dfrac{2}{\sqrt{3}} i_b$ a family of curves may be obtained:

$$\Psi_{\text{qinitial}}\left(i_{qo}, i_{fdo}\right) = L_l I_{qo} + I_{qo} L_{aq}\left(i_{qo}, i_{fdo}\right) \tag{8.95}$$

Considering that the d-axis stator current and field current mmfs are almost equivalent, the cross-coupling saturation in axis q is solved for all purposes. Not so in axis d where no cross-coupling has been explored.

A way out of this situation is to "move" the stator mmf connected as in Figure 8.25 by exchanging phases from a-b&c to b-a&c to c-a&b. This way the rotor is left in the d axis position corresponding to a-b&c (Figure 8.25). Now I_d, I_q, V_d, V_q have to be considered together and thus both families of curves may be obtained simultaneously with

$$i_d = \frac{2}{3}\left[i_a \cos\left(-\theta_{er}\right) + i_b \cos\left(-\theta_{er} + \frac{2\pi}{3}\right) + i_c \cos\left(-\theta_{er} - \frac{2\pi}{3}\right)\right] \tag{8.96}$$

$$i_q = \frac{2}{3}\left[i_a \sin\left(-\theta_{er}\right) + i_b \sin\left(-\theta_{er} + \frac{2\pi}{3}\right) + i_c \sin\left(-\theta_{er} - \frac{2\pi}{3}\right)\right] \tag{8.97}$$

V_d and V_q are obtained with similar formulae but noticing that still $V_b = V_c = -V_a/2$. Then the flux decay equations after integration are

$$\Psi_{\text{dinitial}} = \Psi_{\text{dfinal}} + \int_0^\infty i_d R_a dt + \frac{V_{do}}{V_{abco}} \int_0^\infty V_{\text{diode}}\, dt \tag{8.98}$$

$$\Psi_{\text{qinitial}} = \Psi_{\text{qfinal}} + \int_0^\infty i_q R_a dt + \frac{V_{qo}}{V_{abco}} \int_0^\infty V_{\text{diode}}\, dt \tag{8.99}$$

The final values of flux in the two axes are produced solely by the field current. The flux decay test in the rotor is done again to obtain the following:

$$\Psi_{\text{dinitial}}\left(i_{fo}\right) = \int_0^\infty V_d(t)dt \tag{8.100}$$

$$\Psi_{\text{qinitial}}\left(i_{fo}\right) = \int_0^\infty V_q(t)dt \tag{8.101}$$

Placing the rotor in axis d, then in axis q, and then with $\theta_{er} = \dfrac{\pi}{6}, \dfrac{2\pi}{3}$ will produce plenty of data to document the flux/current families that characterize the SG (Figure 8.24). Flux decay tests results in axis d or q in large SG to determine steady-state parameters as influenced by saturation have been published [9], but still the procedure—standard in principle—has to be further documented by very neat tests with cross-coupling thoroughly considered, and with hysteresis and temperature influence on results eliminated.

The flux decay tests at nonzero or non 90° electric angle, introduced already might be the way to go and get the whole flux/current (or flux/mmf) family of curves that characterizes the SG for various loads.

Note: It may be argued that though practically all values of i_d, i_q, i_{fd} may be produced in flux decay tests, the saturation influence on steady-state parameters may differ under load for same current triplet. This is true because under load (at rated speed), the stator iron is AC magnetized (at frequency f_N) and not DC magnetized as in flux decay tests.

Direct load tests or FEM comparisons with these flux decay tests will tell if this is more than an academic issue. The whole process of standstill flux decay tests may be computerized and thus mechanized, as by now static power switches are available off-the-shelf. The tests take time but apparently notably less than the SSFR. The two types of tests are, in fact, complementary as one produces the steady-state (or static) saturated parameters for specified load conditions, while the other estimates the parameters for transients from such initial on-load steady-state states. The standstill flux decay tests have also been used to estimate the parameters for transients by curve fitting the currents decays versus time [10,11].

8.7 Tests to Estimate the Subtransient and Transient Parameters

Subtransient and transient parameters of the SG manifest themselves when sudden changes at or near stator terminals (short circuits) or in field current occur. Knowing the sudden balanced and unbalanced short-circuit stator current peak value and evolution in time until steady state is useful in the design of the SG protection, calibration of the trip stator and field current circuit breakers and calculation of the mechanical stress in the stator end-turns. Sudden short-circuit tests from no load or load operation have been performed now for more than 80 years and, in general, two stages during these transients have been traditionally identified.

The first one, short in duration, characterized by steep attenuation of stator current I_s, called subtransient stage. The second one, larger, and slower in terms of current decay rate is the so-called transient stage. Phenomenologically, in the transient stage the transient currents in the rotor damper cage (or solid rotor) are already fully attenuated. Subtransient and transient stages are characterized by time constants: the time elapsed for a current decay to $1/e = 0.368$ from its original value. Based on this observation, graphical methods have been developed first to identify the subtransient and transient parameters. As short circuits from no load are typical, only the d-axis parameters are obtained in general (x_d'', T_d'', x_d', T_d').

In acquiring the armature and field current during short circuit, it is important to use a power switch that short circuits all required phases in the same time. Also, noninductive shunts or Hall probe current sensors with leads kept close together, twisted or via optical fiber cables to reduce the induced parasitic voltages, are to be used.

To avoid large errors and high transient voltage in the field circuit, the latter has to be supplied from low-impedance constant voltage supply. For the case of brushless exciters, the field current sensor is placed on the rotor and its output is transmitted through special slip-rings and brushes placed on purpose there, or through telemetry; or an observer is built for the purpose.

8.7.1 Three-Phase-Sudden Short-Circuit Tests

The standard variation of stator terminal rms AC components of current during a three-phase sudden short circuit from no load is

$$I_{ac}(t) = \frac{E}{X_{ds}} + \left(\frac{E}{X_d'} - \frac{E}{X_{ds}} \right) e^{-t/T_d'} + \left(\frac{E}{X_d''} - \frac{E}{X_{ds}'} \right) e^{-t/T_d''} \tag{8.102}$$

$I(t)$—AC rms short circuit current, p.u.

$E(t)$—no-load (initial) AC rms phase voltage in p.u.

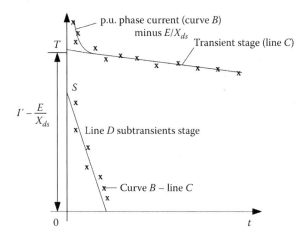

FIGURE 8.27 Analysis of subtransient and transient. Short-circuit current AC components.

If the test is performed below 0.4 p.u., initial voltage X_{ds} is replaced by X_{du} (unsaturated); both taken from the open-circuit saturation curve.

After subtracting the sinusoidal (steady-state) term in Equation 8.93 the second and third terms may be represented in semilogarithmic scales. The rapidly decaying portion of this curve represents the subtransient stage, while the straight line is the transient stage (Figure 8.27).

The extraction of rms values of the stator AC short-circuit current components from its recorded waveform versus time is now straightforward. If the field current is also acquired the stator armature time constant T_a may be determined also. The DC decay stator short circuit current components from all three phases may be extracted from the recorded (acquired) data also.

If a constant voltage low-impedance supply to the field winding is not feasible, it is possible to short circuit simultaneously the stator and the field winding. In this case both the stator current and the field current decay to zero. From the stator view point the constant component in Equation 8.102 has to be eliminated. It is also feasible to re-open the stator circuit (after reaching steady short circuit) and record the stator voltage recovery.

The transient reactance X'_d is

$$X'_d = \frac{E}{I'} \tag{8.103}$$

I' is the "initial AC transient current"—OT Figure 8.27 plus the steady-state short-circuit current E/X_d.
   The subtransient reactance $X_d{}''$ is

$$X''_d = \frac{E}{I''} \tag{8.104}$$

I'' is the total AC current peak at the time of short circuit ($I'' = I' + \overline{OS}$ from Figure 8.27).

The transient reactance is influenced by the saturation level in the machine. If X'_d is to be used to describe transients at rated current, short-circuit tests are to be done for various initial voltages E to plot $X'_d(I')$. The short-circuit time constants T'_d and T''_d are obtained as the slopes of the straight lines C and D in Figure 8.27.

8.7.2 Field Sudden Short-Circuit Tests with Open Stator Circuit

The sudden short circuit stator tests can provide values for x''_d, x'_d, T''_d, T'_d. The stator open-circuit subtransient and transient time constants T''_{do}, T'_{do} are however required to fully describe the operational

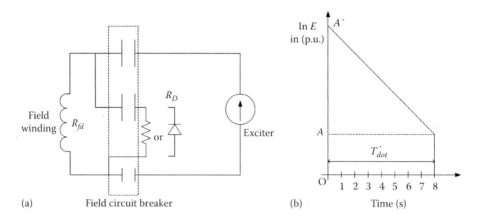

FIGURE 8.28 Field circuit short-circuit tests with open armature (a) the test arrangement and (b) the armature voltage E versus time.

impedance X_d. A convenient test to identify T'_{do} consists of running the machine on armature open circuit with the field circuit provided with a circuit breaker and a field discharge contact with a field discharge resistance (or freewheeling diode) in series. The field circuit is short-circuited by opening the field circuit breaker (Figure 8.28a).

The field current and stator voltage are acquired and represented in a semilogarithmic scale (Figure 8.28b).

The transient open-circuit time constant T'_{dot} in Figure 8.2 accounts also for the presence of additional discharge resistance R_D. Consequently the actual T'_{do} is

$$T'_{do} = T'_{dot} \cdot \left(1 + \frac{R_D}{R_{fd}}\right) \tag{8.105}$$

When a freewheeling diode is used R_D may be negligible.

The subtransient open-circuit time constant T''_{do} may be obtained after a sudden short-circuit test when the stator armature circuit is suddenly opened, and the recovery armature differential voltage is recorded (Figure 8.29).

The value of T''_{do} is obtained after subtracting the line C (Figure 8.29) from the differential voltage ΔE (recovery rms voltage minus rms steady-state voltage) to obtain curve D, approximated to a straight line.

Let us notice that the three-phase short-circuit tests with variants have provided only transient and subtransient parameters in axis d.

8.7.3 The Short-Circuit Armature Time Constant T_a

The stator time constant T_a occurs in the DC component of the three phase short-circuit current:

$$I_{dc}(t) = \frac{E}{X''_d} e^{-t/T_a} \tag{8.106}$$

It may be determined by separating the DC components from of the short-circuit currents of phases a, b, c (Figure 8.30).

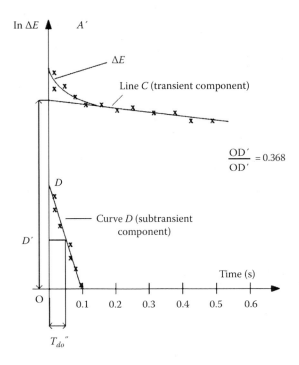

FIGURE 8.29 Difference between recovery voltage and steady-state voltage versus time.

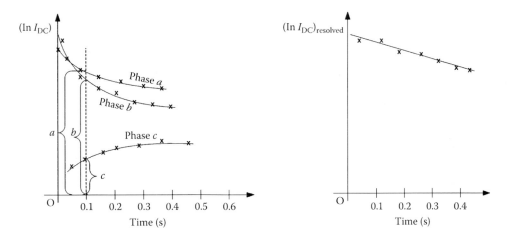

FIGURE 8.30 Direct current components of short-circuit currents.

As it can be seen from Figure 8.30, the DC components of short-circuit currents do not vary in time exactly in the same way. A resolved I_{DC} is calculated, from selected values at same time, as [4] follows:

$$(I_{DC})_{\text{resolved}} = \left(\sqrt{\left(a^2 + b^2 - ab \right)} + \sqrt{a^2 + c^2 - ac} + \sqrt{b^2 + c^2 - cb} \right) \sqrt{\frac{4}{27}} \qquad (8.107)$$

where a, b, and c are the instantaneous values of I_a, I_b, I_c for a given value of time with $a > b > c$. Then, from the semilogarithmic graph of (I_{DC}) resolved versus time, T_a is found as the slope of the straight line.

Note: The above graphical procedure to identify x_d'', x_d', T_d'', T_d', T_a, through balanced short-circuit tests may be computerized in itself to speed up the time to extract these parameters from test data that are today computer-acquired [4].

Various regression methods have also been introduced to fit the test data to the SG model.

8.8 Transient and Subtransient Parameters from *d* and *q* Axis Flux Decay Test at Standstill

The standstill flux decay tests in axes *d* and *q* provide the variation of $i_d(t)$, $i_{fd}(t)$ and, respectively, $i_q(t)$ (Figure 8.31).

This test has been traditionally used to determine—by integration—the initial flux and, thus, the synchronous reactances and the turn ratio *a*.

However, the flux decay transient current responses contain all the transient and subtransient parameters. Quite a few methods to process this information and produce x_d'', x_d', x_q'', T_d'', T_d', T_q'' have been proposed.

Among them, we mention here the decomposition of recorded current in exponential components [10]:

$$I_{d,q}(t) = \sum I_j e^{-t/T_j} \tag{8.108}$$

For the separation of the exponential constants I_j and T_j, a dedicated program based on nonlinear least square analysis may be used. Also, the maximum likelihood estimation (ML) has been applied [11] successfully to process the flux decay data—$i_d(t)$, $i_{fd}(t)$—with two rotor circuits in axes *d* and *q*. To initialize the process, the graphical techniques described in previous paragraphs [12] have also been applied for the scope.

If the tests are performed for low initial currents, there is no magnetic saturation. On the contrary, for large (rated) values of initial currents, saturation is present.

It may be argued that during the transients in the machine at speed and on load, the frequency content of various rotor current differ from the case of standstill flux decay transients. In other words, where the transients and subtransient parameters determined from flux decay tests may be applied safely? At least they may be safely used to evaluate balanced sudden short-circuit transients.

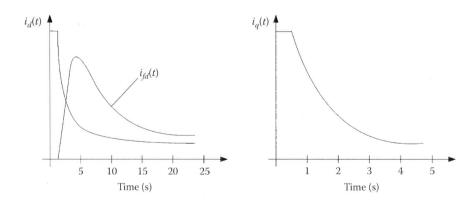

FIGURE 8.31 Flux decay transients at standstill.

8.9 Subtransient Reactances from Standstill Single-Frequency AC Tests

The subtransient reactances X_d'' and X_q'' are associated with fast transients, or large frequency (Figure 8.32). Consequently, at standstill when supplying the stator line from a single-phase AC (at rated frequency) source, the rotor circuits (field winding is short-circuited) experience that frequency. The rotor is placed in axis d by noticing the situation when the AC field current is maximum. For axis q, it is zero.

The voltage, current, and power in the stator are measured and thus:

$$Z_{d,q}'' = \frac{E_{ll}}{2I_a} \tag{8.109}$$

$$R_{d,q}'' = \frac{P_a}{2I_a^2} \tag{8.110}$$

$$X_{d,q}'' = \sqrt{\left(Z_{d,q}''\right)^2 - \left(R_{d,q}''\right)^2} \tag{8.111}$$

The rated frequency f_N is producing notable skin effects in the solid parts of the rotor or in the damper cage. As for the negative impedance Z_2 the frequency in the rotor $2f_N$, it might be useful to do this test again at $2f_N$ and determine again $X_d''(2f_N)$ and $X_q''(2f_N)$ and then use them to define

$$X_2 = \frac{X_d''(2f_N) + X_q''(2f_N)}{2} \tag{8.112}$$

$$R_2 = \frac{R_d''(2f_N) + R_q''(2f_N)}{2} \tag{8.113}$$

It should be noted, however, that the values of X_2 and R_2 are not influenced by DC saturation level in the rotor that is present during SG on load operation. If solid parts are present in the rotor, the skin effect that influences X_2 and R_2 is notably marked in turn by the DC saturation level in the machine at load.

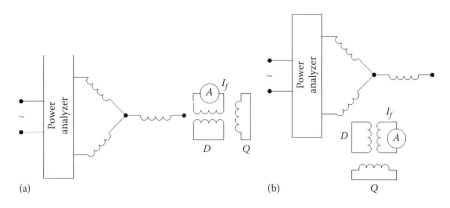

FIGURE 8.32 Line-to-line AC standstill tests (a) axis d and (b) axis q.

8.10 Standstill Frequency Response Tests

Traditionally, short-circuit tests have been performed to check the SG capability to withstand the corresponding mechanical stresses on the one hand, and to provide for subtransient and transient parameter determination to predict transient performance and assist in SG control design, on the other hand.

From these tests, two rotor circuit models are identified: the subtransient and transient submodels—a damper + field winding along d axis and two damper circuits along axis q.

As already illustrated in previous paragraphs, sudden short-circuit tests do not in general produce the transient and subtransient parameters in axis q.

For today's power system dynamics studies, all parameters along axes d and q are required.

Identifying the two rotor circuit models in axis d and separately in axis q may be performed by standstill flux decay tests as illustrated earlier in this chapter.

It seems however that both sudden short-circuit tests and standstill flux decay tests do not completely reflect the spectrum of frequencies encountered by SG under load transients when connected to power system or in stand-alone mode.

This is how standstill frequency tests have come into play. They are performed separately in axes d and q for current levels of 0.5% of rated current and for frequencies from 0.001 to 100 Hz and more.

Not all actual transients in an SG span, this wide-frequency band and thus the identified model from SSFR may be centered on the desired frequency zone.

The frequency effects in solid iron-rotor SGs are very important, and thus the second-order rotor circuit model may not suffice. A third order in both d and q axes proved to be more adequate.

As all standstill tests, the centrifugal effect on the contact resistance of damper bars (or of conducting wedges) to slot walls are not considered though, they may influence notably the identified model, comparisons between SSFR and on load frequency response tests for turbogenerators have spotted such differences. The saturation level is very low in SSFR tests, while in real SG transients the rotor core is strongly DC magnetized and the additional (transient) frequency currents in the solid iron occur in such an iron core. The field penetration depth is increased by saturation and the identified model parameters change.

Running SSFR in the presence of increasing DC pre-magnetization through the field current may solve the problem of saturation influences on the identified model. Direct current premagnetization is required only for frequencies above 1 Hz. Thus the time to apply large DC currents is somewhat limited, to limit the temperature rise during such DC + SSFR tests. Recently, Reference [12] seems to demonstrate such claim. Thus far, however, the pure SSFR tests have been investigated in more detail through a very rich literature, and finally standardized [4].

In what follows the standardized version of SSFR is presented with short notes on latest publications about the subject [4,13].

8.10.1 Background

The basic small perturbation transfer function parameters, as developed in Chapter 5, are as follows:

$$\Delta\Psi_d(s) = G(s)\Delta e_{fd}(s) - L_d(s)\Delta i_d(s) \tag{8.114}$$

$$\Delta\Psi_q(s) = -L_q(s)\Delta i_q(s) \tag{8.115}$$

The "−" signs in Equations 8.114 and 8.115 are common for generators in the United States.

$L_d(s)$—direct axis operational inductance (the Laplace transform of d axis flux divided to i_d with field winding short-circuited $\Delta e_{fd} = 0$)

$L_q(s)$—the quadrature axis operational inductance

$G(s)$—the armature to field transfer function (Laplace transform of the ratio of d axis flux linkage variation to field voltage variation, when the armature is open circuited)

Also,

$$sG(s) = \left(\frac{\Delta i_d(s)}{\Delta i_{fd}(s)} \right)_{\Delta e_{fd}=0} \tag{8.116}$$

Equation 8.116 defines the $G(s)$ for the case when the field winding is short-circuited.

One more transfer function is $Z_{afo}(s)$

$$Z_{afo}(s) = \left(\frac{\Delta e_{fd}(s)}{\Delta i_d(s)} \right)_{\Delta i_{fd}=0} \tag{8.117}$$

It represents the Laplace transform of the field voltage to d-axis current variations when the field winding is open.

Originally, the second-order rotor circuit model (Figure 8.33, Chapter 5) has been used to fit the SSFR.

The presence of the leakage coupling inductance L_{fld} introduced in Reference [14] to better represent the field winding transients, was found positive in cylindrical solid iron rotors and negative in salient-pole SGs.

In general, the stator leakage inductance L_l is considered known.

For high-power cylindrical solid rotor SGs, third-order models have been introduced (Figure 8.34). There is still a strong debate if one or two leakage inductances are required to fully represent such a machine, where the skin effect in the solid iron (and in the possible copper damper strips placed below the rotor field winding slot wedges) is notable. Eventually, they lead to rather complex frequency responses (Figures 8.34 and 8.35) [15].

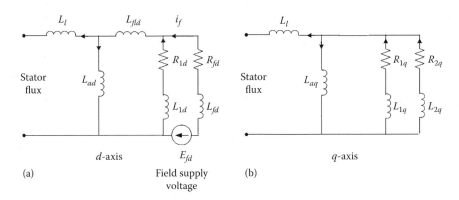

FIGURE 8.33 Second-order SG model: (a) axis d and (b) axis q.

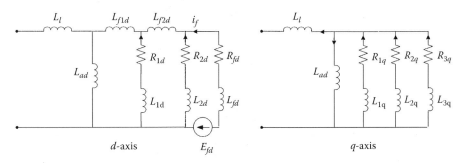

FIGURE 8.34 Third-order model of SG: (a) axis d and (b) axis q.

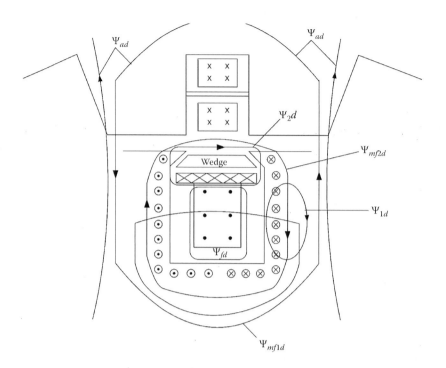

FIGURE 8.35 Third-order model (axis d) with two rotor leakage mutual inductances L_{f12d} and L_{f2d}.

With the leakage inductance L_l given, it is argued if both L_{f12d} and L_{f2d} are necessary to fully represent the actual phenomena in the machine. Though linear circuit theory allows for quite a few equivalent circuits for same frequency response over some given frequency bands, it seems natural to follow the physical phenomena flux paths in the machine (Figure 8.35) [15].

Initially, SSFR methods made use of only stator inductances $L_d(\omega)$, $L_q(\omega)$ responses. It was soon realized that, simultaneously, the rotor response transfer function $G(\omega)$ has to be taken into consideration such that the modeling of the transients in the field winding be adequate. In addition, $Z_{afd}(j\omega)$ is identified also in some processes if the frequency range of interest is below 1 Hz.

The magnitude and the phase of $L_d(j\omega)$, $L_q(j\omega)$, $G(j\omega)$, and $Z_{af0}(j\omega)$ are measured at several frequencies.

The smaller the frequencies, the larger the number of measurements per decade (up to 60) that is required for satisfactory precision [16].

SSFR are done at very low current levels (0.5% of base stator current) in order to avoid overheating as, at least in the low-frequency range, data acquisition for 2–3 periods requires long time internals ($f = 0.001$–1000 Hz).

Although magnetic saturation of the main flux path is avoided, the SSFR [17] makes the stator and rotor iron magnetization process evolve along low amplitude hysteresis cycles where the incremental permeability acts $\mu_i = (100$–$150)\mu_o$.

Consequently, L_{ad}, L_{aq} identified from SSFR are not to be used as unsaturated values with the machine at load.

The SSFR measurable parameters are

$$Z_d(j\omega) = \frac{\Delta e_d(j\omega)}{\Delta i_d(j\omega)}\bigg|_{\Delta e_{fd}=0} \tag{8.118}$$

$$Z_q(j\omega) = \frac{\Delta e_q(j\omega)}{\Delta i_q} \tag{8.119}$$

$$G(j\omega) = \frac{\Delta e_d(j\omega)}{j\omega \Delta i_{fd}(j\omega)}\bigg|_{\Delta i_d = 0} \tag{8.120}$$

$$j\omega G(j\omega) = \frac{\Delta i_{fd}(j\omega)}{\Delta i_d(j\omega)}\bigg|_{\Delta e_{fd} = 0} \tag{8.121}$$

$Z_d(j\omega)$ and $j\omega G(j\omega)$ may be found from same test (in axis d) by additionally acquiring the field current $i_{fd}(j\omega)$.

The measurable parameter $Z_{af0}(j\omega)$ is

$$Z_{afo}(j\omega) = \frac{\Delta e_{fd}(j\omega)}{\Delta i_d(j\omega)}\bigg|_{\Delta i_f = 0} \tag{8.122}$$

or,

$$Z_{afo}(j\omega) = \frac{\Delta e_d(j\omega)}{\Delta i_{fd}(j\omega)}\bigg|_{\Delta i_d = 0} \tag{8.123}$$

The mutual inductance L_{afd} between the stator and field windings is

$$L_{afd} = \frac{2}{3} \lim_{\omega \to o} \frac{Z_{afo}(j\omega)}{j\omega} \tag{8.124}$$

Alternatively from Equation 8.121,

$$L_{afd} = \lim_{\omega \to o} \frac{\Delta i_{fd}(j\omega)}{j\omega \cdot \Delta i_d(j\omega)}\bigg|_{\Delta e_{fd} = 0} \tag{8.125}$$

here R_{fd} is the field resistance plus the shunt and connecting leads.

The typical testing arrangement and sequence is shown in Figure 8.36.

The stator resistance R_a is

$$R_a = \lim_{\omega \to o}|Z_d(j\omega)| \tag{8.126}$$

Though rather straightforward, Equation 8.126 is prone to large errors unless a resolution of 1/1000 is not available at very low frequencies [4].

Fitting a straight line in the very low frequency range is recommended.

If R_a from Equation 8.126 is far away from the manufacturer's data, it is better to use the latter value because the estimation of time constants in $L_d(j\omega)$ would otherwise be compromised.

$$L_d(j\omega) = \frac{Z_d(j\omega) - R_a}{j\omega} \tag{8.127}$$

Typical $L_d(j\omega)$ data are shown in Figure 8.37a and for $j\omega\, G(j\omega)$ in Figure 8.37b.

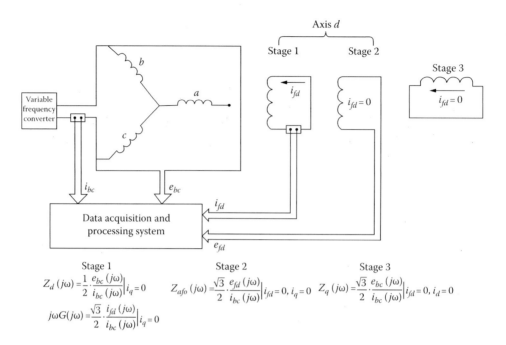

$$Z_d(j\omega) = \frac{1}{2} \cdot \frac{e_{bc}(j\omega)}{i_{bc}(j\omega)}\Big|_{i_q=0}$$

$$Z_{afo}(j\omega) = \frac{\sqrt{3}}{2} \cdot \frac{e_{fd}(j\omega)}{i_{bc}(j\omega)}\Big|_{i_{fd}=0,\, i_q=0}$$

$$Z_q(j\omega) = \frac{\sqrt{3}}{2} \cdot \frac{e_{bc}(j\omega)}{i_{bc}(j\omega)}\Big|_{i_{fd}=0,\, i_d=0}$$

$$j\omega G(j\omega) = \frac{\sqrt{3}}{2} \cdot \frac{i_{fd}(j\omega)}{i_{bc}(j\omega)}\Big|_{i_q=0}$$

FIGURE 8.36 SSFR testing setup stages and computation procedures.

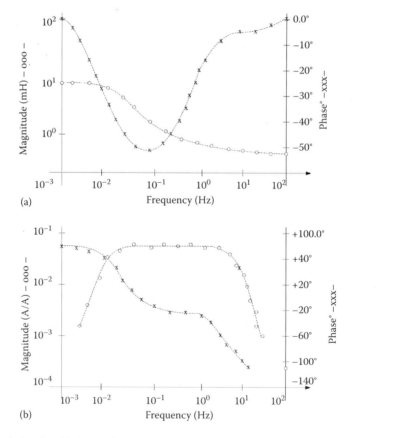

FIGURE 8.37 $L_d(j\omega)$ and $j\omega G(j\omega)$ typical responses: (a) d axis inductance and (b) stator-rotor transfer function.

The $Z_{afo}(j\omega)$ function is computed as depicted in Figure 8.36. The factor $\sqrt{3}/2$ in both $j\omega\, G(j\omega)$ and $Z_{afo}(j\omega)$ expressions is due to the fact that stator mmf is produced with only two phases or because phase b is displaced 30° with respect to field axis.

From quadrature tests, again, R_a is calculated as in axis d, and finally:

$$L_q(j\omega) = \frac{Z_q(j\omega) - R_a}{j\omega} \qquad (8.128)$$

A typical $L_q(j\omega)$ dependence on frequency is shown on Figure 8.38.

Moreover, the test results provide the value of the actual turns ratio $N_{af}(0)$:

$$N_{afo}(0) = \frac{1}{L_{ad}(0)} \lim_{\omega \to 0} \left| \frac{Z_{afo}(j\omega)}{j\omega} \right|_{i_q=0, i_d=0} \qquad (8.129)$$

where $L_{ad}(0)$ is

$$L_{ad}(0) = \lim_{\omega \to 0} \left| L_d(j\omega) \right| \qquad (8.130)$$

The base N_{af} turns ratio is

$$N_{af(base)} = \frac{3}{2} \left(\frac{I_{abase}}{I_{fdbase}} \right) = \frac{2 p_1 N_{fd}}{N_a} k_{w1} k_{f1} \qquad (8.131)$$

where
  p_1—pole pairs
  N_f—turns per filed winding coil
  K_{W1}—total stator winding factor
  K_{f1}—total field form factor

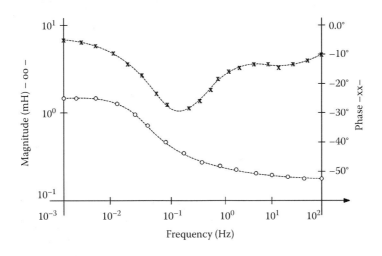

FIGURE 8.38 Typical $L_q(j\omega)$ response.

$N_{af}(0)$ and $N_{af}(\text{base})$ should be very close to each other. The field resistance after reduction to armature winding R_{fd}, is

$$R_{fd} = \frac{\lim_{\omega \to o}\left(j\omega L_{ad} \right)}{\lim_{\omega \to o}\left(\dfrac{\Delta i_{fd}\left(j\omega \right)}{\Delta i_d\left(j\omega \right)} \right) \cdot \dfrac{2}{3} N_{af}\left(0 \right)}; \quad (\Omega) \tag{8.132}$$

The directly measured field resistance R_{fd} may be reduced to armature winding also to yield R_{fd}:

$$R_{fd} = r_{fd} \cdot \frac{3}{2} \cdot \frac{1}{N_{af}^{\;2}\left(0 \right)}; \quad (\Omega) \tag{8.133}$$

Corrections for temperature may be added in Equation 8.133 if necessary.

The base field current may be calculated from Equation 8.131 if $N_{af}(0)$ value is used:

$$i_{fd}\left(\text{base} \right) = \frac{3}{2} I_a\left(\text{base} \right) \cdot \frac{1}{N_{af}\left(0 \right)}; \quad A\left(\text{dc} \right) \tag{8.134}$$

8.10.2 From SSFR Measurements to Time Constants

In a third-order model $L_d(j\omega)$ or $L_q(j\omega)$ are of the form:

$$L_{d,q}\left(j\omega \right) = L_{d,q}\left(0 \right) \cdot \frac{\left(1 + j\omega T_{d,q}''' \right)\left(1 + j\omega T_{d,q}'' \right)\left(1 + j\omega T_{d,q}' \right)}{\left(1 + j\omega T_{d,qo}''' \right)\left(1 + j\omega T_{d,qo}'' \right)\left(1 + j\omega T_{d,qo}' \right)} \tag{8.135}$$

The time constants to be determined through curve fitting, from SSFR are not necessarily the same as those obtained from short-circuit tests (in axis d).

Numerous methods of curve fitting have been proposed [16,18], some requiring the computation of gradients and some avoiding them, like the pattern search described in [4].

The direct ML method combining field short-circuit open SSFR in axis d has been shown to produce not only the time constants, but the very parameters of the multiple-order models.

Rather straightforward analytical expressions to calculate the third-order model parameters from the estimated time constants have been found for the case that $L_{f2d} = 0$ (see Figure 8.33) [19].

To shed more light on the phenomenology within the multiple rotor circuit models, a rather intuitive method to identify the time constants from SSFR, based on phase responses extremes finding [20], is described in what follows.

8.10.3 The SSFR Phase Method

It is widely accepted that second-order circuit models have a bandwidth of up to 5 Hz, and can be easily identified from SSFR tests. But even in this case, some simplifications are required, to enable that the standard set of short-circuit and open-circuit constants be identified.

With third-order models, such simplifications are not anymore indispensable. While curve fitting techniques to match SSFR to the identified model are predominant today, they are not free from shortcomings such as the following:

- They define from the start the order for the model
- Initiate the curve fitting with initial estimates of time constants
- Need to define a cost function and eventually calculate its gradients

Alternatively, it may be possible to identify the time constants pairs $T_1 < T_{10}$, $T_2 < T_{20}$, $T_3 < T_{30}$ Equation 8.136 from the operational inductances $L_d(j\omega)$, $L_q(j\omega)$, $j\omega G(j\omega)$, based on the property of lag circuits [20].

$$L_d(j\omega) = L_d \frac{(1+j\omega T_1)}{(1+j\omega T_{10})} \cdot \frac{(1+j\omega T_2)}{(1+j\omega T_{20})} \cdot \frac{(1+j\omega T_3)}{(1+j\omega T_{30})} \tag{8.136}$$

The R–L branches in parallel that appear in the equivalent circuit of SGs, make pairs of zeros and poles when represented in the frequency domain.

As each pair of zeros and poles form a lag circuit ($T_1 < T_{10}$, $T_2 < T_{20}$, $T_3 < T_{30}$), they may be separated one by one from the phase response. Let us denote $T_1/T_{10} = \alpha < 1$.

The lag circuit main features are as follows:

• It has a maximum phase lag φ at the center frequency of the pole zero pair F_c where

$$\sin\varphi = \frac{\alpha-1}{\alpha+1} \tag{8.137}$$

• The overall gain change due to the zero/pole pair is

$$\text{Gain change} = -20\log\alpha; \quad (\text{dB}) \tag{8.138}$$

• The two time constants T_{pole} and T_{zero} are

$$T_{\text{zero}} = \frac{T_{\text{pole}}}{\alpha}; \quad T_{\text{pole}} = \frac{\sqrt{\alpha}}{2\pi F_c} \tag{8.139}$$

The gain change is, in general, insufficient to be usable in the calculation of α, but Equations 8.127 and 8.139 are sufficient to calculate the zero and pole time constants. Identifying first the maximum phase lag points φ (at frequency F_c) is first operated by Equation 8.137. Then, T_{pole} and T_{zero} are easily determined from Equation 8.139. The number of maximum phase lag points in the SSFR frequency corresponds to the order of the equivalent circuit. The process starts by finding the first T_1, T_{10} pair. Then they are introduced in the experimentally found $L_d(j\omega)$ or $L_q(j\omega)$, and thus eliminated. The order of the circuit is reduced by one unit.

The remaining of the phase will be used to find the second zero-pole pair and so on, until the phase response left does not contain any maximum.

The order of the equivalent circuit is not given initially but claimed at the end, in accordance with the actual SSFR-phase number of maximum phase lag points.

The whole process may be computer programmed easily and ±1 dB gain errors are claimed to be characteristic to this method [20].

To further reduce the errors in determining, the value of the maximum phase lag angle φ and the frequency at which occurs, F_c, sensitive studies have been performed. They show that an error in F_c produces a significant error in the time constants T_{pole} and T_{zero}; α changes the time constants such that ϕ varies notably at same F_c [20].

Therefore, if F_c is varied above and below the firstly identified value F_{ci}, the $L_d(j\omega)$ error varies from positive to negative values. When this error changes sign the correct value of F_c has been reached (Figure 8.39a). For this "correct" value of F_c, the initially calculated value of α is changed up and down until the error changes sign.

The change in sign of phase error (Figure 8.39b) corresponds to the correct value of α. With these correct values, the final values of two time constants are obtained. A reduction in errors to ±0.5° and respectively, ±0.1 dB are claimed by these refinements.

The equivalent rotor circuit resistances may be calculated from the time constants just determined through an analytical solution using a linear transformation [19]. As expected, in such an analytical process the stator leakage inductance L_l plays an important role. There is a value of L_l above which the rotor circuit leakage inductances become negative. Design values of L_l have been shown to produce good results, however.

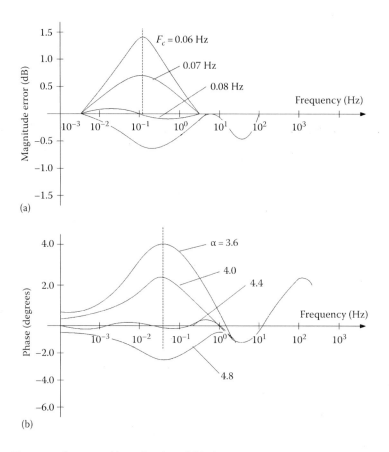

FIGURE 8.39 Variation of errors in (a) amplitude and (b) phase.

A few remarks seem in order:

- The SSFR phase-lag maxima are used to detect the center frequencies of the zero-pole pairs in the multiple circuit model of SG.
- The zero-pole pairs are calculated sequentially and then eliminated one by one from the phase response until no maximum phase lag is apparent. The order of the circuit model appears at the end of the process.
- As the initial values of the time constants are determined from the phase-response maxima, the process of optimizing their values as developed above leads to a unique representation of the equivalent circuit.
- It may be argued that the leakage coupling rotor inductances L_{fl1d}, L_{fl2d} are not identified in the process.
- The phase method may at least be used as a very good starting point for the curve-fitting methods to yield more physically representative equivalent circuit parameters of SGs.

8.11 Online Identification of SG Parameters

As today SGs tend to be stressed to the limit, even predictions of +5% additional stability margin will be valuable as it allows the delivery of about 5% more power safely. To make such predictions in a tightly designed SG and power system requires the identification of generator parameters from nondangerous online tests such that a sudden variation of field voltage for from one or a few active and reactive power levels [21].

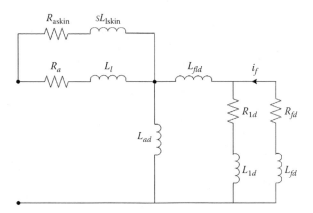

FIGURE 8.40 Addition of fictitious circuit section to account for stator skin effect during fast transients.

This way, the dynamic parameter identification takes place in conditions closer to those encountered in even larger transients. During a specific on-load large transient, the parameters of the SG-equivalent circuit may be considered constant as identified through an estimation method [22] or variable with the level of magnetic saturation and temperature. Intuitively, it seems more acceptable to assume that magnetic saturation influences some (or all) the inductances in the model, while the temperature influences the time constants. The frequency effects are considered traditionally by increasing the order of the rotor circuit model (especially for solid rotors). However, in a fast transient, the skin effect in the stator winding of a large SG may be considerable. Also, during the first milliseconds of such of a transient, the laminated core of stator may enter the model with an additional circuit [23].

When estimating the SG equivalent circuit parameters from rather large online measurements, it may be possible to let the parameters (some of them) vary in time though in reality they vary due to magnetic saturation level that changes in time during transients.

The stator frequency effects may be considered either by adding one cage circuit to model or by letting the stator resistance R_a vary in time during the process [24]. In the postprocessing stage, the variable parameters may be expressed as function of, say, field current, power angle, or stator current. Such functions may be used in other online transients. Alternatively for stator skin effects considerations, an additional fictitious circuit may be connected in parallel with R_a and L_l (Figure 8.40), instead of considering that R_a varies in time during transients. It may be also feasible to adopt a lower-order circuit model (say second order), perform a few representative online tests and, from all experimental data, determine the parameters. Making use of these parameters values through interpolation (artificial neural networks), for example, new transients may be explored safely [25]. Advanced estimation algorithms such as using the concept of synthesized information factor (SIF) [24], artificial neural networks [25] or constraint conjugate gradient methods [22], ML [26] have all been used for more or less successful validation on large machines under load, via transient responses in power angle, filled current, stator current [24]. The jury is still out but, perhaps, soon, an online measurement field strategy, with a pertinent parameter adaptive estimation scheme, will mature enough to enter the standards.

8.11.1 A Small-Signal Injection on Line Technique

A rather recent approach to online parameter identification is related to small-signal (current) injection by a chopper connected between two phases controlled at about 50% modulation index with two frequencies ($\omega_s \pm \omega_e$); $\omega_e \geq 2\pi 200$ rad/s (Figure 8.41) [27]. The 5% p.u. current injection frequency may vary between 0.01 Hz and 1 kHz, to cover all frequency range of interest.

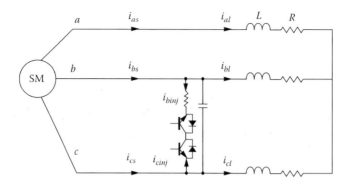

FIGURE 8.41 Online current–injection set up for a synchronous machine with R, L load.

The phase voltages (V_a, V_b) and current (i_a, i_b) are transformed into rotor coordinates and the injected frequencies are separated as ΔV_d, ΔV_q, Δi_d, Δi_q:

$$\begin{vmatrix} \Delta V_q \\ \Delta V_d \end{vmatrix} = \begin{bmatrix} Z_{qq} & Z_{qd} \\ Z_{dq} & Z_{dd} \end{bmatrix} \begin{vmatrix} \Delta I_q \\ \Delta I_d \end{vmatrix} \tag{8.140}$$

The impedances in Equation 8.140 can be traced down to SM dq model parameters and calculated at various injection frequencies in the chopper. Then an optimization algorithm (GOSET in MATLAB®) is used to extract the machine dq model parameters. A comparison with SSFR for the investigated 3.7 kW, 230 V, 1800 rpm SM shows good agreement of key parameters. Operational impedances $x_d(s)$, $x_q(s)$ by GOSET (genetic algorithm) and by SSFR agree satisfactorily over the entire frequency range (Figure 8.42) [27].

Zero sequence test is needed to approximate the stator leakage inductance, but its design value will do, also. Subsequent verification of the method presented here for higher power SGs is still needed before its industrialization.

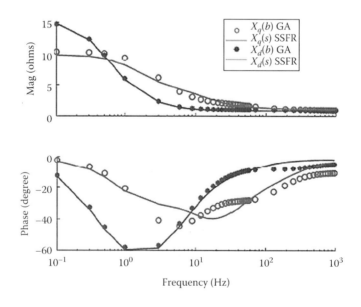

FIGURE 8.42 $X_d(s)$, $Y(s)$ comparisons.

8.11.2 Line Switching (On or Off) Parameter Identification for Isolated Grids

Load rejection method of SG parameter identification (in q axis) is already confirmed in IEEE standard 1110; while, when complemented with field current decay test (for axis d), it offers all machine dq model parameters. Reference [28] develops the dq model of SG with a single input V_f (field winding voltage) by adding the RL equivalent load model (Figure 8.43) [28].

In fact, any power grid may be reduced to an RLC load and an emf, so the method may be extended to grid connected SGs.

The state variables X and the outputs Y are as follows:

$$X = \left[\Psi_D, \Psi_Q, \Psi_F, i_d, i_q \right]^t \quad Y = \left[i_F, V_d V_q, i_d, i_q \right] \tag{8.141}$$

An RL load equivalent model in dq coordinates (p.u.) is as follows:

$$\begin{vmatrix} V_d \\ V_q \end{vmatrix} = R_L \begin{vmatrix} i_d \\ i_q \end{vmatrix} + \frac{1}{\omega_n} X_L \frac{d}{dt} \begin{vmatrix} i_d \\ i_q \end{vmatrix} + \omega_m X_L \begin{vmatrix} i_d \\ i_q \end{vmatrix} \tag{8.142}$$

Marking this model with the standard dq model of SG, a unique model with only one input V_F, is obtained. Magnetic saturation is considered in X_{dm} and X_{qm} by same coefficient K_f [28].

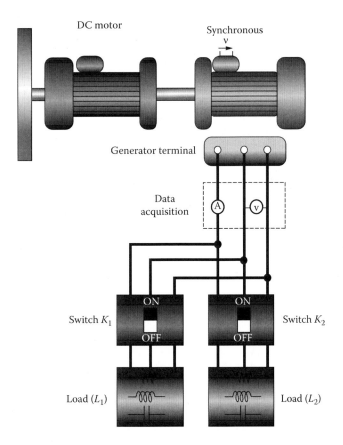

FIGURE 8.43 Online switching and load rejection principle.

This unique model is used to treat online switching of an inductive load, which implies that the steady-state power angle remains zero. No rotor position estimation is thus necessary even if speed varies during online switching, but it is available, as it may be introduced in the SG *dq* voltage/current model, with no need for the motion equation.

From the response of online (or load rejection) switching transients with measured currents, constant V_F, stator voltages, quite a few SG parameters may be obtained by curve fitting (optimization).

An interesting duality between online pure inductive load switching and offline pure capacitive load switching is inferred in [28] and may be instrumental in testing for parameter identification.

A cross-identification (validation) of SG parameters between online switching, offline switching, field short circuit (field current decay) tests is offered in Reference [29] for a small (1.5 kVA) SG. Although the method is easier to apply for small- and medium-power SGs it is worth pursuing it further.

8.11.3 Synthetic Back-to-Back Load Testing with Inverter Supply

Up to 10 MVA SGs are being considered for variable speed wind generators, hydroapplications, and so on. These systems use full-power converters.

Testing them, by using the inverter but not coupling them at shaft it may constitute a good advantage at manufacturers' site.

Figure 8.44a shows a typical such arrangement with two identical SGs and their four-quadrant DC–link AC–AC converters connected to the local power grid.

After both machines are slowly accelerated to speed, to claim small power from the local power grid, their delivered power should be set first on zero.

Subsequently, the reference input powers P_1^* and P_2^* are synchronously switched from + (generator), to − (motoring) such that where one is motor the other is generator. The power switching frequency is high enough such that the machines remain at synchronous speed.

The average power extracted in the process from the local power grid is equivalent to the two lines the losses in machine + converter system: $2\Sigma p$.

Therefore, even the efficiency at different power levels can be obtained:

$$\eta = \frac{2\sum p}{P_1^* + P_2^*} \tag{8.143}$$

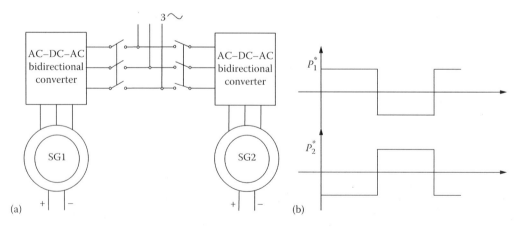

FIGURE 8.44 Synthetic back-to-back loading of SGs.

The heating test may be done implicitly this way with programmed power levels according to given duty cycles are representative for the application.

Note: The above method may be applied also to the PMSGs or DFIGs with PWM bidirectional converters; even for CRIG with full-power PWM bidirectional converters.

Only one machine and dual converter is needed for standard synthetic loading, but then only the average power from the local grid is equal to system losses, while the peak power is experienced by the local power grid.

8.12 Summary

- A rather complete set of testing methods for synchronous generators (SGs) is presented in IEEE Std. 115-1995. IEC has a similar standard.
- The SG testing methods may be classified into standardized and research types. The standardized tests may be performed for acceptance, performance, or parameters.
- Acceptance testing refers to insulation resistance dielectric and partial discharge, resistance measurements, identifying short-circuited field turns, polarity of field poles, shaft currents and bearing insulation, phase sequence TIF (balanced, residual, line-to-neutral) stator terminal voltage waveform deviation and distortion factors, overspeed tests, line discharging (maximum absorbed kVA at zero power factor with zero field current), and acoustic noise. They were presented in some detail in this chapter following, in general, IEEE Std. 115-1996.
- Testing SG for performance refers to saturation curves, segregated losses, power angle, and efficiency.
- The individual loss components are as follows:
  - Friction and windage loss
  - Core loss (on open circuit)
  - Stray load loss (on short circuit)
  - Stator winding loss
  - Field winding loss
- To identify the magnetic saturation curves and then segregate the loss components, to finally calculate efficiency under load, four methods have gained wide acceptance:
  - Separate driving method
  - Electric input method
  - Retardation method
  - Heat transfer method
- The separate driving method is based on the concept of driving the SG at precisely controlled (or rated) speed by an external, low rating (<5% of SG rating) motor, or by the exciter (if it contains an electric machine on shaft) or by the prime mover itself. With the SG driven at rated speed, on open stator circuit, the field current, stator voltage, and the driving motor output are measured for increasing values of field current, until the stator terminal voltage reaches at least 125% of rated value. The open-circuit saturation curve is thus obtained. In addition, as the driving motor output is considered known, the field current may be made zero, and thus the friction and windage losses are measured. From the open-circuit run, where both friction and windage and core losses occur, the core loss for various values of terminal voltage is obtained. The same arrangement may be used to determine the short-circuit saturation curve and then stray load losses for currents levels from 125% to 25% of rated current, if the winding resistance is known from previous measurements. With all loss components known, the efficiency versus power output may be calculated.
- The electrical input method is very similar to the separate driving method, but this time the SG itself works as a synchronous motor without a mechanical load.

- In retardation tests, the uncoupled SG is brought at 105%–110% of rated speed and then left to decelerate in three different situations:
  - (a) Open circuit stator, zero field current, to measure friction and windage loss P_{fw};
  - (b) Open circuit stator, constant field current, to measure friction and windage loss + core loss: $P_{fw} + P_{core}$;
  - (c) Short circuit stator, constant field current, to measure friction and windage + stator winding + stray-load losses: $P_{cul} + P_{strayload} + P_{fw}$.

  The inertia may be obtained from run (a), with P_{fw} already known, from the motion equation, by calculating the speed slowing rate around rated speed. Speed has to be recorded. Then, from runs (b) and (c), P_{core} and $P_{strayload}$ may be segregated.
- The other purpose of steady-state tests is to determine the steady-state parameters of SG and the field current under specified load.
- The leakage reactance X_l is approximated usually to the Potier reactance X_p obtained from a rated current zero-power-factor tests at rated voltage. As $X_p > X_l$ at rated voltage, the same test is performed at 110% rated voltage when X_p approaches X_l.
- A procedure to estimate X_l as the average between the zero sequence (homopolar) reactance $X_o (X_o < X_l)$ and the reactance without the rotor inside the stator bore X_{lair}, is introduced. Alternatively, from X_{lair} the bore air volume reactance X_{air} may be subtracted to obtain the leakage reactance. X_{air} is proportional to the unsaturated uniform air gap reactance X_{adu}; $X_{air} \sim X_{adu} \times gK_c\pi/\tau$: $X_l = X_{lair} - X_{air}$.
- The excitation current at specified load-active and reactive power and voltage is determined by the Potier diagram method. Magnetic saturation is considered the same in both axes. The power angle is needed for the scope and is computed by using the unsaturated values of quadrature axis reactance, X_{qu} of Equation 8.29.
- From the simplified transient SG phasor diagram (Figure 8.14), the excitation current for stability studies is obtained.
- Voltage regulation—the difference between no-load voltage and terminal voltage for specified load and same field current—is calculated based on the computed field current at specified load.
- The power angle may be measured by encoders or other mechanical sensors. The power angle is the electrical angle between the terminal voltage and the field produced emf.
- Temperature tests are required to verify the SG capability to deliver the rated or more power under conditions agreed upon by vendor and buyer. Four methods are described:
  - Conventional (direct) loading
  - Synchronous feedback (back-to-back) loading
  - Zero-power-factor tests
  - Open-circuit running plus short-circuit "loading"
- While direct loading may be done by manufacturer only for small- and medium-power SGs, the same test may be performed after commissioning, with the machine at power grid for all power levels.
- Back-to-back loading implies the presence of two coupled identical SGs with their rotors displaced mechanically by $2\delta_N/p_1$. The stator circuits of the two machines are connected together. One machine acts as a motor, the other as a generator. If an external low rating motor is driving the MS + SG set, then the former covers the losses in the two machines $2\Sigma P$. Care must be exercised to make the losses fully equivalent with those occurring with the SG under specified load.
- The zero-power-factor load tests implies that the SG works as a "synchronous capacitor" (motoring at zero-mechanical-load). Again loss equivalence with actual conditions has to be observed.
- The open-circuit stator test for the field current at specified load is run and the temperature rise until thermal steady state, Δt_{oc}, is measured. Further on, the short-circuit test at rated current is run until new thermal steady state is obtained for a temperature rise of Δt_{sc}. The SG total temperature is $\Delta t_{oc} + \Delta t_{sc}$ minus the temperature differential due to mechanical loss, which is duplicated during the tests.

- All these tests imply efforts and have shortcomings, but the temperature rise for specified load is so important for SG life that it has to be measured even if sophisticated FEM thermal-electromagnetic models of SGs is now available.
- Besides steady-state performance, behavior under transients conditions is also important to predict. To this scope, the SG parameters for transients have to be identified. Power system stability studies rely on SG parameters knowledge.
- PER-UNIT values are used in parameter definitions to facilitate more generality in results referring to SG of various power levels. Only three base independent quantities are in general required: voltage, current, and frequency.
- For the rotor, Rankin has defined a new base field current $i_{fdbase} = I_{fdbase} \times X_{adu}$. I_{fdbase} is the field current (reduced to the stator) that produces the rated voltage on the straight line of no-load saturation curve. X_{adu} is the unsaturated d-axis coupling reactance (in p.u.) between stator and rotor windings. Rankin's reciprocal system produces equal stator to field and field to stator p.u. reactances.
- The steady-state parameters X_d, X_q may be determined by a few carefully designed tests without loading the SG. Quadrature axis reactance X_q is more difficult to segregate, but pure I_q loading of the machine method provides for the X_q values.
- SGs may work with unbalanced load either when connected to the power system or in stand-alone mode. For steady state, by the method of symmetrical components, the value of $Z_2(X_2)$ may be found from a steady-state line to line short circuit. The elimination of the third harmonics from the measured voltage is crucial in obtaining acceptable precision with this testing method.
- The zero sequence (homopolar) reactance X_o is measured from a standstill AC tests with all stator phases in series or in parallel. In general, $X_o \leq X_l$. A two line to neutral short-circuit tests will also provide for X_o.
- The short-circuit ratio SCR $= 1/X_d = I_{FNL}/I_{FSI}$. I_{FSI} id the field current from the three-phase short-circuit (at rated stator current). I_{FNL} is the field current at rated voltage and open-circuit test. SCR has today typical values of 0.4–0.6 for high-power SGs.
- Due to magnetic saturation and its cross-coupling effect—all the methods to determine the steady-state parameters thus far need special corrections to fit the results from direct on load measurements. Families of flux/current curves for axis d and q are required to be obtained from special operation mode tests, in order to have enough data to cope with on-load various situations.
- Standstill flux decay tests may provide the required flux/current families of curves. They imply short-circuiting the stator or field circuits and recording the currents. The tests are done in axis d or q or in a given rotor position.
- Integrating the resistive voltage drop in time, the initial (DC) flux linkage is obtained. Care must be exercised to avoid hysteresis caused errors and to DC measure the resistance before each new test in order to eliminate temperature rise errors.
- To estimate the transient parameters—X_d'', X_d', T_d'', T_d', X_q'', T_q'', X_q', T_q', T_{do}'', T_{do}', T_{qo}''—two main types of tests have been standardized:
  - Sudden three-phase short-circuit tests
  - SSFR
- Three-phase sudden short circuit is used to identify the above parameters of a second-order model of SG in both, axis d and axis q. The methods to extract the parameters, from stator phase currents and field current recording, are essentially either graphical with mechanization by computer programs [27] or regression type.
- Third-order models are required, especially in SGs with solid rotors, due to skin effect strong dependence on frequency in solid bodies.
- Alternatively, there are proposals to determine the parameters for transients from the already mentioned standstill DC flux decay tests, by processing the time variation of currents or voltages during these tests. Though good results have been reported, the method has not met yet worldwide acceptance due to insufficient documentation.

- SSFR tests use quite similar arrangements with DC flux decay tests. The SG is fed with about 0.5% rated currents at frequencies from 0.001 to 100 Hz and more. The input voltage, stator current, field current rms values, and phase lags are measured. The tests take time as at least 2–3 periods have to be recorded at each frequency.
- In general, the amplitude of the operational parameters $L_d(j\omega)$, $L_q(j\omega)$, $G(j\omega)$, $Z_{afo}(j\omega)$ is used to extract the parameters of the third-order model along axes d and q: $X'''_{d,q}$, $X''_{d,q}$, $X'_{d,q}$, $I'''_{d,q}$, $I''_{d,q}$, $I'_{d,q}$ $T'''_{d,qo}$, $T''_{d,qo}$, $T'_{d,qo}$, the stator resistance R_a, field winding resistance R_{fd}, the turns ratio between rotor and stator a. Numerous curve fitting methods have been introduced and improved steadily up to the present time.
- The presence of one or two leakage mutual rotor reactances in the third-order model, with the leakage stator reactance X_l given a priori, is still a matter of debate.
- These reactances seem mandatory (at least one of them) to represent correctly, in the same time, the transient behavior of SGs seen from stator circuit and from the field current side.
- The up-to-now less-favored information from SSFR the phase response versus frequency has been recently put to work to identify the third model of SG [20]. The method is based on the observation that the phase response corresponds to zero/pole pairs in the model's transfer function. By using the center frequency F_C and the value of response phase φ, these maximum phase zero/pole pairs—that is time constants—are calculated rather simply. They are then corrected until only very low errors persist. After the first pair is identified, its circuits are eliminated from the response. The search for phase maximum continues until no maximum persists. The model order comes thus at the end. Determining the resistances and reactances of the SG multiple circuit model, from transient reactances to time constants, may be done by regression methods. Analytical resistance expressions have also been found for the third-order models [19].
- Some verifications of SSFR validity for various large on-load transients have been made. Still, this operation seems yet insufficiently documented, especially for solid rotor SGs.
- Online SG model identification methods, based on same nondamaging transients on load (i.e., up to 30% step field voltage change response at various load conditions) have been introduced recently. Artificial neural networks and other learning methods may be used to extend the model thus obtained to new transients, based on a set of representative on load tests.
- While these very complex online adaptive parameter identification methods go on, there are still hopes that simpler methods such as standstill flux decay tests, standstill DC + SSFR tests, or load recovery tests may be improved to the point to become fully reliable for large on-load transients.
- A small signal injection method between two phases through a chopper, with SG on load has been shown recently to produce good results with reasonable effort [27].
- Similarly, good results are claimed with an on–off line switching method [28].
- Important developments are expected in SG testing in the near future as theory, software, and hardware are continuously upgraded through worldwide efforts by industry and academia. These efforts are driven by power quality ever more demanding standards. As an example, a synthetic back-to-back loading test of SGs with AC–DC–AC converters and low local power grid rating was presented in Section 8.10.3.

References

1. K. Shima, K. Ide, M. Takahashi, Finite-element calculation of leakage inductances of a saturated-pole synchronous machine with damper circuits, *IEEE Transactions*, EC-17(4), 463–471, 2002.
2. K. Shima, K. Ide, M. Takashashi, Analysis of leakage flux distributions in a salient-pole synchronous machine using finite elements, *IEEE Transactions*, EC-18(1), 63–70, 2003.
3. A.M. El-Serafi, J. Wu, A new method for determining the armature leakage reactance of synchronous machines, *IEEE Transactions*, EC-18(1), 80–86, 2003.
4. IEEE Standard 115-1995.

5. A.M. El-Serafi, N.C. Kar, Methods for determining the Q axis saturation characteristics of salient-pole synchronous machines from the measured D-axis characteristics, *IEEE Transactions*, EC-18(1), 80–86, 2003.
6. A.W. Rankin, Per unit impedance of synchronous machines, *AIEE Transactions*, 64, 564–572 and 939–941, 1985.
7. S. Tahan, I. Kamwa, A two factor saturation model for synchronous machine with multiple rotor circuits, *IEEE Transactions*, EC-10(4), 609–616, 1995.
8. M. Biriescu, Gh. Liuba, Identification of reactances of synchronous machine including the saturation influence, *Proceedings of International Conference on Evolution and Modern Aspects of Synchronous Machines*, Zurich, Switzerland, August 27–29, 1991, pp. 55–58.
9. M. Biriescu, G. Liuba, M. Mot, V. Olarescu, V. Groza, Identification of synchronous machine reactances from current decay at standstill test, *Record of ICEM-2000*, Espoo, Finland, 2000, pp. 1914–1916.
10. A. Keyhani, S.I. Moon, A. Tumageanian, T. Leksau, Maximum likelyhood estimation of synchronous machine parameters from flux decay data, *Proceedings of ICEM-1992*, Manchester, UK, Vol. I, 1992, pp. 34–38.
11. N. Dedene, R. Pintelon, Ph. Lataire, Estimation of global synchronous machine model using a MIMO estimator, *IEEE Transactions*, EC-18(1), 11–16, 2003.
12. F.P. de Melle, J.R. Ribeiro, Derivation of synchronous machine parameters from tests, *IEEE Transactions*, PAS-96(4), 1211–1218, 1977.
13. I.M. Canay, Causes of discrepancies in calculation of rotor quantities and exact equivalent diagrams of the synchronous machine, *IEEE Transactions*, PAS-88, 114–1120, 1969.
14. P.L. Dandeno, Discussion. *IEEE Transactions*, EC-9(3), 587–588, 1994.
15. I. Kamwa, P. Viarouge, On equivalent circuit structures for empirical modeling of turbine-generators and discussion. *IEEE Transactions*, EC-9(3), 579–592, 1994.
16. P.L. Dandeno, H.C. Karmaker, Experience with Standstill Frequency Response (SSFR) testing and analysis of salient pole synchronous machines, *IEEE Transactions*, EC-14(4), 1209–1217, 1999.
17. A. Keyhani, H. Tsai, Identification of high order synchronous generator models from SSFR tests data, *IEEE Transactions*, EC-9(3), 593–603, 1994.
18. S. D. Umans, I.A. Mallick, G.L. Wilson, Modelling of solid iron turbogenerators, Part I+II, *IEEE Transactions*, PAS-97(1), 269–298, 1978.
19. Z. Zhao, F. Zheng, J. Gao, L. Xu, A dynamic on-line parameter identification and full system experimental verification for large synchronous machine, *IEEE Transactions*, EC-10(3), 392–398, 1995.
20. A. Watson, A systematic method to the determination of parameters of synchronous machine from results of frequency response tests, *IEEE Transactions*, EC-15(4), 218–223, 2000.
21. C.T. Huang, Y.F. Chen, C.L. Chang, Ch.-Y. Huang, N.D. Chiang, J.Ch. Wang, On line measurement based model parameter estimation for SG model development and identification schemes, Part I+II, *IEEE Transactions*, EC-9(2), 330–343, 1994.
22. I. Boldea, S.A. Nasar, Unified treatment core losses and saturation in the orthogonal axis model of electric machines, *Proceedings of IEEE*, 134-B(6), 355–363, 1987.
23. Z. Zhao, L. Xu, J. Jiang, On line estimation of a variable parameters of synchronous machines using a novel adaptive algorithm—Part I+II and discussion, *IEEE Transactions*, EC-12(3), 193–210, 1997.
24. H.B. Karayaka, A. Keyhani, G.T. Heydt, B.L. Agrawal, D.A. Selin, Neural network based modeling of a large steam turbine-generator rotor body parameters from on-line disturbance data, *IEEE Transactions*, EC-16(4), 305–311, 2001.
25. R. Wamkeua, I. Kamwa, X. Dai-Do, A. Keyhani, Iteratively reweighetd least squares for maximum likelihood identification of synchronous machine parameters from on line tests, *IEEE Transactions*, EC-14(2), 156–166, 1999.

26. I. Kamwa, M. Pilote, H. Corle, P. Viarouge, B. Mpanda-Mabwe, M. Crappe, Computer software to automate the graphical analysis of sudden shortcircuit oscillograms of large synchronous machines, part I+II, *IEEE Transactions*, Vol. EC-10(3), 399–414, 1995.
27. J. Huang, K.A. Corzine, M. Belkhayat, On line synchronous machine parameter extraction from small-signal injection techniques, *IEEE Transactions*, EC-24(1), 43–51, 2009.
28. R. Wamkeue, C. Jolette, I. Kamwa, Advanced modeling of a synchronous generator under line-switching and load rejection for isolated grid applications, *IEEE Transactions*, EC-25(3), 680–689, 2010.
29. R. Wamkeue, C. Jolette, A.B.M. Mabwe, I. Kamwa, Cross-identification of synchronous generator parameters from RTDR test time-domain analytical responses, *IEEE Transactions*, EC-26(3), 776–786, 2011.

Index

F